Contemporary Multivariate Analysis and Design of Experiments

SERIES IN BIOSTATISTICS

Series Editor: Heping Zhang *(Yale University School of Medicine, USA)*

Vol. 1 Development of Modern Statistics and Related Topics
— In Celebration of Prof. Yaoting Zhang's 70th Birthday
edited by Heping Zhang & Jian Huang

Series in Biostatistics – Vol. 2

Contemporary Multivariate Analysis and Design of Experiments

edited by

Jianqing Fan
Princeton University, USA

Gang Li
University of California at Los Angeles, USA

World Scientific

NEW JERSEY • LONDON • SINGAPORE • BEIJING • SHANGHAI • HONG KONG • TAIPEI • CHENNAI

Published by

World Scientific Publishing Co. Pte. Ltd.

5 Toh Tuck Link, Singapore 596224

USA office: 27 Warren Street, Suite 401-402, Hackensack, NJ 07601

UK office: 57 Shelton Street, Covent Garden, London WC2H 9HE

British Library Cataloguing-in-Publication Data
A catalogue record for this book is available from the British Library.

First published Mar 2005
Reprinted Sept 2005

Series in Biostatistics — Vol. 2
CONTEMPORARY MULTIVARIATE ANALYSIS AND DESIGN OF EXPERIMENTS
In Celebration of Professor Kai-Tai Fang's 65th Birthday

ISBN 981-256-120-X

This book is printed on acid-free paper.

Printed in Singapore by Mainland Press

Preface

Technological innovations and exponential increases in computing power have profound impacts on scientific research and development. New technologies allow us to collect data with an unpresredented level of size and complexity. Internet disseminations make large and complex data widely available to statisticians. Statistics has pervaded in every facet of science and engineering. It plays an increasingly important role in computational biology, quantitative finance, information engineering, neuroscience, medicine, and policy making as we enter a new era of quantitative research. These applications pose new challenges to the design of experiments and the analysis of the resulting multivariate data. At the occasion of the 65th birthday of Professor Kai-Tai Fang, we give these two classical fields of statistics a contemporary outlook.

This volume consists of the reviewed papers that contributed by many prominent statisticians, who are the friends and former students of Professor Kai-Tai Fang. It features an exclusive interview with Professor Kai-Tai Fang and seven review articles that are particularly useful to researchers who are new to these areas. It gives an overview of new developments and a contemporary outlook on the analysis of multivariate data, the design of experiment and their related topics. The monograph is dedicated to Professor Fang on his 65th birthday in June 2005. Most of the articles in the book will be presented at the International Conference on Statistics in Honor of Professor Kai-Tai Fang's 65th Birthday, June 20 - 24, 2005, Hong Kong, co-sponsored by the Institute of Mathematical Statistics and the Hong Kong Baptist University, co-chaired by Fred Hickernell and Jianqing Fan.

Professor Kai-Tai Fang, fellow of the Institute of Mathematical Statistics and fellow of the American Statistical Association, is a strong scholar and prolific researcher. He has published over 200 articles, authored or co-authored 17 books, and edited 8 lecture notes and proceedings in a wide range of statistical subjects, including multivariate analysis, design of experiments, and quasi Monte Carlo methods. In addition, Professor Fang has actively participated in a large array of consulting projects, including the designs of chemical and biological experiments, standardization of Chinese garments, and assessment of service of Hong Kong public libraries. As a leading figure in Hong Kong and China, he has greatly popularized the use of statistics in academic research and industry, enthusiastically participated in organizing various professional meetings, and provided conscientious editorial service. He is a strong professional leader and a dedicated educator, who foster a new generation of fertile statisticians worldwide.

Inside this monograph, Ms. Agnes W.L. Loie, former head of Information and Public Relation Office of the Hong Kong Baptist University, presented an interview article with Professor Kai-Tai Fang. It includes many anecdotes of Professor Fang and gives snapshots of Professor Fang's life as a student, researcher and family man. It contains some of his philosophy of life, outlines of his important contributions to statistics, and a list of his publications.

The twenty-three invited papers encompass a wide range of topics and are grouped into four parts. They are independent of each other. Each is dedicated to a specific issue on multivariate analysis, design of experiments, biostatistics, and other statistical issues. This book is targeted to a broad readership. We hope that regardless of their background, readers will find some parts are of their interest and suit their needs.

The first part contains 7 articles on multivariate analysis, studying a number of emerging issues in the field. It begins with Art Owen's introduction of multidimensional variation for quasi-monte Carlo simulation, followed by Rahul Mukerjee's investigation on the higher order power properties for a very general class of empirical discrepancy statistics. Mingjin Wang and Qiwei Yao introduced various methods for modelling multivariate volatilities, a fundamental issue in asset pricing and portifolio management. Some recent advances in two-level structural equation models were surveyed in the paper by Peter Bentler, Jiajuan Liang and Ke-Hai Yuan. Hong-bin Fang, Samuel Kotz and Gang Wei studied the geometric structures of copulas and local dependence patterns, which have wide applications in medical research and risk managements. Driven by the needs of longitudinal studies, Jianxin Pan and Dietrich von Rosen proposed a new data-driven approach for modeling mean-covariance structures in a growth curve model. Tõnu Kollo and Anu Roos enriched the classical multivariate distribution theory by contributing results and simulation methods on Kotz-type elliptical distributions.

The second topic focuses on the design of experiments, consisting of 8 articles, covering a wide array of important subjects. It begins with the construction of the optimal two-step sequential U-type of designs by Peter Winker, followed by the study on the granularity and perfect balance of experimental designs by Fred Hickernell. Scott Beattie and Dennis Lin introduced a new class of Latin hypercube designs for computer experiments. Min-Qian Liu, Hong Qin and Min-Yu Xie gave an overview on some recent developments on the application of the discrete discrepancy to several common experiment designs. Lean designs of orthogonal arrays of 2- and 3-levels were extensively discussed by Chang-Xing Ma and Ling-Yau Chan. Yingcai Su presented quasi-random sampling for estimation of integrals of random fields, including a review of number-theoretic methods. An efficient approach to the probabilistic sensitivity analysis in engineering design was introduced by Agus Sudjianto, Xiaoping Du and Wei Chen. The topic concludes with an investigation on the kernel selection problem for experimental designs by Aijun Zhang.

The third part is on some recent developments in biostatistics with four articles on different subjects. Jianqing Fan, Gang Li and Runze Li presented a contemporary overview on a new class of variable selection techniques for various models in survival analysis with emphasis on oracle properties. Gang Li, Runze Li and Mai Zhou complemented the topic by presenting a review on the empirical likelihood for survival analysis. Ming Tan, Hongbin Fang and Guoliang Tian reviewed the recent advance on statistical analysis for tumor xenograft experiments and developed a new multivariate random effect model. Parameter estimation after termination of a multivariate group sequential test was investigated by Chengqing Wu, Aiyi Liu and Kai F.Yu.

The fourth part of the monograph collects four papers on advance in statistics. The subject begins with a contribution by Yuehua Wu on model selection based on M-estimation and cross-validation, followed by the article by Jinting Zhang on selecting important variables using order-dependent thresholding with emphasis on applications to regression splines. The last two papers in this monograph are on the two-sample t-test by using the Box-Cox transform authored by Hanfeng Chen and Md. Khairul Islam, and on the admissibility of location parameter under a more general class of loss functions contributed by Jian-Lun Xu.

We are most grateful to the enthusiastic support of all of the people who have helped to make this volume possible. We are particularly indebted to Hongbin Fang, Fred Hickernell and Runze Li for their invaluable help. We owe many thanks to Kenny Yeung for providing technological assistance in turning collective contributions into such a wonderful book. We would like to express our gratitude to Heping Zhang, the series editor, and Rok Ting Tan of World Scientific for their valuable assistance and guidance. Each article was reviewed critically by referees. We are grateful to Hongbin Fang, Guoliang Tian, Fred Hickernell, Runze Li, Hua Liang, Jiajuan Liang,

Changxing Ma, Kun Nie, Jianxin Pan, Gang Wei, Yuehua Wu, Jianlun Xu, Ke-Hai Yuan, Aijun Zhang, Jinting Zhang, for their invaluable and conscientious refereeing service.

Finally and most importantly, as his former students, we would like to wholeheartedly thank Professor Kai-Tai Fang for bringing us into the world of statistics, sharing with us his scientific creativity and fertile imagination, teaching us philosophy of sciences, and showing us how to mentor and foster younger generations. Many of our achievements reflect his determination. We are very proud of him, as a teacher and a friend. We wish him the best.

Jianqing Fan, Princeton
Gang Li, Los Angeles
December 5, 2004

Contents

A Conversation with Kai-Tai Fang

Agnes W. L. Loie

Abstract:
Kai-Tai Fang (方開泰) was born in 1940 in Taizhou, Jiangsu province, China. He received his secondary education at the renowned Yangzhou High School in Jiangsu. In 1957, Kai-Tai entered Peking University to read mathematics and in 1963 he undertook graduate studies at the Institute of Mathematics, Academia Sinica, Beijing.

After graduation, Kai-Tai was appointed Assistant Researcher in the Institute of Mathematics, Academia Sinica, a position he held until 1978 when he was promoted to Assistant Professor. The following year, he was transferred to the Institute of Applied Mathematics, Academia Sinica, shortly after which he was promoted to Associate Professor in October 1980. In 1984, Kai-Tai was appointed Associate Director of the Institute and in July 1986, he became Professor.

Fig. 1. Kai-Tai Fang

Kai-Tai has received many awards for his statistical works, which had a profound effect on developments in a wide range of fields. In 1982 he was awarded the special prize of the Ministry of Light Industry for the standardization of apparel sizes for Chinese adults. In 1984, his unified approach to the distribution of restricted occupancy problems won him the second-class prize for Science & Technology from the Academia Sinica. Kai-Tai's precision test methodology and determination gained him second-class

Contemporary Multivariate Analysis and Experimental Design—In Honor Celebration of Professor Kai-Tai Fang's 65th birthday. Edited by Jianqing Fan and Gang Li. The World Scientific Publisher, 2005.

prize for National Standardization in 1988. His solutions to the distribution of some random military coverage problems also won him another Science & Technology prize from the Academia Sinica the following year.

In 1992, Kai-Tai received a first-class award for Most Excellent Textbook from the State Statistical Bureau of the PRC for his authorship of *Statistical Distributions*. In the same year, his book *Generalized Multivariate Analysis* won him a special nationwide award for Most Excellent Book in China by the Government Information and Publication Administration, Beijing. In 1998, the number of citations of Kai-Tai's works reached the ninth highest in the country, according to the *Chinese Science Citation Database*.

In Hong Kong, Kai-Tai was presented with the President's Award for Outstanding Performance in Scholarly Work by Hong Kong Baptist University in 2001. He was made an honorary member of the Hong Kong Statistical Society in 2002.

Kai-Tai has authored and co-authored 16 textbooks and monographs and published more than 200 research papers. He has served on numerous editorial boards and was editor-in-chief of the book series "*Modern Applied Mathematics Methods*" in China from 1990 to 2003. Kai-Tai was instrumental in organizing several influential conferences and workshops, both internationally and nationally. He has supervised the research of many graduate students and provided useful advice, encouragement and collaboration for students and their peers around the world. In honor of his scholarly contributions, Kai-Tai has been elected a Fellow of the Institute of Mathematical Statistics and a Fellow of the American Statistical Association. The latter honored him "for (his) outstanding contributions to multivariate analysis, Quasi-Monte Carlo methods and design of experiments; for (his) leadership in statistical education, consultation and administration; and for (his) editorial service." He has also been elected a Member of the International Statistical Institute.

The following conversation took place in Hong Kong during the fall of 2004.

Loie: When did you start to develop an interest in mathematics and statistics? Did it have anything to do with your family background?

Fang: Not really. Shortly after I was born, World War II broke out and it was an era of complete chaos. My parents took the seven of us to find refuge in rural villages. After the war, things were still chaotic and a formal education system was not in place. The teachers were not serious about teaching and that gave us a perfect excuse to follow suit with regard to learning. I remember that because of a shortage of space and manpower, two classes of different levels shared the same classroom and the same teachers. That meant that the teacher could only devote half his time to teaching us and the other half to teaching the higher level. We lacked interest and the desire to study hard. We aimed only at a mere pass. It was not until headmaster Konghou Wang (王孔厚) stepped into our classroom that I took a positive twist in my learning attitude. It was also then that I began to develop an interest in mathematics.

Fig. 2. Kai-Tai when he was three.

Loie: In what way did he inspire you?

Fang: Every day Mr Wang would give an extra 30-minutes' tuition for our grade six class and he would come up with a list of questions for us to work on. Those who finished first and got all the answers right would be allowed to leave the classroom to play. The questions he set were far from routine and were in fact pretty interesting. There was one that I remember in particular. The question was about a farmer who was selling a basket of eggs. He approached the first family who bought half the basket of eggs plus half an egg. The second family bought half of the remaining total plus half an egg. The third bought the remaining half of the total plus half an egg, then the whole basket of eggs was sold out. How many eggs were there in the basket? It didn't take me long to come up with the answer: seven. I was the first to hand in the answer and was instantly allowed to go out and play. Even my elder sister, who was in the same class and ranked first in class, could not get the answer right. That was the first time in my life that I discovered my strength and competitiveness; it was also the first time I realized I had an edge over my classmates. After that, I was almost always the first to leave the classroom. This self-discovery, coupled with Mr Wang's recognition, worked miracles in building up my confidence and had a far-reaching impact on my self-esteem.

Loie: Would you regard Mr Wang your first mentor?

Fang: Indeed he was. He was instrumental in stimulating my interest in mathematics, an area that I undertook as my lifelong career.

Loie: What happened after that?

Fang: In 1951, I was admitted to Yangzhou High School, one of the most reputable, well-established secondary schools in the whole nation. The school adopted a serious and professional manner and they (the school board) even employed university professors to teach us. The deputy headmaster, for example, was a famous English professor. The quality of the teachers was exceptional and they attached great emphasis to independent thinking. The school had produced many famous graduates such as Chairman Jiang Zemin(江澤民); Hu Jiaomu (胡喬木), Mao Zedong's (毛澤東) secretary; and more than 10 members of the Chinese Academy of Sciences.

Fig. 3. Kai-Tai (front, third from right) and other junior high graduates of Yangzhou High School in 1954.

Loie: Was your interest in mathematics further enhanced here?

Fang: Yes. I met my second teacher who deepened my passion for mathematics. He was Guangzhao Fang (方光照). He adopted an enlightened approach by first asking questions before giving a lecture. This was to inspire us to think. I was always among the first two to answer the questions. His lectures were stimulating and I was captivated by what he had to say. When I was invited by the school to give a talk on how to learn mathematics in the late 1980s, I was so pleased to meet Mr Fang again and I highlighted his teaching approach in my talk.

Loie: I heard that Yangzhou High School was renowned for its whole-person education and its emphasis on encouraging students to develop a wide variety of interest as well as nurturing their psychological strength. How did your secondary education benefit your development as a whole?

Fang: When I entered senior high, I read many books on self-development to boost my psychological quality and I set a number of targets for myself. For example, to increase my perseverance level, I planned a series of target studies for every weekend, a practice that I maintained even when I entered university. Another example is that our school then had a scheme to encourage students to exercise. Those who succeeded in running a certain number of kilometers would be awarded a souvenir. I challenged myself to run every day, even in the severe cold winter climate. All these self-training exercises helped equip me with the determination to overcome future problems, both academic and otherwise; they also gave me the will to succeed. I never give up easily, regardless of the scale of any problem.

Loie: Can you tell me something about your university studies?

Fang: Professors at Peking University had high expectations about their students. Peking University, famous as it was, wanted to do just as well as Moscow University, which then ranked first in a number of areas. Their education strategy was to let the best professors teach first-year students so that the latter would have a solid foundation for their studies. This, I think, was a wise strategy because despite my 10-year stoppage in my studies due to the Cultural Revolution, I still had a firm grasp of mathematical techniques. Because of the keen competition between Peking and Moscow universities and also among students, all of us were under tremendous pressure. Many of my classmates were filled with a sense of negativism even though they performed exceptionally well in their secondary school days.

Fig. 4. Kai-Tai during his university days.

Loie: How did you surpass all the difficulties and keen competition you faced in learning advanced mathematics?

Fang: While at Peking University, I came across a book *How to Solve It - A New Aspect of Mathematical Method*, authored by G. Polya, who was then a professor at the Swiss Federal Institute (where Albert Einstein graduated), and later at Stanford University. This book embodies a wealth of wisdom on thinking skills. In a nutshell, the book establishes a close-knit link between the specific and general and advises readers to be general in order to be specific and vice-versa. It also warns readers that it is better not to have a book at all than to believe all that is written in the book. Professor Polya's book also challenges readers to do something positive to exceed the teachings of books they read. I was deeply moved by Polya's teachings and I put them into practice. I set high expectations of myself and required myself to look for solutions rather than seeking help from my teachers and classmates. Polya's book has had a life-long impact on me and I have applied his teachings to my academic studies and research ever since. It never occurred to me that I would have a chance to thank Professor Polya in person. That chance came when I visited Stanford University in 1982. Professor K.L. Chung (鍾開萊) took me to see him and I told him that to me, he was first and foremost a great educationalist and I trusted that his readers would agree with what I said.

Loie: I understand that you were a student of the renowned P.L. Hsu (許寶騄) and became greatly influenced by his supervision.

Fig. 5. P.L. Hsu (front, second from left) with Kai-tai (back, second from left) and his other graduate students at the Peking University campus in 1963.

Fang: Yes, the next person that impacted me was my supervisor, Professor P.L. Hsu, a UK-educated scholar who laid a solid foundation for multivariate statistical analysis and who had four papers acknowledging his contributions published in the same issue of the prestigious international journal *The Annals of Statistics* in 1980. This was an exceptional treatment by the journal in recognition of his outstanding contributions. Hsu was severely ill in 1962 and was instructed to take full rest by his doctor. Despite his illness, he continued to work full time and take up both research and supervision duties. He required us to study a 50-page book written by a Russian mathematician, a Stalin award winner, and asked us to try to improve his results and make a report in class. After our presentation, Hsu told students that if they followed the Russian approach, they could only come up with a single dimension. He then showed us how to solve the problem using different approaches and came up with a more powerful answer that catered not only to one dimensional statistics but also to high dimensional statistics. This was an eye-opening experience for me. Hsu's insistence in fulfilling his teaching obligations, despite his weak physical condition, and his dedication to research exerted great influence on my future academic career. In fact, my first paper "The Limiting Distribution of Linear Permutation Statistics and its Applications" was completed under his supervision. Hsu said to me that there was a gap in a paper originally published in The *Annals of Statistics* and should I be able to identify and fill in that gap, I would be qualified to graduate. Very soon, I was able to identify the gap and fill it in and even discovered that the paper could well be extended. I put in many new angles which produced some interesting results. After two weeks, I handed in my paper and, after reading it, Hsu told me that I could now graduate. He even recommended that *Acta Peking University* should publish my paper, which it willingly accepted. It was however most unfortunate that before my paper was published, a political movement took place and all publications and newsletters came to a halt. Fortunately, the paper was published 19 years later in *Acta Appl. Math.*

Loie: That was an indelibly dark era for mainland residents. How did you survive this period?

Fang: China then was really shrouded in an intense political climate and people became distant from one another, fearing that any outpouring of genuine feelings would be betrayed, especially if they were about government and policies. Because of this, my years at the university were unhappy - a sharp contrast from my high school days. My dislike for the chaotic political movement in Peking University prompted my decision to pursue further studies at the Institute of Mathematics, Academia Sinica (later changed to Chinese Academy of Sciences) and became the first postgraduate student of Professor Minyi Yue (越民義).

Loie: What did you do in Academia Sinica?

Fang: My first two years at the Institute of Mathematics were fruitful under a favorable academic ambience. In 1965 I was assigned to An Shan Steel and Iron Co and was forced to turn to application instead of just theory. The engineers there treated us nicely and had a high expectation of us. At that time, I used 'non-linear' regression analysis to analyze the data collected. This period signaled a positive change for me in that I could apply my knowl-

Fig. 6. Kai-Tai (front, second from right), other postgraduate students and new staff of Academia Sinica were sent to work in rural villages during the Cultural Revolution.

edge to meet the high expectations people bestowed on us. What I learned in Peking University focused merely on theory and did not touch on any applications, which was the weakness of the Russian system. While I was at An Shan, I was asked to give lectures to the engineers. I covered eight topics in statistics, all of which were published for staff reference, an indication of the high regard in which they held me. However, this favorable situation did not last long as the political movement took shape the following year.

Loie: That was a prelude to a political storm, with the Cultural Revolution just round the corner?

Fig. 7. Grace Yang (centre)

Fang: Indeed. The subsequent years were a complete waste of time. All of us were deprived of the right and opportunity to pursue our studies and research. In 1965 and 1966, I was sent to the villages as a laborer. The following two years came the Cultural Revolution and the political movement lasted until 1976. We were all under tremendous pressure and stress and were uncertain about tomorrow. What was important then was that we could survive today and we did not even dare to think about tomorrow. As much as I hated this period, it did strengthen the psychological side of me, enabling me to face each and every bitter challenge with an unyielding manner. Like Professor Grace Yang (羅昭容) of the University of Maryland once said me: "You have recovered your 10 lost years." She gave me a lot of encouragement on a number of issues.

Loie: When did you learn orthogonal design and start conducting experiments with this method?

Fang: During the early '70s, staff from Peking University and the Institute of Mathematics, Academia Sinica, attempted to promote and apply orthogonal design to the industrial sector. In 1972, I had the opportunity to go to the Tsingdao Beer Factory and other factories. I supervised the engineers there to apply orthogonal design to industrial experiments. It was a precious experience for me to witness the substantial potential of applying orthogonal design to practical industrial use. However, I also detected the considerable difficulties faced by the engineers in understanding statistical methods, especially in calculating the ANOVA Table without the help of computers or calculators. I came to realize the need for statisticians to simplify the complicated statistical theories and methods, and later created "Visualization Analysis" for analytical use on experiment data. Very soon this method was commonly used on the Mainland, triggering a great sense of encouragement and inspiration on my part.

Loie: There were quite a number of contributions that you made to orthogonal design. What were they?

Fang: During my process of promoting the common use of orthogonal design, I encountered quite a number of complicated multi-factor and non-linear issues. The engineers were unable to identify a satisfactory parameter value combination for a long time. An example was a porcelain insulator factory in Nanjing. The factory had a team of staff assigned to conduct experiments continually to identify a satisfactory parameter value combination. Although they had achieved much in their experiments, they still failed to get one of the responses to meet the requirement, thus failing to deliver the glass insulator products. (At that time, the factory received a large number of orders for glass insulators but was unable to deliver the products.) In view of the complexity of the issue, I adhered to the principle of "big net catching big fish". I conducted 25 experiments and arranged six 5-level factors by orthogonal design. From a statistical point of view, the experiment model was non-estimable and was therefore incorrect. However, in those 25 experiments, one had all the responses fulfilling the requirements. That was great news to the factory in-charge. Should one liken the outcome to winning the U.S. lottery or was it significant? In fact, using orthogonal design to conduct 25 experiments actually represented 15,625 experiments, thus greatly increasing the likelihood of attaining an ideal technical/manufacturing condition. In my opinion, the power of fractional factorial design was that the experimental points have a good representation. Since then, I have used the same strategy to solve many of the "lasting, major and difficult" problems of the factories. This success has also injected in me the necessary courage to initiate the Uniform Design theory and method.

Loie: Can you tell me how you came up with uniform design, an approach so well known in the statistical field?

Fang: Using the "big net" approach to get the best combinations, I mustered the courage to create another new approach - the uniform design approach. After I returned to the Academia Sinica in 1970, I came across several occasions which called for the application of a more powerful statistical approach.

For example, in 1975, a factory manufacturing steel for automobiles wanted to come up with a nation-wide standard that needed the numerical calculation of many five-dimension integrals. At a time when computers were much less powerful, it was almost impossible to do so. Luogeng Hua (華羅庚) and Yuan Wang (王元) came up with a method to solve high dimension integral problems. Professor Wang taught me how to use their method and I realized that the method might be applied to experimental design.

Fig. 8. Yuan Wang (right) after being conferred an honorary doctoral degree by Hong Kong Baptist University. From left: Kai-Tai; Daniel Tse, former President and Vice-Chancellor of HKBU; and Lu Ping, another honorary degree recipient in 1998.

In 1978, there were three major missile-related projects covering land, sea and aerospace. A problem-solving approach was needed to tackle all the projects. Again, it was highly challenging. I had to come up with a new method, one that could approximate a complicated system by a simple method with required accuracy. The great challenge was a motivating force to me.

Fig. 9. Kai-Tai with Samuel Kotz and his wife.

I collaborated with Yuan Wang and we worked out the uniform design. This method made possible the calculation of an accurate answer in .00001 seconds with the required accuracy. It was both time- and cost-saving and provided a valuable alternative since it could also be used in computer experiments as well as laboratory experiments. Several years after the uniform design theory was proposed, I discovered that it was being used extensively in the Mainland. Not only was it used for military purposes, it was also adopted by and for civilians.

Loie: The 1980s marked a significant chapter in your life as you started to play a key role in the global scene. Would you consider that as an epoch-making era for you?

Fang: In a way, yes. In 1980 when I had the opportunity for an overseas visit, I did not plan to go initially in view of my wife's illness. She suffered from asthma and, of course, my two daughters were quite young. My standard of English then was low and basically I could not communicate in the language. Besides, we were blocked from the outside world during the 10 years of political movement and were ignorant of what was happening around the world. After much deliberation, I finally visited Yale University with one focus in mind - to learn as much as I could within a limited period of time. The eight months I spent at Yale was, on one hand, tremendously difficult as I was learning from scratch and yet, on the other hand, it was fulfilling as I succeeded in coming up with several papers, one of which was published in a respected journal and one in *Encyclopedia of Statistical Sciences.* While at Yale, Professor Samuel Kotz sent me one of his books which inspired me to come up with a paper - my first paper to be published in a western journal - and which started a series of collaborations between us.

Then in 1981 and 1982, I went to Stanford for a visit and there I met Professor T.W. Anderson, a Princeton graduate who later taught at Columbia and Stanford universities. He asked me to read two papers and then we had some idea about generalized multivariate analysis. Many statisticians wanted to generalize multivariate analysis to non-normal populations, but they failed. The combination of the essence of Polya's and Hsu's teaching - that you need to seek different approaches and you need a powerful tool in order to exceed the work of others - prompted me to undertake a search for such a powerful tool - the $\stackrel{d}{=}$ operator. I systemat-

Fig. 10. T.W. Anderson (second from left) and his wife.

ically developed this tool and wrote two papers on the topic. After publication, Professor Anderson identified a vast potential for further development and subsequently selected a topic for extended research. He engaged more of his students in this research and on my return to the Mainland, I also brought many students into the activity.

This collaborative research of ours continued for eight years, during which more than 50 articles, two monographs and a collection of papers were published. This was a big leap forward compared to the situation prior to my overseas visit when it was a norm for one project to generate only one or two papers. I realized that for people of high standards, the choice of topics was of paramount importance.

Fig. 11. Shiing-Shen Chern (third from left) visiting the Institute of Applied Mathematics, Academia Sinica, in 1986. (From left:) Kai-Tai, Minyi Yue, Wentsun Wu, Yuanshuen Ching and Fang Wu posed a picture with him during the visit.

Our collaboration brought me high international reputation. I was invited by *Encyclopedia* and other journals to act as referee, author or Associate Editor. I was deeply appreciative of Professor Anderson for his identification of the potential and productivity of generalized multivariate analysis as a research topic.

Loie: Did you have the chance to visit other universities?

Fang: Yes, in October 1982, I was invited to give talks at several universities, including Princeton, Yale and Columbia, and the universities of Pennsylvania, Maryland, Rutgers and George Washington. These visits were useful to my work, especially after I was promoted to Associate Director of the Institute of Applied Mathematics, Academia Sinica. Then in 1985-86, upon Professor I. Olkin's recommendation, I taught two subjects in the Swiss Federal Institute as a Guest Professor. It posed another challenge for me as it was the first time I had to teach in

Fig. 12. Dietrich von Rosen (back, centre)

English. During my stay there, I seized the opportunity to visit various prestigious institutions including Oxford, Cambridge, London University College and Imperial College, etc. The visits helped me to "network" which has proved useful throughout my career. In fact, it was during my overseas trips that I met Professors Colin White, C.R. Rao, Norman Lloyd Johnson, D.R. Cox, A.P. Dawid, Y.L. Tong (董永良), Y.S. Chow (周元燊), George Tiao (刁錦寰), Michael Stephens and Dietrich von Rosen. I am glad that our paths crossed as they were all inspirational to me in one way or another.

Loie: I heard that you had a particular "connection" with the University of North Carolina (UNC) at Chapel Hill. How is that so?

Fang: Well, I first went to the University of North Carolina in Chapel Hill - where my former supervisor P.L. Hsu once taught - as visiting professor between 1986 and 1988. I taught generalized multivariate analysis. Since then, I have been linked to the University in different ways. In addition to P.L. Hsu, my former student Jianqing Fan (范劍青), a recipient of the COPSS award and now professor at Princeton, was also a faculty member of the university.

Fig. 13. Kai-Tai fifth from left with C.F. Ng (centre), the Dean of Science and now President and Vice-Chancellor, and departmental colleagues at Hong Kong Baptist University in 1997.

Loie: How would you describe your academic and research pursuits at Hong Kong Baptist University?

Fang: With the encouragement of Kai-

Wang Ng (吳啓宏), I moved to Hong Kong Baptist University (then College) in 1990. My years at Hong Kong Baptist University were the happiest and smoothest of my academic life. Many of my important papers were published during this period in international journals, gaining me global exposure and reputation. The academic ambience here is stimulating and the congeniality among colleagues is notable. The support I gained from the top administrators is keen, facilitating my collaboration with overseas and Mainland academics. The number of PhD and MPhil students supervised by me is comparatively large. All these factors have created an environment conducive to both academic and research developments. I must thank our former and current Presidents Dr Daniel C.W. Tse (謝志偉) and Professor C.F. Ng (吳清輝), as well as Academic Vice-President Professor Herbert H. Tsang (曾憲博) for their encouragement and support. It was at Hong Kong Baptist University that I reaped the most fruitful harvest in terms of academic and research pursuits. It was also here that I received various honors and awards.

Loie: You have developed or further developed quite a number of methods during your academic career here at HKBU. Can you tell me more about them?

Fang: The Quasi-Monte Carlo method was one that was expanded and further developed here in Hong Kong. We started applying the method in Beijing to develop the uniform design. In Hong Kong we continued to apply the Quasi-Monte Carlo Method to experimental design, and also to a variety of statistical problems, including simulation and statistical inference. In 1994, I co-authored a book, *Number-Theoretic Methods in Statistics*, with Yuan Wang to further promote the method and its applications. It was also in the same year that I became President of the newly-established Uniform Design Association of China, a post which I held for 10 years until 2003.

Fig. 14. Peter Winker.

There was a hiccup in the development of uniform design as the uniformity was categorized as a geometric criterion instead of a statistical one. This criticism provided an excuse for people to reject our papers. In view of this, I decided to spend more time on the uniform design theory. In

1992, a participant from North Carolina State University attended my conference in Hong Kong and told me that it was a pity that the Western community did not know about the uniform design. This remark was of great encouragement to me and inspired me to work on more solid basic theory for the uniform design method.

Loie: How did you overcome the technical difficulties of promoting the uniform design theory and method?

Fang: I indeed encountered a number of problems in tackling the issue. First, I was not familiar with the typical tools employed by the Quasi-Monte Carlo method as they were invented by mathematicians such as Luogeng Hua and Yuan Wang. Besides, I am a statistician and not a pure mathematician. One way to solve the problem was for me to learn to use the tools but it would not be effective in light of my age and time.

Second, the uniform design theory in itself was difficult. I therefore spent the first four years, i.e. from 1992-96, working on it. It was like an exploration for me and I made slow progress. It was necessary for me to identify the tools that suited me - on which I spent an enormous amount of time. As the Chinese saying goes: "It is of little use for peonies to blossom only by themselves. They need green leaves to bloom with them." I was stimulated to focus more of my time on the uniform design. In fact, 90 per cent of my academic pursuits have focused on uniform design since then. My collaboration with several scholars led to the discovery of a breakthrough that suited me. I came up with the conjecture that most orthogonal designs were uniform. If that was the case, we could link up orthogonal design with uniform design and obtain a vast development potential for uniform design.

I spent one year with Peter Winker of Germany, a doctoral student then and a professor now, to prove with the computer that my conjecture was true. It was exciting to find that my conjecture was true in that many existing orthogonal designs were also uniform designs. Our result was based on the measure of uniformity proposed by my colleague, Fred Hickernell. This discovery was of mutual benefit to both Fred and myself. For him, his proposed measure of uniformity was initially not appreciated by many but his measure became necessary in uniform design. For me, his measure of uniformity helped prove that many existing

Fig. 15. Rahul Mukerjee.

orthogonal designs were uniform designs. With this, we still had one step to complete - to come up with a mathematical proof.

To achieve this, I invited Rahul Mukerjee, Professor of the India Institute of Management, to collaborate with me. Rahul is a worldwide expert in experimental design. After two weeks, he told me that my conjecture was not always true, even for a two-level factorial case. However, he came up with an excellent result - that we could link up uniformity with orthogonality. A criterion "aberration" was used to measure orthogonal design. For uniform design, the centered discrepancy was used to assess uniform design. With this, Rahul established an analytic relationship between centered discrepancy and aberration.

Fig. 16. From left: A.P. Dawid, Fred Hickernell, Kai-Tai and T.W. Anderson at a conference in 1997. Sitting behind Kai-Tai was C.R. Rao.

This discovery was immediately published in a top journal, *Biometrika*. It opened up an entirely new area that linked up uniform design and factorial design, an area in which I collaborated with C.X. Ma (馬長興) and others, and which resulted in the publication of more than 20 papers since 1999-2000.

Then in 2000, I began collaboration with S.G. Ge (葛根年) from Suzhou University and M.Q. Liu (劉民千) from Nankai University to link up combinatorial design and uniform design. Another new direction was established and this also led to the publication of many research papers.

The breakthrough we achieved in relation to uniform design won international recognition. The *Encyclopedia of Statistics Science* (Second Edition) has chosen uniform design as an entry while the *Handbook of Statistics* Volume 22 (2003) already included uniform design as a chapter. *Springer Handbook of Engineering Statistics* invited us to write a chapter on uniform design for engineers and this too will soon be published.

Uniform design also won national acclaim. The Uniform Design Association of China, for example, reflected the need to conduct national conferences, training courses, workshops and other activities to meet the calls to promote the applications of uniform design in the Mainland.

Application-wise, there were numerous successful applications of uniform design in China and overseas. With the keyword "uniform design", you can call up (on the Internet) hundreds of published case studies. The application of uniform design by Ford Motor Co in the U.S. is exemplary of the applicability of this method. At Ford, under the leadership of Dr. Agus Sudjianto, the technique has become a critical enabler for them to execute "Design for Six Sigma" to support new product development, in particular, automotive engine design. I was told that today, computer experiments using uniform design have become standard practices at Ford Motor Company to support early stage of product design before hardware is available. Uniform design has also been successfully introduced elsewhere. A notable example was the significant contribution made by Professor Dennis Lin (林共進) to promoting the theory and application of uniform design in Taiwan, India and the U.S.

Fig. 17. Kai-Tai poses with Jianqing Fan (third from left) and his other former graduate students during a conference break.

Loie: Apart from research, you also spent much time on statistical education. Can you elaborate on that?

Fang: To promote statistical education, I wrote international monographs, textbooks for undergraduate and postgraduate studies as well as textbooks for engineers in the Mainland and for various targets on different occasions. I was also willing to take up guest professorships. Often, many of the participants who are now professors and industry and university leaders came to me and said they had listened to my lectures on various occasions or studied my textbooks when they were students. I found that quite rewarding.

I have been told that one of my textbooks has been assigned as a compulsory textbook for Analytical Chemistry students. That was beyond my expectations. Although promoting statistical education has increased both my exposure and reputation as a by-product, what I found most gratifying and encouraging was the fact that I can make some contributions to my country.

I also understand that some of my textbooks and articles have been published in layman's terms for different professions so that the non-statistical sectors could also conduct research with statistics including uniform design. For example, the application of multivariate statistics for the standardization of apparel sizes for Chinese adults in 1976-78 was successful and the National Standards Bureau invited me to write a series of lectures. The published articles were collected as a book entitled *Statistics and Standardization.* Another example was the An Shan Steel and Iron Co which I mentioned earlier.

Loie: I know that you have organized or co-organized quite a number of significant conferences, both international and national. Organizing conferences of this scale requires an enormous amount of time and attention. How did you find time to organize these activities amid your already hectic schedule?

Fang: On a national basis, China had fallen behind for at least two decades because of the political turbulence. It was necessary to bring it to par with our counterparts overseas. Collaboration was useful in this regard. I took part in organizing seven nationwide multivariate analysis conferences since 1979, with one part of it theoretical and the other on applications, to provide a platform for establishing collaboration between the two. To attract international collaboration, I organized the Sino-American Statistical Meeting in 1987, which attracted more than 200 participants. In Hong Kong, I organized the International Symposium on Multivariate Analysis and Their Applications in 1992, the International Workshop on Quasi-Monte Carlo Methods and Their Applications in 1995, the International Symposium on Contemporary Multivariate Analysis and Its Applications in 1997, the Symposium on Theory of Uniform Design and Its Applications in 1999, the 4th Monte Carlo and Quasi-Monte Carlo Conference in Scientific Computing in 2000 and the Symposium on the Uniform Experimental Design in 2003.

Fig. 18. Kai-Tai with his parents, sisters and brother in 1957.

Loie: You have been accorded high international reputation as a result of your contributions to the global statistical field. The honors and awards bestowed on you include

Fellow of Institute of Mathematical Statistics (1993), Fellow of American Statistical Association (2001) and numerous awards for your outstanding contributions to multivariate analysis, Quasi-Monte Carlo methods, design of experiments, and for your leadership in statistical education, consultation and administration as well as for your editorial service. Despite all these prestigious honors, you are still a modest man of high integrity, as reflected in the tributes dedicated to you by your peers. How did you manage to always conduct yourself in such a good manner and with such a positive attitude? Was there any advice you took to heart that helped shape you the way you are today?

Fig. 19. A family picture taken at the Summer Palace in 1983.

Fang: My parents set a good example for me. To this day, I still remember vividly the advice my father gave me. He said if you extend your help to others, you should forget about it. On the contrary, if you receive assistance from other people, you should always keep that in mind and return the favor. Before I reached 40, the country was poor. My wife was weak and my daughters were small. Our standard of living was basically minimal. Some of my friends helped me, but I was incapable of paying them back. When my economic situation improved, I paid back all the debts and whenever we came across a friend who needed financial assistance, we never hesitated to lend a helping hand. My mother was exemplary of how one should conduct one's self, even in an adverse environment. She came from a village background and had no educational opportunity or cultural heritage. It was not until after her marriage that she had the chance to learn how to read and write. My father, on the other hand, came from the upper class. My mother kept a low profile but she learned exceptionally fast. And she always presented herself well, regardless of the situation. I learned from my mother that if you want to adapt yourself to a new environment you have to learn to be aware of your surroundings and should not be self-centered. I adhered strictly to this philosophy when I first visited the United States, a country so vastly different from my own in almost every aspect - cultural, logic, systems, terms, etc. I thought of my mother and I began to watch attentively other people's behavior, their culture, their logic, their way of thinking, their strengths and their weaknesses. I became happy when I began to appreciate differences in my environment. I learned the things that were desirable and brought them back to the Mainland. My positive attitude allowed me to keep an open mind in my management style and, because of this, I was promoted to Associate Director of the Institute of Applied Mathematics, Academia Sinica, in a mere one-and-a-half years after my return to the Mainland. Another person who influenced my personal development was Professor K.L. Chung, the first PhD student of P.L. Hsu.

Fig. 20. Kai-Tai and wife Tingmei in 1986.

While I was in the U.S., he told me that many people hid themselves in the office or laboratories to do research and declined to mix with the Americans. He asked me why I went to the States and advised me to go out and mix with people. Following his advice, I joined an activity every two-and-a-half days, be it a seminar or party or social gathering. I benefited greatly by joining these activities and I became aware of a significant improvement in my English communication skills and in my understanding of the Western culture.

Fig. 21. Kai-Tai's family with C.R. Rao's family.

I am a firm believer that great achievements involve great risks. I encountered a dilemma in 1980 when I was offered an opportunity to go abroad as a visiting scholar. My wife then was sick and my two kids still small. This, coupled with my unfavorable financial situation, somehow deterred me from making a positive move. One of my friends said to me that if

I did not go abroad, my career development would be limited. He said every person must face at least one difficult period in his lifespan and that one must face it with bravery and courage in order to overcome it. This remark was inspirational to me. I therefore took the risk of traveling abroad and was psychologically prepared to come back anytime should I receive a telegram with bad news. My wife, Tingmui Li (李庭梅) who was told by some of the neighbors that she might not be able to see me again should I go abroad, supported my move. I deeply appreciate her much-needed understanding and unfailing support.

Loie: Do you have any motto and if yes, could you share it with the younger generation?

Fang: I don't have one in particular but I think that to me, the most important thing is to be a person of integrity. Good character precedes good academic achievements. Also, don't be afraid of difficulties. Face the problems head-on and find a way to solve them. Remember that there is always a way out for those who seek it. I went through the 10-year Cultural Revolution without even knowing whether there would be a tomorrow and I survived. It is important to build a strong psychological shield to shelter yourself from external blow. I also encourage youngsters to work hard. If you decide to go for something, do it with all your might and give the best you can.

Fig. 22. Gang Li (right)

Loie: What post-retirement plans do you have?

Fang: I have been invited to continue supervising postgraduate students, a task which I willingly accept as it has always been my earnest wish to nurture successors for the statistical field. It always gives me great gratification to see my former postgraduate students, such as Jianqing Fan and Gang Li (李剛), now professor at UCLA, doing so well.

I also plan to do some leisure traveling as my previous trips were mostly work-related. It will be nice to have time to listen to music, classical in particular, and to attend to my other interests including playing ping-pong, badminton and Chinese chess, swimming and photography.

I would also like to spend more time with my wife and my two daughters, Ying Fang (方穎) and Yan Fang (方燕), both of whom are now working in the U.S. The time that I spent with them was minimal and I always feel guilty when I come to think of it. Although my wife and I have been married for 36 years, the time that we were actually together was just 19 years. We were seven years apart immediately after our marriage in 1968 and were separated again in 1990 when I came to Hong Kong.

Fig. 23. A happy family reunion in the U.S.

A Tribute to Professor Kai-Tai Fang

Agnes W. L. Loie

Some names of significance popped up during my interview with Professor Kai-Tai Fang. Significant, because they are the names of those whose academic interactions with Kai-Tai have sparked off fruitful collaborations that have impacted on the development of the statistical field in different ways. I took the liberty of contacting some of these scholars, who themselves are key players in the global statistical circle, to invite them to say a few words about Kai-Tai at the onset of his retirement. The invitation was met with immediate and enthusiastic responses. Here is a collective tribute to Kai-Tai by some of his peers, friends and collaborators:

T.W. Anderson, Professor, Stanford University:

Professor Kai-Tai Fang came to the Department of Statistics at Stanford University in the summer of 1981 as a visitor. He was motivated by a desire to develop his knowledge and expertise in multivariate statistical analysis, which is one of the fields of statistics in which I have done a great deal of research and exposition.

A topic that I expected would be challenging and useful for both of us was the statistical methodology appropriate for elliptically contoured distributions.

We worked together during the academic year 1981-82 as well as the summers before and after that academic year. I engaged more of my students in this research; on his return to China Kai-Tai brought many students and associates into our activity.

We continued this collaborative research for a period of years. Many of our studies with those of our students and associates were assembled in a

Contemporary Multivariate Analysis and Experimental Design—In Honor Celebration of Professor Kai-Tai Fang's 65th birthday. Edited by Jianqing Fan and Gang Li. The World Scientific Publisher, 2005.

volume that we edited: *Statistical Inference in Elliptically Contoured and Related Distributions*, 1990.

The Hong Kong Baptist University sponsored several conferences on multivariate analysis; Kai-Tai and I participated in them. One of them was the International Symposium on Multivariate Analysis and its Applications in March 1992 in which nearly 200 statisticians participated. This symposium resulted in a volume co-edited by Kai-Tai, Ingram Olkin, and me and published by the Institute of Mathematical Statistics in 1994.

Kai-Tai has extended our work to multivariate models more general than elliptically contoured ones, particularly *Symmetric Multivariate and Related Distributions* with Samuel Kotz and Kai-Wang Ng (1990) and *Generalized Multivariate Analysis* with Yaoting Zhang (1990). I have reported on some of my extensions of our work in the third edition of my book *An Introduction to Multivariate Statistical Analysis (2003).* Incidentally Kai-Tai and Yaotung Zhang drew heavily on my first edition for their book in Chinese *An Introduction to Multivariate Analysis* (1982).

Kai-Tai has also done significant research in fields of statistics other than multivariate analysis, including design of experiments and clustering. His broad interests include number theory as well.

It has been a pleasure for me to collaborate with Kai-Tai on research, writing, editing, and organizing. I have been impressed again and again by his energy and initiative.

Another feature we share is our birthday: June 5!

Fred J. Hickernell, Professor, Hong Kong Baptist University:

Good statisticians are hard to find; great statisticians are even rarer. Yet, for the past 14 years we have been blessed to have Professor Kai-Tai Fang, one of the most well-known statisticians of China, to be part of our Department of Mathematics. What a tremendous honor this has been!

Kai-Tai joined the Department of Mathematics at Hong Kong Baptist University at a critical time. The Department had recently begun to offer government-approved BSc degrees but had aspirations to engage in internationally recognized research and offer postgraduate degrees. At the time most colleagues were relatively young and inexperienced in doing research. Kai-Tai provided the necessary leadership to the Mathematics Department that spurred our development.

There are many things that I appreciate about Kai-Tai and have learned from him. Any list would be incomplete, but here are a few:

- Kai-Tai leads by example. By publishing important papers in international journals, engaging in various research collaborations and organizing international conferences, he has set a high standard for the rest of us to aspire to.

- Kai-Tai is an expert in involving colleagues and students in his work. One reason that he is so productive is that he knows how to draw on the resources of others. At the same time, he uses these opportunities to mentor other people and help them develop their potential.
- Kai-Tai is generous with his ideas. Choosing good research topics is crucial to success, and some scholars guard their ideas jealously. Not so with Kai-Tai, who freely suggests interesting research problems to his colleagues and students and offers his advice on how best to solve them.
- While some scholars focus on their own interests, Kai-Tai has always strived to further the interests of the whole Mathematics Department, the University and the statistics profession.
- Many mathematicians do not appreciate that statistics is a separate discipline from mathematics, and many statisticians view mathematics as bothersome. Kai-Tai understands how statisticians view problems differently from mathematics, but at the same time recognizes that mathematics provides many useful tools for the statistician.
- Although a world-renowned scholar, Kai-Tai is happy to perform the ordinary teaching and administrative duties expected of a professor. He respects other colleagues, even when they may be junior to him in rank or experience.
- Kai-Tai is hospitable. He loves to receive visitors and students even amidst his hectic schedule. He takes a keen interest in their well-being and makes sure that they enjoy their time at Hong Kong Baptist University.

Probably what I most admire about Kai-Tai, and wish that I could emulate better, is how he can keep his composure and even be gracious in the midst of difficult and emotional situations. I am most grateful to have the opportunity to be his colleague and to learn so much from him. Although he is going on to a well-deserved retirement, it will be a definite and heavy loss for the students and colleagues of the Mathematics Department.

Rahul Mukerjee, Professor, India Institute of Management:

I recall, with much fondness, my first meeting with Professor Kai-Tai Fang in June, 1997. Since then we developed a long-standing friendship and collaboration and I was privileged to visit him many times.

Indeed, even before meeting Kai-Tai, I knew him very well through his path-breaking contributions to such diverse areas as multivariate analysis, experimental design, and so on. After working with him, I was amazed with his profound depth of knowledge, innovative ideas and incisive analytical skill. We collaborated on research problems in the fields of experimental design and empirical likelihood, and on each occasion, it was a most rewarding experience. Certainly, I learned a lot through this association and, just as the other co-workers of Kai-Tai, I hold the highest esteem for him.

In addition to being a scholar of the highest order of eminence, Kai-Tai is an extremely kind person who embodies the Asian human values. I wish him a long, productive life.

Dietrich von Rosen, Professor, Swedish Agriculture University:

Fourteen years ago after a hard day's work at the Tartu Seminar on Multivariate Statistics, Professor Kollo from the University of Tartu and myself discussed about the achievements in multivariate statistical analysis during the last ten years. Perhaps because of our fatigue, we could not identify many new interesting topics with one exception. There were several books on elliptical and spherical distributions and in all of them Professor Kai-Tai Fang was involved. We decided to invite Kai-Tai to our part of the world. Since then Kai-Tai has been a frequent guest in our region, i.e. the triangular region of Estonia, Finland and Sweden.

Our collaboration has been very successful. From my point of view, I cannot overestimate the impact Kai-Tai has had on the statistical community and in particular on my research. I know that Kai-Tai is working extremely hard but at the same time he always has time to share his knowledge and the findings of his research groups. This is how science can and should evolve and to me, Kai-Tai is an excellent ambassador for the international academic society. For example, besides giving several seminars at our university he also conducted courses for non-statisticians. In particular I would like to mention the one on the uniform design which was highly appreciated by many students from different disciplines. Moreover, I remember Kai-Tai once acted as an opponent of a Swedish thesis and he did his duty in such a good and constructive way that everyone was happy after the defence - the thesis defender, the committee and the whole audience including the defender's mother, father and friends.

Something which also characterizes Kai-Tai is effectiveness. Over the years when Kai-Tai visited us, he always brought with him some basic ideas for at least one or two papers. Without these preparations the effectiveness would have been significantly reduced and the output lessened. The fields in which we have been working comprise copulas, generalizations of the complex normal distribution, influential observations, growth curve models and general estimation procedures, etc.

Another reason why things have always become so effective is that there are many excellent students around Kai-Tai, some of whom I still work with. To be a good supervisor and help students open the door to the scientific world is an art. I believe that the research environment created by Kai-Tai and others at Hong Kong Baptist University provides an environment conducive to stimulating students to become excellent researchers. It has always been fruitful for me to visit Hong Kong Baptist University as it was there that some of my research ideas originated.

My participation in several of the conferences organized by Kai-Tai and his colleagues has also proven useful. The conferences were well-organized, the keynote speakers invited were inspiring and the ambience stimulating. Kai-Tai was instrumental in making all these happen.

I am happy to know of such a visionary, warm and friendly scientist and I wish Professor and Mrs. Fang a happy and successful future.

Yuan Wang, Professor, Academy of Mathematics & System Sciences, Chinese Academy of Sciences:

Professor Kai-Tai Fang and I began to cooperate in 1976. Kai-Tai asked me a question on the numerical evaluation of a 5-fold definite integral. I introduced him to use the number theoretic method (or Quasi-Monte Carlo method). He succeeded.

He then knew the number theoretic method might be useful in experimental design. In 1978 a Chinese industrial agency proposed a problem of experimental design to him. As the number of experiments would be too large if the classical methods were used, Kai-Tai turned to the possibility of using the number theoretic method, and he discussed it with me. This was the start of a long cooperation between us.

It has been very nice cooperating with him. We have been very happy and fruitful during these near 20 years of co-operation. While he was working in Beijing, we attended a seminar every week. As most of the time we were not in the same place, we communicated by letters, emails and telephones to exchange our ideas.

Kai-Tai is a very nice man and is very easy to approach. He is very active and often seeks to come up with problems and then look for solutions. In particular, I think he is a genius statistician. When faced with a statistical problem, he can always propose a way to solve it. This leaves me with a deep impression!

I sincerely hope he is healthy and happy.

Peter Winker, Professor, University of Erfurt :

I met Professor Kai-Tai Fang for the first time at a conference in Tartu, Estonia, in 1994. We happened to sit next to each other at the conference banquet, and Kai-Tai introduced me to uniform design theory. I was a complete novice in this field, but discovered that the problem of finding uniform designs or at least low discrepancy designs might be dealt with by methods of integer programming. Given that I had some experience in applying optimization heuristics to problems in statistics and econometrics, we decided to start a joint project on this topic. In fact, within a couple of months, we finished our first joint paper without even meeting for a second time - thanks to e-mail. My next opportunity to meet with Kai-Tai was for the Workshop on Monte Carlo and Quasi-Monte Carlo methods organized at Hong Kong Baptist University in 1995. There, we presented already our second joint paper. Afterwards, the cooperation continued to be very productive and fruitful. We met at different occasions in Heidelberg and Hong Kong, and this summer, Kai-Tai visited Erfurt supported by a grant of a joint Hong Kong - Germany exchange programme.

For me, it is a real pleasure to work with Kai-Tai. He introduced me to a new interesting field and was willing to take over new ideas from optimization heuristics. But besides joint work, it is also a pleasure to meet with Kai-Tai and to enjoy spending some time together. I wish him all the best for his coming anniversary and the time after retirement, and I am quite optimistic that our close collaboration will not end with this event.

Publications

Kai-Tai Fang

Department of Mathematics, Hong Kong Baptist University.
(ktfang@hkbu.edu.hk)

Books:

[1] Fang, K. T. and Dai, S. S. (1973, 1974, 1979). *Basic Methods of Mathematical Statistics.* Science Press, Beijing (in Chinese).

[2] Fang, K. T. (1977, 1981). *The Analysis of Variance.* Science Press, Beijing (in Chinese).

[3] Liu, C. W., Dai, S. S. and Fang, K. T. (1980). *Elements of Probability Papers.* Science Press, Beijing (in Chinese).

[4] Fang, K. T. and Wu, C. Y. (1981). *Mathematical Statistics and Standardization.* Technical Standardization Press, Beijing (in Chinese).

[5] Zhang, Y. T. and Fang, K. T. (1982, 1999, 2003). *An Introduction to Multivariate Analysis.* Science Press, Beijing (in Chinese).

[6] Fang, K. T. and Pan, E. P. (1982). *Clustering Analysis.* Geological Publishing House, Beijing (in Chinese).

[7] Fang, K. T. and Xu, J. L. (1987). *Statistical Distributions.* Science Press, Beijing.

[8] Fang, K. T., Quan, H, and Chen, Q. Y. (1988). *Applied Regression Analysis.* Science Press, Beijing (in Chinese).

[9] Fang, K. T., Xing, K. F. and Liu, G. Y. (1988). *Precision of Test Methods Determination.* Chinese Standardization Press, Beijing (in Chinese).

[10] Fang, K. T. (1989). *Applied Multivariate Analysis.* East China Normal University Press, Shanghai (in Chinese).

Contemporary Multivariate Analysis and Experimental Design—In Honor Celebration of Professor Kai-Tai Fang's 65th birthday. Edited by Jianqing Fan and Gang Li. The World Scientific Publisher, 2005.

[11] Fang, K. T., Kotz, S., and Ng, K. W. (1990). *Symmetric Multivariate and Related Distributions.* Chapman and Hall Ltd., London and New York.

[12] Fang, K. T. and Zhang, Y. T. (1990). *Generalized Multivariate Analysis.* Science Press and Springer-Verlag, Beijing and Berlin.

[13] Fang, K. T. and Wang, Y. (1994). *Number-Theoretic Methods in Statistics.* Chapman and Hall, London.

[14] Fang, K. T. (1994). *Uniform Design and Design Tables.* Science Press, Beijing (in Chinese).

[15] Fang, K.T. and Ma, C.X. (2001). *Orthogonal and Uniform Experimental Designs.* Science Press, Beijing (in Chinese).

[16] Pan, J.X. and Fang, K.T. (2002). *Growth Curve Models and Statistical Diagnostics.* Springer, New York.

Lecture Notes/Proceedings:

[1] Fang, K. T. and Anderson, T. W. (Eds) (1990). *Statistical Inference in Elliptically Contoured and Related Distributions.* Allenton Press Inc., New York.

[2] An, H. Z., Fang, K. T., Hickernell, F. J., and Yang, Z. Q. (1991). *Applied Statistics Lecture Series.* Hong Kong Baptist College.

[3] Chen, X. R., Fang, K. T. and Yang, C. C. (Eds) (1992). *The Development of Statistics: Recent Contributions from China.* Longman Scientific & Technical, London.

[4] Anderson, T. W., Fang, K. T. and Olkin, I. (Eds) (1994). *Multivariate Analysis and Its Applications.* Institute of Mathematical Statistics, Hayward, California.

[5] Fang, K. T. and Lam, P. C. B. (Eds) (1994). *Proceedings 1994 Workshop in Applied Statistics.* Hong Kong Baptist College.

[6] Fang, K. T. and Hickernell, F. J. (Eds) (1995). *Proceedings 1995 Workshop in Applied Statistics.* Hong Kong Baptist University.

[7] Fang, K. T. and Hickernell, F. J. (Eds) (1995). *Proceedings Workshop Quasi-Monte Carlo Methods and Their Applications.* Hong Kong Baptist University.

[8] Fang, K.T., F.J. Hickernell and H. Niederreiter (Eds)(2001). *Monte Carlo and Quasi-Monte Carlo Methods 2000.* Springer, Berlin.

Papers:

1965

[1] Fang, K.T., Dong, Z.Q., and Han, J.Y. (1965). The structure of stationary queue without after-effect. *Acta Mathematicae Applicatae and Computation Sinica* **2** 84–90 (in Chinese).

1972

[2] Fang, K.T. (1972). An introduction to some nonparametric statistical methods. *Mathematics in Practice and Theory.* **5** 58–66 (in Chinese).

1976

[3] Fang, K.T. and Liu, C.W. (1976). The use of range in analysis of variance. *Mathematics in Practice and Theory* No.1 37–51 (in Chinese).

[4] Fang, K.T. (1976). Application of the theory of the conditional distribution for making the standardization of clothes. *Acta Mathematicae Applicatae Sinica* **2** 62–74 (in Chinese).

1977

[5] Sun, S.G. and Fang, K.T. (1977). The test for additional information in multivariate analysis. *Acta Mathematicae Applicatae Sinica.* **3** 81–91 (in Chinese).

[6] Liu, C.W. and Fang,K.T. (1977). Yates' algorithm and its application in 2^n-type orthogonal array. *Mathematics in Practice and Theory* No.3 9–18 (in Chinese).

1978

[7] Fang, K.T. (1978). Clustering analysis. *Mathematics in Practice and Theory* **4** 66–80 (No.1) and 54–62 (No.2) (in Chinese).

1979

[8] Fang, K.T. and Wu, C.Y. (1979). The extreme value problem of some probability functions. *Acta Mathematicae Applicatae Sinica* **2** 132–148 (in Chinese).

1980

[9] Fang, K.T. (1980). Experimental design by uniform distribution. *Acta Mathematicae Applicatae Sinica* **3** 363–372 (in Chinese).

[10] Bai, Z.D. and Fang, K.T., et al. (1980). A problem on independence of random variables. *Kexue Tongbao(Chinese Science Bulletin)* 90–92 (in Chinese).

1981

[11] Wang, Y. and Fang, K.T. (1981). A note on uniform distribution and experimental design. *Kexue Tongbao (Chinese Science Bulletin)* **26** 485–489.
[12] Fang, K.T. (1981). The Limiting distribution of linear permutation statistics and its applications. *Acta Mathematicae Applicatae* **4** 69–82 (in Chinese).
[13] Fang, K.T. (1981). Graphical analysis of multivariate observations. *Mathematics in Practice and Theory* 63–71 (No.3) and 42-48 (No.4) (in Chinese).
[14] Wang, L.H., Fang, K.T. and Zeng, Y.Y. (1981). An application of Bayes discriminant analysis in determining the systematic position of gigantopithecus. *Vertebrata PalAsiatica* **19** 269–275 (in Chinese).

1982

[15] Fang, K.T. (1982). Some clustering methods for a set of ordered observations. *Acta Mathematicae Applicatae Sinica* **5** 94–101 (in Chinese).
[16] Fang, K.T. and Sun, S.G. (1982). Discriminant analysis by distance. *Acta Mathematics Applicatae Sinica* **5** 145–154 (in Chinese).
[17] Fang, K.T. and Ma, F.S. (1982). Splitting in cluster analysis and its applications. *Acta Math. Appl. Sinica* **5** 339–534 (in Chinese).
[18] Fang, K.T., Wang, D.Q. and Wu, G.F. (1982). A class of constrained regression. *Mathematicae Numerica Sinica* **4** 57–69 (in Chinese).
[19] Fang, K.T. (1982). A restricted occupancy problem. *J. Appl. Prob.* **19** 707–711.
[20] Fang, K.T. (1982). Equivalence between Fisher discriminant model and regression model. *Kexue Tongbao (Chinese Science Bulletin)* **27** 803–806 (in Chinese).

1983

[21] Fang, K.T. and Niedzwiecki, D. (1983). A unified approach to distributions in restricted occupancy problem. *Contributions to Statistics, Essays in Honour of Professor Norman Lloyd Johnson (ed. by P.K. Sen)* 147–158. North - Holland.

1984

[22] Fang, K.T and Chen, H.F. (1984). Relationships among classes of spherical matrix distributions. *Acta Mathematicae Applicatae Sinica (English Series)* **1** 139–147.
[23] Fang, K.T. (1984). Some further applications of finite difference operators. *Appl. Math. & Math. Computation* No.4 22–32 (in Chinese).

[24] Fang, K.T. (1984). Theory of majorization and its applications. *Appl. Math. & Math. Computation* No.5 75–86 (in Chinese).
[25] Fang, K.T. and He, S.D. (1984). The problem of selecting a given number of representative points in a normal population and a generalized Mills' ratio. *Acta Math. Appl. Sinica* **7** 293–306 (in Chinese).
[26] Fang, K.T. and Wu, Y.H. (1984). Distributions of quadratic forms and generalized Cochran's Theorem. *Math. in Economics* **1** 29–48 (in Chinese).

1985

[27] Fang, K.T. (1985). Occupancy problems. *Encyclopedia of Statistical Sciences* Vol. 6 (ed. by Kotz, S., Johnson, N.L. and Read, C.B.) 402–406, Wiley.
[28] Fang, K.T. and He, S.D. (1985). Regression models with linear constraints and nonnegative regression coefficients. *Math. Numer. Sinica* **7** 237–246 (in Chinese).
[29] Fan, J.Q. and Fang, K.T. (1985). Minimax estimator and Stein Two-stage estimator of location parameters for elliptically contoured distributions. *Chinese J. Appl. Prob. Stat.* **1** 103–114 (in Chinese).
[30] Zhang, Y., Fang, K.T. and Chen, H.F. (1985). On matrix elliptically contoured distributions. *Acta Math. Scientia* **5** 341–353 (in Chinese).
[31] Fan, J.Q. and Fang, K.T. (1985). Inadmissibility of sample mean and regression coefficients for elliptically contoured distributions. *Northeastern Math. J.* **1** 68–81 (in Chinese).
[32] Fang, K.T. and Xu, J.L. (1985). Likelihood ratio criteria testing hypotheses about parameters of elliptically contoured distributions. *Math. in Economics.* **2** 1–9 (in Chinese).

1986

[33] Fang, K.T. and Xu, J.L. (1986). The direct operations of symmetric and lower-triangular matrices with their applications. *Northeastern Math. J.* **2** 4–16 (in Chinese).
[34] Anderson, T.W., Fang, K.T. and Hsu, H. (1986). Maximum likelihood estimates and likelihood ratio criteria for multivariate elliptically contoured distributions. *The Canadian J. Stat.* **14** 55–59.
[35] Fang, K.T. and Chen, H.F. (1986). On the spectral decompositions of spherical matrix distributions, and some of their subclasses. *J. Math. Res. & Exposition* No.14 147–156 (in Chinese).

1987

[36] Fang, K.T. (1987). A review: on the theory of elliptically contoured distributions. *Advance in Mathematics* **16** 1–15.

[37] Zhang, H.C. and Fang, K.T. (1987). Some properties of left-spherical and right-spherical matrix distributions. *Chinese J. Appl. Prob. Stat.* **3** 97–105 (in Chinese).

[38] Anderson, T.W. and Fang, K.T. (1987). On the Theory of multivariate elliptically contoured distributions. *Sankhya* **49, Series A** 305–315.

[39] Zhang, H.C. and Fang K.T. (1987). The distributions of normal matrix variate distribution. *J. Graduate School* **4** 22–30 (in Chinese).

[40] Fang, K.T. and Xu J.L. (1987). The Mills' ratio of multivariate normal distributions and spherical distributions. *Acta Mathematicae Sinica* **30** 211–220 (in Chinese).

[41] Fang, K.T., Fan, J.Q. and Xu, J.L. (1987). The distributions of quadratic forms of random matrix and applications. *Chinese J. Appl. Prob. Stat.* **3** 289–297 (in Chinese).

[42] Quan, H. and Fang, K.T. (1987). Unbiasedness of some testing hypotheses in elliptically contoured population. *Acta Mathematicae Applicatae Sinica* **10** 215–234 (in Chinese).

[43] Fan, J.Q. and Fang, K.T. (1987). Inadmissibility of the usual estimator for location parameters of spherically symmetric distributions, *Chinese Science Bulletin.* **32** 1361–1364, in English Ser., **34** 533–537.

[44] Fang, K.T. and Chen, Q.Y. (1987). Discussion on the probability method in Statistics, *Mathematical Statistics & Applied Probability* **2** 355–368 (in Chinese).

[45] Fan, J.Q. and Fang, K.T. (1987). Maximum likelihood character of distributions. *Acta Math. Appl. Sinica* (English Ser.) **3** 358–363.

1988

[46] Fang, K.T. and Fang, B.Q. (1988). Some families of multivariate symmetric distributions related to exponential distribution. *J. Multivariate Analysis* **24** 109–122.

[47] Fang, B.Q. and Fang, K.T. (1988). Distributions of order statistics of multivariate l_1–norm symmetric distribution and Applications. *Chinese J. Appl. Prob. Stat.* **4** 44–52 (in Chinese).

[48] Fang, K.T., Xu, J.L. and Teng, C.Y. (1988). Likelihood ratio criteria testing hypotheses about parameters of a class of elliptically contoured distributions. *Northeastern Math. J.* **4** 241–252 (in Chinese).

[49] Fang, K.T. and Fang, B.Q. (1988). Families of Exponential matrix distributions. *Northeastern Math. J.* **4** 16–28 (in Chinese).

[50] Fang, B.Q. and Fang, K.T. (1988). Maximum likelihood estimates and likelihood ratio criteria for location and scale parameters of the multivariate l_1–norm symmetric distributions. *Acta Math. Appl. Sinica* (English Series) **4** 13–22.

[51] Fang, K.T. and Fang, B.Q. (1988). A class of generalized symmetric Dirichlet distributions. *Acta Math. Appl. Sinica* (English Series) **4** 316–322.

[52] Fang, K.T. and Fan, J.Q. (1988). Large sample properties for distributions with rotational symmetries. *Northeastern Math. J.* **4** 379–388 (in Chinese).

[53] Fang, K.T. and Fang, B.Q. (1988). Generalized symmetrized Dirichlet Distributions. *Acta Math. Appl. Sinica* (English Series) **4** 316–323.

1989

[54] Xu, J.L. and Fang, K.T. (1989). Expected values of zonal polynomials of spherical matrix distributions. *Acta Math. Appl. Sinica* (English Series) **5** 6–14.

[55] Fang, K.T. and Xu, J.L. (1989). A class of multivariate distributions including the multivariate logistic. *J. Math. Research and Exposition.* **9** 91–100 (in Chinese).

[56] Fang, K.T. and Fang B.Q. (1989). A characterization of multivariate l_1–norm symmetric distributions. *Statistics & Probability Letters* **7** 297–299 (in Chinese).

[57] Quan, H., Fang, K.T. and Teng , C.Y. (1989). The applications of information function for spherical distributions. *Northeastern Math. J.* **5** 27–32 (in Chinese).

[58] Fang, K.T. and Liang, J.J. (1989). Inequalities for the partial sums of elliptical order statistics related to genetic selection. *The Canadian J. Statistics* **17** 439–446.

1990

[59] Wang, Y. and Fang, K.T. (1990). Number theoretic method in applied statistics. *Chinese Annals of Math. Ser. B* **11** 41–55.

[60] Wang, Y. and Fang, K.T. (1990). Number theoretic method in applied statistics (II). *Chinese Annals of Math. Ser. B* **11** 384–394.

[61] Fang, K.T. and Yuan K.H. (1990). A unified approach to maximum likelihood estimation. *Chinese J. Appl. Prob. Stat.* **7** 412–418 (in Chinese).

[62] Fang, K.T. and Wang, Y. (1990). A sequential algorithm for optimization and its applications to regression analysis. In *Lecture Notes in Contemporary Mathematics* (L. Yang and Y. Wang ed) 17–28. Science Press, Beijing.

1991

[63] Fang, K.T. and Bentler, P.M. (1991). A Largest characterization of spherical and related distributions. *Statistics & Probability Letters* **11** 107–110.

[64] Fang, K.T. and Wang, Y. (1991). A sequential algorithm for solving a system of nonlinear equations. *J. Computational Math.* **9** 9–16.

[65] Liu, S.S.M., Chan, A.K.K. and Fang,K.T. (1991). Evaluation of the promotional tactics for pharmaceuticals in Hong Kong. *Hong Kong Manager* **27** No.6 16–19.

1992

[66] Wang, S.R., Chen X.R. and Fang, K.T. (1992). Statistics in China: A brief account of the past and present, in *The Development of Statistics: Recent Contributions from China* 1–6. Longman, London.

[67] Anderson, T.W. and Fang, K.T. (1992). Theory and Applications elliptically contoured and related distributions. In *The Development of Statistics: Recent Contributions from China* 41–62. Longman, London.

[68] Wang, Y. and Fang, K.T. (1992). A sequential number–theoretic methods for optimization and its applications in statistics. In *The Development of Statistics: Recent Contributions from China* 139–156. Longman, London.

[69] Fang, K.T. and Zhen, H.N. (1992). A maximum symmetric differences principle and its applications in uniform design. *Chinese J. Appl. Prob. Stat.* **8** 10–18 (in Chinese).

[70] Fang, K.T., Kotz, S. and Ng, K.W. (1992). On the L_1–norm distributions in *L_1–Statistical Analysis and Related Methods (Y. Dodge ed.)* 401–413. Elsevier Science Publishers, North Holland, Amsterdam.

[71] Li, G. and Fang, K.T. (1992). The Ramanujan's q–extension of the exponential function and statistical distributions. *Acta Math. Appl. Sinica, (English Ser.)* **8** 264–280.

[72] Fang, K.T., Yuan, K.H. and Bentler, P.M.(1992). Applications of sets of points uniformly distributed on sphere to test multinormality and robust estimation. In *Probability and Statistics* eds by Jiang, Z.P. et al. World Scientific, Singapore, 56–73.

[73] Fang, K.T. and Wang, Y. (1992). Applications of quasi–random sequence in statistics. In *Proceedings of Asian Mathematical Conference 1990 eds by Li, Z. et al.* World Scientific, Singapore, 135–139.

[74] Zhu, L.X., and Fang, K.T. (1992). On projection pursuit approximation for nonparametric regression, In *Order Statistics and Nonparametrics: Theory and Applications, eds by Sen, P.K. and Salama, I.A.*. Elsevier Science Publishers, 455–469.

1993

[75] Fang, K.T. and Yuan K.H. (1993). The limiting distributions of some subclasses of the generalized non–central t–distributions. *Acta Math. Appl. Sinica (English Ser.)* **9** 71–81.

[76] Fang, K.T., Zhu, L.X. and Bentler, P.M. (1993). A necessary test of goodness of fit for sphericity. *J. Multivariate Analysis* **45** 34–55.

[77] Fang, K.T. and Wei, G. (1993). The distribution of a class the first hitting time. *Acta Math. Appl. Sinica* **15** 460–467 (in Chinese).

[78] Fang, K.T. and Zhang, J.T. (1993). A new algorithm for calculation of estimates of parameters of nonlinear regression modeling. *Acta Math. Appl. Sinica* **16** 366–377 (in Chinese).

1994

[79] Zheng, Z. and Fang, K.T. (1994). On Fisher's bound for stable estimators with extension to the case of Hilbert parameter space. *Statistica Sinica* **4** 679–692.

[80] Fang, K.T., Wang, Y. and Bentler, P.M. (1994). Some applications of number–theoretic methods in statistics. *Statistical Science* **9** 416–428.

[81] Fang, K.T., Bentler, P.M. and Yuan, K.H. (1994). Applications of number–theoretic methods to quantizers of elliptically contoured distributions. *Multivariate Analysis and Its Applications, IMS Lecture Notes - Monograph Series* 211–225.

[82] Fang, K. T. and Li, J. K. (1994). Some new results on uniform design. *Chinese Science Bulletin* **21** 1921–24. English version in **40** 268–272, 1995.

[83] Zhu, L. X. and Fang, K. T. (1994). The accurate distribution on the Kolmogorov statistic with applications to bootstrap approximation. *Advance in Applied Mathematics* **15** 476–489.

1995

[84] Pan, J. X. and Fang, K. T. (1995). Multiple outlier detection in growth curve model with unstructured covariance matrix. *Annals Institute of Statistical Mathematics* **47** 137–153.

[85] Zhu, L.X., Wong, H.L., and Fang, K.T. (1995). A test for multivariate normality based on sample entropy and projection pursuit. *J. Statistical Planning and Inference* **45** 373–385.

[86] Wang, Y., Lin, D. K. and Fang, K.T. (1995). Designing outer array points. *J. Quality Technology* **27** 226–241.

[87] Li, R.Z. and Fang, K.T. (1995). Estimation of scale matrix of elliptically contoured matrix distribution. *Statistics & Probability Letters.* **24** 289–297.

[88] Fang, K. T. and Li, R.Z. (1995). Estimation of scatter matrix based on i.i.d. sample from elliptical distributions. *Acta Math. Appl. Sinica, (English Ser)* **11** 405–412.

[89] Fang, K. T. and Hickernell, F. J. (1995). The uniform design and its applications. *Bulletin of The International Statistical Institute, 50th Session, Book 1* 339–349, Beijing.

[90] Zhu, L.X., Fang, K.T., Bhatti, M.I., and Bentler, P.M. (1995). Testing Sphericity of a high–dimensional distribution based on bootstrap approximation. *Pakistan J. Statistics* **11** 49–65.

[91] Zhu, L. X., Fang, K. T., and Zhang, J. T. (1995). A projection NT–type test for spherical symmetry of a multivariate distribution. *New Trends in Probability and Statistics Vol. 3 - Multivariate Statistics and Matrices in Statistics* (E - M Tiit, T. Kollo and H. Niemi eds) 109–122. VSP - TEV, Utrecht, The Netherlands.

[92] Shiu, W. C., Ma, S. L., and Fang, K. T. (1995). On the rank of cyclic Latin squares. *Linear and Multilinear Algebra* **40** 183–188.

[93] Shen, S. Y. and Fang, K.T. (1995). Neural computation on nonlinear regression analysis problems. *International J. Math. & Statist. Sciences* **3** No.2 155–178.

1996

[94] Chen, X.R., Zhu, L.X. and Fang, K.T. (1996). Almost sure convergence of weighted sums. *Statistica Sinica* **6** 499–509.

[95] Fang, K. T. and Hickernell, F.J. (1996). Discussion of the papers by Atkinson, and Bates *et al, J. R. Statist. Soc. B* **58** 103.

[96] Wang, Y. and Fang, K. T. (1996). Uniform design of experiments with mixtures. *Science in China (Series A)* **39** 264–275.

[97] Pan, J. X., Fang, K. T., and Liski, E. (1996). Bayesian local influence in the growth curve model with Rao's simple covariance structure. *J. Multivariate Analysis* **58** 55–81.

[98] Yang, Z. H., Fang, K. T., and Liang, J. J. (1996). A characterization of multivariate normal distribution and its application. *Statistics & Probability Letters* **30** 347–352.

[99] Pan, J. X. and Fang, K. T. (1996). Detecting influential observations in growth curve model with unstructured covariance. *Comput. Statist. and Data Anal.* **22** 71–87.

[100] Zhu, L. X. and Fang, K. T. (1996). Asymptotics for kernel estimate of sliced inverse regression. *The Annals of Statistics* **24** 1053–1068.

[101] Liang, Y. Z. and Fang, K. T. (1996). Robust multivariate calibration algorithm based on least median squares and sequential number theoretic optimization method. *Analyst Chemistry* **121** 1025–1029.

[102] Fang, K. T., Hickernell, F. J. and Winker, P. (1996). Some global optimization algorithms in statistics. In *Lecture Notes in Operations Research*, eds. by Du, D. Z., Zhang, X. S. and Cheng, K. 14–24. World Publishing Corporation.

1997

[103] Chen, X.R., Fang, K.T., Zhang, R.C. and Zhu, L.X. (1997). Asymptotics on combination generators. *Science In China, (Series A)* **27** No.4, 289–297.

[104] Li, R. Z., Fang, K. T. and Zhu, L. X. (1997). Some probability plots to test spherical and elliptical symmetry. *J. Computational and Graphical Statistics* **6** No. 4 1–16.

[105] Fang, K. T. and Li, R. Z. (1997). Some methods for generating both an NT–net and the uniform distribution on a Stiefel manifold and their applications. *Comput. Statist. and Data Anal.* **24** 29–46.

[106] von Rosen, Fang, H. B. and Fang, K. T. (1997). An extension of the complex normal distribution. In *Advances in the Theory and Practice of Statistics: A volume in Honor of Samuel Kotz*, eds. by N. L. Johnson and N. Balakrishnan 415–427. Wiley, New York.

[107] Fang, K. T. (1997). Elliptically contoured distributions. *Encyclopedia of Statistical Science, Update Vol. 1* 212–218. Wiley, New York.

[108] Fang, K. T. and Wang, Y. (1997). Number–theoretic methods. *Encyclopedia of Statistical Science, Update Vol. 2* 993–998. Wiley, New York.

[109] Fang, K. T., Lam, P. C. B. and Wu, Q. G. (1997). Estimation for seemingly unrelated regression equations. *Statistics & Decisions* **15** 183–189.

[110] Zhu, L. X., Fang, K. T., and Li, R. Z. (1997). A new approach for testing symmetry of a high–dimensional distribution. *Bulletin of Hong Kong Math. Society* **1** 35–46.

[111] Pan, J. X., Fang, K. T., and von Rosen, D. (1997). Local influence assessment in the growth curve model with unstructured covariance. *J. Statistical Planning and Inference* **62** 263–278.

[112] Zhu, L. X., Fang, K. T. and Bhatti, M. I. (1977). On estimated projection pursuit–type Cràmer–von Mises statistics. *J. Multivariate Analysis* **63** 1–14.

[113] Winker, P. and Fang, K. T. (1997). Application of Threshold accepting to the evaluation of the discrepancy of a set of points. *SIAM Numer. Analysis* **34** 2038–2042.

[114] Kotz, S., Fang, K. T. and Liang, J. J. (1997). On multivariate vertical density representation and its application to random number generation. *Statistics* **30** 163–180.

[115] Lee, A. W. M., Chan, W. F., Yuen, F. S. Y., Tse, P. K., Liang, Y. Z. and Fang, K. T. (1997). An example of a sequential uniform design:

application in capillary electrophoresis. *Chemometrics and Intelligent Laboratory Systems* **39** 11–18.

[116] Zhang, L., Liang, Y. Z., Yu, R. Q., and Fang, K. T. (1997). Sequential number–theoretic optimization (SNTO) method applied to chemical quantitative analysis. *J. Chemometrics* **11** 267–281.

1998

[117] Winker, P. and Fang, K. T. (1998). Optimal U–type design. In *Monte Carlo and Quasi–Monte Carlo Methods 1996*, eds. by H. Niederreiter, P. Zinterhof and P. Hellekalek 436–448. Springer.

[118] Fang, K. T., Zheng, Z. K. and Lu, W. (1998). Discrepancy with respect to Kaplan–Meier estimator. *Commun. Statist.–Simula.* **27** 329–344.

[119] Pan, J. X., Fang, K. T. and D. von Rosen (1998). On the posterior distribution of the covariance matrix of the growth curve model. *Statist. & Prob. Letters* **38** 33–39.

[120] Zhang, L., Liang, Y. Z., Jiang, J. H., Yu, R. Q. and Fang, K.T. (1998). Uniform design applied to nonlinear multivariate calibration by ANN. *Analytica Chemica Acta* **370** 65–77.

[121] Fang, K. T., Liang, J. J. and Li, R. Z. (1998). A multivariate version of Ghosh's T_3–plot detect non–multinormality. *Computational Stat. and Data Analysis* **28** 371–386.

[122] Tian, G. L. and Fang, K. T. (1998). Stochastic representation and uniform designs for mixture–amount experiments and for mixture experiments under order restrictions. *Science in China* **28** No.12 1087–1101.

1999

[123] Fang, K. T. and Liang, J. J. (1999). Tests of Spherical and elliptical symmetry. *Encyclopedia of Statistical Sciences Update Vol. 3* 686–691. Wiley, New York.

[124] Winker, P. and Fang, K. T. (1999). Randomness and quasi–Monte Carlo approaches, some remarks on fundamentals and applications in statistics and econometrics. *Jahrbücher für Nationalökonomie und Statistics* **218** 215–228 (in German).

[125] Tian, G. L. and Fang, K. T. (1999). Uniform design for mixture–amount experiments and for mixture experiments under order restrictions. *Science in China, Ser. A* **42** No.5 456–470.

[126] Fang, K. T. and Zheng, Z. K. (1999). A two–stage algorithm of numerical evaluation of integrals in number–theoretic methods. *J. Comp. Math.* **17** 285–292.

[127] Fang, K. T. and Li, R. Z. (1999). Bayesian statistical inference on elliptical matrix distributions. *J. Multivariate Anal.* **70** 66–85.

[128] Fang, K. T., Shiu, W. C. and Pan, J. X. (1999). Uniform designs based on Latin squares. *Statistica Sinica* **9** 905–912.

[129] Wu, R. and Fang, K. T. (1999). A risk model with delay in claim settlement. *Acta math. Appl. Sinica* **15** 352–360.

[130] Pan, J. X. and Fang, K. T. (1999). Bayesian local influence in growth curve model with unstructured covariance. *Biometrical J.* **41** 641–658.

2000

[131] Fang, K. T. and Yang, Z. H. (2000). On uniform design of experiments with restricted mixtures and generation of uniform distribution on some domains. *Statist. & Prob. Letters* **46** No.2 113–120.

[132] Xie, M. Y. and Fang, K. T. (2000). Admissibility and minimaxity of the uniform design in nonparametric regression model. *J. Statist. Plan. Inference* **83** 101–111.

[133] Liang, J. J., Li, R., Fang, H. B. and Fang, K. T. (2000). Testing multinormality based low–dimensional projection. *J. Statist. Plan. Infer.* **86** 129–141.

[134] Fang, K.T., Lin, D.K.J. and Ma, C.X. (2000). On the construction of multi-level supersaturated designs. *J. Statist. Plan. Infer.* **86** 239–252.

[135] Fang, K. T., D. K. J. Lin, P. Winker and Y. Zhang (2000). Uniform design: Theory and Applications. *Technometrics* **42** 237–248.

[136] Fang, K. T. and R. Mukerjee (2000). A connection between uniformity and aberration in regular fractions of two–level factorials. *Biometrika* **87** 193–198.

[137] Pan, J. X., Fung, W. K., and Fang, K. T. (2000). Multiple outlier detection in multivariate data using projection pursuit techniques. *J. Statist. Planning and Inference* **83** 153–167.

[138] Fang, K. T., Geng, Z. and Tian, G. L. (2000). Statistical inference for truncated Dirichlet distribution and its application in misclassification. *Biometrical J.* **8** 137–152.

[139] Liang, J. J. and Fang, K. T. (2000). Some applications of Läuter's technique in tests for spherical symmetry. *Biometrical J.* **42** No. 8 923–936.

[140] Ma, C. X., Fang, K. T. and Liski, E. (2000). A new approach in constructing orthogonal and nearly orthogonal arrays. *Metrika* **50** 255–268.

[141] Fang, K. T., Fang, H. B. and von Rosen, D. (2000). A family of bivariate distributions with non–elliptical contours. *Communications in Statistics: Theory and Methods* **29** 1885–1898.

[142] Xu, Q. S., Liang, Y. Z. and Fang, K. T. (2000). The effects of different experimental designs on parameter estimation in the kinetics of a reversible chemical reaction. *Chemometrics and Intelligent laboratory Systems* **52** 155–166.

[143] Mukerjee, R., Chan, L. Y. and Fang, K. T. (2000). Regular fractions of mixed factorials with maximum estimation capacity. *Statistica Sinica* **10** 1117–1132.

[144] Fang, K. T., Kollo, T., and Parring, A. M. (2000). Approximation of the non–null distribution of generalized T^2–statistics. *Linear Algebra and Its Appl.* **321** 27–46.

[145] Fang, K. T. and Ma, C. X. (2000). The usefulness of uniformity in experimental design. In *New Trends in Probability and Statistics* **Vol. 5** (T. Kollo, E.-M. Tiit and Srivastava, M. eds.) 51–59. TEV and VSP, The Netherlands.

[146] Fang, K. T. and Ma, C. X. (2000). Applications of uniformity to factorial designs. *J. Chinese Statist. Assoc.* **38** No.4 441–464 (in Chinese).

[147] Fang, K. T. and Lin, D. K. J. (2000). Theory and applications of the uniform design. *J. Chinese Statist. Assoc.* **38** No.4 331–352 (in Chinese).

2001

[148] Liang, J. J., Fang, K. T., Hickernell, F. J. and Li, R. (2001). Testing multivariate uniformity and its applications. *Math. Computation* **70** 337–355.

[149] Fang, K. T., Wang, S. G. and Wei, G. (2001). A stratified sampling model in spherical feature inspection using coordinate measuring machines. *Statist. & Prob. Letters* **51** 25–34.

[150] Ma, C. X, Fang, K. T. and Lin, D. K. J. (2001). On isomorphism of factorial designs. *J. Complexity* **17** 86–97.

[151] Ma, C. X. and Fang, K. T. (2001). A note on generalized aberration in factorial designs. *Metrika* **53** 85–93.

[152] Shi, P., Fang, K. T., and Tsai, C. L. (2001). Optimal multi-criteria designs for Fourier regression model. *J. Statistical Planning and Inference* **96** 387–401.

[153] Fang, K. T., Yang, Z. H. and Kotz, S. (2001). Generation of multivariate distributions by vertical density representation. *Statistics* **35** 281–293.

[154] Zhao, Y. and Fang, K. T. (2001). Orthogonal exact designs on a sphere, a spherical cap, or a spherical belt. *J. Statist. Plan. Infer.* **98** 279–285.

[155] Chan, L. Y., Fang, K. T. and Mukerjee, R. (2001). A characterization for orthogonal arrays of strength two via a regression model. *Stat. & Prob. Letters* **54** 189–192.

[156] Fang, K. T. and Ma, C. X. (2001). Wrap–around L_2–discrepancy of random sampling, Latin hypercube and uniform designs. *J. Complexity* 608–624.

[157] Fang, K. T. and Ma, C. X. (2001). Relationships Between Uniformity, Aberration and Correlation in Regular Fractions 3^{s--1}. In *Monte Carlo*

and Quasi–Monte Carlo Methods 2000 (Fang, K. T., Hickernell, F. J. and Niederreiter, H., eds.) 213–231. Springer.

[158] Fang, K. T., Ma, C. X. and Mukerjee, R. (2001). Uniformity in Fractional Factorials. In *Monte Carlo and Quasi–Monte Carlo Methods 2000* (Fang, K. T., Hickernell, F. J. and Niederreiter, H., eds.) 232–241. Springer.

[159] Fang, K. T. (2001). Some Applications of Quasi–Monte Carlo Methods in Statistics. In *Monte Carlo and Quasi–Monte Carlo Methods 2000* (Fang, K. T., Hickernell, F. J. and Niederreiter, H., eds.) 10–26. Springer.

[160] Liang, Y. Z., Fang, K. T. and Xu, Q. S. (2001). Uniform design and its applications in chemistry and chemical engineering. *Chemometrics and Intelligent laboratory Systems* **58** 43–57.

2002

[161] Fang, K. T., Ma, C. X. and Winker, P. (2002). Centered L_2–discrepancy of random sampling and Latin hypercube design, and construction of uniform designs. *Math. Computation* **71** 275–296.

[162] Fang, H. B., Fang, K. T. and Kotz, S. (2002). The meta–elliptical distributions with given marginals. *J. Multivariate Analysis* **82** 1–16.

[163] Fang, K. T. (2002). Theory, method and applications of the uniform design. In *Eighth ISSAT International Conference on Reliability and Quality in Design* (H. Pham and M. W. Lu, eds.) 235–239. Anaheim, California.

[164] Fang, K. T. (2002). Experimental designs for computer experiments and for industrial experiments with model unknown. *J. Korean Statist. Society* **31** 1–23.

[165] Deng, S., Liu, W., Fang, K. T. and Ma, C. X. (2002). Sequential number–theoretic algorithm for optimization to the evaluation of form errors. In *Proceedings of The 3rd International Conference on Quality and Reliability* (A. J. Subic et al., eds.) 203–208. RMIT University, Melbourne, Australia.

[166] Fang, K. T., Ge, G. N. and Liu, M. Q. (2002). Construction on $E(f_{NOD})$–optimal supersaturated designs via room squares. In *Calcutta Statistical Association Bulletin* Vol 52 (A. Chaudhuri and M. Ghosh eds.) 71–84.

[167] Fang, K. T., Ge, G. N. and Liu, M. Q. (2002). Uniform supersaturated design and its construction. *Science in China, Der. A* **45** 1080–1088.

[168] Fang, K.T. (2002). Theory, method and applications of the uniform design. *International J. Reliability, Quality, and Safety Engineering* **9** No.4 305–315.

2003

[169] Fang, K. T. and Qin, H. (2003). A note on construction of nearly uniform designs with large number of runs. *Statist. & Prob. Letters* **61** 215–224.

[170] Fang, K. T., Ge, G. N., Liu, M. Q. and Qin, H. (2003). Construction on minimum generalized aberration designs. *Metrika* **57** No.1 37–50.

[171] Ma, C. X., Fang, K. T. and Lin, D. K. J. (2003). A note on uniformity and orthogonality. *J. Statist. Plan. Infer.* **113** 323–334.

[172] Xu, Q. S., Massart, D. L., Liang, Y. Z. and Fang, K. T. (2003). Two–step multivariate adaptive regression spline for modeling a quantitative relationship between gas chromatography retention indices and molecular descriptors. *J. Chromatography A.* **998** 155–167.

[173] Hu, Q. N, Liang, Y. Z., Wang, Y. L., Xu, C. J., Zeng, Z. D., Fang, K. T., Peng, X. L. and Yin, H. (2003). External factor variable connectivity index. *J. Chem. Inf. Comput. Sci.* **43** 773–778.

[174] Wang, X. and Fang, K. T. (2003). The effective dimension and quasi–Monte Carlo integration. *J. Complexity* **19** No.2 101–124.

[175] Fang, K. T., Deng, S. X. and Ma, C. X. (2003). Sequential optimal algorithm for evaluation of form and position error. *Acta Metrologica Sinica* **24** 6–9.

[176] Fang, K. T., Lin, D. K. J. and Qin, H. (2003). A note on optimal foldover design. *Statist. & Prob. Letters* **62** 245–250.

[177] Fang, K. T., Lu, X. and Winker, P. (2003). Lower bounds for centered and wrap–around L_2–discrepancies and construction of uniform designs by threshold accepting. *J. Complexity* **19** 692–711.

[178] Fang, K. T. and Lin, D. K. J. (2003). Uniform designs and their application in industry. In *Handbook on Statistics 22: Statistics in Industry* (R. Khattree and C. R. Rao, Elsevier, eds.) 131–170. North–Holland.

[179] Fang, K. T., Ge, G. N, and Liu, M. Q. (2003). Construction of optimal supersaturated designs by the packing method. *Science in China (Series A)* **33** 446–458, (In Chinese).

[180] Tang, Y., Liang, Y. Z. and Fang, K. T. (2003). Data mining in chemometrics: sub–structures learning via peak combinations searching in mass spectra. *J. Data Science* **1** 481–496.

[181] He, P., Fang, K. T. and Xu, C. J. (2003). The classification tree combined with SIR and its applications to classification of mass spectra. *J. Data Science* **1** 425–445.

[182] Varmuza, K., He, P. and Fang, K. T. (2003). Boosting applied to classification of mass spectral data. *J. Data Science* **1** 391–404.

[183] Hu, Q. N., Liang, Y. Z. and Fang, K. T. (2003). The matrix expression, topological index and atomic attribute of molecular topological structure. *J. Data Science* **1** 361–389.

[184] Fang, K. T. Lin, D. K. J. and Liu, M. Q. (2003). Optimal mixed–level supersaturated design. *Metrika* **58** 279–291.

2004

[185] Fang, K. T. and A. Zhang (2004). Minimum aberration majorization in non–isomorphic saturated designs. *J. Statist. Plan. Infer.* **126** 337–346.

[186] Fang, K. T. and Ge, G. N. (2004). A sensitive algorithm for detecting the inequivalence of Hadamard matrices. *Math. Computation* **73** 843–851.

[187] Fang, K. T., Ge, G. N, and Liu, M. Q. (2004). Construction of optimal supersaturated designs by the packing method. *Science in China Ser. A Math.* **47** 128–143.

[188] Ma, C. X. and Fang, K. T. (2004). A new approach to construction of nearly uniform designs. *International Journal of Materials and Product Technology* **20** 115–126.

[189] He, P., Xu, C. J., Liang, Y. Z. and Fang, K. T. (2004). Improving the classification accuracy in chemistry via boosting technique. *Chemometrics and Intelligent Lab. Systems* **70** 39–46.

[190] Fang, K.T., Lu, X., Tang, Y. and Yin, J. (2004). Construction of uniform design by using resolvable packings and coverings. *Discrete Math.* **274** 25–40.

[191] Fang, K. T., Ge, G. N, Liu, M. Q. and Qin, H. (2004). Combinatorial construction for optimal supersaturated designs. *Discrete Math.* **279** 191–202.

[192] Fang, K. T. and Mukerjee, R. (2004). Optimal selection of augmented pairs designs for response surface modeling. *Technometrics* **46** 147–152.

[193] Fang, K. T. (2004). Theory, method and applications of the uniform experimental design, a historical review. *Application of Statist. and Management* **23** 69–80.

[194] Qin, H. and Fang, K. T. (2004). Discrete discrepancy in factorial designs. *Metrika* **60** 59–72.

[195] Hu, Q. N., Liang, Y. Z., Peng, X. L., Yin, H. and Fang, K. T. (2004). Structural interpretation of a topological index. 1. External factor variable connectivity index (EFVCI). *J. Chem. Inf. Computer. Sci.*. to appear.

[196] Shao, H. Y., Yue, P. Y. K. and Fang, K. T. (2004). Identification of differentially expressed genes with multivariate outlier analysis. *J. Biopharmaceutical Statistics* **14** 629–646.

[197] Fang, K. T., Ge, G. N, Liu, M. Q. and Qin, H. (2003). Construction of uniform designs via super–simple resolvable t–designs. *Utilitas Mathematica.* to appear.

[198] Fang, K.T. (2004). Elliptically contoured distributions. *Encyclopedia of Statistics 2nd Edition.* Wiley, New York, to appear.

[199] Fang, K.T. (2004). Uniform designs, *Encyclopedia of Statistics 2nd Edition.* Wiley, New York, to appear.

[200] He, S. G. L. Yang, K.T. Fang , John F. Widmann (2004). Consistent Estimation of Poisson Intensity in the Presence of Dead Time. *J. American Statist. Assoc.* **99**. to appear.

[201] Fang, K. T., Yin, H. and Liang, Y. Z. (2004). New approach by Kriging methods to problems in QSAR. *J. Chemical Information and Modeling.* to appear.

[202] Fang, K. T. and Mukerjee, R. (2004). Expected lengths of confidence intervals based on empirical discrepancy statistics. *Biometrika.* to appear.

Remark:

The papers [22],[26],[29],[30],[31],[32],[33],[34],[35],[37],[39],[40],[41],[42],[43],[45], [46],[47],[48],[49],[50],[51],[52],[53],[54],[55],[56] were collected in Fang, K. T. and Anderson, T. W. (Eds) (1990). *Statistical Inference in Elliptically Contoured and Related Distributions.* Allenton Press Inc., New York.

Part I

Multivariate Analysis

Multidimensional Variation for Quasi-Monte Carlo

Art B. Owen

Department of Statistics, Stanford University, Stanford, CA 94305, U.S.A. (owen@google.com)

Summary. A pivotal concept in quasi-Monte Carlo (QMC) methods is the total variation, in the sense of Hardy and Krause, of a multidimensional function. The Hardy and Krause variation in turn is based on the Vitali variation. This paper presents an account of variation for QMC. It includes a perhaps surprising find that a bounded function arising in financial valuation has infinite variation. Also included is a previously unpublished low variation function extension method due to Sobol'.

Key words: BVHK, put option, Vitali variation, Sobol' extension

2000 Mathematics Subject Classification: 26B30, 11K38

1 Introduction

This article collects together some properties of multidimensional definitions of the total variation of a real valued function. The subject has been studied for a long time. Many of the results presented here date back at least to the early 1900s.

The main reason for revisiting this topic is that there has been much recent work in theory and applications of Quasi-Monte Carlo (QMC) sampling. For an account of quasi-Monte Carlo integration see Fang & Wang (1994). For integrands with bounded variation in the sense of Hardy and Krause (BVHK) such integrands, QMC attains an error rate of $O(n^{-1}(\log n)^d)$ when using n function evaluations in d dimensions.

Contemporary Multivariate Analysis and Experimental Design—In Honor Celebration of Professor Kai-Tai Fang's 65th birthday. Edited by Jianqing Fan and Gang Li. The World Scientific Publisher, 2005.

When the integrand is in BVHK, then QMC has superior asymptotic behavior, compared to Monte Carlo sampling. Therefore we may like to know when a specific function is in BVHK. Recent introductory text books on real analysis typically cover the notion of total variation for functions of a single real variable. Few of them say much about multidimensional variation. This article fills that gap, providing a machinery for computing or bounding variation. This machinery is then used to show that a widely studied class of financially motivated integrands have infinite variation, so that the QMC rate of convergence for them cannot be found by the usual Koksma-Hlawka bound. This article culminates with a presentation of Sobol's low variation extension method, useful for applying QMC to unbounded integrands.

Discussions of multidimensional variation usually require ungainly expressions that grow in complexity with the dimension d. For this reason, many authors give details for $d = 2$ only. But some results that hold for $d = 2$ do not hold for $d > 2$. For example if $f(x)$ and $g(x)$ are linear functions on the d dimensional cube, then $\min(f(x), g(x))$ is BVHK when $d = 2$ but is not necessarily so when $d > 2$. As a result, we need general tools, like the ones presented here. Some expressions remain complicated, but they do not get more so as d increases.

2 One dimensional variation

Let $f(x)$ be a real valued function defined on $[a, b]$ where $-\infty < a \leq b < \infty$. A "ladder" on $[a, b]$ is a set $\mathcal{Y}$ containing a and finitely many, possibly zero, values from (a, b). The ladder $\mathcal{Y}$ does not contain b except when $a = b$. This case is clearly degenerate, but in some settings below it is simpler to include it than to exclude it. Each element $y \in \mathcal{Y}$ has a successor y_+. If $(y, \infty) \cap \mathcal{Y} = \emptyset$ then $y_+ = b$ and otherwise y_+ is the smallest element of $(y, \infty) \cap \mathcal{Y}$. If the elements of $\mathcal{Y}$ are arranged into increasing order, $a = y_0 < y_1 < \cdots < y_m$, then the successor of y_k is y_{k+1} for $k < m$ and it is b for $k = m$. The value y_+ depends on $\mathcal{Y}$ but this dependence will not be made explicit by the notation.

Let $\mathbb{Y}$ denote the set of all ladders on $[a, b]$. Then the total variation of f on $[a, b]$ is

$$V(f; a, b) = \sup_{\mathcal{Y} \in \mathbb{Y}} \sum_{y \in \mathcal{Y}} |f(y_+) - f(y)|. \tag{1}$$

This variation is written $V(f)$ when $[a, b]$ is understood from context. If $V(f) < \infty$ then f is of bounded variation.

There is no uniquely suitable way to extend the notion of variation to functions of more than one variable. Clarkson & Adams (1933) study six such generalizations, and Adams & Clarkson (1934) mention two more. For quasi-Monte Carlo, the total variation in the sense of Hardy and Krause is the most widely used definition. The early references for that definition are Hardy (1905) and Krause (1903a,1903b), who were studying double Fourier series. That definition of total variation is constructed using the total variation in Vitali's sense. Only these two definitions are considered in this work.

3 Notation

For $x \in \mathbb{R}^d$, write its j'th component as x^j. Thus $x = (x^1, \dots, x^d)$. For $a, b \in \mathbb{R}^d$ write $a < b$ or $a \le b$ if these inequalities hold for all d components. For $a, b \in \mathbb{R}^d$ with $a \le b$, the hyperrectangle $[a, b]$ is the set $\{x \in \mathbb{R}^d \mid a \le x \le b\}$. Also $(a, b) = \{x \in \mathbb{R}^d \mid a < x < b\}$ and $[a, b)$ and $(a, b]$ are defined similarly. The d dimensional volume $\prod_{j=1}^d (b^j - a^j)$ of $[a, b]$ is denoted $\mathrm{Vol}([a, b])$.

For arbitrary points $a, b \in \mathbb{R}^d$, let $\mathrm{rect}[a, b]$ denote the hyperrectangle $[\widetilde{a}, \widetilde{b}]$ with $\widetilde{a}^j = \min(a^j, b^j)$ and $\widetilde{b}^j = \max(a^j, b^j)$. We can think of $\mathrm{rect}[a, b]$ as the "rectangular hull" of $\{a, b\}$.

The subset relation is denoted by $u \subseteq v$ or by $u \subsetneq v$. In the former $u = v$ is allowed, and in the latter $u \ne v$ is assumed.

For $u, v \subseteq \{1, \dots, d\}$ write $|u|$ for the cardinality of u, and $u - v$ for the complement of v with respect to u. For integers $j \le k$, the set $\{j, j+1, \dots, k\}$ is written $j\!:\!k$. A unary minus, denotes the complement with respect to $1\!:\!d$, so that $-u = 1\!:\!d - u$. In expressions such as $1\!:\!d - \{j\}$ and $j\!:\!k \cup u$ the operator ':' has highest precedence. In $-u - v$, the unary minus has higher precedence than the binary minus.

For $u \subseteq 1\!:\!d$, we use x^u to denote the $|u|$-tuple of real values representing the components x^j for $j \in u$. The domain of x^u is the hyperrectangle $[a^u, b^u]$. Suppose that $u, v \subseteq 1 : d$ and $x, z \in [a, b]$ with $u \cap v = \emptyset$. Then $x^u : z^v$ is defined to be the point $y \in [a^{u \cup v}, b^{u \cup v}]$ with $y^j = x^j$ for $j \in u$, and $y^j = z^j$ for $j \notin u$. The expression $x^u\!:\!z^v$ is well defined for $x^u \in [a^u, b^u]$ and $z^v \in [a^v, b^v]$, when $u \cap v = \emptyset$, even if x^{-u} or z^{-v} is left unspecified. We also use the colon to glue together more than two sets of components. For instance $x^u : y^v : z^w \in [a, b]$ is well defined for $x^u \in [a^u, b^u]$, $y^v \in [a^v, b^v]$, and $z^w \in [a^w, b^w]$, when u, v, w are mutually disjoint sets whose union is $1\!:\!d$. It will be clear from context whether a colon glues together a tuple as in $x^u\!:\!x^v$, or denotes a range of integers as in $j\!:\!k$. The main use of the gluing operator is to construct the argument to a function by combining components from multiple sources.

Let $f(x)$ be a real valued function defined on the hyperrectangle $[a, b]$. The function f does not depend on x^u if $f(x^u\!:\!x^{-u}) = f(z^u\!:\!x^{-u})$ holds for all $x^u, z^u \in [a^u, b^u]$ and all $x^{-u} \in [a^{-u}, b^{-u}]$. Similarly, f is a function of x^u alone, if it does not depend on x^{-u}.

For $u \subseteq 1\!:\!d$ and $x^{-u} \in [a^{-u}, b^{-u}]$ we can define a function g on $[a^u, b^u]$ via $g(x^u) = f(x^u : x^{-u})$. We write $f(x^u; x^{-u})$ to denote such a function with the argument x^u on the left of the semi-colon and the parameter x^{-u} on the right.

Many expressions require no special attention for $u = \emptyset$. For instance, when $u = \emptyset$, then the definition of $x^u : z^{-u}$ reduces to z. Sometimes the index set u must be handled specially when it equals $\emptyset$. It is often easier to adopt a convention for $u = \emptyset$ than to explicitly identify it as a special case.

Zero dimensional regions and functions on them are of no direct interest in this work, but they appear as special cases in some derivations. In the sequel, $x^\emptyset$ denotes the "zero-tuple" (), the Cartesian product of zero sets is the set containing the zero-tuple, and the volume of a zero dimensional rectangle is $\prod_{j\in\emptyset}(b^j - a^j) = 1$, just as empty products are conventionally taken to be one. A function f on $[a^\emptyset, b^\emptyset]$ is necessarily constant, with a value denoted by $f()$.

4 Multidimensional variation

The d-fold alternating sum of f over the hyperrectangle $[a, b]$ is

$$\Delta(f; a, b) = \Delta(f; [a, b]) = \sum_{v \subseteq \{1,\dots,d\}} (-1)^{|v|} f(a^v\!:\!b^{-v}). \tag{2}$$

Note particularly that in (2), the coefficient of $f(b)$ is one while that of $f(a)$ is $(-1)^d$. For $u \subseteq 1\!:\!d$, define

$$\Delta_u(f; a, b) = \sum_{v \subseteq u} (-1)^{|v|} f(a^v\!:\!b^{-v}). \tag{3}$$

Here $\Delta_u(f; a, b)$ does not depend on a^{-u} and $\Delta_\emptyset(f; a, b) = f(b)$.

The alternating sums (2) and (3) are well defined even when $a \le b$ does not hold. In general $\Delta(f; a, b) = \pm\Delta(f; \mathrm{rect}[a, b])$. The sign is negative if $a^j > b^j$ holds for an odd number of $j \in 1\!:\!d$.

For each $j = 1, \dots, d$ let $\mathcal{Y}^j$ be a ladder on $[a^j, b^j]$. A (multidimensional) ladder on $[a, b]$ has the form $\mathcal{Y} = \prod_{j=1}^d \mathcal{Y}^j$, and we also use $\mathcal{Y}^u = \prod_{j\in u} \mathcal{Y}^j$.

For $y \in \mathcal{Y}$, the successor point y_+ is defined by taking y_+^j to be the successor of y^j in $\mathcal{Y}^j$. The variation of f over $\mathcal{Y}$ is

$$V_{\mathcal{Y}}(f) = \sum_{y \in \mathcal{Y}} |\Delta(f; y, y_+)|. \tag{4}$$

A ladder is, with minor differences, what Clarkson & Adams (1933) call a "net". Their nets also include upper boundaries from b. Ladders are sets, which allows some manipulations to be economically written. We avoid the term net here, because in quasi-Monte Carlo, a net is a finite list of points satisfying some equidistribution properties.

For the multidimensional setting, let $\mathbb{Y}^j$ denote the set of all ladders on $[a^j, b^j]$, and put $\mathbb{Y} = \prod_{j=1}^d \mathbb{Y}^j$. Then:

Definition 1. *The variation of f on the hyperrectangle $[a,b]$, in the sense of Vitali, is*

$$V_{[a,b]}(f) = \sup_{\mathcal{Y} \in \mathbb{Y}} V_{\mathcal{Y}}(f).$$

When $[a,b]$ is understood, we simply write $V(f)$. The function f is of bounded variation in Vitali's sense (BV) if $V(f) < \infty$.

As described below, variation in the sense of Vitali is not adequate for the study of quasi-Monte Carlo sampling. Instead, the variation in the sense of Hardy and Krause is used. This notion of variation sums the Vitali variations over $[a,b]$ and its "upper faces".

Definition 2. *The variation of f on the hyperrectangle $[a,b]$, in the sense of Hardy and Krause, is*

$$V_{\text{HK}}(f) = V_{\text{HK}}(f; a, b) = \sum_{u \subsetneq 1:d} V_{[a^{-u}, b^{-u}]} f(x^{-u}; b^u). \tag{5}$$

The function f has bounded variation in the sense of Hardy and Krause (BVHK) if $V_{\text{HK}}(f) < \infty$. The definition of bounded variation in Hardy (1905) requires $V_{[a,b]}(f) < \infty$ and $V_{[a^{-u}, b^{-u}]}(f(x^{-u}; z^u)) < \infty$ for all $0 < |u| < d$ and all $z^u \in [a^u, b^u]$. Young (1913) shows that definition to be equivalent to the one above.

The premier use of variation in QMC is in Hlawka's inequality (the Koksma-Hlawka theorem) where the quadrature error has an upper bound equal to $V_{\text{HK}}(f)$ times a discrepancy measure of the points $x_1, \dots, x_n$. See Niederreiter (1992).

If one follows the above definitions literally for a zero dimensional hyperrectangle, then $V_{[a^\emptyset, b^\emptyset]}(f) = |f()|$ and $V_{\text{HK}}(f; a, b) = 0$. The variation $V_{\text{HK}}(f)$ is a semi-norm on functions and not a norm, because it vanishes for constant but non-zero functions. The quantity $V_{\text{HK}}(f) + |f(b)|$ is often used in QMC because it is a norm on functions. This norm can be obtained by adjoining the case $u = 1{:}d$ to the sum in (5).

5 Splits of hyperrectangles

The properties of variation derive from those of alternating sums. Those in turn are based on properties of splits of hyperrectangles.

Definition 3. *A split of the hyperrectangle* $[a,b]$ *is a set* $\{[a_i,b_i] \mid 1 \le i \le m < \infty\}$ *where* $\cup_{i=1}^m [a_i,b_i] = [a,b]$ *and* $[a_i,b_i) \cap [a_j,b_j) = \emptyset$ *when* $i \ne j$.

Note that $[a_i,b_i] \cap [a_j,b_j]$ is not necessarily empty for $i \ne j$. The most basic split is a coordinate split:

Definition 4. *For* $j \in \{1,\dots,d\}$ *and* $c \in [a^j,b^j]$ *the corresponding coordinate split of* $[a,b]$ *is the set* $\{L,R\}$ *of left and right pieces*

$$L = L(j,c) = L(j,c;a,b) = \{x \in [a,b] \mid x^j \le c\}, \quad \text{and,}$$
$$R = R(j,c) = R(j,c;a,b) = \{x \in [a,b] \mid x^j \ge c\},$$

respectively.

Both L and R are closed hyperrectangles: $L = [a,\widetilde{b}]$ where $\widetilde{b}^k = b^k$ for $k \ne j$ and $\widetilde{b}^j = c$, and $R = [\widetilde{a},b]$ where $\widetilde{a}^k = a^k$ for $k \ne j$ and $\widetilde{a}^j = c$. Next we show that the alternating sum over $[a,b]$ is the sum of alternating sums over L and R. Propositions 1 through 4 below recapitulate results from Fréchet (1910).

Proposition 1. *Suppose that the hyperrectangle* $[a,b]$ *is split into* $[a,\widetilde{b}]$ *and* $[\widetilde{a},b]$ *as described above. Then*

$$\Delta(f;a,b) = \Delta(f;a,\widetilde{b}) + \Delta(f;\widetilde{a},b). \tag{6}$$

Proof: Let $c^{\{j\}}$ denote $c \in [a^j,b^j]$ for use as an argument to f. We write a sum over $v \subseteq 1{:}d$ as a double sum. The outer sum is over $u \subseteq 1{:}d - \{j\}$ and the inner sum is over u and $u \cup \{j\}$. Thus $\Delta(f;a,\widetilde{b})$ equals

$$\begin{aligned} &\sum_{u \subseteq -\{j\}} (-1)^{|u|}\big[f(a^u : \widetilde{b}^{-u}) - f(a^{u\cup\{j\}} : \widetilde{b}^{-u-\{j\}})\big] \\ &= \sum_{u \subseteq -\{j\}} (-1)^{|u|}\big[f(a^u : c^{\{j\}} : b^{-u-\{j\}}) - f(a^{u\cup\{j\}} : b^{-u-\{j\}})\big], \end{aligned} \tag{7}$$

and similarly,

$$\Delta(f;\widetilde{a},b) = \sum_{u \subseteq -\{j\}} (-1)^{|u|}\big[f(a^u : b^{-u}) - f(a^u : c^{\{j\}} : b^{-u-\{j\}})\big]. \tag{8}$$

Summing (7) and (8),

$$\sum_{u \subseteq 1:d-\{j\}} (-1)^{|u|}\big[f(a^u : b^{-u}) - f(a^{u\cup\{j\}} : b^{-u-\{j\}})\big] = \Delta(f;a,b). \quad \square$$

Proposition 2. *Suppose that* $\mathcal{Y} = \prod_{j=1}^{d} \mathcal{Y}^j$ *is a ladder on* $[a, b]$. *Then* $\{[y, y_+] \mid y \in \mathcal{Y}\}$ *is a split of* $[a, b]$ *and*

$$\Delta(f; a, b) = \sum_{y \in \mathcal{Y}} \Delta(f; y, y_+). \tag{9}$$

Proof: By construction $a \le y \le y_+ \le b$ for $y \in \mathcal{Y}$, so $\cup_{y \in \mathcal{Y}}[y, y_+] \subseteq [a, b]$. Now suppose that $x \in [a, b]$. Consider $y \in \mathcal{Y}$ where $y^j = a^j$ if $x^j = a^j$, and otherwise $y^j = \max(\mathcal{Y}^j \cap [a^j, x^j))$. Then $y^j \le x^j \le y^j_+$ so that $x \in [y, y_+]$, and hence $\cup_{y \in \mathcal{Y}}[y, y_+] = [a, b]$. Now suppose that $x \in [y, y_+) \cap [\widetilde{y}, \widetilde{y}_+)$ for $y, \widetilde{y} \in \mathcal{Y}$. Then $y^j \le x^j < y^j_+$ and $\widetilde{y}^j \le x^j < \widetilde{y}^j_+$ which implies $y^j = \widetilde{y}^j$, so $y = \widetilde{y}$. Thus $[y, y_+) \cap [\widetilde{y}, \widetilde{y}_+)$ is empty whenever $y \ne \widetilde{y}$, establishing that $\{[y, y_+] \mid y \in \mathcal{Y}\}$ is a split of $[a, b]$. To prove (9), note that the split $\{[y, y_+] \mid y \in \mathcal{Y}\}$ can be obtained by making a sequence of $|\mathcal{Y}| - 1$ coordinate splits of $[a, b]$. □

Proposition 3. *Suppose that* $\{\,[a_i, b_i] \mid 1 \le i \le m < \infty\}$ *is a split of the hyperrectangle* $[a, b]$. *Then*

$$\Delta(f; a, b) = \sum_{i=1}^{m} \Delta(f; a_i, b_i). \tag{10}$$

Proof: Let $\mathcal{Y}^j = \{a^j_1, \ldots, a^j_m, b^j_1, \ldots, b^j_m\} \cap [a^j, b^j)$, define the ladder $\mathcal{Y} = \prod_{j=1}^{d} \mathcal{Y}^j$ and define the split $\mathcal{S} = \{[y, y_+] \mid y \in \mathcal{Y}\}$. Next put $\mathcal{S}_i = \{[y, y_+] \mid y \in \mathcal{Y} \cap [a_i, b_i)\}$. Then $\mathcal{S}_i$ is a split of $[a_i, b_i]$ to which Proposition 2 applies. Also $\mathcal{S}_i$ are mutually disjoint with union $\mathcal{S}$. Therefore $\Delta(f; a, b)$ and $\sum_{i=1}^{m} \Delta(f; a_i, b_i)$ are both equal to $\sum_{i=1}^{m} \sum_{s \in \mathcal{S}_i} \Delta(f; s)$. □

Proposition 4. *Let* $\mathcal{Y}^j$ *and* $\widetilde{\mathcal{Y}}^j$ *be ladders in* $[a^j, b^j]$ *with* $\mathcal{Y}^j \subseteq \widetilde{\mathcal{Y}}^j$ *for* $j = 1, \ldots, d$ *and write* $\mathcal{Y} = \prod_{j=1}^{d} \mathcal{Y}^j$ *and* $\widetilde{\mathcal{Y}} = \prod_{j=1}^{d} \widetilde{\mathcal{Y}}^j$. *Then* $V_{\mathcal{Y}}(f; a, b) \le V_{\widetilde{\mathcal{Y}}}(f; a, b)$.

Proof: The ladder $\mathcal{Y}$ can be changed to $\widetilde{\mathcal{Y}}$ by d steps that each refine just one of the ladders $\mathcal{Y}^j$. Therefore it is sufficient to consider the case where $\widetilde{\mathcal{Y}}^j = \mathcal{Y}^j$ for $j \ne k$ for some $k \in \{1, \ldots, d\}$. Without loss of generality take $k = 1$ and suppose that $\widetilde{\mathcal{Y}}^1 - \mathcal{Y}^1 = \{c\}$ where $y_\ell = \widetilde{y}_\ell < c = \widetilde{y}_{\ell+1} < y_{\ell+1}$ for $0 \le \ell \le m_1$ taking $y_{m_1+1} = b^1$ if $\ell = m_1$. Then $V_{\widetilde{\mathcal{Y}}}(f) - V_{\mathcal{Y}}(f)$ equals

$$\sum_{\widetilde{y}^{2:d}\in\mathcal{Y}^{2:d}}\sum_{\widetilde{y}^1\in\widetilde{\mathcal{Y}}^1}|\Delta(f;\widetilde{y},\widetilde{y}_+)| - \sum_{y^{2:d}\in\mathcal{Y}^{2:d}}\sum_{y^1\in\mathcal{Y}^1}|\Delta(f;y,y_+)|$$

$$= \sum_{y^{2:d}\in\mathcal{Y}^{2:d}}\sum_{y^1\in\{y_\ell\}}|\Delta(f;L(1,c;y,y_+))| + |\Delta(f;R(1,c;y,y_+))|$$

$$- |\Delta(f;L(1,c;y,y_+)) + \Delta(f;R(1,c;y,y_+))|$$

$$\geq 0. \quad \square$$

Proposition 4 allows us to replace the supremum over all ladders in Definition 1 by one over a subset $\widetilde{\mathbb{Y}} \subseteq \mathbb{Y}$ of ladders. If to every $\mathcal{Y} \in \mathbb{Y}$ there is a $\widetilde{\mathcal{Y}} \in \widetilde{\mathbb{Y}}$ with $\mathcal{Y} \subseteq \widetilde{\mathcal{Y}}$, then

$$V(f) = \sup_{\mathcal{Y}\in\widetilde{\mathbb{Y}}} V_{\mathcal{Y}}(f).$$

For instance when $[a,b] = [0,1]^d$ we may suppose that each ladder in $\widetilde{\mathbb{Y}}$ is the d-fold tensor product of some ladder on $[0,1]$. To show this, take $\widetilde{\mathcal{Y}}^k = \cup_{j=1}^d \mathcal{Y}^j$ for $k \in 1\!:\!d$.

A simple ladder that is sometimes useful is one with an equal number of equispaced points in each direction. Let $m \geq 1$ be an integer, set $\mathcal{Y}^j(m) = \mathcal{Y}^j(m,a,b) = \{a^j, a^j + (b^j - a^j)/m, \dots, a^j + (b^j - a^j)(m-1)/m\}$, and put

$$\mathcal{Y}(m) = \mathcal{Y}(m,a,b) = \prod_{j=1}^d \mathcal{Y}^j(m,a,b). \tag{11}$$

Simple ladders can be used to show lower bounds on variation, but we cannot replace the supremum in Definition 1 by the supremum over simple ladders, nor even by the supremum over ladders $\mathcal{Y}$ for which $(y^j - a^j)/(b^j - a^j)$ is always a rational number. We can however restrict attention to ladders for which the cells $[y, y_+]$ are nearly congruent and nearly cubic.

Proposition 5. *Let f be a function on the hyperrectangle $[a,b]$ of positive volume. For $\epsilon > 0$, let $\widetilde{\mathbb{Y}} = \widetilde{\mathbb{Y}}(\epsilon)$ be the set of ladders $\widetilde{\mathcal{Y}}$ for which*

$$\max_{\widetilde{y}\in\widetilde{\mathcal{Y}}}\max_{j\in 1:d}(\widetilde{y}_+^j - \widetilde{y}^j) \leq (1+\epsilon)\min_{\widetilde{y}\in\widetilde{\mathcal{Y}}}\min_{j\in 1:d}(\widetilde{y}_+^j - \widetilde{y}^j). \tag{12}$$

Then $V_{[a,b]}(f) = \sup_{\mathcal{Y}\in\widetilde{\mathbb{Y}}} V_{\mathcal{Y}}(f)$.

Proof: For $\mathcal{Y} \in \mathbb{Y}$, let $\eta = \min_{y\in\widetilde{\mathcal{Y}}}\min_{j\in 1:d}(y_+^j - y^j)$. Because $\mathrm{Vol}([a,b]) > 0$ we have $\eta > 0$. Set $\widetilde{\mathcal{Y}} = \mathcal{Y}$. Then for each $j \in 1\!:\!d$, while there is $y^j \in \mathcal{Y}^j$ with $y_+^j - y^j > 2\epsilon$, replace $\widetilde{\mathcal{Y}}^j$ by $\widetilde{\mathcal{Y}}^j \cup \{(y^j + y_+^j)/2\}$. After a finite number of steps $\mathcal{Y} \subseteq \widetilde{\mathcal{Y}}$ and $\widetilde{\mathcal{Y}}$ satisfies (12) for $\epsilon = 1$.

If $\epsilon \geq 1$, we're done. Otherwise, for an integer $m > 1/\epsilon$ let $k(j, \widetilde{y}^j) = \lfloor m(\widetilde{y}^j_+ - \widetilde{y}^j)/\eta \rfloor$, where $\lfloor z \rfloor$ denotes the greatest integer less than or equal to z. We have $m \leq k < 2m$ because $\widetilde{y}^j_+ - \widetilde{y}^j \in [\eta, 2\eta)$. Next set $\widehat{\mathcal{Y}} = \prod_{j=1}^d \widehat{\mathcal{Y}}^j$, where

$$\widehat{\mathcal{Y}}^j = \bigcup_{\widetilde{y}^j \in \widetilde{\mathcal{Y}}^j} \bigcup_{r=0}^{k(j,\widetilde{y}^j)-1} \left\{ \widetilde{y}^j + r(\widetilde{y}^j_+ - \widetilde{y}^j)/k(j, \widetilde{y}^j) \right\}.$$

The interval $[\widetilde{y}^j, \widetilde{y}^j_+]$ gets split into $k = k(j, \widetilde{y}^j)$ equal width intervals. If an interval in $\widetilde{\mathcal{Y}}$ has been split k ways then its length could have been as small as $k\eta/m$ but not as large as $(k+1)\eta/m$. Thus the shortest interval in the $\hat{\mathcal{Y}}$ ladder has length η/m and the largest has length under $\max_{k \in m:(2m-1)}(k+1)\eta/(km) = (m+1)\eta/m^2$. Now $((m+1)\eta/m^2)/(\eta/m) = (m+1)/m < 1+\epsilon$ because $m > 1/\epsilon$. Thus $\mathcal{Y} \subseteq \widehat{\mathcal{Y}}$ where $\widehat{\mathcal{Y}}$ satisfies (12). □

When computing or bounding $V(f)$ it is often convenient to split the domain of f into hyperrectangular regions and sum the variations from within each of them. The following lemma, stated in Young (1913) justifies such a divisive approach.

Lemma 1. *Let f be defined on the hyperrectangle $[a, b]$. Let $\{[a_i, b_i] \mid 1 \leq i \leq m < \infty\}$ be a split of $[a, b]$. Then*

$$V_{[a,b]}(f) = \sum_{i=1}^m V_{[a_i,b_i]}(f).$$

Proof: Let $\mathcal{Y}$ be a ladder on $[a, b]$. Let $\widetilde{\mathcal{Y}}^j = (\mathcal{Y}^j \cup \{a^j_1, \ldots, a^j_m, b^j_1, \ldots, b^j_m\}) \cap [a^j, b^j)$. Then

$$V_{\mathcal{Y}}(f) \leq V_{\widetilde{\mathcal{Y}}}(f) = \sum_{i=1}^m \sum_{\widetilde{y} \in \widetilde{\mathcal{Y}} \cap [a_i,b_i)} |\Delta(f; \widetilde{y}, \widetilde{y}_+)| \leq \sum_{i=1}^m V_{[a_i,b_i]}(f).$$

Taking the supremum over $\mathcal{Y}$ establishes that $V_{[a,b]}(f) \leq \sum_{i=1}^m V_{[a_i,b_i]}(f)$. Now let $\mathcal{Y}_i$ be ladders on $[a_i, b_i]$ for $i = 1, \ldots, m$. Let $\widetilde{\mathcal{Y}}$ be the ladder with $\widetilde{\mathcal{Y}}^j = \cup_{i=1}^m \mathcal{Y}^j_i$ and let $\widetilde{\mathcal{Y}}_i = \widetilde{\mathcal{Y}} \cap [a_i, b_i) \supseteq \mathcal{Y}_i$. Then

$$\sum_{i=1}^m \sum_{y \in \mathcal{Y}_i} |\Delta(f; y, y_+)| \leq \sum_{i=1}^m \sum_{\widetilde{y} \in \widetilde{\mathcal{Y}}_i} |\Delta(f; \widetilde{y}, \widetilde{y}_+)| = \sum_{\widetilde{y} \in \widetilde{\mathcal{Y}}} |\Delta(f; \widetilde{y}, \widetilde{y}_+)| \leq V(f).$$

Taking the supremum over $\mathcal{Y}_1, \ldots, \mathcal{Y}_m$ yields $\sum_{i=1}^m V_{[a_i,b_i]}(f) \leq V_{[a,b]}(f)$.
□

Suppose that we seek to prove that $V(f) = \infty$. If for $m > 1$ we split $[a, b]$ into m^d congruent hyperrectangles similar in shape to $[a, b]$, then by

Lemma 1, at least one of these smaller hyperrectangles has infinite variation. The proof of infinite variation can therefore always be focussed on an arbitrarily small region within $[a, b]$. Of course, matters would be different had we considered unbounded hyperrectangles.

6 Alternating sums

A function f can be easily recovered from its alternating sums, as follows:

Proposition 6. *Let f be a function on the hyperrectangle $[a, b]$. For $x, c \in [a, b]$*

$$f(x) = f(c) + \sum_{\emptyset \neq u \subseteq 1:d} (-1)^{|u|} \Delta_u(f; x, c). \tag{13}$$

Proof: The right hand side of (13) may be written as

$$\begin{aligned}\sum_{u \subseteq 1:d} (-1)^{|u|} \Delta_u(f; x, c) &= \sum_{u \subseteq 1:d} (-1)^{|u|} \sum_{v \subseteq u} (-1)^{|v|} f(x^v \!:\! c^{-v}) \\ &= \sum_{v \subseteq 1:d} (-1)^{|v|} f(x^v \!:\! c^{-v}) \sum_{u \supseteq v} (-1)^{|u|}.\end{aligned}$$

Next $\sum_{u \supseteq v} (-1)^{|u|} = (-1)^d$ if $v = 1 : d$ and equals 0 otherwise. The sum becomes $(-1)^{2d} f(x) = f(x)$. ☐

For $a, b \in \mathbb{R}^d$, when f has a (Lebesgue) integral over $\mathrm{rect}[a, b]$ then $\int_{[a,b]} f(x)dx = \pm \int_{\mathrm{rect}[a,b]} f(x)dx$. The sign is negative if and only if there are an odd number of indices $j \in 1 : d$ with $a^j > b^j$.

Proposition 7. *Suppose that f is in $L^1[a, b]$ and that $y, y_+, c \in [a, b]$. Then*

$$\sum_{u \subseteq 1:d} (-1)^{|u|} \int_{[y^{-u}:y_+^u, c]} f(x)dx = \int_{[y, y_+]} f(x)dx. \tag{14}$$

Proof: We proceed by induction on d. For $d = 1$ the left hand side of (14) is $\int_{[y,c]} f(x)dx - \int_{[y_+,c]} f(x)dx$ which equals $\int_{[y,y_+]} f(x)dx$. Now suppose that the result holds for dimensions 1 through $d-1$. Then for dimension d the left hand side of (14) is

$$\sum_{v\subseteq -\{d\}} (-1)^{|v|}\Big(\int_{[y^{-v}:y_+^v,c]} f(x)dx - \int_{[y^{-v-\{d\}}:y_+^{v\cup\{d\}},c]} f(x)dx\Big)$$

$$= \int_{[y^d,c^d]} \sum_{v\subseteq -\{d\}} (-1)^{|v|} \int_{[y^v:y_+^{-v-\{d\}},c^{-\{d\}}]} f(x)dx^{-\{d\}}dx^{\{d\}}$$

$$- \int_{[y_+^d,c^d]} \sum_{v\subseteq -\{d\}} (-1)^{|v|} \int_{[y^{v\cup\{d\}}:y_+^{-v-\{d\}},c^{-\{d\}}]} f(x)dx^{-\{d\}}dx^{\{d\}}$$

$$= \int_{[y^{\{d\}},c^{\{d\}}]}\int_{[y^{-\{d\}},y_+^{-\{d\}}]} f(x)dx - \int_{[y_+^{\{d\}},c^{\{d\}}]}\int_{[y^{-\{d\}},y_+^{-\{d\}}]} f(x)dx$$

$$= \int_{[y,y_+]} f(x)dx. \quad \Box$$

The result of Proposition 7 is similar to an inclusion-exclusion familiar in QMC. Let $N([a,b])$ denote the number of points from a list $x_1,\dots,x_n$ that are in $[a,b]$. Then for $a \le x \le y$, $\sum_{u\subseteq 1:d}(-1)^{|u|}N([a,x^u:y^{-u}]) = N([x,y])$.

7 Functions not depending on all variables

The next proposition states a well known deficiency for quasi-Monte Carlo applications, of Vitali's definition of variation:

Proposition 8. *Suppose that $f(x)$ is defined on the hyperrectangle $[a,b]$ and that $f(x)$ does not depend on x^u for non-empty $u \subseteq 1{:}d$. Then $V(f) = 0$.*

Proof: Let $j \in u$. Then for $a \le \widetilde{a} \le \widetilde{b} \le b$,

$$\Delta(f;\widetilde{a},\widetilde{b}) = \sum_{v\subseteq -\{j\}} (-1)^{|v|}\big(f(\widetilde{a}^v:\widetilde{b}^{-v}) - f(\widetilde{a}^{v\cup\{j\}}:\widetilde{b}^{-v-\{j\}})\big) = 0,$$

because f does not depend on whether x^j equals $\widetilde{a}^j$ or $\widetilde{b}^j$. Therefore $V_{\mathcal{Y}}(f) = 0$ for all ladders $\mathcal{Y}$ on $[a,b]$, and so $V(f) = 0$. $\Box$

For the next examples, suppose that $[a,b]$ is a hyperrectangle of positive volume in dimension $d \ge 2$. Let $f_1(x) = 0$ for $x^1 = a^1$ and $f_1(x) = \sin(1/(x^1 - a^1))$ otherwise. Then $V(f_1) = 0$ even though f_1 has infinite variation in the one dimensional sense along the line $a^1 \le x^1 \le b^1$ for any fixed $x^{2:d} \in [a^{2:d}, b^{2:d}]$. Similarly $V(f_2) = 0$, where $f_2(x) = 1$ if x^1 is a rational number and $f_2(x) = 0$ otherwise. Finally, suppose that $f_3(x) = 0$ if $x^1 = a^1 < b^1$ and $f_3(x) = 1/(x^1 - a^1)$ otherwise. Then $V(f_3) = 0$, even though f_3 is unbounded. Example f_3 is given in Fréchet (1910).

8 Invariants and closure

Let $f(x)$ be defined on the hyperrectangle $[a,b]$. Let $\widetilde{f}(x)$ be defined on the hyperrectangle $[\widetilde{a},\widetilde{b}]$ by $\widetilde{f}(x) = f(\widetilde{x})$ where $\widetilde{x}^j = \phi_j(x^j)$ with ϕ_j is a strictly monotone (increasing or decreasing) invertible function from $[\widetilde{a}^j,\widetilde{b}^j]$ onto $[a^j,b^j]$.

Proposition 9. *In the notation above* $V_{[\widetilde{a},\widetilde{b}]}(\widetilde{f}) = V_{[a,b]}(f)$.

Proof: Suppose that $\mathcal{Y}$ is a ladder on $[\widetilde{a},\widetilde{b}]$. For $j = 1,\dots,d$, if ϕ_j is increasing, let $\widetilde{\mathcal{Y}}^j = \{\phi_j(y) \mid y \in \mathcal{Y}^j\}$ and otherwise let $\widetilde{\mathcal{Y}}^j = \{\phi_j(a^j)\} \cup \{\phi_j(y) \mid y \in \mathcal{Y}^j - \{a^j\}\}$. Then $V_{\widetilde{\mathcal{Y}}}(\widetilde{f}) = V_{\mathcal{Y}}(f)$, and so $V_{[a,b]}(f) \le V_{[\widetilde{a},\widetilde{b}]}(\widetilde{f})$. A similar argument using the inverses of ϕ_j yields $V_{[\widetilde{a},\widetilde{b}]}(\widetilde{f}) \le V_{[a,b]}(f)$. □

Proposition 10. *In the notation above, if every* ϕ_j *is increasing, then* $V_{\mathrm{HK}}(\widetilde{f};\widetilde{a},\widetilde{b}) = V_{\mathrm{HK}}(f;a,b)$.

Proof: Because all of the ϕ_j are increasing, the function $\widetilde{f}(\widetilde{x}^{-u};\widetilde{b}^u)$ corresponds to $f(x^{-u};b^u)$. Then Proposition 9 applies to each term in (5). □

Proposition 11. *Let* f *and* g *be functions on the hyperrectangle* $[a,b]$. *If* $f,g \in BVHK$, *then* $f+g$, $f-g$, *and* fg *are in* $BVHK$. *If* $f \in BVHK$ *with* $|f| > C > 0$ *then* $1/f \in BVHK$. *If* $f,g \in BV$, *then* $f+g$ *and* $f-g$ *are in* BV, *but* fg *is not necessarily in* BV. *If for nonempty* $u \subsetneq 1{:}d$ *both* $f \in BV[a^u,b^u]$ *and* $g \in BV[a^{-u},b^{-u}]$ *hold, then* $fg \in BV[a,b]$. *If also* $\alpha,\beta \in \mathbb{R}$, *then* $V_{[a,b]}(\alpha+\beta f) = |\beta|V_{[a,b]}(f)$ *and* $V_{\mathrm{HK}}(\alpha+\beta f) = |\beta|V_{\mathrm{HK}}(f)$.

Proof: The closure rules for BVHK are in Hardy (1905). Those for BV are in Fréchet (1910).Let $y \in \mathcal{Y}$ for a ladder $\mathcal{Y}$ on $[a,b]$. Then $|\Delta(\alpha+\beta f;a,b)| = |\beta||\Delta(f;a,b)|$ from which the variation results for $\alpha+\beta f$ follow easily.

The following example proves the nonclosure of BV under multiplication. Suppose that the dimension of $[a,b]$ is $d=2$ and $\mathrm{Vol}([a,b]) > 0$. Let $f(x) = 1/(x^1-a^1)$ for $x^1 \in (a^1,b^1]$ and $f(x) = 0$ when $x^1 = a^1$. Also let $g(x) = 1 + x^2 - a^2$. Then $V(f) = V(g) = 0$ by Proposition 8. For $\epsilon > 0$ with $\epsilon \le b^1 - a^1$, let $\widetilde{b}^j(\epsilon)$ equal b^j for $j > 1$ and take $\widetilde{b}^1(\epsilon) = a^1 + \epsilon$. Then

$$\Delta(f;a,\widetilde{b}(\epsilon)) = f(\epsilon,b^2) - f(\epsilon,b^1) - f(a^1,b^2) + f(a^1,a^2) = \frac{b^2-a^2}{\epsilon},$$

and so $V(fg) \ge |b^2-a^2|/\epsilon$. □

Proposition 12. *The function f is in BVHK on $[a,b]$ if and only if it can be written $f = f_1 - f_2$ where $\Delta_u(f_i; x, y) \geq 0$ holds for $i = 1, 2$ whenever $x \leq y$ and $u \subseteq 1{:}d$.*

Proof: The "only if" part is due to Hardy (1905). Hardy (1905) and the "if" part is noted in Adams & Clarkson (1934). □

9 Mixed partial derivatives

Vitali's variation is closely connected with the partial derivative of f, taken once with respect to each variable. We write $\partial^{1:d} f(x)$ for $\partial^d f(x)/\prod_{j=1}^{d} \partial x^j$. More generally, for $u \subseteq 1:d$, the mixed partial derivative of f taken once with respect to every x^j for $j \in u$ is denoted $\partial^u f$ and, by convention $\partial^{\emptyset} f(x) = f(x)$. If $\partial^{1:d} f(x)$ exists for all $x \in [a,b]$, then

$$\int_{[a,b]} \partial^{1:d} f(x) dx = \Delta(f; a, b). \tag{15}$$

Equation (15) is immediate for $d = 1$ and follows by induction for $d \geq 1$. Fréchet (1910) used (15) to get the upper bound (17) below, for $V(f)$ from $\partial^{1:d} f$.

Proposition 13. *Suppose that f is a function for which $\partial^{1:d} f$ is defined on the hyperrectangle $[a,b]$. Then*

$$V(f; a, b) \leq \int_{[a,b]} |\partial^{1:d} f(x)| dx, \tag{16}$$

$$V(f; a, b) \leq \mathit{Vol}(\,[a,b]\,) \sup_{x \in [a,b]} \left|\partial^{1:d} f(x)\right|, \quad \textit{and,} \tag{17}$$

$$V(f; a, b) \leq \mathit{Vol}(\,[a,b]\,)^{1/2} \left(\int_{x \in [a,b]} \left(\partial^{1:d} f(x)\right)^2 dx \right)^{1/2}. \tag{18}$$

Proof: For any ladder $\mathcal{Y}$, we find that

$$V_{\mathcal{Y}}(f; a, b) \leq \sum_{y \in \mathcal{Y}} \int_{[y, y_+]} \left|\partial^{1:d} f(x)\right| dx = \int_{[a,b]} \left|\partial^{1:d} f(x)\right| dx,$$

so taking suprema over $\mathcal{Y}$ establishes (16). Applying standard inequalities among L^p norms yields (17) and (18). □

Under mild conditions on $\partial^{1:d}$, equality holds in (16). Continuity of $\partial^{1:d}$ is sufficient, though clearly not necessary. The next result, stated in Fréchet (1910) is better known, though less applicable, than Proposition 13.

Proposition 14. *If $\partial^{1:d}f(x)$ is continuous on the hyperrectangle $[a,b]$ then*

$$V(f) = \int_{[a,b]} \big|\, \partial^{1:d} f(x) \big|\, dx. \tag{19}$$

Proof: Let $\epsilon > 0$. For an integer $m \ge 1$, define the ladder $\mathcal{Y}(m) = \prod_{j=1}^{d} \mathcal{Y}^j$ where $\mathcal{Y}^j = \{a + \ell(b-a)/m \mid \ell = 0,\dots,m-1\}$ for $j = 1,\dots,d$. Because $\partial^{1:d}f(x)$ is continuous on the compact set $[a,b]$, it is uniformly continuous there. Thus there is an integer $m \ge 1$ such that

$$\max_{y\in\mathcal{Y}(m)} \Big(\max_{x\in[y,y_+]} \partial^{1:d} f(x) - \min_{x\in[y,y_+]} \partial^{1:d} f(x) \Big) \le \epsilon.$$

For each $y \in \mathcal{Y}$,

$$\Big| \int_{[y,y_+]} \partial^{1:d} f(x) dx \Big| \ge \int_{[y,y_+]} \Big(\big|\, \partial^{1:d} f(x) \big| - \epsilon \Big) dx \tag{20}$$

holds. Equation (20) is trivial if $\partial^{1:d}f(x)$ has constant sign on $[y, y_+]$, and otherwise, the left and right sides of (20) are positive and negative respectively. Finally,

$$\begin{aligned} V_{[a,b]}(f) &\ge V_{\mathcal{Y}(m)}(f) \\ &= \sum_{y\in\mathcal{Y}(m)} \Big| \int_{[y,y_+]} \partial^{1:d} f(x) dx \Big| \\ &\ge \sum_{y\in\mathcal{Y}(m)} \int_{[y,y_+]} \Big(\big|\, \partial^{1:d} f(x) \big| - \epsilon \Big) dx \\ &= \int_{[a,b]} |\partial^{1:d} f(x)| dx - \epsilon \mathrm{Vol}([a,b]). \quad \square \end{aligned}$$

Proposition 15. *Let f be defined on the hyperrectangle $[a,b]$. Suppose that for some set $u \subseteq 1{:}d$ that $\partial^u f$ exists, and satisfies the Lipschitz-like condition*

$$|\Delta_{-u}(\partial^u f; x, y)| \le A\, \mathit{Vol}(\mathit{rect}[x^{-u}, y^{-u}]), \tag{21}$$

for all $a \le x \le y \le b$ and some $A < \infty$. Then $V(f) \le A\,\mathit{Vol}([a,b])$.

Proof: Let $a \le x \le y \le b$. Then

$$\begin{aligned}\Delta(f;x,y) &= \sum_{v\subseteq -u}\sum_{w\subseteq u}(-1)^{|v|+|w|}f(x^{v\cup w}:y^{-v-w})\\ &= \sum_{v\subseteq -u}(-1)^{|v|}\int_{[x^u,y^u]}\partial^u f(z^u:x^{(-u)\cap(v\cup w)}:y^{-u-v-w})dz^u\\ &= \int_{[x^u,y^u]}\Delta_{-u}(\partial^u f;z^u{:}x^{-u},z^u{:}y^{-u})dz^u,\end{aligned}$$

so that $|\Delta(f;x,y)| \le A|\int_{[x^u,y^u]}\mathrm{Vol}(\mathrm{rect}[x^{-u},y^{-u}])dx^u| \le A\mathrm{Vol}(\mathrm{rect}[x,y])$. Therefore for any ladder $\mathcal{Y}$ on $[a,b]$ we find $V_{\mathcal{Y}}(f) \le A\mathrm{Vol}([a,b])$ and so $V(f) \le A\mathrm{Vol}([a,b])$. □

When $u = 1{:}d$, then the sufficient condition (21) in Proposition 15 reduces to $|\partial^{1:d}f| \le A$. When $u = \emptyset$ then (21) reduces to $|\Delta(f;x,y)| \le A\mathrm{Vol}(\mathrm{rect}[x,y])$, a condition in Fréchet (1910). When $u = \{j\}$ then (21) reduces to a Lipschitz condition for $\partial^{-\{j\}}f$, with respect to x^j, holding uniformly in $x^{-\{j\}}$. When condition (21) holds for u then it also holds for $\widetilde{u} \subseteq u$, so that Fréchet's $u = \emptyset$ condition is the most widely applicable version of (21), and the condition on the full partial derivative $\partial^{1:d}$ is the least widely applicable.

We illustrate the use of the propositions above with an example function having a "cusp" along a hyperplane. For integers $d \ge 1$ and $r \ge 0$ let $f_{d,r}$ be a function on $[0,1]^d$ defined by

$$f_{d,r}(x) = \begin{cases}\max(x^1+\cdots+x^d-1/2,0)^r, & r>0\\ 1_{x^1+\cdots+x^d>1/2}, & r=0.\end{cases}$$

For $u \subseteq 1{:}d$ with $|u| < r$,

$$\partial^u f_{d,r}(x) = r(r-1)\cdots(r-|u|+1)f_{d,r-|u|}(x). \tag{22}$$

If $|u| = r$ then (22) holds everywhere except on the set $E = \{x \mid x^1+\cdots+x^d = 1/2\}$ of d dimensional volume zero. If $|u| > r$ then $\partial^u f_{d,r}(x) = 0$ for $x \notin E$ and is not defined for $x \in E$.

Proposition 16. *$V(f_{d,r};[0,1]^d)$ is finite for $d \le r$ and infinite for $d \ge r+2$.*

Proof: If $r > d$ then $\partial^{1:d}f$ is bounded. If $r = d$ then $\partial^{2:d}f$ exists and is a Lipschitz continuous function in x^1 uniformly in $x^{2:d}$. Therefore $V(f) < \infty$ by Proposition 15 when $d \le r$.

Now suppose that $d \ge r+2$. Let $\mathcal{Y}^j = \{0, 1/(2m), \ldots, (m-1)/(2m)\}$ be a ladder on $[0,1/2]$, and put $\mathcal{Y} = \prod_{j=1}^d \mathcal{Y}^j$. Suppose that $y \in \mathcal{Y}$ with $\sum_{j=1}^d y^j = (m-d+1)/(2m)$. Then $\Delta(f_{d,r};y,y_+) = (2m)^{-r}$. The number of such y is equal to the number of ways to choose d nonnegative integers whose sum is $m-d+1$. Therefore

$$V(f) \ge \frac{1}{(2m)^r}\binom{m}{d-1} \ge \frac{(m-d)^{d-1}}{(d-1)!(2m)^r} \to \infty$$

as $m \to \infty$. Therefore $V(f;[0,1/2]^d) = \infty$ and so $V(f;[0,1]^d) = \infty$ too. □

Taking $r = 0$ in Proposition 16 shows that $V(1_A) = \infty$ for $A = \{x \in [0,1]^d \mid x^1 + \cdots + x^d > 1/2\}$ when $d \ge 2$. Similarly if A is a hyperrectangular region that is not parallel to any of the coordinate axes of $[a,b]$ then 1_A has infinite variation when $d \ge 2$. As d increases, it takes ever greater smoothness along the set E for $f_{d,r}$ to have finite variation.

Proposition 16 has a gap for the case $d = r+1$. Then $\partial^{1:d} f_{d,d-1}$ vanishes for $x \notin E$, but does not exist for $x \in E$. All of the variation of $f_{d,d-1}$ comes from the set E. It is not hard to show that $V(f_{2,1}) = 1$ and that in general $V(f_{d,d-1}) < \infty$. Here is a sketch of the reasoning: By Proposition 5 with $\epsilon = 1$, we need only consider ladders $\mathcal{Y}$ with every $y_+^j - y^j$ in an interval $[\eta, 2\eta)$ where $\eta > 0$. Such a ladder yields fewer than $A\eta^{-d+1}$ hyperrectangles $[y, y_+]$ that intersect the set E, for some $A < \infty$. For each such hyperrectangle, we find that $|\Delta(f_{d,d-1}; y, y_+)| \le B\eta^{d-1}$ for some $B < \infty$. Then $V_{\mathcal{Y}}(f_{d,d-1}) \le AB$.

Proposition 17 considered functions symmetric in their arguments. The infinite variation result is more general:

Proposition 17. *For integer $r \ge 0$, real values $\theta_0, \dots, \theta_d$, and $x \in [a,b]$ let*

$$f(x) = f_{r,\theta}(x) = \begin{cases} \max(\theta_1 x^1 + \cdots + \theta_d x^d - \theta_0, 0)^r, & r > 0 \\ 1_{\theta_1 x^1 + \cdots + \theta_d x^d > \theta}, & r = 0. \end{cases}$$

Let $E = \{x \in [a,b] \mid \theta_1 x^1 + \cdots + \theta_d x^d = \theta_0\}$. If $E \cap [a,b]$ has positive $d-1$ dimensional volume, $d \ge r+2$, and none of $\theta_1, \dots, \theta_d$ is zero, then $V(f) = \infty$.

Proof: The proof follows from two applications of Proposition 9 with linear transformations that reduce the problem to the one handled by Proposition 16. The details are omitted to save space. □

10 Functions vanishing except on one face

The next two propositions consider functions that are zero on all of the hyperrectangle $[a,b]$, except for a boundary face. There are two cases, one for a face that is a single corner of $[a,b]$ and one for a face of positive dimension less than d.

Proposition 18. *Let $a, b \in \mathbb{R}^d$ with $a \le b$ and let $u \subseteq 1\!:\!d$. Suppose that $f(x) = 0$ unless $x^u = a^u$ and $x^{-u} = b^{-u}$. Then*

$$V_{[a,b]}(f) = \begin{cases} |f(a^u : b^{-u})|, & \mathit{Vol}([a,b]) > 0 \\ 0, & \mathit{else.} \end{cases} \tag{23}$$

Proof: If $\mathrm{Vol}([a,b]) = 0$ then $V(f) = 0$ for any real valued f. Assume that $\mathrm{Vol}([a,b]) \neq 0$. Then $V_{\mathcal{Y}}(f; a, b) = |f(a^u : b^{-u})|$ for any ladder $\mathcal{Y}$ on $[a,b]$. $\square$

Proposition 19. *Let $a, b \in \mathbb{R}^d$ with $a \le b$. Let $u, v \subseteq 1\!:\!d$ with $u \cap v = \emptyset$ and $|u \cup v| < d$, and set $w = -u - v \neq \emptyset$. Suppose that $f(x)$ is defined on $[a,b]$ with $f(x) = 0$ unless $x^u = a^u$ and $x^v = b^v$. Then*

$$V_{[a,b]}(f) = \begin{cases} V_{[a^w, b^w]}(f(x^w; a^u\!:\!b^v)), & \mathit{Vol}([a,b]) > 0 \\ 0, & \mathit{else.} \end{cases} \tag{24}$$

Proof: Suppose that $\mathrm{Vol}([a,b]) > 0$. For any ladder $\mathcal{Y}$ on $[a,b]$ and any $y \in \mathcal{Y}$ we find that $\Delta(f; y, y_+) = 0$ if $y^u \neq a^u$ or $y_+^v \neq b^v$. Then $V_{\mathcal{Y}}(f) = V_{\mathcal{Y}^w}(f(x^w; a^u\!:\!b^v))$. $\square$

Proposition 18 is the $w = \emptyset$ version of Proposition 19 if we adopt the convention that the variation of f on $[a^\emptyset, b^\emptyset]$ is $|f()|$.

Proposition 20. *Let $a, b, \widetilde{a}, \widetilde{b} \in \mathbb{R}^d$ with $a \le \widetilde{a} \le \widetilde{b} \le b$. Let $f(x)$ be defined on $[a,b]$ with $f(x) = 1$ for $\widetilde{a} \le x \le \widetilde{b}$ and $f(x) = 0$ otherwise. Then*

$$V_{[a,b]}(f) = \prod_{j=1}^{d} \left(1_{a^j < \widetilde{a}^j} + 1_{\widetilde{b}^j < b^j}\right). \tag{25}$$

Proof: Begin by splitting $[a,b]$ into 3^d hyperrectangles of the form $[a^u, \widetilde{a}^u] \times [\widetilde{a}^v, \widetilde{b}^v] \times [\widetilde{a}^w, \widetilde{b}^w]$, where u, v, w are disjoint subsets of $1:d$ with $u \cup v \cup w = 1:d$. By Lemma 1, $V_{[a,b]}(f)$ is the sum of $V(f)$ taken over these hyperrectangles. Notice that if $v \neq \emptyset$ then f does not depend on x^j over the corresponding hyperrectangle, so $V(f)$ vanishes there. If instead $v = \emptyset$, then f vanishes except at one corner of the hyperrectangle, and so Proposition 18 applies. Therefore

$$\begin{aligned} V_{[a,b]}(f) &= \sum_{u \subseteq 1:d} V_{[a^u, \widetilde{a}^u] \times [\widetilde{b}^u, b^u]}(f) = \sum_{u \subseteq 1:d} \Big(\prod_{j \in u} 1_{a^j < \widetilde{a}^j}\Big)\Big(\prod_{j \notin u} 1_{\widetilde{b}^j < b^j}\Big) \\ &= \prod_{j=1}^{d} \Big(1_{a^j < \widetilde{a}^j} + 1_{\widetilde{b}^j < b^j}\Big). \quad \square \end{aligned}$$

Proposition 20 includes some interpretable special cases. If $a < \widetilde{a} \le \widetilde{b} < b$, then $V_{[a,b]}(f) = 2^d$, so the variation of the indicator function of a hyperrectangle in general position is 2^d. For the indicator function of a single point in general position, we take $\widetilde{a} = \widetilde{b}$, and find again the variation is 2^d. When the boundary of $[\widetilde{a}, \widetilde{b}]$ intersects that of the containing hyperrectangle $[a, b]$, the variation is smaller. If for any j, there is equality at both boundaries, that is $a^j = \widetilde{a}^j$ and $b^j = \widetilde{b}^j$, then $V(f)$ vanishes, reflecting the fact that f does not depend on x^j. If any $a^j = b^j$ then of necessity there is equality at both boundaries so $V(f)$ vanishes, as it must because $a^j = b^j$ implies $\mathrm{Vol}([a, b]) = 0$.

The next result relates Hardy-Krause variation of f to Vitali variation of f, after extending the domain of f to the "upper-right", and filling in constant values.

Proposition 21. *Suppose that $a, b, \widetilde{b} \in \mathbb{R}^d$ with $a < b < \widetilde{b}$. Let $f(x)$ be a real valued function defined on $[a, b]$. Define $\widetilde{f}(x)$ and $\widehat{f}(x)$ on $[a, \widetilde{b}]$ as follows: For $x \in [a, b]$ let $\widetilde{f}(x) = \widehat{f}(x) = f(x)$. For $x \in [a, \widetilde{b}] - [a, b]$ let $\widetilde{f}(x) = 0$ and $\widehat{f}(x) = f(b)$. Then*

$$V_{[a,\widetilde{b}]}(\widetilde{f}) = V_{\mathrm{HK}}(f) + |f(b)|, \quad \textit{and,} \quad V_{[a,\widetilde{b}]}(\widehat{f}) = V_{\mathrm{HK}}(f).$$

Proof: Let $g(x) = f(x)$ on $[a, b]$ and let $g(x) = c$ on $[a, \widetilde{b}] - [a, b]$. Split $[a, \widetilde{b}]$ into 2^d hyperrectangles of the form $[a^u, b^u] \times [b^{-u}, \widetilde{b}^{-u}]$ for $u \subseteq 1:d$. For $u = 1:d$, we find $V_{[a,b]}(g) = V_{[a,b]}(f)$. For $u \ne 1:d$ note that $g(x) - c$ is a function that vanishes except on one face of $[a^u, b^u] \times [b^{-u}, \widetilde{b}^{-u}]$. If $u = \emptyset$, we find $V_{[b,\widetilde{b}]}(g - c) = |f(b) - c|$ by Proposition 18. For $0 < |u| < d$, we apply Proposition 19 to $g - c$ getting the variation of $f - c$ on a face. By Proposition 11 the variation of $f - c$ equals that of f. Then

$$V_{[a,\widetilde{b}]}(g) = \sum_{u \subseteq 1:d} V_{[a^u,b^u] \times [b^{-u},\widetilde{b}^{-u}]}(g) = \sum_{\emptyset \ne u \subseteq 1:d} V_{[a^u,b^u]}(f(x^u, b^{-u})) + |f(b) - c|.$$

The first result follows for $c = 0$, the second for $c = f(b)$. $\square$

11 Variation and ANOVA

If $f \in L^2[a, b]$ where $\mathrm{Vol}([a, b]) > 0$, then there is an analysis of variance (ANOVA) decomposition of f. Liu & Owen (2003) outline properties and references for ANOVA. The ANOVA takes the form

$$f(x) = \sum_{u \subseteq 1:d} f_u(x)$$

where $f_u(x)$ only depends on x^u and $\int_{a^j}^{b^j} f_u(x)dx^j = 0$ whenever $j \in u$.

By Proposition 8, $V(f_u) = 0$ for $|u| < d$ and so $V(f) = V(f_{1:d})$. Let $E(g) = \mathrm{Vol}([a,b])^{-1} \int_{[a,b]} g(x)dx$ denote the expected value of $g(x)$ for random x uniformly distributed in $[a,b]$. Let $\mathrm{Var}(g) = E((g(x) - E(g(x)))^2)$ denote the variance of $g(x)$. Write $\sigma^2 = \mathrm{Var}(f)$ and $\sigma_u^2 = \mathrm{Var}(f_u)$. The ANOVA decomposition is so named because $\sigma^2 = \sum_{u \subseteq 1:d} \sigma_u^2$.

Proposition 22. *If $\sigma_{1:d}^2 > 0$ then $V(f) > 0$. The converse does not hold.*

Proof: Liu & Owen (2003) show that $E(\Delta(f; x, \widetilde{x})^2) = \sigma_{1:d}^2$ for independent random x and $\widetilde{x}$, both uniformly distributed on $[a,b]$. Then if $\sigma_{1:d}^2 > 0$ there exist $x, \widetilde{x} \in [a,b]$ with $|\Delta(f; x, \widetilde{x})| \geq \sigma_{1:d}$ and so $V(f) > 0$. As for the converse, let $f(x) = 1$ if $x = b$ and let $f(x) = 0$ otherwise. Then $0 \leq \sigma_{1:d}^2 \leq \sigma^2 = 0$ but $V(f) = 1$ by Proposition 20. □

12 Indicator functions

Let $[a,b]$ be a d dimensional hyperrectangle and let $A \subseteq [a,b]$. The indicator function of A, also called the characteristic function of A, is given by $1_A(x) = 1$ for $x \in A$ and $1_A(x) = 0$ otherwise. It is clear that $\Delta(1_A; a, b)$ must be an integer and so $V_{\mathcal{Y}}(1_A)$ must also be an integer. Therefore either $V(1_A) = \infty$ or $V(1_A)$ is a nonnegative integer. Also, we easily find that $V_{[a,b]}(1_A) = V_{[a,b]}(1_{[a,b]-A})$ by Proposition 11 because $1_{[a,b]-A} = 1 - 1_A$. □

Proposition 20 gives the variation in Vitali's sense for indicator functions of hyperrectangles. Propositions 16 and 17 show how indicator functions can have infinite variation when $d \geq 2$ and A has a planar boundary. The difference between the cases lies in whether the boundary of A is parallel to any of the coordinate axes of $[a,b]$.

For a more general set A we can for integer $m \geq 1$, split $[a,b]$ into m^d congruent hyperrectangles each similar to $[a,b]$. The variation of f is at least as large as the number of those hyperrectangles with nonzero variation. We anticipate that this number grows in proportion to m^{d-1} for typical sets A of interest. Therefore we first consider when an indicator function has non-zero variation. We know that $V(1_A) = 0$ if 1_A does not depend on x^j for some $j \in 1{:}d$. When $d = 2$ there is a converse as follows:

Proposition 23. *Let $[a,b]$ be a rectangle in $\mathbb{R}^2$ with Vol$([a,b]) > 0$. Let $f : [a,b] \to \{0,1\}$ and suppose that f does depend on x^j for each $j \in \{1,2\}$. Then $V(f; a, b) \geq 1$.*

Proof: Because f depends on x^2 there is a value $y^1 \in [a^1, b^1]$ such that $f(x^{\{2\}}; y^{\{1\}})$ takes both values 0 and 1. Similarly let y^2 be a point in $[a^2, b^2]$ for which $f(x^{\{1\}}; y^{\{2\}})$ takes both values 0 and 1, and put $y = (y^1, y^2)$. Let $\widetilde{y}^1 \in [a^1, b^1]$ and $\widetilde{y}^2 \in [a^2, b^2]$ satisfy $f(\widetilde{y}^1, y^2) = f(y^1, \widetilde{y}^2) = 1 - f(y)$. Let $[\widetilde{a}, \widetilde{b}] = \mathrm{rect}[y, \widetilde{y}]$. Then

$$|\Delta(f; \widetilde{a}, \widetilde{b})| = |f(\widetilde{y}) - f(\widetilde{y}^1, y^2) - f(y^1, \widetilde{y}^2) + f(y)| = |f(\widetilde{y}) - 2 + 3f(y)| \geq 1$$

for $f(\widetilde{y}), f(y) \in \{0, 1\}$. $\square$

The natural analogue of Proposition 23 does not hold true for $d \geq 3$. For $d = 3$, consider $[0, 1]^3$. Let $A_1 = A_2 = A_3 = [0, 1/2)$ and define

$$f(x) = \begin{cases} 1, & x^3 \in A_3 \text{ and } x^2 \in A_2, \\ 0, & x^3 \in A_3 \text{ and } x^2 \notin A_2, \\ 1, & x^3 \notin A_3 \text{ and } x^1 \in A_1, \\ 0, & x^3 \notin A_3 \text{ and } x^1 \notin A_1. \end{cases}$$

The function f depends on each of the 3 components of x. It can be shown that $V(f) = 0$. The result also holds for more general sets A_j.

Suppose that $A \subseteq [a, b]$ is a set, open or closed or neither, with a positive d dimensional volume and a smooth boundary. If a portion of that smooth boundary has positive $d - 1$ dimensional volume and is not parallel to any of the coordinate axes, then for $d \geq 2$, we expect that $V(1_A) = \infty$. For instance if A is the interior of a sphere of positive radius contained inside $[a, b]$ then $V(1_A) = \infty$. Informally, the argument runs as follows. We can find a small hyperrectangle s inside $[a, b]$ with one face in A, the opposite face not in A, and a nearly linear boundary separating $s \cap A$ from $s \cap (-A)$. Then $V(1_A) \geq V_s(1_A)$ and the latter is infinite. The next proposition fills in details.

Proposition 24. *Let A be a subset of the hyperrectangle $[a, b]$ in dimension $d \geq 2$. Suppose that there exists a subhyperrectangle $[\widetilde{a}, \widetilde{b}] \subseteq [a, b]$ of positive volume, an index $j \in 1{:}d$, and a function g defined on $[\widetilde{a}^{-\{j\}}, \widetilde{b}^{-\{j\}}]$ taking values in $(\widetilde{a}^j, \widetilde{b}^j)$ such that either $\widetilde{x} \in A$ when $\widetilde{x}^j > g(\widetilde{x}^{-\{j\}})$ and $\widetilde{x} \notin A$ when $\widetilde{x}^j < g(\widetilde{x}^{-\{j\}})$ or $\widetilde{x} \in A$ when $\widetilde{x}^j < g(\widetilde{x}^{-\{j\}})$ and $\widetilde{x} \notin A$ when $\widetilde{x}^j > g(\widetilde{x}^{-\{j\}})$. Suppose further that $\partial^{\{k\}} g$ is bounded away from zero for each $k \neq j$. Then $V(1_A) = \infty$.*

Proof: For $m \geq 1$, let S_m be the split of $[\widetilde{a}^{-\{j\}}, \widetilde{b}^{-\{j\}}]$ into m^{d-1} congruent hyperrectangles. Let $\widetilde{S}_m = \{s \times [\widetilde{a}^j, \widetilde{b}^j] \mid s \in S_m\}$. Then $\widetilde{S}_m$ is a split of $[\widetilde{a}, \widetilde{b}]$ into m^{d-1} long thin hyperrectangles. For each $s \in S_m$ evaluate g at all 2^{d-1} corners and select a value c strictly between the largest and second largest of these values. From a coordinate split of $s \times [\widetilde{a}^j, \widetilde{b}^j]$ along direction j at point c we find that $V_{s\times[\widetilde{a}^j,\widetilde{b}^j]}(1_A) \geq 1$ and so $V(1_A) \geq m^{d-1}$. □

In Proposition 24 the set A was assumed to be of positive d dimensional volume. Thus for example it does not apply to functions like the indicator of a hypershere that nontrivially intersects the $d \geq 2$ dimensional $[a, b]$. For that case we consider a subhyperrectangle $[\widetilde{a}, \widetilde{b}]$ for which there is an index j and a function g on $[\widetilde{a}^{-\{j\}}, \widetilde{a}^{-\{j\}}]$ with $\widetilde{x} \in A$ if and only if $g(\widetilde{x}^{-\{j\}}) = \widetilde{x}^j$. Once again we can find coordinate splits to show that the variation of 1_A is positive within each of m^{d-1} long thin hyperrectangles constructed as in the proof of Proposition 24.

13 Call and put options

Much work in quasi-Monte Carlo integration has been motivated by some integrands from computational finance. For full details of Monte Carlo applications to computational finance, see Glasserman (2004). Here we present some such integrands, and explain why they are not typically of bounded variation.

For $z \in \mathbb{R}$, let $\varphi(z) = \exp(-z^2/2)/\sqrt{2\pi}$ be the standard normal probability density function, $\Phi(z) = \int_{-\infty}^{z} \varphi(y)dy$ be the corresponding cumulative distribution function, and let Φ^{-1} be the quantile function, mapping $(0, 1)$ to $\mathbb{R}$. We also take $\Phi^{-1}(0) = -\infty$ and $\Phi^{-1}(1) = \infty$.

Many call options have a payoff function that can be expressed in the form:

$$C(x) = \max\left(0, \sum_{r=1}^{R} \alpha_r \exp\left(\beta_{r0} + \sum_{j=1}^{d} \beta_{rj} G^{-1}(x^j)\right) - K\right), \tag{26}$$

for scalars $\alpha_r > 0$ and β_{rj} and a strike price $K > 0$. It is usual to have $G = \Phi$ but sometimes G is an alternative distribution having fatter tails than does the normal. We will assume that $G^{-1}(0) = -\infty$ and $G^{-1}(1) = \infty$. For simplicity some discount factors have been absorbed into the α_r. The value of the option is $\int_{[0,1]^d} C(x)dx$. In cases of interest there exist r and $j \geq 1$ for which $\beta_{rj} \neq 0$ holds. Then C is unbounded on $(0, 1)^d$ and hence cannot be BVHK.

For $f_r(x) = \alpha_r \exp(\beta_{r0} + \sum_{j=1}^{d} \beta_{rj} G^{-1}(x^j))$, let

$$P(x) = \max\left(0, K - \sum_{r=1}^{R} f_r(x)\right). \tag{27}$$

This $P(x)$ is the payoff of a put option whose value $\int_{[0,1]^d} P(x)dx$ is of independent interest. Notice that $C(x)-P(x) = \sum_{r=1}^R f_r(x) - K$. When $G = \Phi$, there is a closed form expression for $\int_{[0,1]^d} f_r(x)dx$ and then an estimate of $P(x)$ can then be easily translated into one for $C(x)$. The function $P(x)$ is bounded because all the $\alpha_r > 0$. When $P(x)$ is BVHK, then quasi-Monte Carlo integration yields an estimate of $\int P(x)dx$ and hence also of $\int C(x)dx$ with error rate $O(n^{-1}\log(n)^d)$.

But $P(x)$ is ordinarily not BHVK. It is continuous but has a cusp along the set $E = \{x \mid \sum_{r=1}^R f_r(x) = 0\}$. As in the proof of Proposition 24 we employ m^{d-1} long thin hyperrectangles that cross E in their long direction. Let j be an index for which $\beta_{rj} \neq 0$ for some $r > 0$. Suppose first that $\beta_{rj} > 0$ so that $f_r(x) \to \infty$ as $x^j \to 1$ for any $x^{-\{j\}}$. The projections of these hyperrectangles in the $-\{j\}$ directions, split a hyperrectangle $[a^{-\{j\}}, b^{-\{j\}}] \subseteq [0,1]^{d-1}$ such that $P(x) > 0$ at every point of $[a^{-\{j\}}, b^{-\{j\}}] \times c^{\{j\}}$ for some $c^{\{j\}} \in (0,1)$. The hyperrectangles extend from $c^{\{j\}}$ to 1 in the x^j direction. When $d \geq 3$, the variation in each long thin hyperrectangle is larger than a fixed multiple of m^{-1} so that the variation of P is infinite.

If instead $\beta_{rj} < 0$, then take long thin hyperrectangles whose $-\{j\}$ projections split a hyperrectangle $[a^{-\{j\}}, b^{-\{j\}}] \subseteq [0,1]^{d-1}$ such that $P(x) = 0$ at every point of $[a^{-\{j\}}, b^{-\{j\}}] \times c^{\{j\}}$ for some $c^{\{j\}} \in (0,1)$. Then take the long direction for the hyperrectangles to be from 0 to $c^{\{j\}}$.

14 Sobol's low variation extensions

Given a function f defined on a subset K of $[a,b]$ we consider ways of extending it to $\widetilde{f}$ defined on all of $[a,b]$ while keeping some control on the size of $V_{[a,b]}(\widetilde{f})$. One application is in proving results like Theorem 2 of Sobol' (1973). Sobol's proof of that theorem was never published. Professor Sobol' kindly described for me the key ideas underlying the proof. See especially equations (29) and (30) below.

The set K is assumed to have some regularity. First we assume that K is a nonempty closed set. Then we designate some point $c \in K$ as an "anchor" for the extensions. This anchor is commonly taken to be a or b or $(a+b)/2$. Then we suppose that

$$x \in K \implies \mathrm{rect}[x,c] \subseteq K. \tag{28}$$

In case $c = b$ then K has the Pareto property. Given $x \in K$ and $y \notin K$ there is at least one j with $y^j < x^j$. The next result appears in Sobol' (1961).

Proposition 25. *Let f be a function on the hyperrectangle $[a,b]$. Suppose that $\partial^{1:d}f$ exists. Then for $x, c \in [a,b]$*

$$f(x) = f(c) + \sum_{\emptyset \neq u \subseteq 1:d} (-1)^{|u|} \int_{[x^u,c^u]} \partial^u f(y^u; c^{-u}) dy^u. \tag{29}$$

Proof: Similarly to equation (15) we find that $\int_{[x^u,c^u]} \partial^u f(y^u; c^{-u}) dy^u = \Delta_u(f; x, c)$. The rest follows from Proposition 6. □

The term $f(c)$ in equation (29) corresponds to the excluded case $u = \emptyset$ under a natural convention.

Next we give a representation of f as a sum of functions of varying dimensionalities, using mixed partial derivatives of f taken once with respect to each x^j for j in a set u. When K contains a d dimensional rectangle of nonzero volume, these derivatives are defined as usual. In particular for points x on the boundary of K, only one sided derivatives defined as limits from within K are used. When K is contained inside a zero volume rectangle there are some coordinate directions from which no meaningful limit can be taken. Let $\nu(K) = \{j \in 1:d \mid \sup_{x\in K} x^j > \inf_{x\in K} x^j\}$ be the set of coordinates that truly vary within K. The formulas below will not depend on the value we give to derivatives with respect to coordinates that do not vary. For definiteness, we take

$$\partial_K^u f(x) = \begin{cases} \partial^u f(x), & u \subseteq \nu(K) \\ 0, & \text{else.} \end{cases}$$

Even when $\nu(K) = \emptyset$, which holds when $K = \{c\}$, we still have $\partial_K^{\emptyset} f(c) = f(c)$.

Definition 5. *Let $c \in [a,b]$ and suppose that K is a nonempty closed subset of $[a,b]$ which satisfies (28), and that $\partial_K^u f(x)$ exists for $x \in K$ and $u \subseteq 1:d$. Then the Sobol' (low variation) extension of f from K to $[a,b]$ with anchor c is*

$$\widetilde{f}(x) = f(c) + \sum_{\emptyset \neq u \subseteq 1:d} (-1)^{|u|} \int_{[x^u,c^u]} 1_{z^u:c^{-u}\in K}\ \partial_K^u f(z^u : c^{-u}) dz^u. \tag{30}$$

To justify the name "extension" requires that $\widetilde{f}(x) = f(x)$ when $x \in K$. Note that $x \in K$ implies $\text{rect}[x,c] \subseteq K$ so that the expression $1_{z^u:c^{-u}\in K}$ can then be removed from equation (30). Next ∂_K^u only differs from ∂^u in cases where $[x^u, c^u]$ has zero volume, and those terms contribute nothing to the sum. Therefore the subscript K can be removed from the partial derivative symbol. Then $\widetilde{f}(x) = f(x)$ by Proposition 25.

Theorem 1. *For $c \in [a,b]$, let K be a nonempty closed subset of $[a,b]$ which satisfies (28). Let f be a function for which $\partial_K^u f(x)$ exists for $x \in K$, and $u \subseteq 1:d$. Let $\widetilde{f}$ be the Sobol' extension of f from K to $[a,b]$ with anchor c. Then*

$$V_{[a,b]}(\widetilde{f}) \leq \int_K |\partial_K^{1:d} f(x)| dx. \tag{31}$$

If $\partial_K^{1:d} f(x)$ is continuous on K, then

$$V_{[a,b]}(\widetilde{f}) = \int_K |\partial_K^{1:d} f(x)| dx. \tag{32}$$

Proof: If $|u| < d$ then the corresponding term in (30) is a function of x that does not depend on x^{-u}, so it has Vitali variation 0. Therefore the Vitali variation of $\widetilde{f}$ equals that of $\widetilde{f}_{1:d}(x) \equiv (-1)^d \int_{[x,c]\cap K} \partial_K^{1:d} f(z) dz$. Let $\mathcal{Y}$ be a ladder on $[a,b]$. For $y \in \mathcal{Y}$,

$$\begin{aligned}
\Delta(\widetilde{f}_{1:d}; y, y_+) &= \sum_{v \subseteq 1:d} (-1)^{|v|} \widetilde{f}_{1:d}(y^v : y_+^{-v}) \\
&= \sum_{v \subseteq 1:d} (-1)^{|v|} (-1)^d \int_{[y^v, c^v] \times [y_+^{-v}, c^{-v}]} 1_{z \in K}\ \partial_K^{1:d} f(z) dz \\
&= \sum_{v \subseteq 1:d} (-1)^{|-v|} \int_{[y^v, c^v] \times [y_+^{-v}, c^{-v}]} 1_{z \in K}\ \partial_K^{1:d} f(z) dz \\
&= \int_{[y, y_+]\cap K} \partial_K^{1:d} f(z) dz,
\end{aligned}$$

so that $V_{\mathcal{Y}}(\widetilde{f}_{1:d}) \leq \int_K |\partial_K^{1:d} f(x)| dx$. Taking the supremum over $\mathcal{Y}$ proves (31). To prove (32), note that K is compact, so $\partial_K^{1:d} f$ is uniformly continuous on K. We may split $[a,b]$ into a regular grid of m^d hyperrectangles, sum $V(f)$ over those hyperrectangles that are contained within K, and let $m \to \infty$. □

For the Hardy-Krause variation of $\widetilde{f}$ we need to consider which x^{-v} values when glued to b^v give a point in K. Let $K(b^v) = K_{-v}(b^v) = \{x^{-v} \in [a^{-v}, b^{-v}] \mid x^{-v} : b^v \in K\}$.

Theorem 2. *Assume the conditions of Theorem 1 and that $c = b$. Then*

$$V_{\mathrm{HK}}(\widetilde{f}) \leq \sum_{v \subsetneq 1:d} \int_{K(b^v)} |\partial_K^{-v} f(x^{-v}, b^v)| dx^{-v}. \tag{33}$$

Proof: From the definition, $V_{\mathrm{HK}}(\widetilde{f}) = \sum_{v\subsetneq 1:d} V_{[a^{-v},b^{-v}]}(\widetilde{f}(x^{-v};b^v))$. If $-v \subseteq \nu(K)$ then $\widetilde{f}(x^{-v};b^v)$ is also the Sobol' extension of $f(x^{-v};b^v)$ from $K(b^v)$ to $[a^{-v},b^{-v}]$ with anchor b^{-v}, and so

$$V_{[a^{-v},b^{-v}]}(\widetilde{f}(x^{-v};b^v)) \le \int_{K(b^v)} |\partial^{-v} f(x^{-v};b^v)|dx^{-v}. \tag{34}$$

Now suppose that $j \in -v$ and $j \notin \nu(K)$. Then $\widetilde{f}(x^{-v};b^v)$ does not depend on x^j, so that $V(\widetilde{f}(x^{-v};b^v)) = 0$ and again (34) holds. Summing (34) over $v \subsetneq 1\!:\!d$ establishes (33). $\square$

Acknowledgments

It is a pleasure to thank: Ilya M. Sobol' for communicating how he proved Theorem 2 of Sobol' (1973), Linda Yamamoto, Stanford's Mathematics and Computer Science librarian, for help in locating some articles, an anonymous reviewer for helpful comments, and Professors Gang Li and Jianqing Fan for inviting this contribution. Finally I thank Fang Kai-Tai for his many contributions to quasi-Monte Carlo sampling.

References

Adams, C. R. & Clarkson, J. A. (1934). Properties of functions $f(x, y)$ of bounded variation. *Transactions of the American Mathematical Society* **36** 711.

Clarkson, J. A. & Adams, C. R. (1933). On definitions of bounded variation for functions of two variables. *Transactions of the American Mathematical Society* **35** 824–854.

Fang, K. T. & Wang, Y. (1994). *Number-Theoretic Methods in Statistics.* Chapman and Hall, London.

Fréchet, M. (1910). Extension au cas des intégrales multiples d'une définition de l'intégrale due à Stieltjes. *Nouvelles Annales de Mathématiques* **10** 241–256.

Glasserman, P. (2004). *Monte Carlo methods in financial engineering.* Springer, New York.

Hardy, G. H. (1905). On double Fourier series, and especially those which represent the double zeta-function with real and incommensurable parameters. *Quarterly Journal of Mathematics* **37** 53–79.

Krause, M. (1903a). Über mittelwertsätze im gebiete der doppelsummen and doppelintegrale. *Leipziger Ber.* **55** 239–263.

Krause, M. (1903b). Über Fouriersche reihen mit zwei veränderlichen grössen. *Leipziger Ber.* **55** 164–197.

Liu, R. & Owen, A. B. (2003). Estimating mean dimensionality, Technical report, Stanford University, Department of Statistics.

Niederreiter, H. (1992). *Random Number Generation and Quasi-Monte Carlo Methods. S.I.A.M.*, Philadelphia, PA.

Sobol'. I. M. (1961). An exact estimate of the error in multi-dimensional quadrature formulas for function classes $\widetilde{w}_1^0$ and $\widetilde{h}_1^0$. *USSR, Comput. Maths and Math Physics* pp. 228–240.

Sobol'. I. M. (1973). Calculation of improper integrals using uniformly distributed sequences. *Soviet Math Dokl* **14**(3) 734–738.

Young, W. H. (1913). *Proceedings of the London Mathematical Society (2)* **11** 142.

Higher Order Power Properties of Empirical Discrepancy Statistics

Rahul Mukerjee

Indian Institute of Management Calcutta, Joka, Diamond Harbour Road, Kolkata 700 104, India (rmuk1@hotmail.com)

Summary. We investigate the higher order power properties for a very general class of empirical discrepancy statistics. This class includes the Cressie-Read discrepancy statistics and, in particular, the empirical likelihood ratio statistic. Under the criterion of average local power along spherical contours, it is seen that these competing statistics can be discriminated at the third order of comparison and that the Pearsonian chi-square statistic tends to have an edge over others.

Key words: Average power, contiguous alternatives, Cressie-Read discrepancy, Edgeworth expansion, empirical likelihood, Pearsonian chi-square statistic, third-order.

2000 Mathematics Subject Classification: 62G10

1 Introduction

Empirical likelihood (Owen, 1988) has been of significant interest in the statistics and econometrics literature; accounts of recent developments are available in Owen (2001) and Mittelhammer, Judge & Miller (2000). Corcoran (1998) introduced a very general class of empirical discrepancy statistics. This class includes the Cressie-Read discrepancy statistics (Baggerly, 1998) as an important subclass and, in particular, the empirical likelihood ratio (ELR) and Pearsonian chi-square statistics. Recently, Bravo (2003) reported illuminating results on second order power of Cressie-Read discrepancy statistics under contiguous alternatives and showed that the ELR

Contemporary Multivariate Analysis and Experimental Design—In Honor Celebration of Professor Kai-Tai Fang's 65th birthday. Edited by Jianqing Fan and Gang Li. The World Scientific Publisher, 2005.

statistic enjoys an optimality property in terms of second-order local maximinity along spherical contours. He also observed that if instead one works with the criterion of average local power along such contours, then all statistics become equivalent at the second order of comparison. Thus under the latter criterion, a third-order comparison of the statistics is warranted. This problem is addressed in the present article with reference to the general class of empirical discrepancy statistics (Corcoran, 1998).

As one can anticipate, the algebra for the third-order comparison is significantly more challenging than that at the second order. In order to alleviate this difficulty to some extent, we consider an approach, akin to that in comparing parametric likelihood-based tests (Mukerjee, 1990a, b), which avoids the direct computation of approximate cumulants. Even then, the present nonparametric setting entails extra terms that need careful attention. The final formula for third-order local average power reveals that the Pearsonian chi-square statistic tends to have an edge over others under the present criterion. This may be contrasted with the findings in Bravo (2003) on the second-order local maximinity of the ELR statistic.

After presenting the preliminaries in Section 2, we obtain an expression for the third-order power under contiguous alternatives in Section 3. This, in turn, leads to a formula for the third-order local average power. The formula as well as its implications are discussed in Section 4.

2 Preliminaries and a stochastic expansion

Let $X_1, \ldots, X_n$ be independent $q \times 1$ random vectors from an unknown common distribution with mean θ and a positive definite covariance matrix Γ. Interest lies in the null hypothesis $H_0 : \theta = \theta_0$. Let $\{p_i\}$ be a nonparametric likelihood supported on the observed data, and $\hat{p_i} = n^{-1}$ be the nonparametric maximum likelihood estimator of p_i. Following Corcoran (1998), an empirical discrepancy statistic for $H_0 : \theta = \theta_0$ is given by

$$T(\theta_0) = (2n^2/J_2) inf \sum_{i=1}^{n} h(p_i, \hat{p_i}),$$

where the infimum is over $p_i (1 \leq i \leq n)$, subject to $\sum p_i = 1$ and $\sum p_i(X_i - \theta_0) = 0$. Here $h(u, v)$ is a smooth discrepancy measure that satisfies $h(u, u) = 0$, but need not be symmetric in its arguments. Also, J_m is the mth derivative of $h(p_i, \hat{p_i})$ with respect to p_i, evaluated at $p_i = \hat{p_i} = n^{-1}$. As in Corcoran (1998), it is supposed that $J_m/(n^{m-2}J_2) = O(1)$, for sufficiently many values of $m = (1, 3, 4, \ldots)$. Thus $\rho_3 = J_3/(nJ_2)$ and $\rho_4 = J_4/(n^2 J_2)$ are $O(1)$ constants.

The choice $h(u,v) = \lambda^{-1}\{1-(u/v)^{-\lambda}\}$ leads to the important subclass of Cressie-Read discrepancy measures and entails $\rho_3 = -(\lambda+2)$. The cases $\lambda = 0, -1, -2, -\frac{1}{2}$ and 1 correspond to the empirical likelihood, Kullback-Leibler, Euclidean, Hellinger and Pearsonian chi-square discrepancies respectively (the first two are defined in a limiting sense).

We work under the same conditions as in Bravo (2003); these conditions justify the Edgeworth expansion in Section 3. Contiguous alternatives of the form $H_n : \theta = \theta_n$ are considered, where $\theta_n = \theta_0 + n^{-1/2}\Gamma^{1/2}\gamma$ and $\gamma = (\gamma_1, \ldots, \gamma_q)'$ is free from n. As in Bravo (2003), we note that in the present context of nonparametric inference, the population distribution functions under the null and contiguous alternative hypotheses are related in the sense that they are both assumed to belong to the same class of distributions, indexed by the mean θ. Thus the distribution of $X_i - \theta_n$, under θ_n, is the same as that of $X_i - \theta_0$, under θ_0 (the difference with a parametric location model is that the form of the distribution is now unknown). Consequently, defining

$$\begin{aligned} Z_i &= (Z_{i1}, \ldots, Z_{iq})' = \Gamma^{-1/2}(X_i - \theta_0), \\ Z_i^* &= (Z_{i1}^*, \ldots, Z_{iq}^*)' = \Gamma^{-1/2}(X_i - \theta_n), \end{aligned} \tag{1}$$

the standardized moments

$$\begin{aligned} \beta_{m(1)\ldots m(u)} &= E_{\theta_0}(Z_{im(1)} \cdots Z_{im(u)}) \\ &= E_{\theta_n}(Z_{im(1)}^* \cdots Z_{im(u)}^*) \quad (1 \le m(1), \ldots, m(u) \le q) \end{aligned} \tag{2}$$

do not depend on γ. In particular, $\beta_m = 0$ and $\beta_{ms} = \delta_{ms}$ $(1 \le m, s \le q)$, where δ_{ms} is Kronecker delta. Let

$$A_{m(1)\ldots m(u)} = n^{-1/2} \sum_{i=1}^{n} \{Z_{im(1)} \ldots Z_{im(u)} - \beta_{m(1)\ldots m(u)}\}, \tag{3}$$

$$A_{m(1)\ldots m(u)}^* = n^{-1/2} \sum_{i=1}^{n} \{Z_{im(1)}^* \ldots Z_{im(u)}^* - \beta_{m(1)\ldots m(u)}\}. \tag{4}$$

By (1) - (4),

$$A_m = A_m^* + \gamma_m (1 \le m \le q),$$

$$A_{ms} = A_{ms}^* + n^{-1/2}(\gamma_m A_s^* + \gamma_s A_m^* + \gamma_m \gamma_s) \qquad (1 \le m, s \le q), \tag{5}$$

$$A_{mst} = A_{mst}^* + \gamma_m \delta_{st} + \gamma_s \delta_{mt} + \gamma_t \delta_{ms} + o_p(1) \qquad (1 \le m, s, t \le q), \tag{6}$$

are stochastically bounded under θ_n. Hence following Bravo (2004), with further simplification, a stochastic expansion for $T(\theta_0)$ under θ_n is given by

$$T(\theta_0) = W'W + o_p(n^{-1}), \tag{7}$$

where

$$W = A + n^{-1/2}W_1 + n^{-1}W_2, \tag{8}$$

with $A = (A_1, \ldots, A_q)'$, $W_j = (W_{j1}, \ldots, W_{jq})'$ $(j = 1, 2)$, and using the summation convention,

$$W_{1m} = -\frac{1}{2}A_{ms}A_s - \frac{1}{6}\rho_3\beta_{mst}A_sA_t \qquad (1 \le m \le q), \tag{9}$$

$$\begin{aligned} W_{2m} = &\frac{1}{2}(1 + \frac{1}{2}\rho_3)^2(A'A)A_m - \frac{1}{24}(3\rho_3^2 - \rho_4)\beta_{mstu}A_sA_tA_u \\ &+ \frac{1}{9}\rho_3^2\beta_{mst}\beta_{uvt}A_sA_uA_v + \frac{3}{8}A_{mt}A_{st}A_s - \frac{1}{6}\rho_3A_{mst}A_sA_t \\ &+ \frac{1}{2}\rho_3\beta_{mst}A_{tu}A_sA_u - \frac{1}{12}\rho_3\beta_{stu}A_{mu}A_sA_t \qquad (1 \le m \le q). \end{aligned} \tag{10}$$

The following notation will be used later. Let $G_m(\gamma)$, $G_{ms}(\gamma)$, $G_{mst}(\gamma)$ etc. be multivariate Hermite polynomials in γ, i.e., $G_m(\gamma) = \gamma_m$, $G_{ms}(\gamma) = \gamma_m\gamma_s - \delta_{ms}$, and so on. Also, let $Q_m(\gamma)$, $Q_{ms}(\gamma)$, $Q_{mst}(\gamma)$, ... represent the first, second, third, ... raw moments of a q-variate normal distribution with mean vector γ and covariance matrix I_q. Thus $Q_m(\gamma) = \gamma_m$, $Q_{ms}(\gamma) = \gamma_m\gamma_s + \delta_{ms}$, and so on.

3 Third-order power

3.1 Approximate characteristic function

We first obtain the approximate characteristic function of W under θ_n. This approach yields the corresponding Edgeworth expansion without explicit evaluation of approximate cumulants of W. Let $\xi = (\xi_1, \ldots, \xi_q)' = (-1)^{1/2}\tau$, where τ is a q-dimensional auxiliary variate. By (8),

$$\begin{aligned} E_{\theta_n}\{exp(\xi'W)\} = &E_{\theta_n}[\{1 + n^{-1/2}\xi_mW_{1m} + n^{-1}(\xi_mW_{2m} \\ &+ \frac{1}{2}\xi_m\xi_uW_{1m}W_{1u})\}exp(\xi'A)] + o(n^{-1}). \end{aligned} \tag{11}$$

From (2),(4) and (5), it is easy to obtain the cumulants of A under θ_n. Hence

$$\begin{aligned} E_{\theta_n}\{exp(\xi'A)\} = &\{1 + \frac{1}{6}n^{-1/2}\beta_{mst}\xi_m\xi_s\xi_t + n^{-1}U(\xi)\} \\ &\times exp(\xi'\gamma + \frac{1}{2}\xi'\xi) + o(n^{-1}), \end{aligned} \tag{12}$$

where $U(\xi)$ is a polynomial in ξ, whose coefficients are constants free from n. Clearly, $U(\xi)$ is the same for all statistics under consideration. Its detailed form of will not be needed in what follows.

In view of (11), we next consider $E_{\theta_n}\{W_{1m}exp(\xi' A)\}$. Equation (12) yields an Edgeworth expansion for A, under θ_n, with margin of error $o(n^{-1/2})$. This helps in handling the second term of W_{1m} in (9). Note also that the first term of W_{1m} is the same for all statistics under consideration. Thus, after considerable algebra, one obtains

$$\begin{aligned} E_{\theta_n}\{W_{1m}exp(\xi' A)\} = {} & [C_m^{(0)}(\xi+\gamma) + U_m^{(0)}(\xi,\gamma) + n^{-1/2}\{C_m^{(1)}(\xi+\gamma) \\ & + C_m^{(2)}(\xi,\gamma) + U_m^{(1)}(\xi,\gamma)\}] \\ & \times exp(\xi'\gamma + \frac{1}{2}\xi'\xi) + o(n^{-1/2}), \end{aligned} \tag{13}$$

where

$$\begin{aligned} C_m^{(0)}(\gamma) &= -\frac{1}{6}\rho_3\beta_{mst}Q_{st}(\gamma), \\ U_m^{(0)}(\xi,\gamma) &= -\frac{1}{2}\beta_{mst}\{Q_{st}(\tilde{\gamma}) - \tilde{\gamma}_s\gamma_t\}, \end{aligned} \tag{14}$$

$$\begin{aligned} C_m^{(2)}(\xi,\gamma) = {} & \frac{1}{36}\rho_3\beta_{mst}\beta_{uvw}\{G_{uvw}(\gamma)Q_{st}(\tilde{\gamma}) \\ & -3\gamma_v\gamma_w Q_{stu}(\tilde{\gamma}) + 3\gamma_w Q_{stuv}(\tilde{\gamma})\}, \end{aligned} \tag{15}$$

and $\tilde{\gamma} = (\tilde{\gamma}_1, \ldots, \tilde{\gamma}_q)' = \xi + \gamma$. Here $C_m^{(1)}(\gamma)$ is a polynomial in γ and $U_m^{(1)}(\xi,\gamma)$ is a polynomial in ξ and γ. The coefficients in these polynomials are $O(1)$ constants. The detailed forms of $C_m^{(1)}(\gamma)$ and $U_m^{(1)}(\xi,\gamma)$ are not needed in the sequel. It may be noted that the first term of W_{1m} entails $U_m^{(0)}(\xi,\gamma)$ and $U_m^{(1)}(\xi,\gamma)$ in (13) while the second term of W_{1m} accounts for the other terms in (13). Thus $U_m^{(0)}(\xi,\gamma)$ and $U_m^{(1)}(\xi,\gamma)$ remain the same for all statistics that are being considered.

Remark 1. One of the two leading terms within squared brackets in (13), namely $U_m^{(0)}(\xi,\gamma)$, does not depend on ξ and γ only through $\xi+\gamma$. This may be contrasted with what happens for parametric likelihood-based tests (Mukerjee 1990a, b). The same phenomenon is seen to persist with $E_{\theta_n}\{W_{2m}exp(\xi' A)\}$. Then, unlike $U_m^{(0)}(\xi,\gamma)$, some of the resulting additional terms involve ρ_3 and influence the comparison in which we are interested here. Hence these additional terms, shown explicitly in (20) and (21) below (see also (19)), need careful attention.

Remark 2. An explanation for $U_m^{(0)}(\xi,\gamma)$ helps in understanding the subsequent development. By (2), (4) and (5), up to the first order of approximation, the limiting distribution of $(A_{ms}, A')'$, under θ_n, is $(q+1)$-variate normal with mean $(0, \gamma')'$, and covariance matrix

$$\begin{bmatrix} \sigma^2_{ms} & \psi'_{ms} \\ \psi_{ms} & I_q \end{bmatrix},$$

where $\psi_{ms} = (\beta_{ms1}, \ldots, \beta_{msq})'$ and σ^2_{ms} is a constant. Hence, under appropriate moment assumptions (Bravo, 2003), via a conditioning argument,

$$\begin{aligned} E_{\theta_n}\{A_{ms}A_s exp(\xi' A)\} &= E_{\theta_n}[\{\psi'_{ms}(A-\gamma)\}A_s exp(\xi' A)] + o(1) \\ &= \beta_{mst}\{Q_{st}(\tilde{\gamma}) - \tilde{\gamma}_s\gamma_t\}exp(\xi'\gamma + \frac{1}{2}\xi'\xi) + o(1). \end{aligned}$$

In view of the first term of W_{1m} in (9), the form of $U^{(0)}_m(\xi, \gamma)$ is now evident.

Precisely the same arguments as in Remark 2 help in handling the subsequent terms in the right-hand side of (11). Thus by (2), (4)-(6), (9) and (10), at the expense of a heavy algebra one can obtain $E_{\theta_n}\{W_{2m}exp(\xi' A)\}$ and $E_{\theta_n}\{W_{1m}W_{1u}exp(\xi' A)\}$ with margin of error $o(1)$. Using these calculations together with (12)-(15) in (11), we eventually get

$$\begin{aligned} E_{\theta_n}\{exp(\xi' W)\} &= \{1 + n^{-1/2}L_1(\xi,\gamma) + n^{-1}L_2(\xi,\gamma)\} \\ &\quad \times exp(\xi'\gamma + \frac{1}{2}\xi'\xi) + o(n^{-1}), \end{aligned} \tag{16}$$

where

$$L_1(\xi,\gamma) = \frac{1}{6}\beta_{mst}\xi_m\xi_s\xi_t + \xi_m\{C^{(0)}_m(\xi+\gamma) + U^{(0)}_m(\xi,\gamma)\}, \tag{17}$$

$$L_2(\xi,\gamma) = L_{20}(\xi,\gamma) + \xi_m L^{(m)}_{21}(\xi+\gamma) + L_{22}(\xi,\gamma), \tag{18}$$

$$L_{22}(\xi,\gamma) = \rho_3\sum_{j=1}^{5} M_j(\xi,\gamma) - \rho_3^2 M_6(\xi,\gamma), \tag{19}$$

$$M_1(\xi,\gamma) = -\frac{1}{2}\beta_{mst}\beta_{uvt}\xi_m\gamma_v Q_{su}(\tilde{\gamma}),$$

$$M_2(\xi,\gamma) = \frac{1}{12}\beta_{stu}\beta_{muv}\xi_m\gamma_v Q_{st}(\tilde{\gamma}), \tag{20}$$

$$M_3(\xi,\gamma) = \frac{1}{6}(\beta_{mstu}\gamma_u - \gamma_m\delta_{st} - \gamma_s\delta_{mt} - \gamma_t\delta_{ms})\xi_m Q_{st}(\tilde{\gamma}), \tag{21}$$

$$M_4(\xi,\gamma) = \frac{1}{36}\beta_{mst}\beta_{uvw}\xi_m\{G_{uvw}(\gamma)Q_{st}(\tilde{\gamma}) - 3\gamma_v\gamma_w Q_{stu}(\tilde{\gamma})\}, \tag{22}$$

$$M_5(\xi,\gamma) = -\frac{1}{12}\beta_{mst}\beta_{uvw}\xi_m\xi_u\gamma_w Q_{stv}(\tilde{\gamma}),$$

$$M_6(\xi,\gamma) = \frac{1}{72}\beta_{mst}\beta_{uvw}\xi_m\gamma_u Q_{stvw}(\tilde{\gamma}). \tag{23}$$

In (18), $L_{20}(\xi, \gamma)$ is a polynomial in ξ and γ, and $L_{21}^{(m)}(\gamma)$ is a polynomial in γ for each m. The coefficients in these polynomials are $O(1)$ constants. In particular, $L_{20}(\xi, \gamma)$ is the same for all statistics under consideration. The detailed forms of $L_{20}(\xi, \gamma)$ and $L_{21}^{(m)}(\gamma)$ will not be needed later.

The terms in $L_{22}(\xi, \gamma)$ will play a crucial role in Section 4. Of these, $M_j(\xi, \gamma)$ $(j = 1, 2, 3)$ arise from $E_{\theta_n}\{W_{2m}exp(\xi' A)\}$ as hinted in Remark 2, while the remaining $M_j(\xi, \gamma)$ arise from (15) and $E_{\theta_n}\{W_{1m}W_{1u}exp(\xi' A)\}$, on simplification.

3.2 Third-order power function

For $j \geq 1$ and $\Delta \geq 0$, let $K_{j,\Delta}(.)$ and $k_{j,\Delta}(.)$ stand respectively for the distribution and density functions of a possibly non-central chi-square variate with j degrees of freedom and non-centrality parameter Δ. Write $\overline{K}_{j,\Delta}(.) = 1 - K_{j,\Delta}(.)$, and given $\alpha (0 < \alpha < 1)$, define z^2 via $\overline{K}_{q,0}(z^2) = \alpha$. Let $S = \{w = (w_1, \ldots, w_q)' : w'w > z^2\}$, and $D = (D_1, \ldots, D_q)'$, where $D_m = \partial/\partial w_m$. Write $\phi_q(.)$ for the q-variate normal density with null mean and covariance matrix I_q. Define

$$R_j(\gamma, z^2) = \int_S L_j(-D, \gamma)\phi_q(w - \gamma)\, dw \qquad (j = 1, 2). \tag{24}$$

The approximate characteristic function (16) readily yields an Edgeworth expansion for W under θ_n. Using such an expansion, from (7) it follows that, for any $O(1)$ constant b,

$$\begin{aligned} P_{\theta_n}\{T(\theta_0) > z^2 + n^{-1}b\} &= \overline{K}_{q,\Delta}(z^2) + n^{-1/2}R_1(\gamma, z^2) \\ &\quad + n^{-1}\{R_2(\gamma, z^2) - bk_{q,\Delta}(z^2)\} + o(n^{-1}), \end{aligned} \tag{25}$$

where $\Delta = \gamma'\gamma$. From (14), (17) and (24), one can explicitly obtain $R_1(\gamma, z^2)$. The resulting expression agrees with with Bravo (2003) and vanishes at $\gamma = 0$. Hence by (25), $P_{\theta_0}\{T(\theta_0) > z^2 + n^{-1}b\} = \alpha + o(n^{-1})$ holds, provided $b = b_o$, where $b_o = R_2(0, z^2)/k_{q,0}(z^2)$. Let $\hat{b_o}$ be obtained by replacing any population moment in b_o by its sample analog. Then $\hat{b_o} = b_o + o_p(1)$, and the critical region $T(\theta_0) > z^2 + n^{-1}\hat{b_o}$ has size $\alpha + o(n^{-1})$. By (25), the third-order power function for this critical region is given by

$$\begin{aligned} P_{\theta_n}\{T(\theta_0) > z^2 + n^{-1}\hat{b_0}\} &= P_{\theta_n}\{T(\theta_0) > z^2 + n^{-1}b_o\} + o(n^{-1}) \\ &= \overline{K}_{q,\Delta}(z^2) + n^{-1/2}R_1(\gamma, z^2) \\ &\quad + n^{-1}R_2^*(\gamma, z^2) + o(n^{-1}), \end{aligned} \tag{26}$$

where

$$R_2^*(\gamma, z^2) = R_2(\gamma, z^2) - R_2(0, z^2)\{k_{q,\Delta}(z^2)/k_{q,0}(z^2))\}. \tag{27}$$

4 Average power

4.1 Expression for average power

We now consider the average power along spherical contours of the form $\gamma'\gamma = \mu(> 0)$. Let $V_1(\mu, z^2)$ and $V_2(\mu, z^2)$ be the averages of $R_1(\gamma, z^2)$ and $R_2^*(\gamma, z^2)$ respectively along such contours. Following Bravo (2003), $V_1(\mu, z^2) = 0$. Hence, by (26), the third-order average power function is given by, say,

$$P(\mu) = \overline{K}_{q,\mu}(z^2) + n^{-1}V_2(\mu, z^2) + o(n^{-1}). \qquad (28)$$

In order to examine the third-order term $V_2(\mu, z^2)$, note that by (18), (24) and (27)

$$R_2^*(\gamma, z^2) = R_{20}^*(\gamma, z^2) + R_{21}^*(\gamma, z^2) + R_{22}^*(\gamma, z^2), \qquad (29)$$

where

$$R_{2j}^*(\gamma, z^2) = R_{2j}(\gamma, z^2) - R_{2j}(0, z^2)\{k_{q,\Delta}(z^2)/k_{q,0}(z^2)\} \ (j = 0, 1, 2), \qquad (30)$$

with $\Delta = \gamma'\gamma$, and

$$R_{2j}(\gamma, z^2) = \int_S L_{2j}(-D, \gamma)\phi_q(w - \gamma)\, dw \qquad (j = 0, 2), \qquad (31)$$

$$R_{21}(\gamma, z^2) = \int_S (-D_m)L_{21}^{(m)}(-D + \gamma)\phi_q(w - \gamma)\, dw. \qquad (32)$$

For $j = 0, 1, 2$, let $V_{2j}(\mu, z^2)$ be the average of $R_{2j}^*(\gamma, z^2)$ along $\gamma'\gamma = \mu$. Following Mukerjee (1990a), by(30) and (32), $V_{21}(\mu, z^2) = 0$. Hence by (28) and (29),

$$P(\mu) = \overline{K}_{q,\mu}(z^2) + n^{-1}\{V_{20}(\mu, z^2) + V_{22}(\mu, z^2)\} + o(n^{-1}). \qquad (33)$$

Since $L_{20}(\xi, \gamma)$ is the same for all statistics under consideration, it is evident from (30) and (31) that the same holds for $V_{20}(\mu, z^2)$. Therefore, by (33), in order to study third-order average power, hereafter we consider only $V_{22}(\mu, z^2)$.

Now, from (19)-(23), $L_{22}(\xi, 0) = 0$, so that by (30) and (31), $V_{22}(\mu, z^2)$ reduces to the average of $R_{22}(\gamma, z^2)$ along $\gamma'\gamma = \mu$. From (19) and (31), it now follows that

$$V_{22}(\mu, z^2) = \rho_3 \sum_{j=1}^{5} B_j(\mu, z^2) - \rho_3^2 B_6(\mu, z^2), \qquad (34)$$

where, for each j, $B_j(\mu, z^2)$ is the average of

$$\int_S M_j(-D,\gamma)\phi_q(w-\gamma)\,dw$$

along $\gamma'\gamma = \mu$. For instance, from (20), proceeding along the line of Mukerjee (1990b),

$$\int_S M_1(-D,\gamma)\phi_q(w-\gamma)\,dw = -\beta_{mst}\beta_{uvt}\gamma_v\{\gamma_m\gamma_s\gamma_u k_{q+6,\Delta}(z^2) + (\gamma_m\delta_{su} + \gamma_s\delta_{mu} + \gamma_u\delta_{ms})k_{q+4,\Delta}(z^2)\},$$

where $\Delta = \gamma'\gamma$, so that

$$B_1(\mu, z^2) = -(\mu/q)(2\beta_{mst}\beta_{mst} + \beta_{mmt}\beta_{sst})k_{q+4,0}(z^2) + O(\mu^2).$$

Expressions for $B_j(\mu, z^2)$, $2 \le j \le 6$, can be obtained similarly from (20)-(23). Using these in (34), upon simplification,

$$V_{22}(\mu, z^2) = (\mu/q)k_{q+4,0}(z^2)a(z^2) + O(\mu^2),$$

where

$$a(z^2) = \rho_3\{\beta_{mmss} - q(q+2) - \beta_{mst}\beta_{mst} - \frac{1}{2}\beta_{mmt}\beta_{sst}\} - (\rho_3 + \frac{1}{6}\rho_3^2)(\beta_{mst}\beta_{mst} + \frac{3}{2}\beta_{mmt}\beta_{sst})\{z^2/(q+4)\}. \qquad (35)$$

Hence, in order to perform well with regard to third-order local average power, a statistic should keep $a(z^2)$ large.

4.2 Implications

The expression for $a(z^2)$ does not involve ρ_4. Since $\rho_3 = -(\lambda + 2)$ for Cressie-Read discrepancy statistics, it is clear that this subclass is essentially complete under the present criterion within the class of empirical discrepancy statistics.

Note that $\beta_{mst}\beta_{mst} + \frac{3}{2}\beta_{mmt}\beta_{sst} > 0$, whenever the β_{mst} are not all zeros. In this situation, the coefficient of z^2 in $a(z^2)$ is maximum when $\rho_3 + \frac{1}{6}\rho_3^2$ is minimum, i.e., uniquely when $\rho_3 = -3$, which corresponds to the Pearsonian chi-square statistic. Thus, if the β_{mst} are not all zeros, then any statistic with $\rho_3 \neq -3$ will have a smaller $a(z^2)$ than the Pearsonian chi-square statistic, and hence will be dominated by the latter under the present criterion, for sufficiently large z^2, i.e., for sufficiently small α. This is comparable with an optimality property of Rao's score statistic in the parametric setting (Mukerjee, 1990b).

For a given α or equivalently z^2, the formula (35) for $a(z^2)$ can be employed to compare standard choices of ρ_3 over various possibilities for the β_{mst} and β_{mmss}. To illustrate this, consider the univariate case, $q = 1$, and write $\kappa_1 = \beta_{111}^2$, $\kappa_2 = \beta_{1111}$. Then (35) reduces to

$$a(z^2) = \rho_3(\kappa_2 - 3 - \frac{3}{2}\kappa_1) - \frac{1}{2}(\rho_3 + \frac{1}{6}\rho_3^2)z^2\kappa_1. \tag{36}$$

Let $\alpha = 0.05$, i.e., $z^2 = 3.8415$. If $\kappa_2 > 3$, then by (36), the Pearsonian chi-square statistic (for this, $\rho_3 = -3$) has a larger $a(z^2)$ than the empirical likelihood, Kullback-Leibler, Euclidean or Hellinger discrepancy statistics (for these, $\rho_3 = -2$, -1, 0 and $-\frac{3}{2}$ respectively) when $\kappa_2 - 3$ is less than $1.82\kappa_1$, $2.14\kappa_1$, $2.46\kappa_1$ or $1.98\kappa_1$ respectively. On the other hand, if $\kappa_2 = 3$, then the Pearsonian chi-square statistic has a larger $a(z^2)$ than each of its aforesaid rivals whenever $\kappa_1 > 0$, and if $\kappa_2 < 3$, then the same holds for every κ_1.

Alternatively, one can consider expected $a(z^2)$ under suitable prior specification for the β_{mst} and β_{mmss}, and then optimize with respect to ρ_3. To illustrate this in the univariate case, we assign a uniform prior on (κ_1, κ_2) over

$$\{(\kappa_1, \kappa_2) : 0 \le \kappa_1 \le 1.8,\ \kappa_1 + 1 < \kappa_2 < \frac{15}{8}\kappa_1 + \frac{9}{2}\},$$

which is the range of (κ_1, κ_2) for the Pearsonian system of distributions; see Pearson & Hartley (1958, p. 210). By (36), the expected $a(z^2)$ with respect to such a uniform prior equals,

$$-\frac{607}{1960}\rho_3 - \frac{117}{245}(\rho_3 + \frac{1}{6}\rho_3^2)z^2,$$

which is maximum at $\rho_3 = -(3 + \frac{607}{312}z^{-2})$. Furthermore, for every z^2, the Pearsonian chi-square statistic has a larger expected $a(z^2)$ than each of its four rivals mentioned in the last paragraph.

Acknowledgments

Thanks are due to Professor K.T. Fang for a very helpful discussion. This work was supported by a grant from the Center for Management and Development Studies, Indian Institute of Management Calcutta.

References

Baggerly, K. A. (1998). Empirical likelihood as a goodness-of-fit measure. *Biometrika* **85** 535–547.

Bravo, F. (2003). Second-order power comparisons for a class of nonparametric likelihood-based tests. *Biometrika* **90** 881–890.

Bravo, F. (2004). Bartlett-type adjustments for empirical discrepancy test statistics. Preprint.

Corcoran, S. A. (1998). Bartlett adjustment of empirical discrepancy statistics. *Biometrika* **85** 967–972.

Mittelhammer, R., Judge, G. & Miller, D. (2000). *Econometric Foundations.* Cambridge University Press, Lodon.

Mukerjee, R. (1990a). Comparison of tests in the multiparameter case i. second order power. *J. Mult. Anal* **33** 17–30.

Mukerjee, R. (1990b). Comparison of tests in the multiparameter case ii. a third order optimality property of Rao's test. *J. Mult. Anal* **33** 31–48.

Owen, A. B. (1988). Empirical likelihood ratio confidence intervals for a single functional. *Biometrika* **75** 237–249.

Owen, A. B. (2001). *Empirical Likelihood.* Chapman and Hall, London.

Pearson, E. S. & Hartley, H. O. (1958). *Biometrika Tables for Statisticians*, **Volume I**. Cambridge University Press, London.

Modelling Multivariate Volatilities: An Ad Hoc Method

Mingjin Wang[1] and Qiwei Yao[2]

[1] Department of Statistics, London School of Economics and Political Science, Houghton Street, London, WC2A 2AE, U.K. (m.wang1@lse.ac.uk)
[2] Department of Business Statistics and Econometrics, Guanghua School of Management, Peking University 100871, Beijing, China (q.yao@lse.ac.uk)

Summary. The goal of the paper is two-fold. We first survey the available methods for modelling multivariate volatility processes. We then propose a new and simple method with numerical illustration.

Key words: Conditional covariance matrix, financial returns, multivariate GARCH, statistical estimation.

2000 Mathematics Subject Classification: 62M10,62H99,62P05

1 Introduction

Volatility plays an important role in controlling and forecasting risks in various financial operations. For a univariate return series, volatility is often represented in terms of conditional variances or conditional standard deviations. Many statistical models have been developed for modelling univariate conditional variance processes. While univariate descriptions are useful and important, problems of risk assessment, asset allocation, hedging in futures markets and options pricing require a multivariate framework, since high volatilities are often observed in the same time periods across different assets. Statistically this boils down to model time-varying conditional variance and covariance matrices of a vector-valued time series. Section 2 below lists

Contemporary Multivariate Analysis and Experimental Design—In Honor Celebration of Professor Kai-Tai Fang's 65th birthday. Edited by Jianqing Fan and Gang Li. The World Scientific Publisher, 2005.

some existing statistical models for multivariate volatility processes. We refer to Bauwens, Laurent & Rombouts (2003) for a more detailed survey on this topic. We propose a new and ad hoc method with numerical illustration in section 3. We concludes in section 4 with a brief summary.

2 Existing methods

Let $\mathbf{x}_t = (x_{1,t}, \cdots, x_{d,t})^\tau$ be a $d \times 1$ return series of d assets. Let $\mathcal{F}_t$ be the σ-algebra generated by $\{\mathbf{x}_k, k \le t\}$, which represents the information set at time t. We assume

$$E(\mathbf{x}_t|\mathcal{F}_{t-1}) = 0, \qquad \mathrm{Var}(\mathbf{x}_t|\mathcal{F}_{t-1}) = \boldsymbol{\Sigma}_t = \big(\sigma_{ij,t}\big). \tag{1}$$

The goal is to model the conditional variance-covariance matrix $\boldsymbol{\Sigma}_t$ which is a $d \times d$ non-negative definite matrix. Different models for $\boldsymbol{\Sigma}_t$ have been proposed over the last two decades. We review some of the models drawing major attraction in the literatures below, and refer to Bauwens, Laurent & Rombouts (2003) for a more extensive survey.

2.1 The BEKK GARCH models

One of the most general forms, proposed by Engle & Kroner (1995), is the BEKK representation of a multivariate GARCH(p, q) process

$$\boldsymbol{\Sigma}_t = \mathbf{C}_0'\mathbf{C}_0 + \sum_{k=1}^{K}\sum_{i=1}^{q} \mathbf{A}_{ik}\mathbf{x}_{t-i}\mathbf{x}_{t-i}'\mathbf{A}_{ik}' + \sum_{k=1}^{K}\sum_{j=1}^{p} \mathbf{B}_{jk}\boldsymbol{\Sigma}_{t-j}\mathbf{B}_{jk}', \tag{2}$$

where $\mathbf{C}_0, \mathbf{A}_{ik}, \mathbf{B}_{jk}$ are $d \times d$ matrices and $\mathbf{C}_0$ is upper triangular.

Although the form of the above model is quite general especially when K is reasonably large, it suffers from the problems due to overparametrization. See Engle & Kroner (1995) for more discuss on the identification problem of this model.

Similar to univariate GARCH models, the standard estimation method for the BEKK model (2) is the quasi-maximum likelihood estimation (qMLE) facilitated by assuming $\mathbf{x}_t|\mathcal{F}_{t-1} \sim N(\mathbf{0}, \boldsymbol{\Sigma}_t)$. The consistency and the asymptotic normality of the qMLE have been established by Comte & Lieberman (2003). Note that even for moderately large d, the qMLE is a solution of a high-dimensional nonlinear optimization problem. Therefore in practice some approximate and iterative estimation methods are often more efficient.

2.2 Factor and orthogonal models

In order to reduce the number of parameters in modelling multivariate volatilities, different types of decompositions for $\boldsymbol{\Sigma}_t$ are often employed in model-specifications. For instance, writing $\boldsymbol{\Sigma}_t$ as the sum of a time-varying part (usually with reduced rank) and a homoscedastic part, Engle, Ng & Rothschild (1990) proposed a factor multivariate GARCH model as follows:

$$\boldsymbol{\Sigma}_t = \boldsymbol{\Omega} + \sum_{k=1}^{K} \mathbf{g}_k \mathbf{g}_k' \Big(\sum_{i=1}^{q} \alpha_{ik} \mathbf{f}_k' \mathbf{x}_{t-i} \mathbf{x}_{t-i}' \mathbf{f}_k + \sum_{j=1}^{p} \beta_{jk} \mathbf{f}_k' \boldsymbol{\Sigma}_{t-j} \mathbf{f}_k \Big). \tag{3}$$

where α_{ik}, β_{jk} are non-negative constants, $\boldsymbol{\Omega}$ is a time-invariant non-negative definite constant matrix, and $\mathbf{g}_k, \mathbf{f}_k$ are $d \times 1$ constant vectors satisfying the constraints $\sum_{\ell=1}^{d} f_{k\ell} = 1$ for $k = 1, \cdots, K$ and $\mathbf{f}_k' \mathbf{g}_i = 0$ for $k \neq i$, and 1 for $k = i$.

Model (3) is called a Factor-GARCH$(p, q; k)$ model. The K linear combinations $\theta_{k,t} = \mathbf{f}_k' \mathbf{x}_t$, $k = 1, 2, \cdots, K$, represent K common factors of which the conditional variances are specified as K different univariate GARCH(p, q) models. It is easy to see that (3) is a special case of the BEKK model if we put $\mathbf{A}_{ik} = \sqrt{\alpha_{ik}} \mathbf{f}_k \mathbf{g}_k'$ and $\mathbf{B}_{jk} = \sqrt{\beta_{jk}} \mathbf{f}_k \mathbf{g}_k'$.

Factor-GARCH models can be estimated with qMLE method. In practice, the factor representation portfolios $\theta_{k,t}$ are usually set as known and a two-stage estimation scheme can then be invoked. See Lin (1992) for other estimation procedures.

Orthogonal GARCH (hereafter "O-GARCH") model, which is based on the principal components of $\mathbf{x}_t$, can virtually be viewed as a special case of Factor-GARCH model (Alexander & Chibumba (1997)). Let the unconditional covariance matrix of $\mathbf{x}_t$ be $\boldsymbol{\Sigma}$. Based on the eigen-decomposition $\boldsymbol{\Sigma} = \mathbf{W} \boldsymbol{\Lambda} \mathbf{W}'$, where $\mathbf{W}'\mathbf{W} = \mathbf{I}_d$ and $\boldsymbol{\Lambda}$ is a diagonal matrix. O-GARCH specification first fits the conditional variance of each principal component $\boldsymbol{\zeta}_t \equiv (\zeta_{1,t}, \cdots, \zeta_{d,t})' = \mathbf{W}' \mathbf{x}_t$ with a univariate GARCH model:

$$\zeta_{i,t} | \mathcal{F}_{t-1} \sim N(0, \lambda_{i,t}),$$
$$\lambda_{i,t} = \omega_i + \sum_{u=1}^{q_i} \alpha_{iu} \zeta_{i,t-u}^2 + \sum_{v=1}^{p_i} \beta_{iv} \lambda_{i,t-v},$$

and then take $\boldsymbol{\Sigma}_t = \mathbf{W} \boldsymbol{\Lambda}_t \mathbf{W}'$ as the conditional variance matrix of $\mathbf{x}_t$, where $\boldsymbol{\Lambda}_t = \mathrm{diag}(\lambda_{1,t}, \cdots, \lambda_{d,t})$. This effectively assumes that the principal components are also *conditionally* uncorrelated.

Obviously, O-GARCH model is easy to fit in practice even when d is large or very large. However, it treats unconditionally uncorrelated principal components as conditionally correlated as well, which is typically untrue.

This may lead to nonsensical or even wrong results. See Fan, Wang & Yao (2004).

Recently, Fan, Wang & Yao (2004) proposed to model multivariate volatilities in terms of a decomposition based on the so-called conditionally uncorrelated components (CUC) of $\mathbf{x}_t$. It overcomes the aforementioned shortcoming of the O-GARCH models.

2.3 Conditional correlation models

It always holds that

$$\mathbf{\Sigma}_t = \mathbf{D}_t \mathbf{\Gamma}_t \mathbf{D}_t, \tag{4}$$

where $\mathbf{D}_t = \text{diag}(\sqrt{\sigma_{11,t}}, \cdots, \sqrt{\sigma_{dd,t}})$, and $\mathbf{\Gamma}_t$ is the conditional correlation matrix $\mathbf{x}_t$.

Assuming that $\mathbf{\Gamma}_t$ does not change over time t and modelling each $x_{j,t}$ with a univariate GARCH model, (Bollerslev 1990) proposed a constant conditional correlation (CCC) framework which simplified the estimation and inference procedures substantially. However, it is questionable if the time-invariant conditional correlation is a realistic assumption in practice.

The dynamic conditional correlation (DCC) model of Engle (2002) computes the time changing conditional correlation matrix from the standardized residuals series

$$\mathbf{\Gamma}_t = \text{diag}\{\mathbf{Q}_t\}^{-1/2} \mathbf{Q}_t \text{diag}\{\mathbf{Q}_t\}^{-1/2}, \tag{5}$$

where

$$\mathbf{Q}_t = \mathbf{S}(1 - \theta_1 - \theta_2) + \theta_1(\boldsymbol{\xi}_{t-1}\boldsymbol{\xi}'_{t-1}) + \theta_2 \mathbf{Q}_{t-1}, \tag{6}$$

and $\xi_{k,t}$ are the standardized residuals obtained from the raw residuals $x_{k,t}/\{\hat{\sigma}_{kk,t}\}^{1/2}$, and $\mathbf{S}$ is the sample covariance matrix of $\{\boldsymbol{\xi}_t\}_{t=1}^n$.

A slightly different formulation was suggested by Tse & Tsui (2002):

$$\mathbf{\Gamma}_t = \mathbf{\Gamma}(1 - \theta_1 - \theta_2) + \theta_1 \mathbf{\Gamma}_{t-1} + \theta_2 \mathbf{\Psi}_{t-1} \tag{7}$$

to fit the correlation process. Here $\mathbf{\Gamma} = \{\rho_{ij}\}$ is a time-invariant $d \times d$ positive definite parametric matrix with unit diagonal elements and $\mathbf{\Psi}_{t-1}$ is, for example, the sample correlation matrix of $\{\boldsymbol{\xi}_t\}_{t-M}^{t-1}$. This specification is called varying correlation multivariate GARCH model or simply VC-MGARCH model.

Although qMLE method is available in principle for all these conditional correlation models, some two-stage estimation schemes have been developed to increase the computational efficiency, and have apparently been used more often in practice.

3 A new ad hoc method

3.1 Method

Note in (1), $\sigma_{ii,t} = \text{Var}(x_{i,t}|\mathcal{F}_{t-1})$. We may model $\sigma_{i,t}^2 \equiv \sigma_{ii,t}$ using any appropriate univariate volatility models based on univariate time series $\{x_{ti}\}$.

To model the off-diagonal elements $\sigma_{ij,t}$ with $i < j$, put

$$y_{ij,t} = (x_{i,t} + x_{j,t})/2. \tag{8}$$

We may model its conditional variance $\omega_{ij,t} = \text{Var}(y_{ij,t}|\mathcal{F}_{t-1})$ again by a simple univariate model. Note that for $1 \leq i < j \leq d$,

$$\sigma_{ij,t} = 2\omega_{ij,t} - \frac{\sigma_{i,t}^2 + \sigma_{j,t}^2}{2}. \tag{9}$$

Hence once we have derived univariate volatility models for each component $x_{i,t}$ and the combined series $y_{ij,t}$, the models for the conditional covariances is implied by (9) above.

In practice, we may use simple GARCH(1,1) models for modelling both $\sigma_{i,t}^2$ and $\omega_{ij,t}$, namely

$$\sigma_{i,t}^2 = \alpha_i + \beta_i x_{i,t-1}^2 + \gamma_i \sigma_{i,t-1}^2, \tag{10}$$

$$\omega_{ij,t} = \alpha_{ij} + \beta_{ij} y_{ij,t-1}^2 + \gamma_{ij}\omega_{ij,t-1}. \tag{11}$$

It is clear that the above proposal overcomes the difficulties due to overparametrization, and can be implemented in a computationally efficient manner since all the components of $\boldsymbol{\Sigma}_t$ are practically fitted separately. Furthermore, we have the flexibility in choosing appropriate univariate models for $\sigma_{i,t}^2$ and $\omega_{ij,t}$, which may be GARCH, stochastic volatility models, semi-parametric or nonparametric volatililty models, or some empirical methods such as rolling exponential smoothing. However the simplicity in both the structure and the feasibility does come with a price unfortunately. First the implied estimator for the conditional variance $\boldsymbol{\Sigma}_t$ may not necessarily be a non-negative definite matrix. (A quick remedy may be to shrink the negative eigenvalues of the estimated $\boldsymbol{\Sigma}_t$ to 0). Furthermore, the approach suffers from a kind of innate inconsistency in model specification. For example, under the GARCH(1,1) specification of (10) and (11), the conditional variance of a portfolio $\mathbf{a}'\mathbf{x}_t$ is not necessarily GARCH(1,1). Also note that we may define $y_{ij,t}$ differently from the form (8), still $\sigma_{ij,t}$ may be uniquely determined by $\sigma_{i,t}^2, \sigma_{j,t}^2$ and $\omega_{ij,t}$. However the estimator for $\sigma_{ij,t}$ implied may be different.

3.2 Numerical illustration

We illustrate the new method with two real data sets. The first one consists of the daily log returns (in percentages) of two exchange rate series, namely, the Deutsche mark (D) and the Japanese yen (J) versus U.S. dollar. It covers the period of 3 January 1990 — 23 June 1998, for a total of 2131 observations. The data was downloaded from the website of the Federal Reserve Bank of New York and has been analyzed by Tse & Tsui (2002) using the VC-MGARCH model. See Figures 1(a) & (b) for the time series plots of these two series. The second data set contains the four indices from Asian stock markets, i.e. the Hang Seng index of Hong Kong (HS), the Japan Nikkei 225 index (JN), the Shanghai Composite index (SH) and the Taiwan Weighted index (TW). Daily close prices adjusted for dividends and splits are obtained directly from the website of Yahoo!Finance. We applied log-difference transformation to convert them into continuously compounded returns. Adjustment was also made to account for the effect of different holidays of these four markets. The data consist of 1507 observations covering the period of 1 August 1997 — 31 July 2004. The time series plots for the second data set are omitted to save the space.

Descriptive statistics for all the six series are reported in Table 1. All the series are leptokurtic and the nulls of normal distribution can be rejected based on the Jarque-Bera test for all series. The Ljung-Box portmanteau statistics of the two exchange rates series suggest that there exists no significant evidence for the autocorrelation structures in both the series. We extract the mean values from these two series and focus our attention to their covariance matrix modelling. For the Asian market data, the portmanteau statistics reveal some autocorrelation structure in HS and TW series. Accordingly we fit an AR(5) model for each of these four series first. The analysis reported below was conducted with the filtered series.

For the first data set, a univariate GARCH(1,1) model is fitted to D and J, respectively, using qMLE method subject to the "variance targeting" constraint in the sense that the long run variance is just the sample variance (see Engle (2002)). In order to obtain an estimator for conditional correlation between D and J, another univariate GARCH(1,1) model is fitted to (D+J)/2 using the same method. Table 2 presents the estimated parameters. Standard errors are omitted to save space. See Figure 1(c) and (d) for the fitted volatility for these two return series and (e) and (f) for the fitted covariance and correlation between D and J, respectively. A horizontal line in Figure 1(f) is drawn to show the level of unconditional correlation between these two series.

Table 1. Summary Statistics of the Two Data Sets

	D	J	HS	JN	SH	TW
Mean	0.0025	-0.0023	-0.0193	-0.0388	0.0101	-0.0411
Stdev	0.6746	0.6750	2.0974	1.6868	1.5109	1.9432
Min	-2.8963	-4.5228	-14.7346	-9.0145	-8.7277	-9.9360
Max	3.1030	3.2269	20.2083	8.8876	8.8491	9.7871
Skewness	0.0197	-0.5065	0.6226	0.0107	0.1881	-0.0199
Kurtosis	4.7731	6.5508	14.9947	5.2678	8.2629	5.2284
J-B	279.29	1210.62	9131.42	322.97	1748.09	311.91
$Q_1(10)$	13.9104	16.0221	14.8263	6.5265	8.2794	20.9968
$Q_1(20)$	21.8039	27.4127	30.1055	10.7677	16.9162	38.4476
$Q_2(10)$	287.0760	83.8406	226.8334	74.5163	131.2148	88.4256
$Q_2(20)$	460.5919	111.9082	242.1270	100.1453	192.3026	103.3381

Note: J-B stands for the Jarque-Bera statistics. $Q_1(k)$ and $Q_2(k)$ represents the Ljung-Box portmanteau statistics of the original and squared return series, respectively.

It is interesting to compare the fitted conditional correlation in Figure 1(f) with those in Figure 4 of Tse & Tsui (2002). The later was obtained using BEKK model and VC-MGARCH model, both of them needed an intensive searching method to maximize the corresponding likelihood functions. The magnitudes and the time-varying patterns in these two figures are very similar. This suggests that our ad hoc method is as capable as those more sophisticated models in representing dynamic correlation structure at least for this data set. Furthermore, the fitted conditional correlation process always stays between -1 and 1. Hence the corresponding conditional covariance is automatically a non-negative definite matrix.

To further check the possible misspecification of the fitted model, we use the Ljung-Box Q portmanteau statistics of the cross-product of the standardized error series.

More specifically, we use the $Q(k)$ statistics of $\hat{u}_{i,t}^2 - 1, t = 1, 2, \cdots, T$ to check adequacy of the volatility model for the i-th series and the $Q(k)$ statistics of $\hat{u}_{i,t}\hat{u}_{j,t} - \hat{\rho}_{ij,t}, t = 1, 2, \cdots, T$ to check the correlation modelling between i-th and j-th series, where $\hat{u}_{i,t} = x_{i,t}/\hat{\sigma}_{i,t}$ is the standardized residuals. χ_k^2 is selected as a null reference distribution[3]. The two columns on the right in Table 2 list the values of $Q(10)$ and $Q(20)$. Apparently, at any conventional level of significance, there is no evidence to indicate the remnant

[3]Although there is no rigorous theory for such a test so far, the simulation study in Tse & Tsui (1999) suggests that it indeed provides a reasonable test with good size and power.

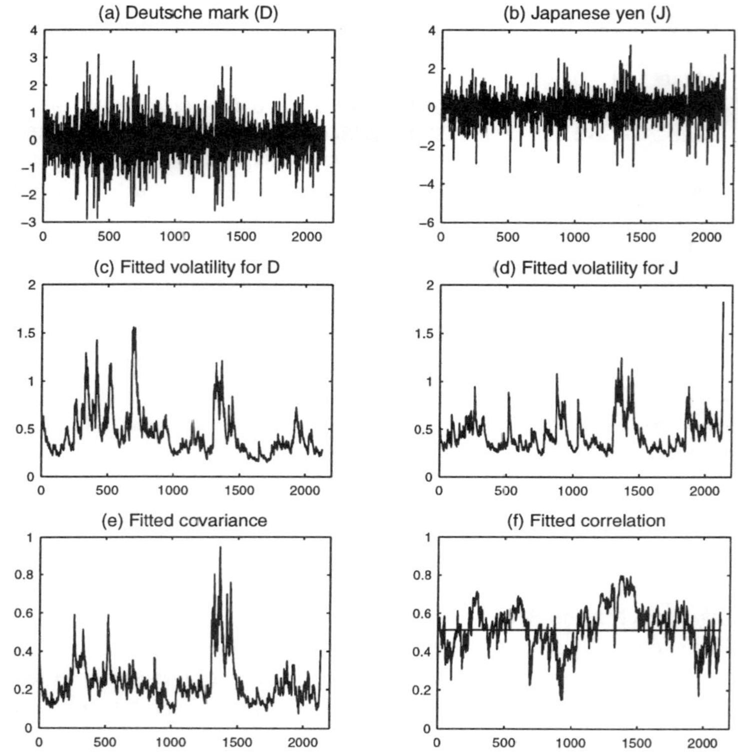

Fig. 1. *Time series plot of daily log-return (in percentage) of (a) Deutsche mark (D) and (b) Japanese yen (J) versus US dollar; the fitted volatility of the return series of (c) Deutsche mark and (d)Japanese yen; and the fitted (e) conditional covariance and (f) conditional correlation between D and J using the ad hoc method.*

autocorrelation structure in the residuals. This confirms quantitatively that the new method works well for such a data set.

The intractability of estimating a high-dimensional volatility model is a notorious fact in modelling multivariate volatility processes. For instance, for a four-dimension BEKK model it requires to solve an optimization problem with at least 42(=10+16+16) parameters. However, our ad hoc method can handle such a situation in a pretty easy manner. As a matter of fact, we need to estimate 10 univariate GARCH(1,1) models only, and the six conditional covariance can be derived according to (9). Table 2 lists the estimated coefficients for the Asian market data set. The six fitted conditional correlation series are plotted in Figure 2, where the horizontal line in each

Table 2. Estimation and Diagnostic Checking Results

	α	β	γ	$Q(10)$	$Q(20)$
D	0.0061	0.0509	0.9357	11.5085	15.9470
J	0.0108	0.0439	0.9325	3.6654	10.7579
(D,J)	0.0066	0.0432	0.9376	15.8036	23.5034
HS	0.0731	0.0946	0.8887	7.9466	14.3098
JN	0.2203	0.0777	0.8447	5.0772	9.1901
SH	0.1056	0.1322	0.8214	7.3288	20.1829
TW	0.3347	0.0862	0.8242	5.8506	15.4980
(HS,JN)	0.1228	0.0860	0.8674	5.0718	16.5563
(HS,SH)	0.0562	0.0956	0.8737	12.4900	23.6990
(HS,TW)	0.1952	0.0902	0.8385	7.0120	13.5878
(JN,SH)	0.0916	0.0673	0.8655	9.1555	12.9979
(JN,TW)	0.2163	0.0559	0.8423	2.1990	4.5946
(SH,TW)	0.1785	0.1015	0.7867	7.4842	19.4355

panel is the corresponding unconditional correlation. Furthermore, the values of the Q portmanteau statistic in the two very-right columns of Table 2 suggest the adequacy of the fitting. Note the (global) unconditional correlations are pretty close to 0 in Figures 2(b), (d) and (f), it seems reasonable to observe some negative conditional correlations in those plots. Note that the estimated conditional variances are not guaranteed to be non-negative definite. We calculate the eigenvalues for each fitted covariance matrix and the negative values only occur at the smallest eigenvalues of 23 points over the whole 1502(=1507-5) observations.

4 Conclusion

After reviewing some of the major multivariate volatility models, we put forward a new ad hoc method to model the conditional covariance process. Numerical results based on two real data sets suggest that a practically meaningful fitting may be obtained in a computationally efficient manner from applying the proposed new method. Therefore it might be worthwhile to investigate the theoretically properties of this ad hoc method more thoroughly.

Acknowledgements

The research was partially supported by an EPSRC research grant. Mingjin Wang was also supported by the Natural Science Foundation of China (grant 70201007).

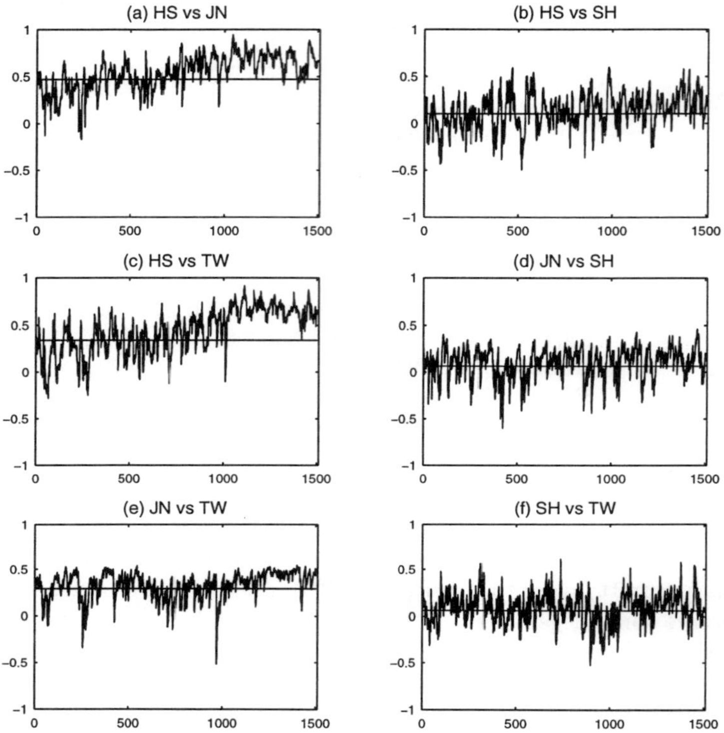

Fig. 2. *The fitted conditional correlations between (a) HS and JN, (b) HS and SH, (c) HS and Tw, (d) JN and SH, (e) JN and TW, (f) SH and TW for the Asian Stock Market data using the ad hoc method.*

References

Alexander, C., & Chibumba, A. (1997). *Multivariate orthogonal factor GARCH.* University of Sussex, Mimeo.

Bauwens, L., Laurent, S. & Rombouts, J. V. K. (2003). *Multivariate GARCH models: a survey.* CORE, Universit'e catholique de Louvain, Louvain-la-Neuve, Belgium.

Bollerslev, T. (1990). Modelling the coherence in short-run nominal exchange rates: a multivariate generalized ARCH model. *Reveiw of Economics and Statistics* **72** 498–505.

Comte, F. & Lieberman, O. (2003). Asymptotic theory for multivariate GARCH processes. *Journal of Multivariate Analysis* **84** 61–84.

Engle, R. F., Ng, V. K. & Rothschild, M. (1990). Asset pricing with a factor-ARCH covariance structure: empirical estimates from treasury bills. *Journal of Econometrics* **45** 213–238.

Engle, R. F. (2002). Dynamic conditional correlation — a simple class of multivariate GARCH models. *Journal of Business of Economic Statistics* **20** 339–350.

Engle, R. F. & Kroner, K. F. (1995). Multivariave simultaneous generalised ARCH. *Econometric Theory* **11** 122–150.

Fan, J., Wang, M. & Yao, Q. (2004). Modelling multivariate volatilities via conditional uncorrelated components. *Under preparation.*

Lin, W. (1992). Alternative estimators for factor GARCH models - a Monte Carlo comparison. *Journal of Applied Econometrics* **7** 259–279.

Magnus, J. R. & Neudecker, H. (1999). *Matrix Differential Calculus with Applications in Statistics and Econometrics.* (Revised edition). Wiley, New York.

Tse, Y. K. & Tsui, A. (1999). A note on diagnosing multivariate conditional heteroscedasticity models. *Journal of Time Series Analysis* **20** 679–691.

Tse, Y. K. & Tsui, A. (2002). A multivariate GARCH modle with time-varying correlations. *Journal of Business and Economic Statistics* **20** 351–362.

Some Recent Advances in Two-level Structural Equation Models: Estimation, Testing and Robustness

Peter M. Bentler[1], Jiajuan Liang[2] and Ke-Hai Yuan[3]

[1] University of California, Los Angeles, Department of Psychology, Box 951563, Los Angeles, CA 90095-1563, U.S.A. (bentler@ucla.edu)
[2] University of New Haven, School of Business, 300 Boston Post Road, West Haven, CT 06516, U.S.A. (jliang@newhaven.edu)
[3] University of Notre Dame, Department of Psychology, Notre Dame, IN 46556, U.S.A. (kyuan@nd.edu)

Summary. Estimation, testing and robustness are three common problems in two-level structural equation models. Liang & Bentler (2004a) developed the EM approach to estimation of model parameters and studied the asymptotic properties of the estimators. In a series of articles, Yuan and Bentler studied the problems of testing model fit and the robustness of tests and standard errors. This paper reviews some recent advances in research on the three problems, and provides some numerical comparisons between our methodology and selected existing ones.

Key words: Mean and covariance structure, maximum likelihood estimation, robustness, two-level structural equation model, testing hypothesis

2000 Mathematics Subject Classification: 62F35, 62H12, 62H15, 62P15, 62P25

1 Introduction

Two-level structural equation models (SEM for simplicity) have been used to analyze data collected from a hierarchical sampling scheme. For example, in evaluating students' performance in an educational program, test scores may be col-

Contemporary Multivariate Analysis and Experimental Design—In Honor Celebration of Professor Kai-Tai Fang's 65th birthday. Edited by Jianqing Fan and Gang Li. The World Scientific Publisher, 2005.

lected from students nested in schools; in studying some psychological behavior of children, data may be collected from children nested in households; in evaluating the effectiveness of a new drug, data may be collected from patients divided into groups according to their age; or in longitudinal studies, data may be collected from the same group of individuals at different time points. In collecting this type of sample data, the hierarchical sampling scheme consists of two steps: 1) randomly select some groups (such as schools and households); and 2) randomly select some individuals from each of the selected groups. The data collected by this two-step sampling are usually called two-level (or clustered, hierarchically structured) data. The individuals (such as students) nested in different groups (such as schools) are called level-1 units, and the groups the level-2 units. In some practical situations, data with more than two levels can be encountered. For example, students can be considered to be nested in different classes, while the classes are nested in various schools. Such data collected from students have a three-level structure. In general, the more levels that are utilized, the more complicated the model and its associated methodologies will become. However, most existing studies on multilevel models mainly focus on the case of two levels, and similarly, we will focus on two-level structural equation models. Comprehensive discussions of multilevel models can be found in books such as Goldstein (1995), Heck & Thomas (2000), Hox (2002), Kreft & de Leeuw (1998), Raudenbush & Bryk (2002), Reise & Duan (2003), and Snijders & Bosker (1999).

A general two-level SEM may contain observations from both level-1 and level-2 units. They are called level-1 observations and level-2 observations, respectively. For example, scores collected from students' tests are level-1 observations; financial resources and teaching facilities such as the number of labs in a school are level-2 observations, which are collected only through the level-2 units. It is assumed that level-1 observations are influenced by two sources: 1) variables and factors characterizing the differences among individuals, which are called level-1 influences; and 2) variables and factors characterizing the differences across the level-2 units, which are called level-2 influences. A model for characterizing the effect of level-1 influences is called the level-1 model, and a model for characterizing the effect of level-2 influences, the level-2 model. A basic assumption of two-level SEM is that the effects of various variables or factors on the response (indicator) variables are additive. In structural equation models, there is a strong preference to consider observed variables as indicators of hypothesized latent variables or factors, and hence models may include specifications that relate latent to observed variables, and latent variables to each other. Such latent variables or factors may exist at one or both levels in the case of two-level SEM. The most widely known latent variable model is the factor analysis model, in which the latent variables generate the observed variables and are simply correlated among themselves. Because of the wide applications of factor analysis models in various fields, especially in the behavioral and social sciences, e.g., education, medicine, psychology, and sociology, two-level SEM with a factor analysis model as its level-1 or level-2 (or at both levels) model have been very common in practical applications.

Two-level SEM can be developed in a variety of ways. Some earlier discussions on the formulation of two-level SEM can be referred to (to name a few): Goldstein

& McDonald (1988), Muthén (1989, 1994, 1997) McDonald & Goldstein (1987), Lee (1990), Lee and Poon (1992, 1998), McDonald (1993) and du Toit & du Toit (2002). Based on this earlier work, Liang & Bentler (2004a) proposed the following formulation for analysis of two-level SEM:

$$\begin{pmatrix} \boldsymbol{z}_g \\ \boldsymbol{y}_{gi} \end{pmatrix} = \begin{pmatrix} \boldsymbol{z}_g \\ \boldsymbol{v}_g \end{pmatrix} + \begin{pmatrix} \mathbf{0} \\ \boldsymbol{v}_{gi} \end{pmatrix}, \tag{1}$$

where $\{\boldsymbol{z}_g : \ q \times 1, \ g = 1, \ldots, G\}$ is a set of i.i.d. (independently identically distributed) level-2 observations whose level-1 component can be regarded as a constant (zero, for simplicity), and $\{\boldsymbol{y}_{gi} : \ p \times 1, \ i = 1, \ldots, N_g; \ g = 1, \ldots, G\}$ is a set of level-1 observations that are assumed to be decomposed as the sum of level-1 effect ($\boldsymbol{v}_{gi}$) and level-2 effect ($\boldsymbol{v}_g$). For each fixed g, $\{\boldsymbol{y}_{gi} : \ i = 1, \ldots, N_g\}$ are i.i.d. observations from the same group (level-2 unit), while for all i and g, $\{\boldsymbol{y}_{gi}\}$ are not independent. $\boldsymbol{v}_g$ is the level-2 latent random vector characterizing the level-2 effects, and $\boldsymbol{v}_{gi}$ the level-1 latent random vector characterizing the level-1 effects. It is typically assumed that both $\boldsymbol{z}_g$ and $\boldsymbol{v}_g$ are independent of $\boldsymbol{v}_{gi}$, but $\boldsymbol{z}_g$ and $\boldsymbol{v}_g$ are generally correlated. The zero vector ($q \times 1$) in (1) implies that the level-2 observation $\boldsymbol{z}_g$ has a constant effect (assuming zero without loss of generality) on all level-1 units nested in the same level-2 unit. The set of level-1 sample sizes $\{N_g : \ g = 1, \ldots, G\}$ is called the level-1 sample design and G the level-2 sample size. When there are no level-2 observations $\boldsymbol{z}_g$'s (or $\boldsymbol{z}_g \equiv \mathbf{0}$), formulation (1) reduces to

$$\boldsymbol{y}_{gi} = \boldsymbol{v}_g + \boldsymbol{v}_{gi}. \tag{2}$$

For example, Lee (1990), Lee and Poon (1992, 1998) and Lee & Tsang (1999) studied two-level SEM by using formulation (2).

Within the setup of (1), we further denote $\boldsymbol{\mu}_z = E(\boldsymbol{z}_g)$, $\boldsymbol{\mu}_y = E(\boldsymbol{y}_{gi}) = E(\boldsymbol{v}_g)$, $\boldsymbol{\Sigma}_{zz} = \text{cov}(\boldsymbol{z}_g)$, $\boldsymbol{\Sigma}_B = \text{cov}(\boldsymbol{v}_g)$, $\boldsymbol{\Sigma}_W = \text{cov}(\boldsymbol{v}_{gi})$ and $\boldsymbol{\Sigma}_{zy} = \text{cov}(\boldsymbol{z}_g, \boldsymbol{y}_{gi}) = \text{cov}(\boldsymbol{z}_g, \boldsymbol{v}_g)$. It is typically assumed that $E(\boldsymbol{v}_{gi}) = \mathbf{0}$. With $p + q$ means and $p(p+1)/2 + (p+q)(p+q+1)/2$ variances and nonduplicated covariances, there are a total of

$$R = p + q + \frac{p(p+1)}{2} + \frac{(p+q)(p+q+1)}{2} \tag{3}$$

free parameters in the saturated model, which implies that all means, variances and nonduplicated covariances are free parameters. In a two-level SEM, the means $\boldsymbol{\mu}_z$, $\boldsymbol{\mu}_y$ and variance-covariance matrices $\boldsymbol{\Sigma}_{zz}$, $\boldsymbol{\Sigma}_B$, $\boldsymbol{\Sigma}_W$ and $\boldsymbol{\Sigma}_{zy}$ can be further structured. For example, a substantive theory might represent $\boldsymbol{v}_{gi}$ as generated by a factor model with $\boldsymbol{\Sigma}_W = \boldsymbol{\Lambda}_W \boldsymbol{\Phi} \boldsymbol{\Lambda}_W' + \boldsymbol{\Psi}_W$. Similarly, $\boldsymbol{\Sigma}_B$ may also have a factor structure like that of $\boldsymbol{\Sigma}_W$. The variables in $\boldsymbol{z}_g$ might predict the between level (level-2) factors, which will lead to structured $\boldsymbol{\Sigma}_{zy}$ and $\boldsymbol{\mu}_y$. The means, variances and nonduplicated covariances of the observed data are no longer the model's free parameters in the structured model. They are actually functions of a more basic set of free parameters such as factor loadings, factor covariances, error variances, the regression coefficients and residual error variances. Denote the set of free parameters as a parameter vector $\boldsymbol{\theta}$. The structured model or the null hypothesis can be represented as

$$\boldsymbol{\mu} = \begin{pmatrix} \boldsymbol{\mu}_z \\ \boldsymbol{\mu}_y \end{pmatrix} = \boldsymbol{\mu}(\boldsymbol{\theta}) = \begin{pmatrix} \boldsymbol{\mu}_z(\boldsymbol{\theta}) \\ \boldsymbol{\mu}_y(\boldsymbol{\theta}) \end{pmatrix}, \quad (p+q) \times 1$$

$$\boldsymbol{\Sigma}_W = \boldsymbol{\Sigma}_W(\boldsymbol{\theta}), \quad p \times p \tag{4}$$

$$\widetilde{\boldsymbol{\Sigma}}_B = \widetilde{\boldsymbol{\Sigma}}_B\,(\boldsymbol{\theta}) = \begin{pmatrix} \boldsymbol{\Sigma}_{zz}(\boldsymbol{\theta}) & \boldsymbol{\Sigma}_{zy}(\boldsymbol{\theta}) \\ \boldsymbol{\Sigma}_{yz}(\boldsymbol{\theta}) & \boldsymbol{\Sigma}_B(\boldsymbol{\theta}) \end{pmatrix}, \quad (p+q) \times (p+q).$$

Of course, model (4) has to be identified. There is no simple rule for deciding an identified model. Let r be the number of parameters in $\boldsymbol{\theta}$, a necessary condition for model identification is $r \leq R$. A similar inequality has to hold for the means and mean parameters, as well as for the covariances and covariance parameters. A detailed discussion of identification on conventional (one-level) SEM is given by Bollen (1989). This can be applied to each level of a two-level SEM and covers most instances. As in modeling with independent groups (see e.g., Bentler (2004)), sometimes cross-level constraints on parameters can serve to identify a model that would otherwise be underidentified at a given level.

As discussed in Liang & Bentler (2004a), formulation (1) covers a variety of formulations for two-level SEM in the literature. The complexity of the structured model (4) may come from the level-1 model for $\boldsymbol{v}_{gi}$, or from the level-2 model for $\boldsymbol{v}_g$ or from both. By setting up formulation (1), we want to answer the questions: Q1) how do the level-1 factors in $\boldsymbol{v}_{gi}$ influence one another at the within level (level-1), how do the level-2 factors in $\boldsymbol{v}_g$ influence one another at the between level (level-2), and what are the effects from level-1 and level-2 factors on the observations $\{\boldsymbol{z}_g, \boldsymbol{y}_{gi}\}$? Q2) how well does model (4) fit the observed data $\{\boldsymbol{z}_g, \boldsymbol{y}_{gi}\}$ after the model parameters are estimated? and Q3) What assumptions have to be imposed on formulation (1) for evaluating the methods for answering Q1) and Q2), and what is the sensitivity of the methods to violation of assumptions? Methods for addressing Q1) require adequate approaches to parameter estimation. Methods for addressing Q2) involve testing hypotheses related to model fit, and those for addressing Q3) are concerned with the robustness of statistics. Thus in the next sections, we will discuss the recent literature with regard to formulation (1) with the structured model (4): parameter estimation, hypothesis testing, and robustness of statistics.

2 Parameter Estimation

The most commonly used method for parameter estimation in multilevel models is maximum likelihood (ML) based on the normal distributional assumption. A competitive alternative is the generalized least squares (GLS) procedure. We will mainly discuss these two with an emphasis on the ML method for continuous variables.

2.1 Maximum likelihood estimation based on Gauss-Newton type or Fisher scoring algorithms

In the context of multilevel models, maximum likelihood is based on the assumption that the variables $\boldsymbol{z}_g$, $\boldsymbol{v}_g$ and $\boldsymbol{v}_{gi}$ are multivariate normally distributed. Based on this, the log likelihood can be obtained. The associated first- and second-order partial derivatives provide various algorithms for parameter estimation. An advantage of employing the Gauss-Newton or Fisher scoring algorithms is that the Hessian matrix is not required, that is, the second order derivatives of the ML function are not needed. The expected information matrix typically used in these methods is substantially easier to compute than the observed information matrix. With regard to specific implementations, Goldstein & McDonald (1988) pointed out that two-level SEM is a special case of their general two-level model, which is different from (1). McDonald & Goldstein (1987) proposed a formulation for analysis of two-level SEM and provided derivatives for implementing the quasi-Newton or Fisher scoring algorithms for ML estimation. McDonald & Goldstein (1987) model can be formulated in the form of (1) (Liang & Bentler (2004a)). Muthén (1994, 1997) proposed the so-called MUML approach to parameter estimation in two-level SEM. Muthén's (1994, 1997) model can also be formulated in the form of (1). MUML is an approximate ML approach that reduces to full ML estimation for a balanced level-1 sample design $\{N_g \equiv n,\ g = 1, \ldots, G\}$. For unbalanced designs, some studies found that MUML provides similar parameter estimates as ML (Muthén (1991)); Hox (1993); McDonald (1994); Hox & Maas (2001). Yuan & Hayashi (2004) studied conditions for MUML inference to be asymptotically valid. Lee (1990) studied the asymptotics of parameter estimates for formulation (2) and pointed out the possible implementation of a Fisher scoring algorithm for obtaining the ML estimate (MLE) of $\boldsymbol{\theta}$. du Toit & du Toit (2002) provided the technical derivation for the ML method and model implemented in LISREL 8.5 (du Toit & du Toit (2001)).

2.2 Maximum likelihood estimation based on EM-type algorithms

The Expectation Maximization (EM) algorithm (Dempster, Laird & Rubin (1977)) has been successfully used to solve ML estimation problems with missing data. Observing the complexity that can result from an unbalanced sample design, Raudenbush (1995) constructed the "pseudo-balanced" sample design from an unbalanced sample design by conceiving some observations as "missing". Under his pattern of missing data, Raudenbush developed an EM algorithm for obtaining the ML estimator (MLE) of two-level SEM. Raudenbush (1995) two-level SEM can be formulated in the form of (1) (Liang & Bentler (2004a)). Lee & Poon (1998) considered another pattern of missing data in two-level SEM by conceiving of the level-2 latent variables as missing observations. Without involving the mean structure, Lee and Poon developed an EM algorithm for ML analysis of formulation (2). An advantage of Lee & Poon (1998) EM algorithm is that the E-step function has the same form as the ML function from a conventional (one-level) multiple-group SEM. As a result, the M-step of Lee & Poon (1998) EM algorithm can be realized by employing some standard SEM packages such as EQS (Bentler (2004)), LISREL (du Toit & du Toit (2001)), and M*plus* (Muthén & Muthén (2004)). This can save some computational effort in implementing the algorithm.

To compensate the disadvantages of no mean structure and the computation of a large number of inverse matrices in Lee & Poon (1998) approach, Bentler & Liang (2003a) generalized Lee & Poon (1998) methodology to the case of mean structures and improved their algorithm by using a matrix decomposition that avoids the computation of a large number of inverse matrices. Bentler & Liang (2003b) illustrated the practical implementation of their EM algorithm by setting up an EQS model for simultaneous two-level mean and covariance structure analysis. Liang & Bentler (2004a) proposed (1) as a formulation for general two-level SEM and pointed out that its coverage includes many existing formulations in the literature. Following the same line as in Lee & Poon (1998), Liang & Bentler (2004a) developed an EM algorithm for fitting formulation (1) with the structural model (4). Their algorithm has been coded in EQS Version 6.0 (Bentler (2004)). This new version provides structured or nonsaturated simultaneous two-level mean and covariance structure analysis that is yet to be made available in parallel packages such as LISREL and M*plus*. An example based on Liang & Bentler's algorithm in EQS is given in Section 5.

2.3 Estimation based on generalized least squares

It is well-known that the least squares (LS) method has been used in regression analysis for parameter estimation without any distributional assumption on the variables in the underlying model. The LS method has been generalized to various weighted versions, and the resulting versions are usually called generalized least squares (GLS). GLS can be implemented under distributional assumptions or without imposing any distributional assumption on the underlying model. The GLS estimator is taken as that based on a minimization of the sum of generalized distances between the observed data and the modeled data. The various ways for measuring distance result in various GLS estimators. Browne (1974, 1984) developed the GLS method for analysis of conventional (one-level) SEM. One of his weighting methods (1984) provides an asymptotically distribution free (ADF) method. His techniques for developing the GLS and ADF methods have been generalized to the case of two-level SEM. For example, Lee (1990) studied the GLS estimator for two-level SEM and obtained some asymptotic results. Lee (1990) pointed out that the Gauss-Newton algorithm or the Fisher scoring algorithm can be employed to obtain the GLS estimator for parameters in formulation (2). Lee & Poon (1992) proposed both ML and GLS analysis with small level-1 samples for formulation (2). Poon & Lee (1992) proposed another way to do GLS estimation for two-level SEM. Lee's (1990) proposal is still used today, as computational methods for practical implementation of existing GLS theory in two-level SEM remain based on the Gauss-Newton or Fisher scoring algorithms.

3 Testing Model Fit

Testing model fit in two-level SEM amounts to testing the null hypothesis of the structured model against the alternative hypothesis of the unstructured model. When the two-level SEM is formulated as (1), the null hypothesis can be stated as

$$\begin{aligned} H_0 &: \text{ the structured model (4) is true,} \\ \text{versus } H_1 &: \boldsymbol{\mu},\ \boldsymbol{\Sigma}_W \text{ and } \widetilde{\boldsymbol{\Sigma}}_B \text{ in (4) are unstructured.} \end{aligned} \tag{5}$$

The primary method for testing (5) under the assumption of multivariate normality of the involved variables is the likelihood ratio (LR) statistic defined by

$$T_{LR} = l(\hat{\boldsymbol{\theta}}) - l(\hat{\boldsymbol{\theta}}_s), \tag{6}$$

where $l(\cdot)$ is the negative twice of log-likelihood function, $\hat{\boldsymbol{\theta}}$ is the MLE for $\boldsymbol{\theta}$ from the structured model (H_0 is true), and $\hat{\boldsymbol{\theta}}_s$ the MLE for $\boldsymbol{\theta}$ from the saturated (unstructured) model (H_1 is true). The LR statistic in (6) has an asymptotic chi-square χ^2_{R-r} (R is given by (3)) under the normal distributional assumption as described in McDonald & Goldstein (1987), Lee & Poon (1998), and Liang & Bentler (2004a). The degrees of freedom $R - r$ is the difference between number of free parameters in the saturated model and in the structured model.

It has been found that the direct LR statistic (6) usually rejects the true model (H_0 is true) too often (with a higher type I error rate) in the case of conventional (one-level) SEM. For example, Hu, Bentler & Kano (1992), and Bentler & Yuan (1999) studied the effect of small sample size on test statistics for conventional SEM. In fitting two-level SEM, there are two sets of sample sizes: the set of level-1 sample sizes $\{N_g : g = 1, \ldots, G\}$ and the level-2 sample size G. It is mainly the level-2 sample size G that influences the convergence of T_{LR} to χ^2_{R-r} (Yuan & Bentler (2002)). Due to the sensitivity of T_{LR} to small G and to the normal distributional assumption, various types of corrections or adjustments to T_{LR} and related induced statistics for testing model fit for conventional SEM have been proposed, see, for example, Satorra and Bentler (1988, 1990, 1994), Yuan and Bentler (1997, 1998a, 1998b, 1999a). Some of these corrected chi-square and induced statistics have been generalized to fitting two-level SEM. We summarize some recent results as follows.

3.1 Rescaled likelihood ratio statistics

Following the approach to rescaling the LR statistics for fitting conventional SEM, Yuan & Bentler (2002) developed the rescaled LR statistic

$$T_{RLR} = \frac{T_{LR}}{\hat{\lambda}} \tag{7}$$

for fitting two-level SEM, where T_{LR} is defined by (6) and $\hat{\lambda}$ is a rescaling factor that is a function of an estimator of $\boldsymbol{\theta}$ and it has the property that for normally distributed data $\hat{\lambda} \to 1$ as $G \to \infty$. Thus, under the normal distributional assumption on (1), T_{RLR} is asymptotically equivalent to T_{LR}. The Monte Carlo study in Yuan & Bentler (2002) implies that, with finite samples, T_{RLR} performs almost as well as T_{LR} in controlling type I errors with normally distributed data. When the normal distributional assumption is violated, T_{RLR} performs much better than T_{LR}. T_{RLR} also tends to over-reject the correct covariance structure for small sample sizes. Thus T_{RLR} has better performance than T_{LR} in protecting against a violation of the normal distributional assumption on (1).

3.2 Corrected ADF-type statistics

A variety of ADF (Asymptotically Distribution Free)-type statistics have been constructed to fit conventional SEM. In addition to the ADF test mentioned above, Browne (1984) also proposed a residual-based ADF statistic that can be applied to any inefficient but consistent estimator. Yuan and Bentler (1997, 1998b, 1999a) proposed corrected versions of Browne's (1984) ADF statistics, as well as some new F-statistics. These statistics have been generalized to fitting two-level SEM by Yuan & Bentler (2003a). They proposed six ADF-type statistics and studied the empirical performance of eight test statistics (including two existing LR-type statistics) for fitting two-level SEM. These statistics are (Yuan & Bentler (2003a)):

1. the LR statistic T_{LR} in (6);
2. the rescaled LR statistic T_{RLR} in (7);
3. the ADF-type χ^2-statistic T_{ADF};
4. the residual-based ADF-type χ^2-statistic T_{RADF};
5. the corrected ADF-type χ^2-statistic T_{CADF};
6. the corrected residual-based ADF-type χ^2-statistic T_{CRADF};
7. the ADF-type F-statistic F_{ADF};
8. the residual-based ADF-type F-statistic F_{RADF}.

The ADF-type statistics in 3)–6) have an asymptotic χ^2-distribution, and the ADF-type statistics in 7)–8) have an asymptotic F-distribution. Based on their empirical study, Yuan & Bentler (2003a) commented that the six ADF-type statistics in 3)–8) have similar performance in a multilevel context as they do in the conventional one-level SEM context as studied in Yuan and Bentler (1997, 1998b, 1999a, 1999b). The T_{ADF} in 3) and the T_{RADF} in 4) tend to reject the correct models too often unless the sample size is huge. T_{CADF} in 5) and T_{CRADF} in 6) slightly under-reject the correct models for a small level-2 sample size G. F_{ADF} in 7) and F_{RADF} in 8) moderately over-reject correct models for a small level-2 sample size G, and both are not significantly influenced by the level-1 sample sizes. The commonly used LR-type statistic T_{LR} in 1) performs acceptably only when data are multivariate normal. The rescaled LR-type statistic T_{RLR} in 2) over-rejects correct models too often for a small level-2 sample size G, thus performing not as desirably as it does in the conventional SEM context (Hu, Bentler & Kano (1992)).

Yuan and Bentler's (2003a) study shows that there are no ideal statistics for fitting two-level SEM if the level-2 sample size G is too small. The bootstrap or resampling-based procedure has been suggested for dealing with small samples in SEM (e.g., Yung & Chan (1999)). However, bootstrap also faces problems with small samples (e.g., Polansky (2000)), especially when not every replication enjoys a converged solution in the context of SEM (Yuan & Hayashi (2003); Yung & Bentler (1996)). For a moderate or a large G, Yuan & Bentler (2003a) recommended using the regular LR-type statistic T_{LR} if the data are known to be multivariate normal, otherwise using T_{CRADF} or F_{RADF} if no distributional information is known. A simple example on the performance of the eight statistics is given in Section 5.

4 Robustness

Reliable statistical analysis of two-level SEM can be accomplished in three major ways: use a statistical machinery whose assumptions, such as multivariate normality in the case of ML, are justified in practice; use a method such as an ADF estimator or a correction to a test statistic that makes minimal assumptions; or use a method under violation of assumptions where theory can guarantee that the violation will have no serious effect on statistical inferences. Experience indicates that normality is an unlikely assumption in practice (e.g., Micceri (1989)). Further, ADF-type statistics are quite slow to converge to their limiting distributions, and hence larger level-2 sample sizes may be needed than are available. As a result, the study of robustness is important.

The purpose of studying robustness is to find out how a violation of the normal distributional assumption affects the validity of some existing methodologies for ML analysis of two-level SEM. This has been studied in conventional SEM for a long time (e.g., Anderson & Amemiya (1988); Browne & Shapiro (1988); Satorra & Bentler (1990) and remains an active area of research (e.g., Satorra (2002); Yuan and Bentler, 1998a, 1998b, 1999b, 2000a, 2000b). In this section we will summarize some recent results on the robustness of test statistics and standard errors of parameter estimators in two-level SEM. We will find that asymptotic robustness theory developed for one-level SEM models can be extended to two-level SEM models. This means that some test statistics and the standard errors of certain parameters can be trusted under violation of assumptions. However, as in the case of one-level models, there are conditions on the model and its parameters that remain hard to evaluate in practice.

4.1 Robustness of test statistics

A slight violation of the multivariate normal distribution is the family of elliptically contoured distributions (ECD for simplicity), which includes the normal as its special case. Comprehensive studies of ECD and its related distributions were done by Fang, Kotz & Ng (1990), Fang & Zhang (1990), and Gupta & Varga (1993). Having many properties that are similar to those of the normal distribution, particularly noteworthy is that an ECD maintains distributional symmetry

(Kano (1994)). For example, an ECD has no skewness (zero skewness) in all of its univariate marginals. The multivariate t-distribution is perhaps the most familiar ECD that has been used for statistical modeling (Lange, Little & Taylor (1989)). Lee & Poon (1992) proposed a GLS approach to parameter estimation for formulation (2) within ECD. A study on the robustness of the LR-type statistics in two-level SEM can help to understand their sensitivity to violations of the normal distributional assumption.

Yuan & Bentler (2003b) studied the asymptotic distribution of LR-type statistics within the class of ECD. They found that, by proper rescalings, two LR-type statistics still have an asymptotic χ^2-distribution under certain conditions. Those conditions include large level-1 and level-2 sample sizes and that the within and between model parameters are separable. Their results show that the two rescaled LR-type statistics under study are robust within the class of ECD under the required conditions. Under a set of similar conditions to those in Yuan and Bentler (2003b, 2003c) proved that the regular LR-statistics T_{LR} in (6) is asymptotically robust (keeps the same χ^2-distribution as under the normal distributional assumption) within a large class of nonnormal distributions including the pseudo normal distributions proposed in Yuan & Bentler (1999b). Yuan & Bentler (2003c) also implies that rescaled LR-type statistics are asymptotically robust within the class of pseudo elliptical distributions (PED for simplicity, see Yuan & Bentler (1999b)). Because the family of PED includes the multivariate normal and the ECD as its special cases, Yuan & Bentler (2003c) results show that the rescaled LR-statistic T_{RLR} is asymptotically robust against a large class of nonnormal distributions.

4.2 Robustness of standard errors

Under the normal distributional assumption on formulation (1), the MLE $\hat{\boldsymbol{\theta}}$ of the model parameter vector $\boldsymbol{\theta}$ in model (4) has an asymptotic multivariate normal distribution (Lee (1990); Liang & Bentler (2004a)). The standard errors of the components in $\hat{\boldsymbol{\theta}}$ can be obtained by inverting the corresponding information matrix. These standard errors are used as measures for accuracy of the MLE's of the model parameters. When there is a violation of the normal distributional assumption, it is unknown whether the standard errors are still valid for measuring accuracy of the MLE's. Yuan & Bentler (2003d) studied the robustness of standard errors in multilevel models. Under certain conditions, they concluded that for some parameters the standard errors computed under the normal distributional assumption are robust (remain unchanged) within a large class of nonnormal distributions.

5 Examples

In this section we will provide some numerical examples that illustrate the three aspects of recent advances in two-level SEM as discussed in Sections 2-4.

Example 5.1 *Estimation of two-level SEM with simultaneous mean and covariance structure using EQS* (excerpted from Liang & Bentler (2004a)).

The practical data set school.dat is from the National Education Longitudinal Study (NELS: 88) and can be downloaded from the M*plus* website http://www.statmodel.com/mplus/examples. The data set contains many academic measurements of $N = 5198$ students (level-1 units) nested in $G = 235$ schools (level-2 units). We only take the data in columns 7-10 which are students' scores from four courses, and the data in columns 20-21, which are the school-level (level-2) observations. Then we have level-1 observations $\{\boldsymbol{y}_{gi} : 4 \times 1\}$ and level-2 observations $\{\boldsymbol{z}_g : 2 \times 1\}$. Formulation (1) with the structured model (4) is set up for the ML analysis. In the two-level SEM under analysis, there is only one level-1 factor F_W =math-ability of students, which is measured by four indicator variables Y_1, Y_2, Y_3 and Y_4, representing the students' scores from four math tests. Similarly, there is only one level-2 factor F_B =general background, which is also measured by the same four indicator variables. The factor loading for Y_1 is fixed as the constant "1" as a reference at both levels. The level-2 factor F_B is assumed to be further predicted by two level-2 observable variables Z_1 (minority) and Z_2 (school type) with a random error. In addition, it is assumed that the mean $E(F_B)$ cannot represent the mean of Y_3 completely and a free mean parameter has to be added in the prediction of Y_3 by F_B in the level-2 model. The model can be easily specified by a path diagram. Readers who are familiar with EQS can easily see the measurement relationships from the EQS input program in the Appendix. Here we present expressions for the mean and covariance structure from the measurement relationships. The covariance matrices have a factor structure given by

$$\boldsymbol{\Sigma}_W = \boldsymbol{\Lambda}_W \boldsymbol{\Phi}_W \boldsymbol{\Lambda}_W' + \boldsymbol{\Psi}_W, \ \boldsymbol{\Sigma}_B = \boldsymbol{\Lambda}_B \boldsymbol{\Phi}_B \boldsymbol{\Lambda}_B' + \boldsymbol{\Psi}_B, \ \widetilde{\boldsymbol{\Sigma}}_B = \begin{pmatrix} \boldsymbol{\Sigma}_{zz} & \boldsymbol{\Sigma}_{zy} \\ \boldsymbol{\Sigma}_{yz} & \boldsymbol{\Sigma}_B \end{pmatrix},$$

where

$$\boldsymbol{\Lambda}_W = (1, \theta_1, \theta_2, \theta_3)', \quad \boldsymbol{\Phi}_W = \mathrm{var}(F_W) = (\theta_4), \quad \boldsymbol{\Psi}_W = \mathrm{diag}(\theta_5, \theta_6, \theta_7, \theta_8),$$

$$\boldsymbol{\Lambda}_B(\boldsymbol{\phi}) = (1, \phi_1, \phi_2, \phi_3)', \quad \boldsymbol{\Phi}_B(\boldsymbol{\phi}) = \mathrm{var}(F_B), \quad \boldsymbol{\Psi}_B(\boldsymbol{\phi}) = \mathrm{diag}(\phi_4, \phi_5, \phi_6, \phi_7),$$

$$\boldsymbol{\Sigma}_{zz} = \begin{pmatrix} \phi_8 & \phi_9 \\ \phi_9 & \phi_{10} \end{pmatrix}, \quad \boldsymbol{\Sigma}_{zy} = \boldsymbol{\Sigma}_{yz}{}' = \Big(\mathrm{cov}(Z_i, Y_j)\Big), \ (i = 1, 2; j = 1, 2, 3, 4)$$

with

$$\mathrm{var}(F_B) = v_b = \phi_8 \phi_{11}^2 + \phi_{10} \phi_{12}^2 + 2\phi_9 \phi_{11} \phi_{12} + \phi_{13},$$

$$\mathrm{cov}(Z_1, Y_1) = v_1 = \phi_8 \phi_{11} + \phi_9 \phi_{12}, \quad \mathrm{cov}(Z_1, Y_2) = \phi_1 v_1,$$

$$\mathrm{cov}(Z_1, Y_3) = \phi_2 v_1, \quad \mathrm{cov}(Z_1, Y_4) = \phi_3 v_1,$$

$$\mathrm{cov}(Z_2, Y_1) = v_2 = \phi_9 \phi_{11} + \phi_{10} \phi_{12}, \quad \mathrm{cov}(Z_2, Y_2) = \phi_1 v_2,$$

$$\mathrm{cov}(Z_2, Y_3) = \phi_2 v_2 + \phi_{14}, \quad \mathrm{cov}(Z_2, Y_4) = \phi_3 v_2.$$

The mean structure is given by

$$\boldsymbol{\mu}_z = (\mu_1, \mu_2)', \quad \boldsymbol{\mu}_y = (\mu_b, \phi_1\mu_b, \phi_2\mu_b + \mu_4, \phi_3\mu_b)', \quad \mu_b = \phi_{11}\mu_1 + \phi_{12}\mu_2 + \mu_3.$$

In this example, there are six observable variables (Z_1, Z_2, Y_1, Y_2, Y_3, Y_4) but only four free mean parameters μ_i ($i = 1, 2, 3, 4$). So we have a nonsaturated mean structure. The EQS input program is provided in the Appendix for analysis of this simultaneous mean and covariance structure. The results are given in Table 1, where the S.E. (standard error) and the model chi-square are computed by the formulas in Liang & Bentler (2004a). The p-value=0.33 shows that this is an acceptable mean and covariance structure for the selected data. By referring to the LISREL manual (du Toit & du Toit (2001)) and the M*plus* manual (Muthén & Muthén (2004)), we can observe that the current versions of LISREL and M*plus* do not provide an option for the analysis of a nonsaturated mean structure.

Table 1. Parameter estimates, standard errors and model chi-square for Example 5.1 (S.E.=Standard Error

	θ_1	θ_2	θ_3	θ_4	θ_5	θ_6	θ_7	θ_8
ESTIMATE	1.046	0.682	1.024	0.751	0.341	0.255	1.348	2.482
S.E.	0.021	0.022	0.031	0.024	0.015	0.015	0.028	0.053

	ϕ_1	ϕ_2	ϕ_3	ϕ_4	ϕ_5	ϕ_6	ϕ_7	ϕ_8
ESTIMATE	1.052	0.579	1.177	0.014	0.009	0.033	0.051	4.437
S.E.	0.006	0.043	0.012	0.005	0.004	0.009	0.017	0.409

	ϕ_9	ϕ_{10}	ϕ_{11}	ϕ_{12}	ϕ_{13}	ϕ_{14}
ESTIMATE	-.366	0.142	-.175	-.025	0.132	-.023
S.E.	0.057	0.013	0.015	0.083	0.017	0.007

	μ_1	μ_2	μ_3	μ_4
ESTIMATE	4.630	1.170	3.262	0.728
S.E.	0.137	0.025	0.145	0.105

TEST	chi-square= 12.51, d.f.= 11, p-value= 0.33

Example 5.2 *Robustness of eight test statistics for two-level SEM* (excerpted from Yuan & Bentler (2003a)).

Yuan & Bentler (2003a) proposed six test statistics for fitting two-level SEM and carried out an empirical study on the performance of eight statistics that include their six new ones and the two existing ones T_{LR} and T_{RLR} in (6) and (7), respectively. The eight statistics are summarized in subsection 3.2. They used four types of distributions as the underlying distributions of the two-level SEM under study: condition I) multivariate normal (MVN); condition II) elliptically contoured distribution (ECD); condition III) log multivariate normal (LMVN); and condition IV) rescaled log multivariate normal (RLMVN). Formulation (1) is employed to generate the two-level data with

$$\boldsymbol{v}_g = \boldsymbol{\lambda}_b f_b + \boldsymbol{e}_b, \qquad f_b = \alpha + \boldsymbol{\beta}' \boldsymbol{z}_g + \epsilon,$$

where $\boldsymbol{\lambda}_b$ is a 8×1 vector, $\boldsymbol{\mu}_z = E(\boldsymbol{z}_g)$ is a 3×1 vector, $\boldsymbol{\Sigma}_{zz} = \text{cov}(z_g)$ is a 3×3 matrix, $E(\boldsymbol{e}_b) = \mathbf{0}$, $\text{cov}(\boldsymbol{e}_b) = \boldsymbol{\Psi}_b$ is a 8×8 diagonal matrix, $E(\epsilon) = 0$, $\psi = \text{var}(\epsilon)$ is a scalar; and

$$\boldsymbol{v}_{gi} = \boldsymbol{\Lambda}_w \boldsymbol{f}_w + \boldsymbol{e}_w, \quad E(\boldsymbol{v}_{gi}) = \mathbf{0}, \quad \boldsymbol{\Sigma}_W = \boldsymbol{\Lambda}_w \boldsymbol{\Phi}_w \boldsymbol{\Lambda}_w' + \boldsymbol{\Psi}_w,$$

where $\boldsymbol{\Lambda}_w$ is a 8×2 matrix, $\boldsymbol{\Phi}_w$ is a 2×2 correlation matrix and $\boldsymbol{\Psi}_w$ is a 8×8 diagonal matrix. Referring to (4), we have

$$\boldsymbol{\mu}_y = \boldsymbol{\lambda}_b(\alpha + \boldsymbol{\beta}'\boldsymbol{\mu}_z), \quad \boldsymbol{\Sigma}_{zy} = \boldsymbol{\Sigma}_{zz}\boldsymbol{\beta}\boldsymbol{\lambda}_b', \quad \boldsymbol{\Sigma}_B = \boldsymbol{\lambda}_b(\boldsymbol{\beta}'\boldsymbol{\Sigma}_{zz}\boldsymbol{\beta} + \psi)\boldsymbol{\lambda}_b' + \boldsymbol{\Psi}_b.$$

In generating the two-level data, a balanced level-1 sample design ($N_g \equiv n$) is employed for each case. Simulation was carried out for 500 replications and

$$\text{Empirical type I error rate} = \frac{\text{Number of rejections}}{\text{Number of replications}}.$$

The results are summarized in Table 2. The type I error rates for the six χ^2-type statistics T_{LR}, T_{RLR}, T_{ADF}, T_{RADF}, T_{CADF}, and T_{CRADF} are computed by referring to their asymptotic chi-square distribution χ^2_{67}, and those for the two F-type statistics F_{ADF} and F_{RADF} are computed by referring to their asymptotic F-distribution $F_{67,G-67}$ according to the choice of level-2 sample size G in Table 2. All type I error rates are computed with significance level $\alpha = .05$.

The following empirical conclusions can be summarized from Table 2.

1. Under the normal distributional assumption on model (4), the two LR-type χ^2-statistics T_{LR} and T_{RLR} seem to have similar performance (in controlling type I error rates). Both of them tend to over-reject the true model for small level-2 sample size G, and their performance is hardly influenced by the level-1 sample size. T_{LR} is more sensitive to normal data than T_{RLR}: it rejects the true model with higher (much higher for ECD and RLMVN) probability than that of T_{RLR};
2. The two ADF-type χ^2-statistics T_{CADF} and T_{CRADF} seem to have similar performance. Both of them tend to under-reject the true model for all cases;
3. The two ADF-type F-statistics F_{ADF} and F_{RADF} seem to have similar performance. They tend to over-reject the true model for small level-2 sample size G and are more like their asymptotic F-statistics when G increases (e.g., $G \geq 500$);

Table 2. Empirical type I error rates of the eight statistics for four types of underlying distributions in Example 5.2 (significance level $\alpha = .05$)(Level-2 sample size $G = 150$ with balanced level-1 sample size n)

		Test Statistics							
Distribution	n	T_{LR}	T_{RLR}	T_{ADF}	T_{RADF}	T_{CADF}	T_{CRADF}	F_{ADF}	F_{RADF}
MVN	50	.062	.072	.948	.950	.012	.020	.078	.090
	100	.080	.112	.950	.956	.020	.026	.106	.106
	200	.062	.080	.954	.954	.010	.014	.078	.102
	500	.092	.120	.938	.942	.022	.028	.096	.098
ECD	50	.982	.238	.968	.976	.012	.034	.070	.082
	100	.986	.230	.956	.954	.012	.016	.052	.074
	200	.998	.224	.968	.962	.012	.018	.058	.090
	500	.996	.228	.958	.960	.004	.006	.064	.066
LMVN	50	.134	.116	.971	.976	.014	.032	.076	.124
	100	.146	.124	.959	.964	.010	.018	.066	.110
	200	.138	.114	.963	.970	.029	.050	.092	.146
	500	.136	.102	.961	.968	.010	.018	.077	.104
RLMVN	50	.820	.232	.983	.988	.010	.020	.046	.080
	100	.882	.226	.955	.974	.002	.012	.041	.100
	200	.898	.202	.967	.976	.014	.036	.078	.124
	500	.920	.218	.968	.980	.016	.034	.057	.114

(Continued) (Balanced level-1 sample size $n = 50$)

		Test Statistics							
Distribution	G	T_{LR}	T_{RLR}	T_{ADF}	T_{RADF}	T_{CADF}	T_{CRADF}	F_{ADF}	F_{RADF}
MVN	150	.062	.072	.948	.950	.012	.020	.078	.090
	200	.058	.068	.812	.822	.046	.046	.088	.094
	500	.036	.044	.268	.270	.042	.042	.058	.056
	1000	.044	.054	.130	.130	.040	.040	.044	.044
ECD	150	.982	.238	.968	.976	.012	.034	.070	.082
	200	.980	.218	.816	.836	.018	.022	.060	.066
	500	.994	.126	.270	.278	.028	.030	.040	.048
	1000	.998	.090	.086	.092	.036	.038	.040	.042
LMVN	150	.134	.116	.971	.976	.014	.032	.076	.124
	200	.138	.102	.842	.870	.035	.042	.069	.088
	500	.094	.060	.284	.290	.048	.048	.050	.054
	1000	.080	.032	.098	.098	.022	.022	.028	.028
RLMVN	150	.820	.232	.983	.988	.010	.020	.046	.080
	200	.846	.164	.845	.870	.020	.040	.057	.074
	500	.934	.108	.269	.296	.036	.056	.070	.076
	1000	.964	.090	.110	.118	.028	.034	.040	.040

4. The two ADF-type χ^2-statistics T_{ADF} and T_{RADF} seem to require a huge level-2 sample size G to keep the type I error rates under control. For example, even for $G = 1000$, both of them still reject the true model quite often (their type I error rates are much larger than the significance level $\alpha = .05$). So one should be careful when using T_{ADF} and T_{RADF} to test the model fit of two-level SEM without any distributional assumption.

Example 5.3 *Robustness of standard errors* (excerpted from Yuan & Bentler (2003d)).

Yuan & Bentler (2003d) studied the asymptotic robustness of the standard errors of the MLE $\hat{\boldsymbol{\theta}}$ for general multilevel models including formulation (1) associated with model (4). One of their conclusions is that the standard errors of the MLE $\hat{\boldsymbol{\theta}}$ for model (4) are asymptotically robust within a large family of nonnormal distributions. Such a family is generated by the stochastic representation (SR) (see Fang, Kotz & Ng (1990); Yuan & Bentler (1999b))

$$\boldsymbol{x} = r\boldsymbol{L}\boldsymbol{\xi}, \tag{8}$$

where r is a scalar random variable independent of the random vector $\boldsymbol{\xi} = (\xi_1, \ldots, \xi_m)'$, $\boldsymbol{L}$ is a $d \times m$ constant matrix of rank d such that $\boldsymbol{L}\boldsymbol{L}' = \boldsymbol{\Sigma}$. Both r and $\boldsymbol{\xi}$ have finite fourth-order moments. Note that ECD (MVN is included in ECD) can be generated from the SR (8) when suitably choosing r and $\boldsymbol{\xi}$ (Fang, Kotz & Ng (1990)). But the family represented by (8) is much larger than that of ECD (Yuan & Bentler (1999b)).

Yuan & Bentler (2003d) constructed an example of two-level SEM of the form of formulation (2) (a special case of (1)). In the example, both the between and the within covariance matrices have a factor structure:

$$\boldsymbol{\Sigma}_B = \boldsymbol{\Sigma}_B(\boldsymbol{\theta}_b) = \boldsymbol{\Lambda}_b\boldsymbol{\Phi}_b\boldsymbol{\Lambda}_b' + \boldsymbol{\Psi}_b, \quad \boldsymbol{\Sigma}_W = \boldsymbol{\Sigma}_W(\boldsymbol{\theta}_w) = \boldsymbol{\Lambda}_w\boldsymbol{\Phi}_w\boldsymbol{\Lambda}_w' + \boldsymbol{\Psi}_w,$$

where the parameter vectors $\boldsymbol{\theta}_w$ and $\boldsymbol{\theta}_b$ do not share any common parameter, $\boldsymbol{\Psi}_b$ and $\boldsymbol{\Psi}_w$ are diagonal matrices, and $\boldsymbol{\Lambda}_b$ and $\boldsymbol{\Lambda}_w$ are matrices of factor loadings with the structure that each indicator variable only measures one factor. Now assume that the level-2 random component $\boldsymbol{v}_g$ and the level-1 random component $\boldsymbol{v}_{gi}$ in (2) are generated by SR (8) with

$$\boldsymbol{L} = \boldsymbol{L}_b = (\boldsymbol{\Lambda}_b\boldsymbol{\Phi}_b^{\frac{1}{2}}, \boldsymbol{\Psi}_b^{\frac{1}{2}}), \quad \text{and} \quad \boldsymbol{L} = \boldsymbol{L}_w = (\boldsymbol{\Lambda}_w\boldsymbol{\Phi}_w^{\frac{1}{2}}, \boldsymbol{\Psi}_w^{\frac{1}{2}}),$$

respectively. The scalar variable r and the random vector $\boldsymbol{\xi} = (\xi_1, \ldots, \xi_m)'$ in SR (8) satisfy the condition

$$E(r^4) = 1, \quad E(\xi_i^4) \equiv 3.$$

Yuan & Bentler (2003d) concluded that the standard errors of the ML estimates of $\hat{\boldsymbol{\theta}}_b$ and $\hat{\boldsymbol{\theta}}_w$ are asymptotically robust (level-2 sample size $G \to \infty$) within a subfamily of (8), which is still much large than ECD. This implies that the standard errors of the MLE for model (4) under the normal distributional assumption can be still valid within a wide class of nonnormal distributions.

6 Comments and Supplements

In this paper we focused on summarizing certain selected recent advances in estimation, testing, and robustness in two-level SEM with continuous data. Of course, there exist many other recent developments in analysis of two-level SEM besides those methodologies summarized in this paper. In this section we provide a few examples of other research directions and/or advances in the study of two-level SEM. For example, in the discussion of estimation in Section 2, we only considered the case where model parameters have no constraints. Lee & Tsang (1999) proposed an EM algorithm for ML estimation with constraints for formulation (2). Bentler, Liang & Yuan (2004) generalized Lee and Tsang's (1999) model and algorithm to formulation (1). Zhang & Lee (2001) studied the asymptotic theory of two-level SEM with constraints. Lee & Song (2001) studied ML estimation of two-level latent variable models (including two-level SEM) with mixed continuous and polytomous data. In our discussion of testing model fit in Section 3, we only considered the case of fitting two-level SEM without any prior information. Lee & Song (2001) studied the problem of testing hypothesis in two-level SEM using a Bayesian approach.

In both Section 2 and Section 3, we considered the problem of estimation and testing in two-level SEM using the standard one-step framework. That is, we estimate the level-1 and level-2 parameters, and fit the level-1 and level-2 models, simultaneously. This is optimal if the model is correct, but may not be optimal if parts of the model are incorrect as a misspecification in one part of the model may propagate to other parts of the model. Yuan & Bentler (2003e) proposed an approach to stepwise analysis of two-level SEM. They divided the process of fitting a two-level SEM into the steps: 1) fitting the saturated model and obtaining its parameter estimates (e.g., by the ML method); 2) fitting the within-level (level-1) structural model in the framework of conventional (one-level) SEM; 3) formulating the between-level (level-2) structural model and checking the common parameters of the within-level and the between-level by a Wald-type statistic; and 4) fitting the between-level structural model in the framework of conventional SEM by using the parameter estimates obtained at the within-level in step 2). This stepwise approach can be employed to analyze covariance structure models with more than two levels.

Many additional problems in two-level SEM need tackling or remain open. For example, when a two-level SEM is perturbed by a random factor, Song & Lee (2004) gave a method for assessing the local influence of a minor perturbation in the proposed model. Liang & Bentler (2004b) (under review) proposed a heterogeneous two-level latent variable model and its associated algorithm for practical data analysis. It remains unclear how the estimates of model parameters and test statistics are affected when two-level data contain outliers or influential cases at one or another level. In the case of conventional SEM, Yuan & Bentler (2001) studied the effects of outliers on estimators and tests, and Yuan, Marshall & Weston (2002) gave a method for handling influential cases. In still another direction, Muthén (2001) pointed out new opportunities in the study of latent variable models using a combination of categorical and continuous latent variables. It remains open whether Muthén's methodologies for dealing with SEM with categorical latent variables can be effectively generalized to the case of two-level SEM. Also in the case of conventional SEM, besides the χ^2-type and the F-type statistics, other fit indices are available (Yuan & Marshall (2004)). An open question is whether those fit indices can be effectively generalized to the case of two-level SEM. Clearly this paper has been able to shed some light on only limited issues and advances in the huge field of multilevel modeling.

Acknowledgments

This research is supported by National Institute on Drug Abuse grants DA01070, DA00017, and a UNH 2004 Summer Faculty Fellowship & Research Award.

Appendices

EQS input program for the model in Example 5.1

```
/TITLE
EQS Input Program for Analysis of the School Data Set
Within Model
/SPECIFICATIONS
DATA='SCHOOL.DAT'; CASES=5198; VARIABLES=21; METHOD=ML;
MATRIX=RAW; GROUPS=2; ANALYSIS=COV; MULTILEVEL=ML;
CLUSTER=V19;
/LABELS
V7=Y6; V8=Y7; V9=Y8; V10=Y9; V19=SCHOOL; V20=X3; V21=X4;
F1=FW;
/EQUATIONS
Y6=1FW+E6;
Y7=*FW+E7;
Y8=*FW+E8;
Y9=*FW+E9;
/VARIANCES
FW=*;
E6-E9=*;
/END

/TITLE
Between Model
/LABELS
V7=Y6; V8=Y7; V9=Y8; V10=Y9; V19=SCHOOL; V20=X3; V21=X4;
F1=FB;
/EQUATIONS
```

```
Y6=1FB+E6;
Y7=*FB+E7;
Y8=*V999+*FB+E8;
Y9=*FB+E9;
FB=*V999+*X3+*X4+D1;
X3=*V999+E3;
X4=*V999+E4;
/VARIANCES
E6-E9=*;
D1=*;
E3-E4=*;
/COVARIANCES
E3,E4=*;
E8,E4=*;
/TECHNICAL
ITR=200; CON=.000001;
/END
```

References

Anderson, T. W. & Amemiya, Y. (1988). The asymptotic normal distribution of estimators in factor analysis 122.

Bentler, P. M. (2004). *EQS 6 Structural Equations Program Manual.* Encino, CA: Multivariate Software (www.mvsoft.com).

Bentler, P. M. & Liang, J. (2003a). Two-level mean and covariance structures: maximum likelihood via an EM algorithm. In *Multilevel Modeling: Methodological Advances, Issues, and Applications* (S. Reise & N. Duan, eds.) 53–70. Mahwah, NJ: Erlbaum.

Bentler, P. M. & Liang, J. (2003b). Simultaneous mean and covariance structure analysis for two-level structural equation models on EQS. In *New Developments in Psychometrics* (H. Yanai, A. Okada, K. Shigemasu, Y. Kano, & J. J. Meulman, eds.) 123–132. Tokyo: Springer-Verlag.

Bentler, P. M., Liang, J. & Yuan, K. H. (2004). Constrained ML estimation for two-level mean and covariance structure models: an extension to Lee and Tsang's model and algorithms. Under review.

Bentler, P. M. & Yuan, K. H. (1999). Structural equation modeling with small samples: test statistics. *Multivariate Behavioral Research* **34** 181–197.

Bollen, K. A. (1989). *Structural Equations with Latent Variables.* New York: Wiley.

Browne, M. W. (1974). Generalized least squares estimators in the analysis of covariance structures. *South Africa Statistical Journal* **8** 1–24.

Browne, M. W. (1984). Asymptotic distribution-free methods in the analysis of covariance structures. *British J. Math. and Statist. Psych.* **37** 62–83.

Browne, M. W. & Shapiro, A. (1988). Robustness of normal theory methods in the analysis of linear latent variate models. *British J. Math. and Statist. Psych.* **41** 193–208.

Dempster, A. P., Laird, N. M. & Rubin, D. B. (1977). Maximum likelihood from incomplete data via EM algorithm (with discussion). *J. Roy. Statist. Soc.* lB **39** 1–38.

du Toit, M. & du Toit, S. (2001). *Interactive LISREL: User's Guide.* Scientific Software International, Inc.

du Toit, M. & du Toit, S. (2002). Multilevel structural equation modeling. In *Handbook of Quantitative Multilevel Analysis* (J. de Leeuw & I. Kreft, eds.) Chapter 13. Kluwer Publications. To appear.

Fang, K. T., Kotz, S. & Ng, K. W. (1990). *Symmetric Multivariate and Related Distributions.* London: Chapman and Hall.

Fang, K. T. & Zhang, Y. (1990). *Generalized Multivariate Analysis.* Berlin, Beijing: Springer-Verlag and Science Press.

Goldstein, H. (1995). *Multilevel Statistical Models*, 2nd Edition. London: Edward Arnold.

Goldstein, H. & McDonald, R. P. (1988). A general model for the analysis of multilevel data. *Psychometrika* **53** 455–467.

Gupta, A. K. & Varga, T. (1993). *Elliptically Contoured Models in Statistics.* Dordrecht: Kluwer Academic.

Heck, R. H. & Thomas, S. L. (2000). *An Introduction to Multilevel Modeling Techniques.* Mahwah, NJ: Erlbaum.

Hox, J. J. (1993). Factor analysis of multilevel data. Gauging the Muthén model. In *Advances in Longitudinal and Multilevel Analysis in the Behavioural Sciences* (J. H. L. Oud & R. A. W. van Blokland-Vogelesang, eds.) 141–156. Nijmegen, NL: ITS.

Hox, J. J. (2002). *Multilevel Analysis: Techniques and Applications.* Mahwah, NJ: Erlbaum.

Hox, J. J. & Maas, C. J. M. (2001). The accuracy of multilevel structural equation modeling with pseudobalanced groups and small samples. *Structural Equation Modeling* **8** 157–174.

Hu, L., Bentler, P. M. & Kano, Y. (1992). Can test statistics in covariance structure analysis be trusted?. *Psychological Bulletin* **112** 351–362.

Kano, Y. (1994). Consistency property of elliptical probability density functions. *J. Multivariate Anal.* **51** 139–147.

Kreft, I. & de Leeuw, J. (1998). *Introducing Multilevel Modeling.* London: Sage.

Lange, K. L., Little, R. J. A. & Taylor, J. M. G. (1989). Robust statistical modeling using the *t*-distribution. *J. Amer. Statist. Assoc.* **84** 881–896.

Lee, S. Y. (1990). Multilevel analysis of structural equation models. *Biometrika* **77** 763–772.

Lee, S. Y. & Poon, W. Y. (1992). Two-level analysis of structural equation model for unbalanced designs with small level one samples. *British J. Math. and Statist. Psych.* **45** 109–123.

Lee, S. Y. & Poon, W. Y. (1998). Analysis of two-level structural equation models via EM type algorithms. *Statistica Sinica* **8** 749–766.

Lee, S. Y. & Shi, J. (2001). Maximum likelihood estimation of two-level latent variable models with mixed continuous and polytomous data. *Biometrics* **57** 787–794.

Lee, S. Y. & Song, X. Y. (2001). Hypothesis testing and model comparisons in two-level structural equation models. *Multiv. Behav. Res.* **36** 639–655.

Lee, S. Y. & Tsang, S. Y. (1999). Constrained maximum likelihood estimation of two-level covariance structure model via EM type algorithms. *Psychometrkia* **64** 435–450.

Liang, J. & Bentler, P. M. (2004a). A new EM algorithm for fitting two-level structural equation models. *Psychometrika* **69** No. 1 101–122.

Liang, J. & Bentler, P. M. (2004b). Heterogeneous mean and covariance structure analysis for two-level latent variable models. In *The 2003 Joint Statistical Meeting, August 3-7, 2003.* San Francisco, CA. Under review.

McDonald, R. P. (1993). A general model for two-level data with responses missing at random. *Psychometrika* **58** 575–585.

McDonald, R. P. (1994). The bilevel reticular action model for path analysis with latent variables. *Sociological Methods and Research* **22** 399–413.

McDonald, R. P. & Goldstein, H. (1989). Balanced versus unbalanced designs for linear structural relations in two-level data. *British J. Math. and Statist. Psych.* **42** 215–232.

Micceri, T. (1989). The unicorn, the normal curve, and other improbable creatures. *Psychological Bulletin* **105** 156–166.

Muthén, B. O. (1989). Latent variable modeling in heterogeneous populations. *Psychometrika* **54** 557–585.

Muthén, B. O. (1991). Multilevel factor analysis of class and student achievement components. *Journal of Educational Measurement* **28** 338–354.

Muthén, B. O. (1994). Multilevel covariance structure analysis. *Sociological Methods and Research* **22** 376–398.

Muthén, B. O. (1997). Latent variable modeling of longitudinal and multilevel data. In *Sociological Methodology* (A. Raftery, ed.) 453–481. Boston: Blackwell Publishers.

Muthén, B. O. (2001). Second-generation structural equation modeling with a combination of categorical and continuous latent variables: New opportunities for latent class/latent growth modeling. In *New Methods for the Analysis of Change* (L. M. Collins, & A. Sayer, eds.) 291–322. Washington, D.C.: APA.

Muthén, L. K. & Muthén, B. O. (2004). *Mplus Version 3 User's Guide.* Los Angeles: Muthén and Muthén.

Polansky, A. M. (2000). Stabilizing bootstrap-t confidence intervals for small samples. *Canadian J. Statist.* **28** 501–516.

Poon, W. Y. & Lee, S. Y. (1992). Maximum likelihood and generalized least squares analyses of two-level structural equation models. *Statist. and Prob. Letters* **14** 25–30.

Raudenbush, S. W. (1995). Maximum likelihood estimation for unbalanced multilevel covariance structure models via the EM algorithm. *British J. Math. and Statist. Psych.* **48** 359–370.

Raudenbush, S. W. & Bryk, A. S. (2002). *Hierarchical Linear Models*, 2nd Edition. Newbury Park: Sage.

Reise, S. & Duan, N. eds. (2003). *Multilevel Modeling: Methodological Advances, Issues, and Applications.* Mahwah, NJ: Erlbaum.

Satorra, A. (2002). Asymptotic robustness in multiple group linear-latent variable models. *Econometric Theory* **18** 297–312.

Satorra, A. & Bentler, P. M. (1988). Scaling corrections for chi-square statistics in covariance structure analysis. American Statistical Association 1988 Proceedings of Business and Economics Sections 308–313. Alexandria, VA: American Statistical Association.

Satorra, A. & Bentler, P. M. (1990). Model conditions for asymptotic robustness in the analysis of linear relations. *Comput. Statist. and Data Anal.* **10** 235–249.

Satorra, A. & Bentler, P. M. (1994). Corrections to test statistics and standard errors in covariance structure analysis. In *Latent Variable Analysis: Applications for Developmental Research* (A. von Eye & C. C. Cloggs, eds.) 399–419. Newbury Park, CA: Sage.

Snijders, T. & Bosker, R. (1999). *Multilevel Analysis: An Introduction to Basic and Advanced Multilevel Modeling.* Thousand Oaks, CA: Sage.

Song, X. Y. & Lee, S. Y. (2004). Local influence analysis of two-level latent variable models with continuous and polytomous data. *Statistica Sinica* **14** 317–332.

Yuan, K. H. & Bentler, P. M. (1997). Mean and covariance structure analysis: theoretical and practical improvements. *J. Amer. Statist. Assoc.* **92** 767–774.

Yuan, K. H. & Bentler, P. M. (1998a). Robust mean and covariance structure analysis. *British J. Math. and Statist. Psych.* **51** 63–88.

Yuan, K. H. & Bentler, P. M. (1998b). Normal theory based test statistics in structural equation modeling. *British J. Math. and Statist. Psych.* **51** 289–309.

Yuan, K. H. & Bentler, P. M. (1999a). *F*-tests for mean and covariance structure analysis. *J. Educ. and Behav. Statist.* **24** 225–243.

Yuan, K. H. & Bentler, P. M. (1999b). On normal theory and associated test statistics in covariance structure analysis under two classes of nonnormal distributions. *Statistica Sinica* **9** 831–853.

Yuan, K. H. & Bentler, P. M. (2000a). Robust mean and covariance structure analysis through iteratively reweighted least squares. *Psychometrika* **65** 43–58.

Yuan, K. H. & Bentler, P. M. (2000b). Robust transformation with applications to structural equation modeling. *British J. Math. and Statist. Psych.* **53** 31–50.

Yuan, K. H. & Bentler, P. M. (2001). Effect of outliers on estimators and tests in covariance structure analysis. *British J. Math. and Statist. Psych.* **54** 161–175.

Yuan, K. H. & Bentler, P. M. (2002). On normal theory based inference for multilevel models with distributional violations. *Psychometrika* **67** 539–561.

Yuan, K. H. & Bentler, P. M. (2003a). Eight test statistics for multilevel structural equation models. *Comput. Statist. and Data Anal.* **44** 89–107.

Yuan, K. H. & Bentler, P. M. (2003b). On the asymptotic distributions of two statistics for two-level covariance structure models within the class of elliptical distributions. *Psychometrika.* To appear.

Yuan, K. H. & Bentler, P. M. (2003c). Asymptotic robustness of the normal theory likelihood ratio statistic for two-level covariance structure models. *J. Multivariate Anal.* To appear.

Yuan, K. H. & Bentler, P. M. (2003d). Asymptotic robustness of standard errors in multilevel structural equation models. Under review.

Yuan, K. H. & Bentler, P. M. (2003e). Stepwise analysis of multilevel covariance structure models. Under review.

Yuan, K. H. & Hayashi, K. (2003). Bootstrap approach to inference and power analysis based on three statistics for covariance structure models. *British J. Math. & Statist. Psych.* **56** 93–110.

Yuan, K. H. & Hayashi, K. (2004). On Muthén's maximum likelihood for two-level covariance structure models. *Psychometrika.* To appear.

Yuan, K. H. & Marshall, L. L. (2004). A new measure of misfit for covariance structure models. *Behaviormetrika* **31** 67–90.

Yuan, K. H., Marshall, L. L. & Weston, R. (2002). Cross-validation by downweighting influential cases in structural equation modeling. *British J. Math. and Statist. Psych.* **55** 125–143.

Yung, Y. F. & Bentler, P. M. (1996). Bootstrapping techniques in analysis of mean and covariance structures. In *Advanced structural equation modeling: Techniques and issues* (G. A. Marcoulides & R. E. Schumacker, eds.) 195–226. Hillsdale, NJ: Lawrence Erlbaum.

Yung, Y. F. & Chan, W. (1999). Statistical analyses using bootstrapping: Concepts and implementation. In *Statistical strategies for small sample research* (R. Holye, eds.) 81–105. CA: Sage Publications, Inc.

Zhang, W. & Lee, S. Y. (2001). Asymptotic theory of two-level structural equation models with constraints. *Statistica Sinica* **11** 135–145.

Dependence Patterns of Random Variables: Geometric Properties of Copulas

Hong-Bin Fang[1], Samuel Kotz[2] and Gang Wei[3]

[1] Division of Biostatistics, University of Maryland Greenebaum Cancer Center, 22 South Greene Street, Baltimore, MD 21201, U.S.A. (hfang@umm.edu)
[2] Department of Engineering Management, School of Engineering and Applied Science, The George Washington University, Washington DC 20052, U.S.A. (kotz@gwu.edu)
[3] Department of Mathematics, Hong Kong Baptist University, Kowloon Tong, Hong Kong (gwei@math.hkbu.edu.hk)

Summary. Copulas, as dependence measures of random variables, have wide applications in distribution theory, medical research, multivariate survival analysis, risk management, and other fields. The basic properties of copulas have been studied extensively in the literature. However, their geometric and topological properties, which are very important for properly characterizing the dependence patterns of random variables, have not as yet caught statisticians' attention. In this paper, we shall study the geometric structures of copulas and the local dependence pattern of random variables. Important classes of copulas, such as the polynomial copulas, the piecewise linear and quadratic copulas are also investigated.

Key words: Copulas, Dependence measures, Dependence index; Local dependence function; Multivariate distributions.

2000 Mathematics Subject Classification: 62E10, 62H10, 62H20

1 Introduction

Professor Kai Tai Fang has been a leading innovator in the area of dependence and multivariate distributions for the last 30 years and we are honored to provide this paper on the occasion of his 65th birthday.

Contemporary Multivariate Analysis and Experimental Design—In Honor Celebration of Professor Kai-Tai Fang's 65th birthday. Edited by Jianqing Fan and Gang Li. The World Scientific Publisher, 2005.

Copulas were first introduced by Sklar (1959) as dependence functions of random variables. A historical review and description of the major developments of copula analysis are available in Dall'Aglio, Kotz & Salinetti (1991), Schweizer (1991) and Nelsen (1999), among other sources. Copulas contain all the dependence information of corresponding random variables and have been applied as an effective construction of multivariate distributions with given marginals. They have been widely applied in various fields of medicine, science and technology. To mention a few recent examples, Jouini & Clemen (1996) proposed a class of copula models for aggregating expert opinions. Frees, Carriere & Valdez (1996) and Embrechts, Lindskog & McNeil (2003) considered the dependence analysis with copulas in risk management. Fan, Prentice & Hsu (2000) discussed the dependence measurement of copulas in bivariate survival analysis. Fang, Fang & Kotz (2002) constructed meta-elliptical distributions based on elliptical copulas. Basrak, B., Klaassen, Beekman, Martin & Boomsma (2004) analyzed human genetic linkages using copulas.

Definition 1.1 An $n(\geq 2)$-dimensional copula is an n-dimensional cumulative distribution function (cdf), denoted by $C_n(u_1, u_2, \cdots, u_n)$, whose support is the n-dimensional hypercube $[0,1]^n$ and whose univariate marginal distributions are uniformly distributed on $[0,1]$.

Let $X_1, \cdots, X_n$ be random variables, defined on a common probability space, with the individual distribution functions $F_1(x_1), \cdots, F_n(x_n)$ and the joint distribution function $H(x_1, \cdots, x_n)$. Sklar (1996) shows that there exists an n-dimensional copula $C_n(u_1, \cdots, u_n)$ such that, for all $(x_1, \cdots, x_n) \in \mathcal{R}^n$,

$$H(x_1, \cdots, x_n) = C_n(F_1(x_1), \cdots, F_n(x_n)). \tag{1.1}$$

If $F_1(x_1), \cdots, F_n(x_n)$ are continuous, then $C_n(u_1, \cdots, u_n)$ is unique; otherwise $C_n(u_1, \cdots, u_n)$ is uniquely determined on (Range F_1)$\times \cdots \times$(Range F_n).

Consequently, for continuous multivariate distribution functions, the univariate marginals and the multivariate dependence structure can be separated, and the dependence structure can be represented by a copula. Assume furthermore that the random vector $\mathbf{x} = (X_1, \cdots, X_n)^T$ has the density function $h(x_1, \cdots, x_n)$ and marginal density functions, $f_1, \ldots, f_n$, respectively. Then the copula density function of $\mathbf{x} = (X_1, \cdots, X_n)^T$ exists and moreover

$$c(u_1, \cdots, u_n) \equiv \frac{\partial^n C(u_1, \cdots, u_n)}{\partial u_1 \cdots \partial u_n} = \frac{h(F_1^{-1}(u_1), \cdots, F_n^{-1}(u_n))}{\prod_{i=1}^n f_i(F_i^{-1}(u_i))}; \tag{1.2}$$

alternatively

$$h(x_1, \cdots, x_n) = c(F_1(x_1), \cdots, F_n(x_n)) \prod_{i=1}^n f_i(x_i). \tag{1.3}$$

The function $c(F_1(x_1), \cdots, F_n(x_n))$ in the above equation is referred to as the *density weighting function*. For a given copula $C(u_1, \cdots, u_n)$ and marginal distribution functions $F_1(x_1), \cdots, F_n(x_n)$ of $\mathbf{x}$, the multivariate distribution of $\mathbf{x}$ can be constructed from (1.2) or (1.3). Hence, copulas allow us to construct models which go beyond the standard ones as far as the level of dependence is concerned. They yield a powerful tool for testing a wide variety of dependence measurements. The study of copulas have attracted substantial attention of statisticians in the last 20 years (see *e.g.* Drouet-Mari & Kotz (2001)). The basic properties of copulas have been summarized by Nelsen (1999) and Embrechts, Lindskog & McNeil (2003). However, the geometric and topological properties of copulas are quite important for characterizing dependence patterns of random variables and based on our knowledge has been scarcely addressed in the literature. In this paper, we intend to investigate the simple fundamental properties of copulas and the dependence among random variables. Some geometric features of copulas will be considered. Specially, the piecewise linear and piecewise quadratic copulas will be studied and their probabilistic interpretations will be provided. The notion of the "holes" in the domains of multivariate distributions will be introduced. It is shown that "holes" in the copulas are essential for characterizing dependence patterns. Based on these results, some interesting multivariate distributions could be constructed.

Without loss of generality, we will focus in this paper on bivariate copulas. The corresponding results for multivariate copulas can straightforwardly be obtained. This article is organized as follows. In Section 2 polynomial copulas being an important family of copulas are studied. The necessary conditions for a polynomial to be a copula are given. The piecewise linear and quadratic copulas are discussed in Section 3. Section 4 provides the topological invariance properties of copulas. In Section 5, we characterize the local dependence with standardized cross difference ratio. Section 6 presents our conclusions.

2 Polynomial Copulas

Numerous copulas of various types have been constructed in the past decades. A comprehensive review in this connection are given in Drouet-Mari & Kotz (2001) and Embrechts, Lindskog & McNeil (2003). Many copulas can be obtained from the well known multivariate distributions by means of formulas (1.1) or (1.2) (see *e.g.* Fang, Fang & Kotz (2002). However, most of the copulas or copula densities obtained in this manner could not be written in explicit forms, for example the copula of a normal distribution. Kimeldorf & Sampson (1975) constructed a family of piecewise-uniform copulas. Johnson & Kotz (1998) studied the nicked and notched square distributions which are simple piecewise uniform distribution on the unit square ($[0,1]^2$). These distributions provide a method for constructing simple copulas.

An important family of explicit copulas is the polynomial copulas. Given any non-negative integer m, by *power-m polynomial* we mean a polynomial of the following form:

$$p(u,v) = \sum_{i,j} a_{ij} u^i v^j, \tag{2.1}$$

where $0 \le i, j \le m$, $i + j \le m$ and there is at least one $a_{ij} \neq 0$ with $i + j = m$. If $p(u, v)$ is also a copula (see below), we call it a (non-trivial) *polynomial copula of power-m*. By the definition of copulas we have immediately the following lemma.

Lemma 2.1 *The necessary conditions for a power-m polynomial $p(u, v)$ to be a copula are:*

(i) *$ij = 0$, $a_{ij} = 0$, i.e. the polynomial $p(u, v)$ is divisible by uv and $C(u, v) = p(u, v)$ possesses a density.*

(ii) *$\{a_{ij},\ ij \neq 0,\ i + j \le m\}$ are solutions of the linear system:*

$$\begin{cases} \sum_{i=1}^{m-j} a_{ij} = 0, & j = 2, 3, \cdots, m-1 \\ \sum_{i=1}^{m-1} a_{i1} = 1, & \\ \sum_{j=1}^{m-j} a_{ij} = 0, & i = 2, 3, \cdots, m-1 \\ \sum_{j=1}^{m-1} a_{1j} = 1. & \end{cases} \tag{2.2}$$

In the system (2.2) there are $N_1 = \frac{1}{2}m(m-1)$ variables and $2(m-1)$ equations. We could list the rows of these variables in a vector to form the N_1-dimensional vector

$$\boldsymbol{a} = (a_{11}, \ldots, a_{1,m-1}, a_{2,1}, \ldots, a_{2,m-2}, \ldots, a_{m-2,1}, a_{m-2,2}, a_{m-1,1})^T. \tag{2.3}$$

We have the following lemma dealing with the solution space of the system (2.2).

Lemma 2.2 *The solution space of the linear system (2.2) is a linear space topologically isomorphic to $\mathbb{R}^{N_2}$, where $N_2 = \frac{1}{2}(m-2)(m-3)$.*

Proof. Utilizing the definition of $\boldsymbol{a}$ given in (2.3) the coefficient matrix $\boldsymbol{A}$ of the system (2.2) takes on the following form:

$$\boldsymbol{A} = \left[\begin{array}{ccccc} \overbrace{1\ 1 \ldots 1}^{m-1} & \overbrace{0\ 0 \ldots 0}^{m-2} & \cdots & \overbrace{0\ 0}^{2} & \overbrace{0}^{1} \\ 0\ 0 \cdots 0 & 1\ 1 \cdots 1 & \cdots & 0\ 0 & 0 \\ \vdots & \vdots & \ddots & \vdots & \vdots \\ 0\ 0 \cdots 0 & 0\ 0 \cdots 0 & \cdots & 1\ 1 & 0 \\ 0\ 0 \cdots 0 & 0\ 0 \cdots 0 & \cdots & 0\ 0 & 1 \\ \hdashline 1\ 0 \cdots 0 & 1\ 0 \cdots 0 & \cdots & 1\ 0 & 1 \\ 0\ 1 \cdots 0 & 0\ 1 \cdots 0 & \cdots & 0\ 1 & 0 \\ \vdots & \vdots & \ddots & \vdots & \vdots \\ 0\ 0 \cdots 1 & 0\ 0 \cdots 0 & \cdots & 0\ 0 & 0 \end{array}\right]$$

The first $m - 1$ rows (block 1) and the last $m - 1$ rows (block 2) are two sets of linearly independent vectors. The summation of all the vectors in each one of the two blocks equal to the same vector $\mathbf{1} = (1, 1, \ldots, 1)^T$ (of N_1-dimension). Hence $\text{Rank}(\boldsymbol{A}) \le 2(m-1) - 1$.

However, if we get rid of the first row of A, then any row of the remaining matrix cannot be reduced to a zero vector by elementary row operations. Thus we have $\text{Rank}\boldsymbol{A} = 2(m-1) - 1$.

Finally the existence of the simple copula uv means that there exists always a solution $a_{11} = 1$, $a_{ij} = 0$, for all $ij \neq 1$. The dimension of the solution space is

$$\begin{aligned} N_2 &= N_1 - \text{Rank}\boldsymbol{A} \\ &= \tfrac{1}{2}m(m-1) - (2(m-1)-1) = \tfrac{1}{2}(m-2)(m-3). \end{aligned}$$

This completes the proof.

Below we shall state and prove a general theorem dealing with polynomial copulas.

Theorem 2.3 *Polynomial copulas have the following properties:*

(i) *There are no linear and cubic copulas.*
(ii) *There is one and only one quadratic copula: $C(u,v) = uv$.*
(iii) *For any $m > 3$ there are infinitely many non-trivial polynomial copulas of power-m. The coefficient set of all the polynomial copulas of power-m is a convex hull in a $\frac{1}{2}(m-2)(m-3)$-dimensional space. In particular, any non-trivial polynomial copula of power-m does not contain the power terms $u^{m-1}v$ and uv^{m-1}.*

Proof. Assume that the quadratic copula takes the following general forms

$$C_2(u,v) = \alpha u^2 + \beta uv + \gamma v^2 + \lambda u + \eta v + \delta.$$

To determine the unknown coefficients, we shall apply the four boundary conditions. For example, for $C_2(u,v)$, we have

$$\begin{aligned} &C_2(0,0) = 0 \Rightarrow \delta = 0, \\ &C_2(0,v) = 0 \Rightarrow \gamma v^2 + \eta v = 0, \quad i.e., \ \gamma = 0, \ \eta = 0, \\ &C_2(u,0) = 0 \Rightarrow \alpha u^2 + \lambda u = 0, \quad i.e., \ \alpha = 0, \ \lambda = 0, \\ &C_2(1,1) = 0 \Rightarrow \beta = 1. \end{aligned}$$

Assertion (ii) of the theorem follows immediately. Assertion (i) can be proved in the same manner. Actually if we attempt to determine the coefficients of a cubic polynomial to form a copula, we shall end up with uv only which is just quadratic. Thus, we only need to prove the part (iii) of the theorem.

Denote by $p(u,v,\boldsymbol{a})$ a polynomial $p(u,v)$ with the coefficient vector $\boldsymbol{a}$. From Lemmas 2.1 and 2.2 we know that the solution of the linear system (2.2) leads to a polynomial $p(u,v,\boldsymbol{a})$ satisfying all the boundary conditions of a copula. The polynomial $p(u,v,\boldsymbol{a})$ has density over $[0,1]^2$,

$$q(u,v,\boldsymbol{a}) = \frac{\partial^2 p}{\partial u \partial v}, \quad \text{and} \quad p(u,v,\boldsymbol{a}) = \int_0^u \int_0^v q(s,t,\boldsymbol{a}) ds dt. \tag{2.4}$$

The function $q(u,v,\boldsymbol{a})$ depends continuously on the parameter vector $\boldsymbol{a}$. With $\boldsymbol{a}_0 = (1,0,...,0)^T$, $q(u,v,\boldsymbol{a}_0) = 1$ is a particular density function corresponding to the copula uv. Therefore, there exists an open neighborhood S_0 of $\boldsymbol{a}_0$ in the solution space of the system (2.2), such that $\forall \boldsymbol{a} \in S_0$, the corresponding $q(u,v,\boldsymbol{a})$ is non-negative over $[0,1]^2$, and is therefore a copula density function. Hence the polynomial $p(u,v,\boldsymbol{a})$ is a polynomial copula. Moreover, since S_0 is an open set, there are infinitely non-trivial many polynomial copulas of power-m. The convexity of the parameter set is due to the convexity of the copula set, i.e. any convex linear combination of copulas is again a copula. The absence of the terms $u^{m-1}v$ and uv^{m-1} is due to the fact from (2.2) that $a_{m-1,1} = a_{1,m-1} = 0$. This completes the proof of the theorem.

Corollary 2.4 *A power-4 polynomial $p(u,v)$ is a copula iff it is of the form*

$$p_4(u,v) = uv[\alpha + (1-\alpha)(u+v-uv)], \qquad (u,v) \in [0,1]^2, \tag{2.5}$$

where $0 \le \alpha \le 2$.

Proof. We know from Theorem 3.3 that the solution space of the linear system (2.2) (for $m = 4$) is one dimensional ($(m-2)(m-3)/2 = 1$). So we use $\alpha = a_{11}$ as the independent parameter. The general solution of the system (2.2) is $a_{13} = a_{31} = 0$ and $a_{12} = a_{21} = -a_{22} = 1 - \alpha$. Hence the general polynomial copula of power-4 must be of the form:

$$C(u,v) = \alpha uv + (1-\alpha)(u^2v + uv^2) - (1-\alpha)u^2v^2. \tag{2.6}$$

Since $C(u,v)$ is a copula, $\partial^2 C(u,v)/\partial u \partial v$ is non-negative over $[0,1]^2$, that is

$$\begin{array}{rl} \alpha + 2(1-\alpha)(u+v) - 4(1-\alpha)uv \ge 0, & (u,v) \in [0,1]^2 \\ \iff \quad 2(1-\alpha)(u+v-2uv) \ge -\alpha. & \end{array} \tag{2.7}$$

With $(u,v) \in [0,1]^2$, we always have

$$1 \ge (u+v-2uv) \ge u^2 + v^2 - 2uv \ge 0.$$

It is easy to verify that (2.7) is valid if and only if $2(1-\alpha) \ge -\alpha$ and $\alpha \ge 0$. Namely, $0 \le \alpha \le 2$. This completes the proof.

Corollary 2.5 *A power-5 polynomial $p(u,v)$ is a copula iff it is of the form*

$$\begin{aligned} p_5(u,v) = {} & \alpha uv + \beta uv^2 + (1-\alpha-\beta)uv^3 + (1-\alpha-\beta-\gamma)u^2v \\ & + \gamma u^2v^2 - (1-\alpha-\beta)u^2v^3 + (\beta+\gamma)u^3v - (\beta+\gamma)u^3v^2, \end{aligned} \tag{2.8}$$

where (α,β,γ) are parameters such that the density $\partial^2 p_5(u,v)/\partial u \partial v$ is non-negative over $[0,1]^2$. The range of (α,β,γ) is a convex hull which is included in the polyhedron:

$$\mathcal{P} = \{(\alpha,\beta,\gamma):\ \alpha \ge 0,\ 3 - 2\alpha - \beta \ge 0,\ 2 - \alpha + \beta + \gamma \ge 0,\ 2\alpha - \gamma - 1 \ge 0\}.$$

Proof. Analogously to the previous corollary, with $m = 5$, the number of free parameters is $(m-2)(m-3)/2 = 3$. We choose $a_{11} = \alpha$, $a_{12} = \beta$, and $a_{22} = \gamma$ as the free parameters. The general solution of the system (2.2) is then

$$\begin{aligned} a_{13} &= 1 - (\alpha + \beta), \\ a_{23} &= -a_{13} = -1 + \alpha + \beta, \\ a_{32} &= -(\beta + \gamma), \\ a_{31} &= -a_{32} = \beta + \gamma, \\ a_{21} &= -\gamma - a_{23} = 1 - (\alpha + \beta + \gamma). \end{aligned}$$

Hence the power-5 polynomial copulas must be of the form in (2.8). The polyhedron bounds of the parameters can be checked since the density $\partial^2 p_5(u,v)/\partial u \partial v$ is non-negative over $[0,1]^2$.

The high power polynomial copulas have much more degrees of freedom to be used as an approximation to other copulas. As a matter fact even a small subset of this large family would be sufficient for this purpose. For example, Li, Mikusinski, Sherwood & Taylor (1997) have proved that Bernstein polynomials could be used to approximate any continuous copula with the convergence rate proportional to the power of the polynomial.

3 Piecewise Linear/Quadratic Copulas

The piecewise linear copulas have received special attention in the literature. Together with the piecewise quadratic copulas, they have the very simple and direct probabilistic meaning of the dependence pattern between two marginal random variables, i.e. the singularity and local independence. By *singularity* we mean a probability measure is concentrated on the lower dimensional manifold (one dimensional for the case of bivariate distributions). *Local independence* means that the two marginal random variables are independent conditional on a subset of the supporting set of a probability space. Also, we restrict the idea of *piecewise* to the partition of the $[0,1]^2$ into finite or countably infinite union of connected open sets with boundaries as the one dimensional curves of finite length.

Example 3.1. Let (X, Y) be uniformly distributed on the L_1-norm unit sphere $\{(x,y) : |x| + |y| = 1\}$ (Figure 3.1(a)). Then, $|X| + |Y| = 1$. Its marginal distribution function is given by

$$Q(x) = \begin{cases} 0 & \text{if } x < -1, \\ (1+x)/2 & \text{if } |x| \le 1, \\ 1 & \text{if } x > -1. \end{cases} \tag{3.1}$$

Then, the copula of (X, Y) is as follows,

$$C(u,v) = \begin{cases} u + v/2 - 1/4, & \text{if } |u-1/2| + |v-1/2| \le 1/2, \\ & \text{for } |u-1/2| + |v-1/2| > 1/2 : \\ 0, & 0 \le u < \frac{1}{2},\ 0 \le v < \frac{1}{2}, \\ u, & 0 \le u < \frac{1}{2},\ \frac{1}{2} \le v \le 1, \\ v, & \frac{1}{2} \le u \le 1,\ 0 \le v < \frac{1}{2}, \\ u + v - 1, & \frac{1}{2} \le u \le 1,\ \frac{1}{2} \le v \le 1, \end{cases} \tag{3.2}$$

which is piecewise linear. It is composed of five flats and symmetric on $u = v$ (Figure 3.1(b)).

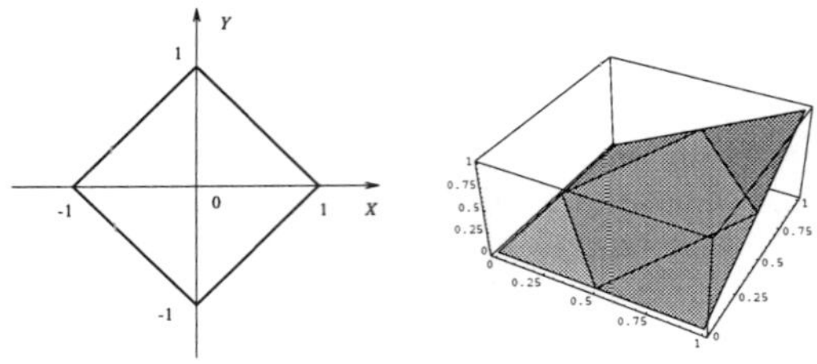

(a) (b)

Figure 3.1 A piecewise linear copula.

Theorem 3.1 *Let $C(u,v)$, $F(x)$ and $G(y)$ be the copula and marginal distribution functions of (X,Y), respectively. The joint distribution of X and Y is singular iff the corresponding copula is* **piecewise additive**, *i.e. there exists a partition of the $[0,1]^2$ such that over each of the partition cells A_i, $i = 1, 2, ...,$*

$$C(u,v)|_{A_i} = C_1(u) + C_2(v), \qquad i = 1, 2, \cdots, \tag{3.3}$$

where $C_1(u)$ and $C_2(v)$ are increasing functions.

Proof. *Necessity:* Since (X,Y) is singular, there is an one-dimensional manifold on which the probability measure concentrated. Under the transformation $\{u = F(x), v = G(y)\}$, there is a corresponding one-dimensional manifold in $[0,1]^2$. Then $[0,1]^2$ can be partitioned along the one-dimensional manifolds. Thus within each partition set the increment of cdf is the sum of the increments of marginal distributions, i.e.

$$\begin{aligned} C(u+\Delta u, v+\Delta v) - C(u,v) &= [C(u+\Delta u, v) - C(u,v)] \\ &\quad + [C(u, v+\Delta v) - C(u,v)] \\ &= C_1(\Delta u) + C_2(\Delta v), \end{aligned}$$

as expected.

Sufficiency: Evidently an additive copula has no density. Due to the increment structure of the cdf, the probability measure must be located along the boundaries of the partition cells. This completes the proof.

If the random vector (X,Y) is singular, its probability measure is concentrated on an one-dimensional manifold. It should be noted the one-dimensional manifold may be a genuine subset of the union of boundaries of the partition sets used to define a piecewise additive copula.

Corollary 3.2 *The copula of X and Y is piecewise linear iff (X,Y) is singular and there is a partition of the probability space given by $\bigcup A_i$, such that all the conditional cdfs: $F(x|Y \le y)|A_i$ and $G(y|X \le x)|A_i$ are proportional to $F(x)$ and $G(y)$ respectively.*

Example 3.2. Consider the bivariate random vector (X,Y) with its probability mass concentrated on two linear sections with the joint density is given by

$$h(x,y) = \begin{cases} \sqrt{2}/4 & -1 \le x \le 0,\ y = -x; \\ \frac{1}{2}\sqrt{x^2+y^2} & 0 \le x \le 1, \quad y = x. \end{cases}$$

The marginal distributions of X and Y are respectively,

$$F(x) = \begin{cases} \frac{1}{2} + \frac{1}{2}x, & -1 \le x \le 0, \\ \frac{1}{2} + \frac{1}{2}x^2, & 0 \le x \le 1, \end{cases} \qquad F^{-1}(u) = \begin{cases} 2u - 1, & 0 \le u \le \frac{1}{2}, \\ \sqrt{2u-1}, & \frac{1}{2} \le u \le 1, \end{cases}$$

and

$$G(y) = \frac{1}{2}y + \frac{1}{2}y^2, \quad 0 \le y \le 1, \quad G^{-1}(v) = \frac{1}{2}(-1 + \sqrt{1+8v}),$$

and the copula of (X, Y) is

$$C(u,v) = \begin{cases} 0, & 0 \le u \le \frac{1}{2},\ 2u - \frac{3}{2} + \frac{\sqrt{1+8v}}{2} \le 0, \\ v, & \frac{1}{2} \le u \le 1,\ \frac{\sqrt{1+8v}-1}{2} - \sqrt{2u-1} \le 0 \\ u - \frac{3}{4} + \frac{\sqrt{1+8v}}{4}, & \text{elsewhere}. \end{cases}$$

Thus, the copula is piecewise additive but not piecewise linear even though the probability mass is totally concentrated on two line sections.

The following Example 3.3 gives a non-trivial piecewise quadratic copula. It follows from Theorem 2.3, if a random vector (X, Y) has a quadratic copula, then X and Y are independent. However, if (X, Y) has a piecewise quadratic copula, then X and Y are locally independent but not necessarily independent.

Example 3.3. We provide an example of a non-trivial piecewise quadratic copula. The unit square $[0,1]^2$ is partitioned into three parts with the two characterizing boundaries:

$$6v = 17 - 8u - \sqrt{289 - 308u + 28u^2} \quad \text{and} \quad 2v + u + 1 = \sqrt{1 + 18u - 3u^2}.$$

Then the three parts are (cf. Figure 3.2(b))

$$\begin{aligned} A &= \left\{(u,v) \in [0,1]^2 : \ 2v \ge \sqrt{1+18u-3u^2} - u - 1\right\}, \\ B &= \left\{(u,v) \in [0,1]^2 : \ v \ge \tfrac{1}{6}(17 - 8u - \sqrt{289 - 308u + 28u^2}), \right. \\ &\qquad \left. v \le \tfrac{1}{2}(\sqrt{1+18u-3u^2} - u - 1)\right\}, \end{aligned}$$

and

$$D = \left\{(u,v) \in [0,1]^2 : \ 6v \le 17 - 8u - \sqrt{289 - 308u + 28u^2}\right\}.$$

The copula defined over the three sets is given by (see Figure 3.2)

$$C(u,v) = \begin{cases} u(1+v)/2, & (u,v) \in A, \\ (u^2 + 6uv + v^2 + u + v)/10, & (u,v) \in B, \\ v(2+u)/3, & (u,v) \in D. \end{cases}$$

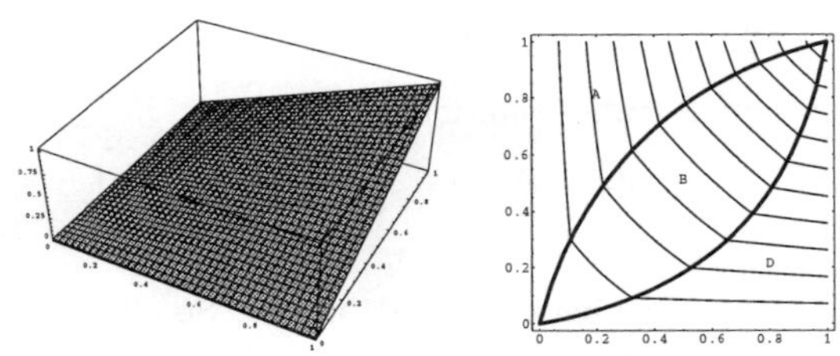

(a) (b)

Figure 3.2 A piecewise quadratic copula.

Since a convex linear combination of the copulas is still a copula, many other quadratic copulas could easily be obtained from the known ones.

4 The "Holes" in the Domain

In this section we shall explore a very simple and quite elementary topological nature of the dependence pattern: the number and the shape of the "holes" in the domains of bivariate copula density functions.

Let (X, Y) be a bivariate random vector. Stoyanov (1997) constructed a classical counterexample involving an interesting non-normal bivariate distribution of (X, Y) such that X, Y,and $X + Y$ are normal but not the joint distribution of (X, Y). The basic idea is to *punch* four square holes A_i $(i = 1, 2, 3, 4)$, symmetrically in the support of a bivariate normal density function, and to move the probability masses in $\boldsymbol{A}_i$ to the other square holes with symmetrical position, B_i $(i = 1, 2, 3, 4)$, respectively. This procedure ensures that the marginals are not affected. It is clear that after *punching* the four holes and rearranging the probability mass, the joint probability density function will no longer be normal over the supporting set.

In practice, the models of multivariate distributions with "holes" in the supporting sets occur quite often. For example, the distribution of number of fishes in an area of sea surrounded with a number of islands. More examples emerge in biology, meteorology, communications and medicine. General geometrical problems arise naturally from probabilistic considerations. Could we construct a non-normal bivariate pdf which is continuous over its support, yet with as many holes as one desires? If so, could the holes have any particular boundaries? At first sight these problems do not seem to be easy. However we shall demonstrate that, though the explicit expression might not be easily available, the numerical solutions for these problems are quite simple and straightforward using the copula analysis and a computer.

Before proceeding further we shall describe rigorously the concept of *holes*. Let (X, Y) be a bivariate continuous random vector with the pdf $h(x, y)$ and the marginal distributions $F(x)$ and $G(y)$. $\mathcal{D} = [a, b] \times [c, d]$ is called a *square domain* of (X, Y), where

$$\begin{aligned} a &= \inf\{x : h(x, y) > 0\},\ c = \inf\{y : h(x, y) > 0\}, \\ b &= \sup\{x : h(x, y) > 0\},\ d = \sup\{y : h(x, y) > 0\}. \end{aligned} \tag{4.1}$$

The invertible continuous transformations

$$\begin{cases} u = F(x) \\ v = G(y) \end{cases} \qquad \begin{cases} x = F^{-1}(u) \\ y = G^{-1}(v) \end{cases} \tag{4.2}$$

are topological isomorphisms between $[0, 1]^2$ and $\mathcal{D}$. So all the topological structures of the domain $\mathcal{D}$, such as the number of holes and the connectedness are valid in the corresponding copula density function.

Definition 4.1 Let (X, Y) be a continuous random vector with the support S and the square domain $\mathcal{D}$. Suppose that $A \subset \mathcal{D}$ is a connected open set and its boundary consisting of piecewise differentiable curves of finite length. Let $S \bigcap A = \phi$. If there exists a open neighborhood of A in $\mathcal{D}$, A_ε such that $A_\varepsilon \backslash A \subset S$, then A is called a *hole*.

Since the pdf is equal to 0 at the holes, the number of holes for the pdf and those for the corresponding copula density function are same by (1.3). For given marginals, the construction of bivariate pdf's with holes can be based on copulas, and the number of holes are invariant under the transformation (4.2). Furthermore, since the marginal cdf's of (X, Y) are strictly increasing, some properties of hole boundaries are consistent with those for the pdf and the copula density, such as piecewise monotonicity and differentiability. In this subsection, we shall study the shapes and bounds of holes by means of a copula density.

Now we use the terminology *rectangular holes* to denote those with boundaries that are parallel to the coordinate axes. Particularly a *single rectangular hole* also includes the situation when the left and the right, or upper and lower sides of the unit square (the domain of copula density) are taken as equivalent.

From (1.3), if there is a rectangular hole in the pdf of (X, Y), then there is also a rectangular hole in its copula density and conversely. The following theorem offers exact upper bounds on the width, height, and the area of a single rectangular hole in the supporting set of a copula density function.

Theorem 4.1 *The exact upper bounds on the width* (α), *height*(β), *and area*$(\alpha\beta)$ *of a single rectangular hole in the copula* $C(u, v)$ *are*

$$1 - \alpha - \beta > 0 \text{ and } \alpha\beta < \max\{\alpha(1-\alpha), \beta(1-\beta)\} < 1/4, \tag{4.3}$$

Furthermore, the bounds can be arbitrarily closely approached as $\{\beta \to (1-\alpha)_-\}$, $\{\alpha \to (1-\beta)_-\}$, *and* $\{\alpha \to \frac{1}{2}-, \beta \to \frac{1}{2}-\}$ *respectively.*

Proof. Consider a typical single rectangular hole A surrounded by the supporting set of copula density, as shown in Figure 4.1. Assume the hole has width α and height β. Let d_1, d_2, b_1, and b_2 be the probability measure of the regions D_1, D_2, B_1, and B_2 respectively. From the definition of a hole these measures must be positive. Since the marginal of any copula is the uniform distribution on $(0, 1)$, we must have that

$$\alpha = d_1 + d_2, \text{ and } \beta = b_1 + b_2.$$

But none of T_1, T_2, T_3, and T_4 is of a zero measure, thus $d_1 + d_2 + b_1 + b_2 < 1$. Hence, $\alpha > 0$, $\beta > 0$ and $\alpha + \beta > 1$. The area of the hole has bounds:

$$\alpha\beta < \max\{\alpha(1-\alpha), \beta(1-\beta)\} < 1/4.$$

Exactness of the bounds is obvious.For any α and β satisfying the above conditions, a copula density function with rectangular hole (α wide and β high) could be directly constructed by assigning the probability mass (α, β, $1 - \alpha - \beta$) uniformly to $\{(D_1, D_2), (B_1, B_2), (T_1, T_2, T_3, T_4)\}$ respectively. The proof is thus completed.

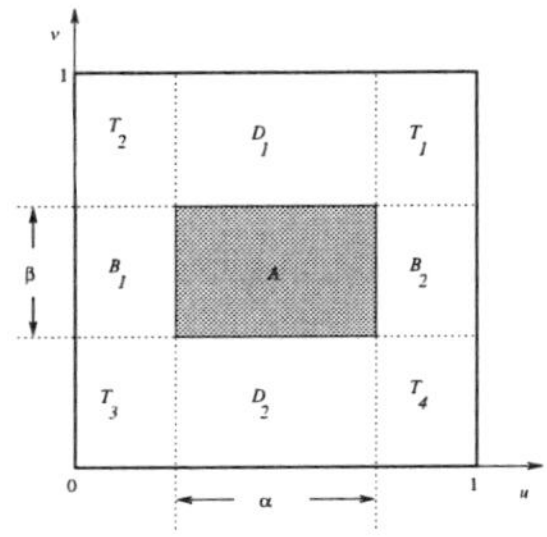

Figure 4.1 A typical rectangular hole within the domain of a copula

All the rectangles that could be used to construct a single hole of a copula density will be now called *admissible rectangles.*

Corollary 4.2 *Let $h(x, y)$ be a bivariate pdf with the marginal distributions $F(x)$ and $G(y)$. There exists a single rectangular hole $[a, b] \times [c, d]$ within the pdf $h(x, y)$ if and only if*

$$1 - (F(b) - F(a)) - (G(d) - G(c)) > 0.$$

(If this inequality is satisfied, the hole $[a, b] \times [c, d]$ will also be called admissible). A bivariate pdf $h(x, y)$ with non-rectangular holes exists provided the holes are strictly covered by an admissible rectangle.

The algorithm for constructing the rectangular hole in above proof of the theorem is a modification of the procedure used by Stoyanov (1997) in his counterexample for normal distributions. We may call it a "mass swapping" algorithm since we could proceed by starting from the uniform distribution over the unit square and then swapping/redistributing the probability mass in the nine rectangular regions to obtain a copula with a single rectangular hole. With some modification this method could also be used to construct other holes with non-rectangular shapes as long as the holes are covered by an admissible rectangular hole.

However we have noticed that a copula density with holes constructed by the mass swapping algorithm is usually piecewise continuous over the supporting region with jumps appearing along the boundaries of the eight rectangular regions. It seems plausible that we could construct infinitely many copula densities with holes, which are yet continuous everywhere over the supporting region. For this construction another algorithm will be introduced.

Theorem 4.3 *For a set of holes (which could be countably infinite) whose area is bounded by an admissible rectangle* **A**, *there exist copula densities possessing these holes.*

Proof. We intend to prove the theorem by constructing directly the copula densities with desired holes. The proof is rather lengthy but quite intuitive. Since the basic idea of the algorithm is to redistribute the measure mass along the vertical/horizontal directions as if the measure mass squeezed along the orthogonal tubes, we shall simply call the algorithm the *squeeze algorithm.* Now we consider a typical situation shown in Figure 4.2, where the holes are bounded by the admissible rectangle $\mathbf{A}= [a, b] \times [c, d]$. The four major steps of the squeeze algorithm are

(i) Starting from the uniform distribution over $[0, 1]^2$. Suppose the mass measure (area) of the holes is S. Set the density to be 0 over the holes.
(ii) Distribute the mass S over $\mathbf{D}_1 \bigcup \mathbf{D}_2$ such that the marginal density in the u-direction still remains 1 for $u \in [0, 1]$.
(iii) Take mass S from $\bigcup_{i=1}^4 \mathbf{T}_i$ such that the marginal density in the v-direction for $v \in [0, c] \bigcup [d, 1]$ is still 1.
(iv) Distribute mass S over $\mathbf{B}_1 \bigcup \mathbf{B}_2$ such that the marginal densities in the u-direction for $u \in [0, a] \times [b, 1]$ and in the v-direction for $v \in [c, d]$ are 1.

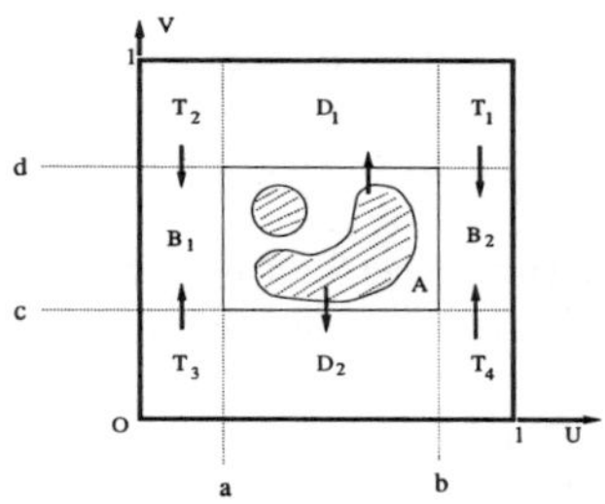

Figure 4.2 Two holes in an admissible rectangle **A**.

We provide further clarification. After step (i), mass S is being lost from $\mathbf{A}$. The new marginal mass densities are $f^*(u)$ and $g^*(v)$. The functions $\theta(u) = 1 - f^*(u)$ and $\eta(v) = 1 - g^*(v)$ are the density functions used as a standard to redistribute the mass S on $\mathbf{A}^c$. Evidently

$$\int_a^b \theta(u)du = \int_c^d \eta(v)dv = S.$$

We shall use the capital Roman letters to represent the area of the 9 different rectangular regions, e.g., T_1 will stand for the area of $\mathbf{T}_1$, etc. In steps (ii) and (iii) the mass S is to be distributed over the eight regions in the following proportion:

$$\begin{cases} \mathbf{D}_1 : r_{D_1} = (1-d)/(c+(1-d)) \\ \mathbf{D}_2 : r_{D_2} = c/(c+(1-d)) \\ \mathbf{B}_1 : r_{B_1} = a/(a+(1-b)) \\ \mathbf{B}_2 : r_{B_2} = (1-b)/(a+(1-b)) \end{cases} \qquad \begin{cases} \mathbf{T}_1 : r_{T_1} = -r_{B_2} r_{D_1} \\ \mathbf{T}_2 : r_{T_2} = -r_{B_1} r_{D_1} \\ \mathbf{T}_3 : r_{T_3} = -r_{B_1} r_{D_2} \\ \mathbf{T}_4 : r_{T_4} = -r_{B_2} r_{D_2}. \end{cases}$$

Step (ii): Typically in $\mathbf{D}_2$, we have to distribute the extra mass $Q = r_{D_2} S$. Admissibility of $\mathbf{A}$ ensures that $Q < T_3 + T_4$. Thus we could distribute the mass Q over $\mathbf{D}_2$ with a density function $q(u, v) \geq 0$ satisfying

$$\begin{aligned} \int_0^c q(u,v)dv &= r_{D_2}\theta(u), \quad u \in [a, b], \\ \int_a^b q(u,v)du &< a + (1-b), \quad v \in [0, c]. \end{aligned} \tag{4.4}$$

In particular, we could choose $q^*(u,v) = r_{D_2}\theta(u)/c$. A similar procedure is applied to $\mathbf{D}_1$.

Step (iii): Typically in $\mathbf{T}_3$ and $\mathbf{T}_4$ masses $-r_{T_3}S$ and $-r_{T_4}S$ have subtracted respectively. The simplest method is to choose constant densities. Similar procedures are applied to $\mathbf{T}_1 \bigcup \mathbf{T}_2$.

Step (iv): Now we have only $\mathbf{B}_1$ and $\mathbf{B}_2$ to deal with. Typically in $\mathbf{B}_1$, mass $r_{B_1}S$ should be distributed with the density function $w(u,v) \geq 0$ such that

$$\begin{aligned} &\int_0^a w(u,v)du = r_{B_1}\eta(v), \quad v \in [c,d], \\ &\int_c^d w(u,v)dv = r(u) \quad u \in [0,a], \end{aligned} \tag{4.5}$$

where $r(u)$ is the mass subtracted from $\mathbf{T}_2 \bigcup \mathbf{T}_4$ along the vertical line section with a fixed u. For the particular situation when constant densities are taken for $\mathbf{T}_2 \bigcup \mathbf{T}_4$ in step (iii) we could simply take $w(u,v) = r_{B_1}\eta(v)/a$. Similar procedures are applied to $\mathbf{B}_2$.

Using the squeeze algorithm described in detail above we have actually constructed a particular piecewise continuous copula density with desired holes given by

$$\frac{\partial^2 C(u,v)}{\partial u \partial v} = \begin{cases} I_{\{(u,v)\notin \text{holes}\}}, & (u,v) \in \mathbf{A} \\ 1 + r_{D_1}\theta(u)/(1-d), & (u,v) \in \mathbf{D}_1 \\ 1 + r_{D_2}\theta(u)/c, & (u,v) \in \mathbf{D}_2 \\ 1 + r_{B_1}\eta(v)/a, & (u,v) \in \mathbf{B}_1 \\ 1 + r_{B_2}\eta(v)/(1-b), & (u,v) \in \mathbf{B}_2 \\ 1 + r_{T_1}S/[(1-b)(1-d)], & (u,v) \in \mathbf{T}_1 \\ 1 + r_{T_2}S/[a(1-d)], & (u,v) \in \mathbf{T}_2 \\ 1 + r_{T_3}S/(ac), & (u,v) \in \mathbf{T}_3 \\ 1 + r_{T_4}S/[c(1-b)], & (u,v) \in \mathbf{T}_4 \end{cases} \tag{4.6}$$

There are infinitely many choices for the density functions $q(u,v)$ and $w(u,v)$ satisfying the conditions (4.4) and (4.5) in steps (ii) and (iv). With minor modifications for the constant densities in $\bigcup_{i=1}^4 \mathbf{T}_i$, we could make the final copula density with holes to be continuous across the boundaries $\mathbf{T}_i \bigcap \mathbf{B}_j$ and $\mathbf{T}_i \bigcap \mathbf{D}_j$ $(i = 1,2,3,4,\ j = 1,2)$ and therefore continuous everywhere over the supporting set. This completes the proof of the theorem.

For a given the marginal distributions, a bivariate pdf with specified holes can be constructed by the squeeze algorithm. A further study is given Fang (1998).

5 Local Dependence Measurement

Although copulas contain all the information concerning the dependence patterns of random variables, the local dependence pattern seems to be seldom explored. Practitioners are usually satisfied with the average information about dependence over the whole domain of distributions, such as the Spearman coefficient, Kendall coefficient and etc. However, important and indicative local properties could be totally filtered out by these average indices. In this section, we shall study the local dependence index via the cross difference and describe local dependence properties such that some fundamental features of the bivariate distributions could be captured either in the limit or at different stages.

The basic operation of cross differencing has been mentioned in the literature over 50 years ago (see *e.g.* Kendall (1938); Kruskal (1958)). It was averaged by the integration over the whole domain of the random variables and used as an ordinal measure of association. Now we shall examine this operation for local conditional distributions over small rectangular domains.

Definition 5.1 Let **W** be a rectangle within the domain of a copula $[0,1]^2$. **W** is partitioned into four smaller rectangles **A**, **B** diagonally and **C**, **D** "counter diagonally" and probability masses over them are denoted by A, B, C and D respectively. The quantity $(A+B)-(C+D)$ is called the *cross difference* over rectangle **W** respect to this partition.

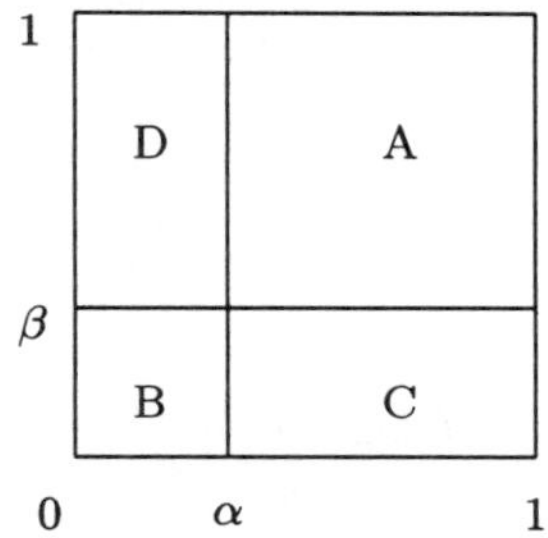

Figure 5.1 The general probability assignments.

Let $C(u,v)$ be the copula of the random variables X and Y. Consider the cross difference $\Delta(\alpha,\beta)$ of $C(u,v)$ over $[0,1]^2$ with the partition lines coinciding with the coordinates α and β respectively (Figure 5.1). Denoting the probability mass over the rectangle **C** by γ (≥ 0), we have

$$\Delta(\alpha,\beta) = 1 - 2(\alpha - \beta + 2\gamma). \tag{5.1}$$

In particular, if X and Y are independent, the cross difference is

$$\Delta_0(\alpha,\beta) = 1 - 2(\alpha+\beta) + 4\alpha\beta. \tag{5.2}$$

If X and Y are perfectly positively dependent, the cross-difference becomes

$$\Delta_1(\alpha,\beta) = 1 - 2|\alpha-\beta|. \tag{5.3}$$

Furthermore, for all α, $\beta \in (0,1)$, we have $\Delta(\alpha,\beta) \leq \Delta_1(\alpha,\beta)$ and

$$\Delta_1(\alpha,\beta) - \Delta_0(\alpha,\beta) = 4\min(\alpha,\beta)\min(1-\alpha,1-\beta). \tag{5.4}$$

Definition 5.2 For $0 < \alpha, \beta < 1$, the value

$$\tau = \frac{\Delta(\alpha,\beta) - \Delta_0(\alpha,\beta)}{\Delta_1(\alpha,\beta) - \Delta_0(\alpha,\beta)} = \frac{\Delta(\alpha,\beta) - (1 - 2(\alpha+\beta) + 4\alpha\beta)}{4\min(\alpha,\beta)\min(1-\alpha, 1-\beta)}$$

is defined as the *dependence index* with respect to a partition (α, β).

Equation (5.4) shows that $-1 \leq \tau \leq 1$, and the equality holds if and only if the two marginal random variables possess perfect positive/negative dependence. With the dependence index τ, we can obtain a general picture of the local dependence patterns of a bivariate distribution (or copula) at different stages. Let us partition the domain $[0,1]^2$ of a copula $C(u,v)$ into $n = 2^{2m}$ identical small squares. The probability mass over a typical square $[u, u+du] \times [v, v+dv]$ is

$$C(u+du, v+dv) - C(u, v+dv) - C(u+du, v) + C(u,v),$$

which is also a form of cross-difference. Then over every square consisting of four neighboring small squares we could evaluate the cross difference and the dependence index and obtain $2^{2(m-1)}$ such indices. Plotting them over the domain we arrive at the dependence picture at the 2^{-m}-th stage.

In the last decade the study of local dependence between two continuous random variables X and Y has attracted some interest. Holland & Wang (1987) and Wang (1993) introduced the local dependence function given by the formula (see also Bairamov, Kotz, & Kozubowski (2003) for a slightly different definition of local dependence):

$$\lambda(x,y) = \frac{\partial^2 \log f(x,y)}{\partial x \partial y} = \frac{f f_{xy} - f_x f_y}{f^2} \tag{5.5}$$

where f denotes the joint pdf of (X, Y) and f_x is the partial derivative at (x, y) respect to x. The authors motivate this as the "limit of the local cross-ratio defined for adjacent cell probabilities, formed by a two-dimensional rectangular grid, when the length and width of the rectangles shrink to zero." A relationship between the dependence index and the dependence function is given by the following theorem.

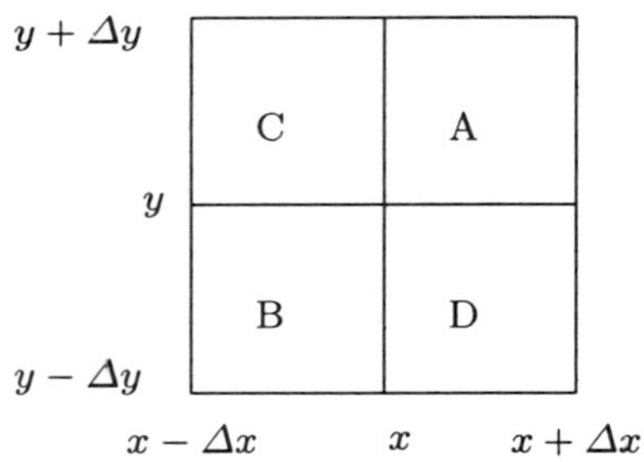

Figure 5.2 Small rectangles around a typical point (x, y).

Theorem 5.1 *Consider a typical point (x, y) and a small rectangle $T = [x - \Delta x, x + \Delta x] \times [y - \Delta y, y + \Delta y]$ within the domain of a smooth pdf $f(x, y)$ (see Figure 5.2). Suppose that $f(x, y) \neq 0$, then as Δx and Δy approach zero, the dependence index on T is given by*

$$\tau = \frac{1}{4}\lambda(x,y)\Delta x\Delta y + o(\Delta x\Delta y). \tag{5.6}$$

Proof. For notational convenience we shall use Q_2 to represent homogeneous quadratic polynomials in $(\Delta x, \Delta y)$. We shall simply carry out Taylor expansion for each configuration. With the notation defined above we have

$$\begin{aligned}
A &= \int_0^{\Delta x}\int_0^{\Delta y} f(x+u, y+v)dudv \\
&= \left(f + \tfrac{f_x\Delta x + f_y\Delta y}{2} + \tfrac{f_{xx}(\Delta x)^2 + f_{yy}(\Delta y)^2}{6} + \tfrac{f_{xy}\Delta x\Delta y}{4}\right)\Delta x\Delta y + o(Q_2) \\
B &= \left(f - \tfrac{f_x\Delta x + f_y\Delta y}{2} + \tfrac{f_{xx}(\Delta x)^2 + f_{yy}(\Delta y)^2}{6} + \tfrac{f_{xy}\Delta x\Delta y}{4}\right)\Delta x\Delta y + o(Q_2) \\
C &= \left(f - \tfrac{f_x\Delta x - f_y\Delta y}{2} + \tfrac{f_{xx}(\Delta x)^2 + f_{yy}(\Delta y)^2}{6} - \tfrac{f_{xy}\Delta x\Delta y}{4}\right)\Delta x\Delta y + o(Q_2) \\
D &= \left(f + \tfrac{f_x\Delta x - f_y\Delta y}{2} + \tfrac{f_{xx}(\Delta x)^2 + f_{yy}(\Delta y)^2}{6} - \tfrac{f_{xy}\Delta x\Delta y}{4}\right)\Delta x\Delta y + o(Q_2).
\end{aligned}$$

Then,

$$\alpha = \tfrac{B+C}{A+B+C+D} = \frac{1}{2} - \frac{f_x\Delta x}{4f} + \frac{f_{xx}(\Delta x)^2 + f_{yy}(\Delta y)^2}{12f} + o(Q_2)$$

$$\beta = \tfrac{B+D}{A+B+C+D} = \tfrac{1}{2} - \tfrac{f_y\Delta y}{4f} + \tfrac{f_{xx}(\Delta x)^2 + f_{yy}(\Delta y)^2}{12f} + o(Q_2).$$

The cross-difference becomes

$$\Delta(\alpha,\beta) = \frac{(A+B)-(C+D)}{A+B+C+D} = \frac{f_{xy}\Delta x\Delta y}{4f} + o(Q_2), \tag{5.7}$$

and the cross-difference in the independent case is

$$\Delta_0(\alpha,\beta) = \frac{f_x f_y \Delta x\Delta y}{4f^2} + o(Q_2). \tag{5.8}$$

Using (5.7) and (5.8), we obtain

$$\tau = \frac{\Delta(\alpha,\beta) - \Delta_0(\alpha,\beta)}{4\min(\alpha,\beta)\min(1-\alpha,1-\beta)} = \frac{1}{4f^2}(f_{xy}f - f_x f_y)\Delta x\Delta y + o(Q_2),$$

and the proof is completed.

Corollary 5.2 *A bivariate distribution with smooth pdf $f(x,y)$ has a constant local dependence function λ if and only if the pdf is of the following form*

$$f(x,y) = \exp\{\lambda xy + A(x) + B(y)\}, \tag{5.9}$$

where $A(x)$ and $B(y)$ are differentiable functions.

The bivariate normal distribution with the covariance matrix

$$\boldsymbol{\Sigma} = \begin{pmatrix} \sigma_1^2 & \sigma_1\sigma_2\rho \\ \sigma_1\sigma_2\rho & \sigma_2^2 \end{pmatrix},$$

belongs to the exponential family (5.8), and the corresponding local dependence function takes the constant form $\lambda = \rho/((1-\rho^2)\sigma_1\sigma_2)$. Kotz & Nadarajah (2003) studied local dependence functions of elliptically symmetric distributions.

6 Discussion

In statistical modeling, it is useful to have tractable multivariate distributions with given marginals in order to be able to quantify the effect of dependence of the variables contained in the model. The copula analysis is a efficient technique to describe the dependence patterns of random variables. Without referring to the copulas, we can hardly visualize and distinguish between the random variables depending upon each other in the same manner with different marginal distributions. The finite set of joint moments and correlation coefficients of two random variables contain insufficient information about their dependence structure. In particular, the number and the shape of holes in the supporting set of the joint pdf is essentially impossible to describe by global "average indices", such as moments and correlation coefficients.

In this article, we studied the basic geometric properties of copulas and defined a version of local dependence index of random variables. An analysis of geometric structures of copulas is an effective tool to characterize dependence structures. The local dependence analysis allows us to capture all of the dependence information among random variables. Additional copulas and certain non-classical multivariate distributions could be obtained utilizing the structures of copulas. The applications of copula techniques in data analysis can refer to Fang, Fang & von Rosen (2000) and Basrak, Klaassen, Beekman, Martin & Boomsma (2004).

Acknowledgments

The authors thank Professor Kai-Tai Fang for his constructive comments. This research was supported in part by the UGC grant of Hong Kong Government RGC/94-95/38 and an FRG grant of Hong Kong Baptist University.

References

Bairamov, I., Kotz, S. & Kozubowski, T. F. (2003). A new measure of linear local dependence. *Statistics* **37** No. 13 243–258.

Basrak, B., Klaassen, C. A., Beekman, M., Martin, N. G. & Boomsma, D. I. (2004). Copulas in QTL mapping. *Behaviour Genetics* **34** 161–171.

Dall'Aglio, G., Kotz, S. & Salinetti, G. (1991). *Advance in Probability Distributions with Given Marginals.* Kluwer Academic, Dordrecht.

Drouet-Mari, D. & Kotz, S. (2001). *Correlation and Dependence.* World Scientific Publish., Singapore.

Embrechts, P., Lindskog, F. & McNeil, A. (2003). Modelling dependence with copulas and applications to risk management. In *Handbook of Heavy Tailed Distributions in Finance* (S. T. Rachev, eds.) 329–384. Elsevier Science B.V., Amsterdam.

Fan, J. J., Prentice, R. L. & Hsu, L. (2000). A class of weighted dependence measures for bivariate failure time data. *Journal of Royal Statistical Society B* **62** 181–190.

Fang, H. B. (1998). *Some Non-classical Multivariate Distributions.* Ph.D. Thesis, Hong Kong Baptist University.

Fang, H. B., Fang, K. T. & Kotz, S. (2002). The meta-elliptical distributions with given marginals. *Journal of Multivariate Analysis* **82** 1–16.

Fang, K. T., Fang, H. B. & von Rosen, D. (2000). A family of bivariate distributions with non-elliptical contours. *Communication in Statistics– Theory and Methods* **29** 1885–1898.

Frees, E. W., Carriere, J. & Valdez, E. (1996). Annuity valuation with dependent mortally. *Journal of Risk and Insurance* **63** 229–261.

Holland, P. W. & Wang, Y. J. (1987). Dependence function for continuous bivariate densities. *Communication in Statistics– Theory and Methods* **16** 863–876.

Johnson, N. L. & Kotz, S. (1998). Nicked and notched square distributions. *TR dated 2-10-98*, Dept. of Operations Research, The George Washington University, Washington DC.

Jouini, M. N. & Clemen, R. T. (1996). Copula models for aggregating expert opinions. *Operations Research* **44** 444–457.

Kendall, M. G. (1938). A new measure of rank correlation. *Biometrika* **30** 81–93.

Kimeldorf, G. & Sampson, A. (1975). Uniform representations of bivariate distributions. *Communication in Statistics– Theory and Methods* **4** 617–627.

Kotz, S. & Nadarajah, S. (2003). Local dependence functions for the elliptically symmetric distributions. *Sankhyā (Ser. B): The Indian Journal of Statistics* **65** Part I 207–223.

Kruskal, W. H. (1958). Ordinal measures of association. *Journal of the American Statistical Association* **53** 814–861.

Li, X., Mikusinski, P., Sherwood, H. & Taylor, M. D. (1997). On approximation of copulas. In *Distributions with Given Marginals and Moment Problems* (V. Bene**v**s & J. **v**St**v**epán, eds.) 107–116. Kluwer Academic Publishers, Dordrecht, The Netherlands.

Nelsen, R. B. (1999). *An Introduction to Copulas.* Lecture Notes in Statistics **139**. Springer, New York.

Schweizer, B. (1991). Thirty years of copulas. In *Advance in Probability Distributions with Given Marginals* (G. Dall'Aglio, S. Kotz & G. Salinetti, eds.) 13–50. Kluwer Academic, Dordrecht.

Sklar, A. (1959). Fonctions de répartition á n dimensions et leurs marges. *Publications de l'Institut de Statistique de l'Université de Paris* **8** 229–231.

Sklar, A. (1996). Random variables, distribution functions, and copulas – a personal look backward and forward. In *Distributions with Fixed Marginals and Related Topics* (L. Rüschendorff, B. Schweizer & M. D. Taylor, eds.) 1–14. Institute of Mathematical Statistics, Hayward, CA.

Stoyanov, J. M. (1997). *Counterexamples in Probability.* Wiley, New York.

Wang, Y. J. (1993). Construction of continuous bivariate density functions. *Statistica Sinica* **3** 173–187.

Modelling Mean-Covariance Structures in the Growth Curve Model

Jianxin Pan[1] and Dietrich von Rosen[2]

[1] Department of Mathematics, University of Manchester, Manchester M13 9PL, U.K. (jpan@maths.man.ac.uk)
[2] Department of Biometry and Engineering, Swedish Agriculture University, Box 7013, S-750 07 Uppsala, Sweden (dietrich@math.uu.se)

Summary. The growth curve model (GCM) has been widely used in longitudinal studies and repeated measures. Most existing approaches for statistical inference in the GCM assume a specific structure on the within-subject covariances, e.g., compound symmetry, AR(1) and unstructured covariances. This specification, however, may select a suboptimal or even wrong model, which in turn may affect the estimates of regression coefficients and/or bias standard errors of the estimates. Accordingly, statistical inferences of the GCM may be severely affected by mis-specification of covariance structures. Within the framework of the GCM in this paper we propose a data-driven approach for modelling the within-subject covariance structures, investigate the effects of mis-specification of covariance structures on statistical inferences and study the possible heterogeneity of covariances between different treatment groups.

Key words: Covariance structures, growth curve models, heterogeneity of covariances, joint mean-covariance modelling, maximum likelihood estimation, mis-specification of covariance structures.

2000 Mathematics Subject Classification: 62F15, 62F30, 62H12, 62H15

1 Introduction

The *growth curve models* (GCM) are generalized multivariate analysis-of-variance models that are useful especially in longitudinal studies and repeated measures (Potthoff & Roy (1964)). The GCM is defined by

Contemporary Multivariate Analysis and Experimental Design—In Honor Celebration of Professor Kai-Tai Fang's 65th birthday. Edited by Jianqing Fan and Gang Li. The World Scientific Publisher, 2005.

$$Y_{p\times n} = X_{p\times m} B_{m\times r} Z_{r\times n} + \epsilon_{p\times n} \tag{1}$$

where Y is the response matrix of n subjects measured at p time points, and X and Z are within- and between-subject design matrices with ranks m and r, respectively. Typically, the columns of X are the powers of time at which repeated measures are made when polynomials of time are used to model the mean structures. The rows of Z are the indicators of treatment groups, i.e., the ith row of Z is given by $z_i' = (0, ..., 0, 1_{n_i}', 0, ..., 0)$ where 1_{n_i} is the $(n_i \times 1)$ vector with all components being one and n_i is the sample size of the ith treatment group $(i = 1, 2, ..., r;\ \sum_{i=1}^{r} n_i = n)$. The columns of the error matrix ϵ are assumed to be independent p-variate Normal with $(p \times 1)$ mean vector 0 and $(p \times p)$ covariance matrices Σ_i, depending on the treatment group of which the responses are generated $(i = 1, 2, ..., r)$. We denote this by $\epsilon \sim N_{p\times n}(0; \Sigma_1, \Sigma_2, ..., \Sigma_r; I_n)$ where I_n is the identity matrix with size n.

When the covariances are homegenerous, i.e., $\Sigma_1 = \Sigma_2 = ... = \Sigma_r \equiv \Sigma$, the estimates of the parameters B and Σ were discussed in the literature by many authors including Potthoff & Roy (1964), Khatri (1966) and von Rosen (1989) among others. Particularly, if the homegenerous covariance matrix Σ is given then the maximum likelihood estimate (MLE) of B must have an explicit form in terms of *generalized weighted least squares* (GWLS)

$$\widehat{B}(\Sigma) = (X'\Sigma^{-1}X)^{-1}X'\Sigma^{-1}YZ'(ZZ')^{-1}. \tag{2}$$

When Σ is unknown, a *two-step* estimation strategy is commonly used to calculate the estimate of B. In other words, we first find an *appropriate* estimate of Σ and then plug it into the GWLS in (2) (e.g., Potthoff & Roy (1964); Gleser & Olkin (1970)). In particular, when the MLE of Σ is utilized then the resulting GWLS estimate is the MLE of B (Rao (1965); von Rosen (1989)). Obviously, the GWLS in (2) says that the estimate of B may depend on the estimate of Σ, of which an exception is that Σ has the so-called Rao's *simple covariance structure* (SCS) :

$$\Sigma = X\Gamma X' + Q\Theta Q' \tag{3}$$

where Γ and Θ are $(m \times m)$ and $((p-m) \times (p-m))$ positive definite and Q is orthogonal to X, i.e., $Q'X = 0$ (Rao (1966)). In fact, it can be shown $\widehat{B}(\Sigma) = \widehat{B}(I_p)$ *if and only if* Σ is of Rao' SCS given in (3) (Kariya (1985); Pan & Fang (2002)). Typical examples of the SCS include *compound symmetry* and *random regression coefficients* structures (Lee (1988); Pan & Fang (2002)). Accordingly, the estimate of B is affected by the estimate of Σ unless Σ is within the space of the SCS. On the other hand, since the estimated covariance of $\widehat{B}$ is given by

$$\widehat{\mathrm{Cov}}(\widehat{B}) \equiv \widehat{\mathrm{Cov}}(\mathrm{vec}(\widehat{B})) = c\ [(ZZ')^{-1} \otimes (X'\widehat{\Sigma}^{-1}X)^{-1}] \tag{4}$$

(e.g., von Rosen (1989)) where c is a constant and $\otimes$ denotes the Kronecker product of two matrices, it is obvious that $\widehat{\mathrm{Cov}}(\widehat{B})$ depends on the estimate of Σ even if Σ falls into the space of the SCS. Accordingly, correct estimate of the covariances plays an important role in statistical inference of the GCM.

In the statistical literature the GCM was studied under a variety of assumptions of covariance structures, for example, unstructured covariance (UC) by Potthoff & Roy (1964) and von Rosen (1989), the SCS by Rao (1966) and Lee (1988), compound symmetry structure by Lee (1988), AR(1) by Fujikoshi, Kanda & Tanimura (1990) and Lee (1988), random regression coefficients structure by Rao (1966), etc.. With the specification of SCS and UC, statistical diagnostics including outlier and influential observation detections was addressed within likelihood and Bayesian framework by Pan & Fang (2002) and Pan, Fang & von Rosen (1997, 1998, 1999). From inferential and predictive points of view, Lee (1991) and Keramidas & Lee (1995) suggested several selection criteria to choose an appropriate covariance structure from a menu of candidates. This kind of *menu-selection* procedures, however, may not be optimal. For example, when the true covariance structure is not contained in the menu the selected covariance structure, though "best" in some sense, may not be close to the true value. Consequently, statistical inference may be badly affected by the mis-specification of covariance structures. On the other hand, the assumption of homogenerous covariances, i.e., $\Sigma_1 = \Sigma_2 = ... = \Sigma_r$, might not be true in practice. For example, in many biological and medical problems the homogeneity assumption does not hold because different treatment groups may have different variations over time. Also, it is not uncommon that within-subject correlation structures may vary from group to group. Accordingly, we hope to establish a mechanism to test whether or not the assumption of homogenerous covariances is true.

In this paper we propose a *data-driven* approach to jointly model the mean and covariance structures for all treatment groups. The approach is based on a modified Cholesky decomposition advocated by Pourahmadi (1999; 2000) for modelling homogenerous covariance structures. We extend Pourahmadi's (1999) approach to model heterogenerous covariances and in the modelling approach the homogeneity assumption becomes testable. We also investigate the effects of mis-specification of covariance structures on statistical inferences in the GCM. This paper is organized as follows. In Section 2 the modified Cholesky decomposition is briefly reviewed and models for mean-covariance structures are proposed. In Section 3 maximum likelihood estimation is developed and in Section 4 principle of testing homogenerous covariances is described. In Section 5 a real data set, Cattle data (Kenward (1987)), is analyzed for illustration. Numerical comparisons between the data-driven and menu-selection approaches are made as well. In Section 6 we discuss some further issue and in the Appendix we give the technical details of the proposed approach.

2 Regression models for mean-covariance structures

For illustration, let us look at the homogenerous GCM $Y \sim N_{p\times n}(XBZ, \Sigma, I_n)$ first. We assume the $(p \times p)$ covariance matrix Σ is positive definite in this paper. Accordingly, there is a unique lower triangular matrix T with 1's as diagonal entries and a unique diagonal matrix D with positive diagonal entries such that $T\Sigma T' = D$. This modified Cholesky decomposition has a transparent statistical interpretation: the below-diagonal entries of T are the negatives of the *autoregressive coefficients*, ϕ_{jk}, in

$$\hat{y}_j = \mu_j + \sum_{k=1}^{j-1} \phi_{jk}(y_k - \mu_k),$$

the linear least squares predictor of y_j based on its predecessors $y_{(j-1)}, \ldots, y_1$, where $\mu_j = E(y_j)$ and y_j is the jth component of the $(p \times 1)$ response y, the column random variable of Y $(j = 1, 2, \ldots, p)$. It can be shown that the diagonal entries of D are the *innovation variances* $\sigma_j^2 = \mathrm{Var}(y_j - \hat{y}_j)$ (Pourahmadi (2000)). Obviously, it follows that $\Sigma^{-1} = T'D^{-1}T$.

For the heterogeneous GCM $Y \sim N_{p\times n}(XBZ, \Sigma_1, \Sigma_2, \ldots, \Sigma_r; I_n)$, we take the modified Cholesky decomposition for each covariance matrix Σ_i, i.e., $T_i\Sigma_i T_i' = D_i$, and then obtain the autoregressive coefficients ϕ_{jki} from the lower triangular matrices T_i and the innovation variances σ_{ji}^2 from the diagonal matrices D_i $(j = 1, 2, \ldots, p;\ k = 1, 2, \ldots, j-1;\ i = 1, 2, \ldots, r)$. In a spirit of Pourahmadi (1999), we propose the following regression models to model the mean and covariance structures, simultaneously,

$$\mu_{ji} = x_j'\beta_i, \qquad \phi_{jki} = a_{jk}'\gamma_i \qquad \text{and} \qquad \log\sigma_{ji}^2 = h_j'\lambda_i \tag{5}$$

where μ_{ji} is the mean of the responses in the ith group measured at the jth time point, β_i, γ_i and λ_i are $(m \times 1)$, $(q \times 1)$ and $(d \times 1)$ regression coefficients for the ith group, and β_i is actually the ith column of B. The covariates x_j (i.e., the transpose of the jth row of X), a_{jk} and h_j are associated with the powers of time when using polynomials of time to model the mean and covariance structures for growth data. For example, we may choose

$$\begin{aligned} x_j &= (1, t_j, t_j^2, \ldots, t_j^{m-1})' \\ a_{jk} &= (1, (t_j - t_k), (t_j - t_k)^2, \ldots, (t_j - t_k)^{q-1})' \\ h_j &= (1, t_j, t_j^2, \ldots, t_j^{d-1})' \end{aligned} \tag{6}$$

if the within-subject correlation only depends on the elapsed time, where t_j is the jth time point at which observations are made. In the literature a Brownian motion specified to covariance structures of the GCM was considered by Lundbye-Christense (1991), which is a special case of the mean-covariance models (5) with the structures (6).

The advantages of the joint regression modelling of mean-covariance structures in (5) are multi-folds, for example, a) it is a data-driven approach that is capable to capture the true structures for mean and covariance, b) the resulted estimates of covariance matrices $\widehat{\Sigma}_i$ are guaranteed to be positive definite, c) the reparameterized regression coefficients have transparent statistical interpretations in terms of autoregressive coefficients and innovation variances (Pourahmadi, 1999, 2000), and d) the assumption of homogenerous covariances becomes testable. We will discuss these issues in more details in the following sections.

3 Maximum likelihood estimation

Denote $Y = (Y_1, Y_2, ..., Y_r)$ and $Z = (Z_1, Z_2, ..., Z_r)$ where Y_i and Z_i are the $(p \times n_i)$ responses and $(r \times n_i)$ between-subject design matrices of the ith group $(i = 1, 2, ..., r)$, respectively. Similar to Pourahmadi (1999), it can be shown that the log-likelihood function $\ell \equiv \ell(B; \gamma_1, ..., \gamma_r; \lambda_1, ..., \lambda_r)$ of the heterogeneous GCM $Y \sim N_{p\times n}(XBZ, \Sigma_1, \Sigma_2, ..., \Sigma_r; I_n)$ modelled with the regression models (5), except a constant being $-(pn/2)\log(2\pi)$, has the following three representations corresponding to B, $(\gamma_1, ..., \gamma_r)$ and $(\lambda_1, ..., \lambda_r)$, respectively,

$$\begin{aligned} \ell &= -\tfrac{1}{2}\sum_{i=1}^{r} n_i \log|\Sigma_i| - \frac{1}{2}\sum_{i=1}^{r} \mathrm{tr}\{\Sigma_i^{-1}(Y_i - XBZ_i)(Y_i - XBZ_i)'\} \\ &= -\tfrac{1}{2}\sum_{i=1}^{r} n_i \log|D_i| - \frac{1}{2}\sum_{i=1}^{r}\sum_{j=1}^{p}(e_{ji} - [A_{ji}^e]'\gamma_i)' D_i^{-1}(e_{ji} - [A_{ji}^e]'\gamma_i) \\ &= -\tfrac{1}{2}\sum_{i=1}^{r}\sum_{j=1}^{p} n_i h_j'\lambda_i - \frac{1}{2}\sum_{i=1}^{r}\sum_{j=1}^{p}(e_{ji} - \hat{e}_{ji})'(e_{ji} - \hat{e}_{ji})/\exp\{h_j'\lambda_i\} \end{aligned} \tag{7}$$

where

$$\begin{aligned} E_i &= Y_i - XBZ_i = (e_{1i}, e_{2i}, ..., e_{pi})', \\ A_{ji}^e &= \sum_{k=1}^{j-1} a_{jk} e_{ki}' \qquad \text{and} \qquad \hat{e}_{ji} = \sum_{k=1}^{j-1} \phi_{jki} e_{ki} \end{aligned}$$

for $i = 1, 2, ..., r$ and $j = 1, 2, ..., p$. Note that when $j = 1$ the sum notation $\sum_{k=1}^{0}$ means zero here.

Let $\mathrm{vec}(B) = (\beta_1', \beta_2', ..., \beta_r')'$ be the $mr \times 1$ vector by vectorizing the matrix B through column by column. Taking differentiation of ℓ in (7) with respect to $\mathrm{vec}(B)$, γ_i and λ_i leads to the following estimating equations, respectively,

$$\begin{aligned} \partial\ell/\partial\mathrm{vec}(B) &= \sum_{i=1}^{r} \mathrm{vec}[X'\Sigma_i^{-1}(Y_i - XBZ_i)Z_i'] = 0 \\ \partial\ell/\partial\gamma_i &= \sum_{j=1}^{p} A_{ji}^e(e_{ji} - [A_{ji}^e]'\gamma_i)/\sigma_{ji}^2 = 0 \\ \partial\ell/\partial\lambda_i &= -\tfrac{n_i}{2}\sum_{j=1}^{p} h_j + \frac{1}{2}\sum_{j=1}^{p}(e_{ji} - [A_{ji}^e]'\gamma_i)'(e_{ji} - [A_{ji}^e]'\gamma_i)h_j/\sigma_{ji}^2 = 0 \end{aligned} \tag{8}$$

The estimating equations above in general have no explicit solutions and certain numerical optimization procedures such as the Nowton-Raphson algorithm and Fisher-scoring algorithm are used instead. In the Appendix we show that the Fisher information matrix of the parameter $\theta = (\mathrm{vec}(B)'; \gamma_1', ..., \gamma_r'; \lambda_1', ..., \lambda_r')'$ must have the form

$$\mathcal{I} \equiv E(-\partial^2\ell/\partial\theta\partial\theta') = \begin{pmatrix} \mathcal{I}_{11} & 0 & 0 \\ 0 & \mathcal{I}_{22} & \mathcal{I}_{32}' \\ 0 & \mathcal{I}_{32} & \mathcal{I}_{33} \end{pmatrix} \tag{9}$$

where

$$\mathcal{I}_{11} \equiv E(-\partial^2\ell/\partial \text{vec}(B)\partial \text{vec}(B)') = \sum_{i=1}^{r} \left[(Z_i Z_i') \otimes (X' \Sigma_i^{-1} X)\right] \tag{10}$$

and the matrices $\mathcal{I}_{22}$ $(rq \times rq)$, $\mathcal{I}_{33}$ $(rd \times rd)$ and $\mathcal{I}_{32}$ $(rd \times rq)$ are block-diagonal with ith block being non-zero $(i = 1, 2, ..., r)$. Their detailed matrix forms are provided in the Appendix.

Based on the above equations, we propose the following Fisher-scoring algorithm to calculate the MLEs of the parameters in the GCM.

Algorithm:

Step 1: Given a starting value of θ, say $\theta_0 = ((\text{vec}(B^0)'; \gamma_1^{0'}, ..., \gamma_r^{0'}; \lambda_1^{0'}, ..., \lambda_r^{0'})'$, we form the covariance matrices $\Sigma_i^0 = \Sigma_i(\gamma_i^0, \lambda_i^0)$ using the modified Cholesky decomposition where $i = 1, 2, ..., r$.

Step 2: Use the following procedure

$$\begin{aligned} \text{vec}(B^1) = \text{vec}(B^0) &+ \left\{ \sum_{i=1}^{r} \left[(Z_i Z_i') \otimes (X'[\Sigma_i^0]^{-1} X)\right] \right\}^{-1} \\ &\times \left\{ \sum_{i=1}^{r} \text{vec}\left[X'[\Sigma_i^0]^{-1}(Y_i - XB^0 Z_i) Z_i'\right] \right\} \end{aligned}$$

and

$$\begin{pmatrix} \gamma^1 \\ \lambda^1 \end{pmatrix} = \begin{pmatrix} \gamma^0 \\ \lambda^0 \end{pmatrix} + \begin{pmatrix} \mathcal{I}_{22} & \mathcal{I}_{32}' \\ \mathcal{I}_{32} & \mathcal{I}_{33} \end{pmatrix}^{-1}_{\theta=\theta_0} \begin{pmatrix} \partial\ell/\partial\gamma \\ \partial\ell/\partial\lambda \end{pmatrix}_{\theta=\theta_0} \tag{11}$$

to update the parameter estimates of B, $\gamma = (\gamma_1', ..., \gamma_r')'$ and $\lambda = (\lambda_1', ..., \lambda_r')'$, respectively.

Step 3: Use the updated value $\theta_1 = ((\text{vec}(B^1)'; (\gamma^1)'; (\lambda^1)')'$ in Step 2 to replace θ_0 and then repeat Steps 1 and 2 above. These procedures are repeated until convergence for θ.

A by-product of the algorithm above is the asymptotic variance-covariance matrix of the MLE $\hat{\theta} = (\text{vec}(\widehat{B})'; \hat{\gamma}_1', ... \hat{\gamma}_r'; \hat{\lambda}_1', ..., \hat{\lambda}_r')'$, which is obtained by simply calculating the inverse of the Fisher information matrix (9), evaluated at the MLE $\hat{\theta}$. Regarding the starting values γ_i^0 and λ_i^0 $(i = 1, 2, ..., r)$, a convenient choice is $\gamma_1^0 = \cdots = \gamma_r^0 = 0$ and $\lambda_1^0 = \cdots = \lambda_r^0 = 0$ $(i = 1, 2, ..., r)$. In other words, the starting values of covariance matrices in all groups are chosen to be an identity matrix $\Sigma_1 = \cdots = \Sigma_r = I_p$. Alternatively, those can be chosen from the sample covariance matrices (Pourahmadi (2000)). Similarly, the regression coefficients B may start from the sample mean of each group.

4 Hypothesis tests and model selection

As mentioned in Section 1, most literature work in the GCM assumes a homogeneous covariance across all the groups, i.e., $\Sigma_1 = \Sigma_2 = \cdots = \Sigma_r$. Within the framework of the mean-covariance models in (5), this becomes a testable assumption. In fact, testing the homogeneity is equivalent to testing the following hypothesis

$$H_0 : \gamma_1 = \gamma_2 = \cdots = \gamma_r \qquad \text{and} \qquad \lambda_1 = \lambda_2 = \cdots = \lambda_r \tag{12}$$

where the parameters in the regression coefficients B are arbitrary. The likelihood ratio test statistic for testing the homogeneity (12) can be computed straightforwardly but its exact distribution is difficult to obtain. Instead, we could use the asymptotic likelihood ratio test. Let $\hat{\ell}_0$ and $\hat{\ell}_1$ be the maximized log-likelihoods under the null hypothesis H_0 and the alternative hypothesis H_1 of which H_0 is not true, respectively. The homogeneity hypothesis (12) can then be tested using $-2(\hat{\ell}_0 - \hat{\ell}_1) \sim \chi^2$ on $(r-1)(q+d)$ degrees of freedom. When H_0 is rejected, the usual hypothesis of covariance homogeneity is not true and heterogeneous covariances exist across the treatment groups. On the other hand, acceptance of H_0 in (12) implies no evidence to against the homogeneous covariance assumption. Note in this case the homogeneous covariance is modelled jointly with the mean without any specifications of structures. The mean structure, however, may vary from group to group in this case.

When the null hypothesis H_0 in (12) is rejected, we may need to further identify the type of dependence present by investigating the following hypotheses

$$H_0 : \lambda_1 = \lambda_2 = \cdots = \lambda_r \qquad \text{and} \qquad H_0 : \gamma_1 = \gamma_2 = \cdots = \gamma_r \tag{13}$$

The first hypothesis in (13) indicates that the innovation variances are the same across the treatment groups, while the second implies there is no difference for within-subject correlation among groups. Again, we can test the null hypotheses in (13) using the asymptotic likelihood ratio tests $-2(\hat{\ell}_0 - \hat{\ell}_1) \sim \chi^2$ on appropriate degrees of freedom.

In the GCM we may also be interested in testing whether or not both the mean and covariance structures are the same across the treatment groups. In other words, we want to test the following hypothesis

$$\begin{aligned} H_0 : \ & \beta_1 = \beta_2 = \cdots = \beta_r, \ \gamma_1 = \gamma_2 = \cdots = \gamma_r \text{ and} \\ & \lambda_1 = \lambda_2 = \cdots = \lambda_r \end{aligned} \tag{14}$$

where β_i is the ith column of the regression coefficients matrix B $(i = 1, 2, ..., r)$. Similarly, the hypothesis in (14) can be tested using the asymptotic likelihood ratio statistic $-2(\hat{\ell}_0 - \hat{\ell}_1) \sim \chi^2$ on $(r-1)(m+q+d)$ degrees of freedom.

When using polynomials of time to model the mean and covariance structures, obviously we need to choose the appropriate degrees of polynomials m, q and d in (6). In a spirit of Pan & MacKenzie (2003), we propose to use the following Baysian Information Criterion (BIC)

$$\text{BIC}(m, q, d) = -(2/n)\hat{\ell}_{\max} + (m+q+d)\{(\log n)/n\} \tag{15}$$

to choose the most appropriate degrees of polynomials, where $\hat{\ell}_{\max} = \ell(\hat{B}; \hat{\gamma}_1, ..., \hat{\gamma}_r; \hat{\lambda}_1, ..., \hat{\lambda}_r)$ is the maximized log-likelihood for the models with the specific degree trip (m, q, d) and $m + q + d$ is the number of parameters in the associated models, including polynomials of degree zero (i.e., intercept). The best triple of degrees, say (m^*, q^*, d^*), satisfies

$$(m^*, q^*, d^*) = \arg \min_{(m,q,d)} \{\text{BIC}(m, q, d)\} \tag{16}$$

where m, q and d lie in the range from 1 to p. The global search of the best triple, however, is computationally intensive because the number of maximizations required to find the best triple (m^*, q^*, d^*) is as large as p^3. Even if the number of repeated measurements, p, is mediate, the search for (m^*, q^*, d^*) may be highly computationally time-consuming.

Within the framework of linear regression models, Pan & MacKenzie (2003) proposed a profile search strategy that saturates the degrees m, q and d in pairs. Their study shows that the profile search is able to capture the best triple (m^*, q^*, d^*) in most circumstances. A significant advantage is that the number of maximizations for searching for (m^*, q^*, d^*) reduces to $3p+1$. For more details one can refer to Pan & MacKenzie (2003). In the real data analysis presented in the next section we adopt this strategy to locate the degree triple in the modelling of the mean and covariance structures. Our analysis confirms that the profile search does lead to the global best triple (m^*, q^*, d^*).

So far we have assumed that the degree triple (m, q, d) of polynomials is chosen to be the same across all treatment groups. It is not uncommon, however, that different treatment group may have a different degree triple. In principle the above parameter estimation procedure and model selection strategy are also suitable to this case but the search of the optimal degrees of polynomials is more computational intensive. On the other hand, testing the hypothesis of homogeneous covariance $H_o : \Sigma_1 = \Sigma_2 = \cdots = \Sigma_r$ no longer reduces to testing of the hypothesis (12) in this case because the dimension of the parameters γ_1, γ_2, ..., γ_r or λ_1, λ_2, ..., λ_r may not be the same. However, the asymptotic likelihood ratio test can be still applied to this case as long as the MLEs of the covariance matrices under the null and alternative hypotheses are obtained.

5 An Example

In this section we analyze Kenward's Cattle data (1987) using the joint mean-covariance modelling strategy. We also compare the data-driven approach to menu-selection methods through the data analysis.

Kenward (1987) analyzed an experiment in which cattle were assigned randomly to two treatment groups A and B, and their weights were recorded to study the effect of treatment on intestinal parasites. Thirty animals received treatment A and another thirty received treatment B. The animals were weighted 11 times over 133-day period at 0, 14, 28, 42, 56, 70, 84, 98, 112, 126 and 133 in days. Pourahmadi (2000) analyzed the data in treatment group A, modelling the covariance structure by adopting a saturated mean model and employing two cubic polynomials of time in the augmented regression model defined in (5), one for the autoregressive coefficients and another for the innovation variances. Below we analyze the two group data simultaneously using the proposed mean-covariance modelling strategy within the framework of growth curve models.

Firstly, we adopt Pan & MacKenzie (2003) BIC-based profile search strategy to select the best degree triple (m^*, q^*, d^*) of polynomials used in the modelling. We find that $(m^*, q^*, d^*) = (11, 5, 4)$, i.e., the mean has a saturated structure, and

Table 1. The maximum likelihood estimates of parameters involved in the autoregressive coefficients and innovation variances, i.e., $\hat{\gamma}_i' = (\hat{\gamma}_{i1}, ..., \hat{\gamma}_{i5})$ and $\hat{\lambda}_i' = (\hat{\lambda}_{i1}, ..., \hat{\lambda}_{i4})$ (estimated standard errors in parentheses)

Group	Parameter	$l = 1$	$l = 2$	$l = 3$	$l = 4$	$l = 5$
B	γ_{1l}	0.185(.006)	-1.628(.104)	1.568(.158)	-1.137(.188)	0.694(.231)
	λ_{1l}	3.518(.077)	0.672(.258)	2.229(.258)	-0.185(.258)	
A	γ_{2l}	0.182(.003)	-1.671(.061)	1.497(.106)	-1.031(.147)	0.365(.164)
	λ_{2l}	3.488(.078)	-1.172(.258)	0.234(.258)	-0.988(.258)	

the autoregressive coefficients and innovation variances are modelled in terms of quartic and cubic polynomials of lag/time, respectively. The minimum value of the BIC is BIC$(11, 5, 4) = 72.468$. Table 1 above reports the parameter estimates and the associated standard errors as well, while Figure 1 below gives the sample regressograms (solid points) and the fitted polynomial curves (solid curves) for the autoregressive coefficients and innovation variances in both groups. Note that the estimated coefficients presented in Table 1 are those pertaining to the orthogonal polynomials in order to avoid singularity of the design matrices.

Secondly, we study whether or not the covariances in Groups A and B are homogeneous, which is equivalent to testing if the null hypothesis (12) is true where $r = 2$. We therefore maximize the log-likelihood functions under the null and alternative hypotheses and obtain $\hat{\ell}_0 = -2120$ and $\hat{\ell}_1 = -2092.167$ so that $-2(\hat{\ell}_0 - \hat{\ell}_1) = 56$. We then compare this value to the Chi-square distribution with $(r-1)(q+d) = 9$ degrees of freedom, i.e., χ_9^2, and conclude that there is a highly significant evidence to against the null hypothesis. In other words, heterogeneous covariances exist for the two group cattle data. Furthermore, we may be concerned with whether either the autoregressive coefficients or innovation variances vary from group to group. We therefore test the hypotheses presented in (13) where $r = 2$. For the innovation variances, we test the null hypothesis $H_0 : \lambda_1 = \lambda_2$ against the alternative hypothesis $H_1 : \lambda_1 \neq \lambda_2$. Under the null and alternative hypotheses, the maximized log-likelihood functions are $\hat{\ell}_0 = -2118$ and $\hat{\ell}_1 = -2092.167$, respectively, and hence the testing statistic value is given by $-2(\hat{\ell}_0 - \hat{\ell}_1) = 58$. When comparing to χ_4^2, we know that the null hypothesis H_0 is rejected and conclude that the innovation variances are different in the two treatment groups. For the autoregressive coefficients, we test the null hypothesis $H_0 : \gamma_1 = \gamma_2$ versus the alternative hypothesis $H_1 : \gamma_1 \neq \gamma_2$. Under H_0 and H_1, the maximized log-likelihood functions are $\hat{\ell}_0 = -2093.113$ and $\hat{\ell}_1 = -2092.167$, respectively, so that the testing statistic takes the value $-2(\hat{\ell}_0 - \hat{\ell}_1) = 4$. When comparing it to χ_5^2, this time we have no evidence to against the null hypothesis H_0. In other words, the autoregressive coefficients are not significantly different in the two treatment groups, which confirms Pan and MacKenzie's (2003) finding where a group indicator is incorporated into covariance modelling.

Thirdly, in order to gain an insight of merit on the mean-covariance modelling we compare this strategy to several menu-selection approaches. Table 2

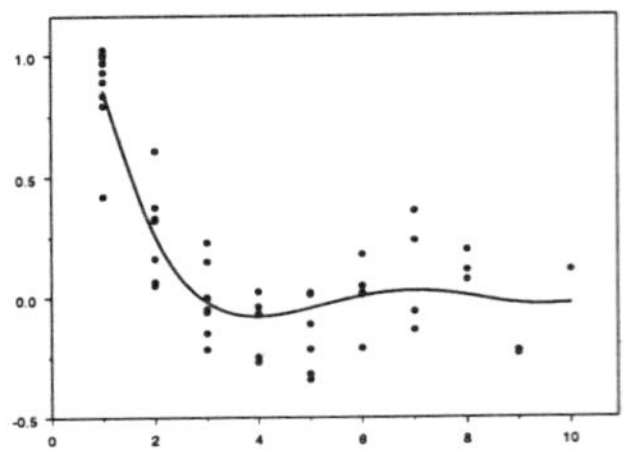

(a) auto. coeff. for Group A

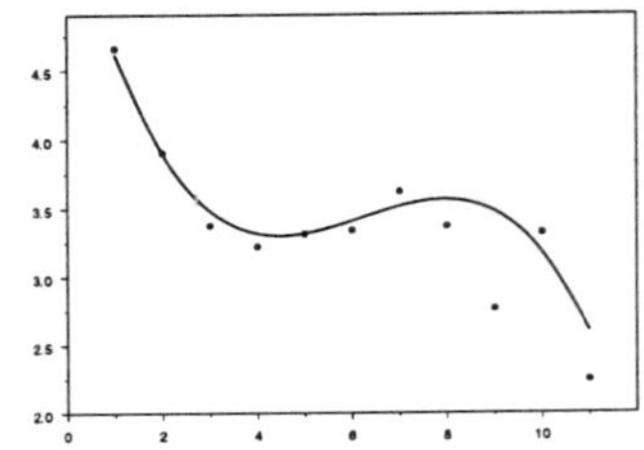

(b) log-innov. var. for Group A

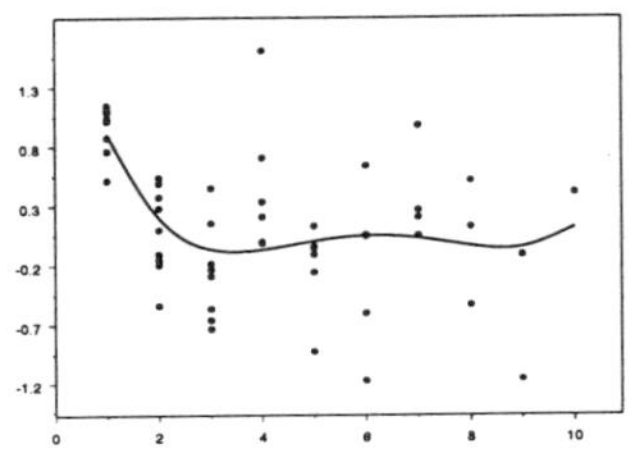

(c) auto. coeff. for Group B

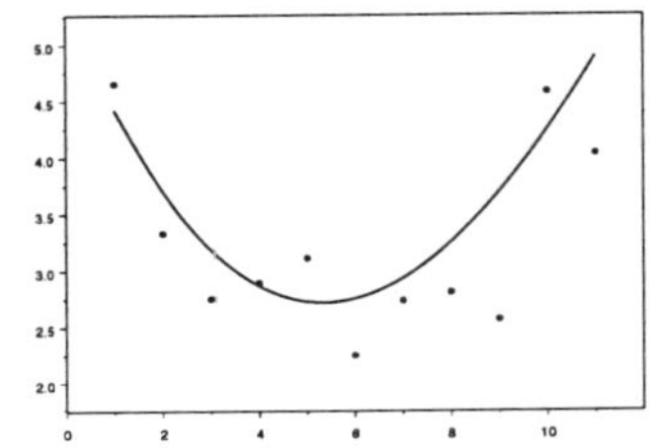

(d) log-innov. var. for Group B

Fig. 1. The Sample regressograms (solid points) and the fitted polynomials of lag/time with the best degree triple $(m^*, q^*, d^*) = (11, 5, 4)$. The panels (a) and (b) are those for Group A and (c) and (d) for Group B, respectively.

Table 2. Comparison between mean-covariance modelling and mean-selections where all models assume a saturated mean

Covariance	Parameter No.	log-likelihood	BIC
SCS	154	-2018.396	77.789
AR(1)	26	-2161.371	73.520
CS	26	-2409.231	81.782
Modelling	40	-2092.167	72.468

above presents the numbers of parameters, the maximized log-likelihood functions and the BIC values for the mean-covariance modelling and several specifications of covariance structures, including the Rao's simple covariance structure (SCS), AR(1) and and compound symmetry (CS). Note that since the mean structure is saturated, i.e., $m = 11$, the Rao's SCS is completely identical to the unstructured covariance (UC). In Table 2, when AR(1) or CS is used to specify the covariance structures it is possible that the two treatment groups may have different variances and correlation coefficients. Therefore the BIC values presented in Table 2 are the average in the two groups under these two circumstances. From Table 2 it is obvious that the mean-covariance modelling approach proposed in this paper performs better than the menu-selection approaches in terms of BIC model selection criterion. Figure 2 below compares those modelling approaches through different curves fitted to the sample autoregressive coefficients and innovation variances (dot points) for the two treatment groups, where the solid curve represents the fitting using the mean-covariance modelling technique, while dot, dash and dash-dot curves are the fitting with Rao's SCS, CS and AR(1) covariance specifications, respectively.

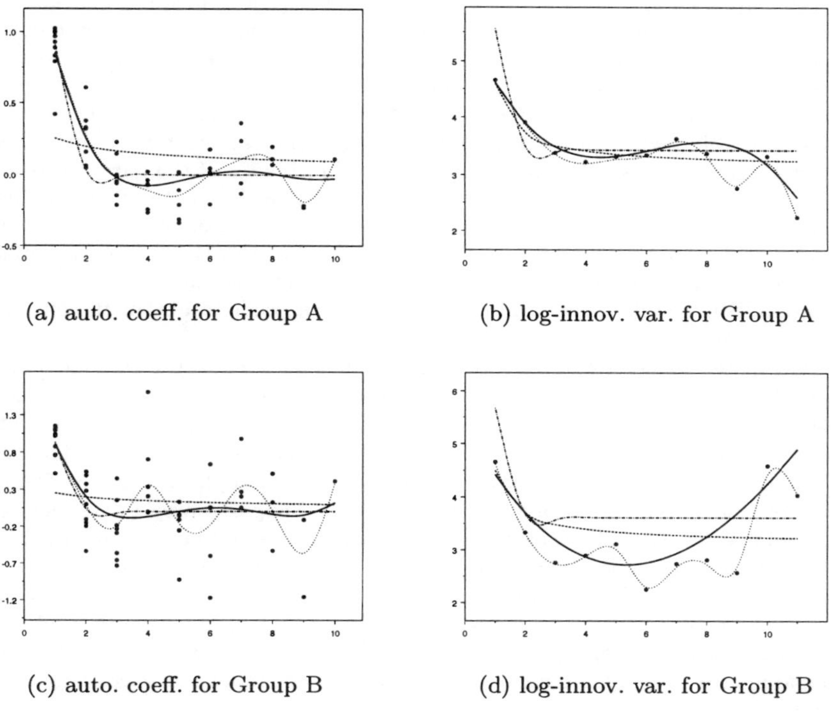

(a) auto. coeff. for Group A

(b) log-innov. var. for Group A

(c) auto. coeff. for Group B

(d) log-innov. var. for Group B

Fig. 2. The Sample regressograms (solid points), the fitted curves using the mean-covariance modelling technique (solid curve) and with covariance specification being Rao's SCS (dot curve), Compound Symmetry (dash curve) and AR(1) (dash-dot curve). The panels (a) and (b) are those for Group A while (c) and (d) are for Group B.

Again, Figure 2 shows that the mean-covariance modelling approach fits the data well. It also clearly shows the menu-selection approach may mis-specify the covariance structures. For example, neither the CS nor AR(1) is able to capture the true covariance structure, while Rao's SCS tends to over-fit the covariance structure. This, in turn, may influence the standard deviation of the estimated regression coefficients and accordingly may bias the statistical inferences of the GCM.

6 Discussion

In this paper we propose a data-driven approach to jointly model the mean and covariance structures for longitudinal data within the framework of growth curve models. The covariance matrices of repeated measures are reparameterized in terms of the modified Cholesky decomposition and the reparameterized parameters have a transparent statistical interpretation - autoregressive coefficients and innovation variances. These reparameterized parameters are further fitted using regression models. The maximum likelihood estimates of the parameters are obtained using the Fisher-scoring algorithm. Based on the joint models, the homogeneous covariance assumption becomes testable. The optimal joint model can be obtained by searching for the most appropriate degree triple of polynomials used for modelling the mean, autoregressive coefficients and innovation variances. A profile BIC-based search strategy is proposed in order to obtain the optimal degree triple.

Compared to menu-selection approaches, the joint mean-covariance modelling strategy specifies no structures on the covariance matrices of within-subject correlation. In contrast, menu-selection approaches assume a specific structure to the covariance matrices. When the structure is misspecified, statistical inferences of the regression coefficients may be incorrect. For example, within the framework of generalized estimating equations (GEE) Wang & Carey (2003) showed that misspecification of covariance structures produces too large standard deviations for regression coefficients and hence results in inefficient estimates. Ye & Pan (2004a) further modelled the mean and covariance structures in GEE using regression models. Very recently they (Ye & Pan (2004b)) proposed to use local-likelihood estimation approach developed by Fan, Farmen & Gijbels (1998) to nonparametrically model the mean and covariance structures for large longitudinal data.

For the growth curve models, Rao's simple covariance structure plays a special role in the sense that within this sub-covariance space the MLE of regression coefficients no longer depends on the choice of covariance structures. It is more interesting to see how this specific covariance structure can be characterized in terms of the autoregressive coefficients and innovation variances. In other words, we want to know under which condition satisfied by the autoregressive coefficients and innovation variances the covariance falls into Rao's simple covariance space. This issue awaits for further exploitation.

To our knowledge, this is the first article that addresses the joint model of mean-covariance structures in the growth curve models in terms of the data-driven regression technique. Of course, under certain specifications of covariance structures statistical modelling was widely discussed in the literature for the GCM, multilevel models, structural equation models, etc. Modelling mean-covariance structures without any specifications of covariance structure distinguishes our approach from the literature work. In addition, our previous experience on statistical diagnostics in the GCM (e.g., Pan & Fang (2002); Pan *et al*, 1997, 1998, 1999) shows that the covariance structure plays an important role in outlier detection and influential observation identification. The diagnostics issue studied within the framework of joint mean-covariance modelling will be reported in a follow-up paper.

Acknowledgments

We would like to thank Prof Kaitai Fang for his nice collaboration in the area of multivariate analysis, growth curve models and statistical diagnostics during the past decade. Our joint work with him already led to a series of publications including papers (e.g., Pan *et al*, 1997, 1998, 1999) and a Springer Series book (Pan & Fang (2002)). We also thank the editors and a referee for their helpful comments on the manuscript. This research work is funded by a EPSRC grant of the UK (GR/R69884).

Appendices: Derivation of the Fisher Information matrix

First, based on (8) we have the second-order derivatives of the log-likelihood with respect to vec(B), γ_i and λ_i as given below

$$\begin{aligned} \partial^2\ell/\partial\text{vec}(B)\partial\text{vec}'(B) &= -\sum_{i=1}^{r}\left[(Z_iZ_i')\otimes(X'\Sigma_i^{-1}X)\right] \\ \partial^2\ell/\partial\gamma_i\partial\gamma_i' &= -\sum_{j=1}^{p}[A_{ji}^e][A_{ji}^e]'/\sigma_{ji}^2 \qquad\qquad (A.1) \\ \partial^2\ell/\partial\lambda_i\partial\lambda_i' &= -\tfrac{1}{2}\sum_{j=1}^{p}(e_{ji}-[A_{ji}^e]'\gamma_i)'(e_{ji}-[A_{ji}^e]'\gamma_i)(h_jh_j')/\sigma_{ji}^2 \end{aligned}$$

where $A_{ji}^e = \sum_{k=1}^{j-1} a_{jk}e_{ki}'$ and e_{ji}' is the kth row of the residual matrix $E_i = Y_i - XBZ_i$ ($j = 1, 2, ..., p$; $i = 1, 2, ..., r$). Similarly, the second-order mixed derivatives of ℓ with respect to vec(B), γ_i and λ_i can be written into

$$\begin{aligned}
\partial^2\ell/\partial\gamma_i\partial\text{vec}'(B) &= -\sum_{j=1}^{p}\Big\{[A^x_{ji}][(e_{ji}-[A^e_{ji}]'\gamma_i)'Z_i'\otimes I_m] \\
&\qquad + [A^e_{ji}][Z_i'\otimes(x_j-[A^x_{ji}]'\gamma_i)]\Big\}/\sigma^2_{ji} \\
\partial^2\ell/\partial\lambda_i\partial\text{vec}'(B) &= -\sum_{j=1}^{p}h_j\text{vec}'\Big[(x_j-[A^x_{ji}]'\gamma_i)(e_{ji}-[A^e_{ji}]'\gamma_i)'Z_i'\sigma^2_{ji}\Big] \\
\partial^2\ell/\partial\lambda_i\partial\gamma_i' &= -\sum_{j=1}^{p}h_j(e_{ji}-[A^e_{ji}]'\gamma_i)'[A^e_{ji}]'/\sigma^2_{ji}
\end{aligned} \tag{A.2}$$

where $A^x_{ji}=\sum_{k=1}^{j-1}a_{jk}x_k'$ and $X=(x_1,...,x_p)'$.

Second, when taking expectation to the first equation of (A.1) it obviously results in $\mathcal{I}_{11}$ as given by (10) because the second-order derivative of ℓ with respect to vec(B) is a constant. On the other hand, by noting that $E(e_{ji})=0$ and $E(A^e_{ji})=\sum_{k=1}^{j-1}a_{jk}E(e_{ki}')=0$ we have $\mathcal{I}_{21}=0$ and $\mathcal{I}_{31}=0$. Therefore, the Fisher information matrix must have the form given in (9), in which the matrices $\mathcal{I}_{22}$, $\mathcal{I}_{32}$ and $\mathcal{I}_{33}$ are obviously block-diagonal due to the forms of score functions given in (8).

The ith diagonal block of $\mathcal{I}_{22}$ $(i=1,2,...,r)$ can be calculated through

$$\begin{aligned}
\mathcal{I}_{22}(i,i) &= E[-\partial^2\ell/\partial\gamma_i\partial\gamma_i'] = \sum_{j=1}^{p}E(A^e_{ji}[A^e_{ji}]')/\sigma^2_{ji} \\
&= \sum_{j=1}^{p}E\Big([\sum_{k=1}^{j-1}a_{jk}e_{ki}'][\sum_{l=1}^{j-1}a_{jl}e_{l_i}']'\Big)/\sigma^2_{ji} \\
&= \sum_{j=1}^{p}\sum_{k=1}^{j-1}\sum_{l=1}^{j-1}a_{jk}a_{jl}'E(e_{ki}'e_{li})/\sigma^2_{ji} \\
&= \sum_{j=1}^{p}\sum_{k=1}^{j-1}\sum_{l=1}^{j-1}a_{jk}a_{jl}'(n_i\sigma_{kli})/\sigma^2_{ji} \\
&= n_i\sum_{j=1}^{p}W_{ji}/\sigma^2_{ji} = n_iW_i
\end{aligned} \tag{A.3}$$

where

$$W_{ji}=\sum_{k=1}^{j-1}\sum_{l=1}^{j-1}a_{jk}a_{jl}'\sigma_{kli} \qquad \text{and} \qquad W_i=\sum_{j=1}^{p}W_{ji}/\sigma^2_{ji} \tag{A.4}$$

where σ_{kli} is the (k,l)th element of the matrix Σ_i. On the other hand, since $T_iE_i=E_i-\widehat{E}_i\sim N_{p\times n_i}(0,D_i,I_{n_i})$ where $\widehat{E}_i=(\hat{e}_{1i},\hat{e}_{2i},...,\hat{e}_{pi})'$ and $\hat{e}_{ji}=\sum_{k=1}^{j-1}\phi_{jki}e_{ki}$ we have $e_{ji}-\hat{e}_{ji}\sim N_{n_i}(0,\sigma^2_{ji}I_{n_i})$ so that

$$(e_{ji}-\hat{e}_{ji})'(e_{ji}-\hat{e}_{ji})/\sigma^2_{ji}\sim\chi^2_{n_i}$$

which implies that $E[(e_{ji}-\hat{e}_{ji})'(e_{ji}-\hat{e}_{ji})/\sigma^2_{ji}]=n_i$. Accordingly, the ith diagonal block of $\mathcal{I}_{33}$ $(i=1,2,...,r)$ can be expressed as

$$\begin{aligned}
\mathcal{I}_{33}(i,i) &= E[-\partial^2\ell/\partial\lambda_i\partial\lambda_i'] \\
&= \tfrac{1}{2}\sum_{j=1}^{p} E\Big[(e_{ji}-[A_{ji}^e]'\gamma_i)'(e_{ji}-[A_{ji}^e]'\gamma_i)/\sigma_{ji}^2\Big](h_jh_j') \\
&= \tfrac{1}{2}\sum_{j=1}^{p} E\Big[(e_{ji}-\hat{e}_{ji})'(e_{ji}-\hat{e}_{ji})/\sigma_{ji}^2\Big](h_jh_j') \qquad (A.5)\\
&= \tfrac{n_i}{2}\sum_{j=1}^{p}(h_jh_j') = \frac{n_i}{2}H'H
\end{aligned}$$

where $H = (h_1, h_2, ..., h_p)'$ is the design matrix involved in the modelling of innovation variances.

Finally, the ith diagonal block of $\mathcal{I}_{32}$ is given by

$$\begin{aligned}
\mathcal{I}_{32}(i,i) &= E[-\partial^2\ell/\partial\lambda_i\partial\gamma_i'] \\
&= \sum_{j=1}^{p} h_j E\Big\{(e_{ji}-[A_{ji}^e]'\gamma_i)'[A_{ji}^e]'\Big\}/\sigma_{ji}^2 \\
&= \sum_{j=1}^{p} h_j\Big(E[A_{ji}^e e_{ji}] - E([A_{ji}^e][A_{ji}^e]')\gamma_i\Big)'/\sigma_{ji}^2 \qquad (A.6)\\
&= n_i\sum_{j=1}^{p} h_j(A_{ji}^\sigma - W_{ji}\gamma_i)'/\sigma_{ji}^2
\end{aligned}$$

where $E([A_{ji}^e][A_{ji}^e]') = n_iW_{ji}$ is already showed in (A.3) while

$$\begin{aligned}
E[A_{ji}^e e_{ji}] &= E\Big(\sum_{k=1}^{j-1} a_{jk}e_{ki}'e_{ji}\Big) = \sum_{k=1}^{j-1} a_{ji}E(e_{ki}'e_{ji}) \\
&= \sum_{k=1}^{j-1} a_{ji}(n_i\sigma_{jki}) = n_i\sum_{k=1}^{j-1} a_{jk}\sigma_{jki} \equiv n_iA_{ji}^\sigma.
\end{aligned}$$

In summary, the Fisher information matrix has the form in (9) in which the block forms of $\mathcal{I}_{11}$, $\mathcal{I}_{22}$, $\mathcal{I}_{33}$ and $\mathcal{I}_{32}$ are provided by (10), (A.3), (A.5) and (A.5), respectively.

References

Fan, J., Farmen, M. & Gijbels, I. (1998). Local maximum likelihood estimation and inference. *Journal of the Royal Statistical Society* B **60** 591–608.

Fujikoshi, Y., Kanda, T. & Tanimura, N. (1990). The growth curve model with an autoregressive covariance structure. *Annals of Institute of Statistical Mathematics* **42** 533–542.

Gleser, L. J. & Olkin, I. (1970). Linear models in multivariate analysis. In *Essays in Probability and Statistics* (R. C. Bose, I. M. Chakravarti, P. C. Mahalanobis, C. R. Rao & K. J. C. Smith, eds.) 267–292. Chapel Hill, North Carolina: University of North Carolina Press.

Kariya, T. (1985). *Testing in the multivariate general linear model*, Kinokuniya Printing Company, Tokyo, Japan.

Kenward, M. G. (1987). A method for comparing profiles of repeated measurements. *Applied Statistics* **36** 296–308.

Keramidas, E. M. & Lee, J. C. (1995). Selection of a covariance structure for growth-curves. *Biometrical Journal* **37** 783–797.

Khatri, C. G. (1966). A note on a MANOVA model applied to problems in growth curve. *Annals of Institute of Statistical Mathematics* **18** 75–86.

Lee, J. C. (1988). Prediction and estimation of growth curve with special covariance structure. *Journal of the American Statistical Association* **83** 432–440.

Lee, J. C. (1991). Tests and model selection for the general growth curve model. *Biometrics* **47** 147–159.

Lundbye-Christense S. (1991). A multivariate growth curve model for pregnancy. *Biometrics* **47** 637–657.

Pan, J. & Fang, K. T. (2002) *Growth curve models and statistical diagnostics.* New York: Springer Series in Statistics.

Pan, J., Fang, K. T. & von Rosen, D. (1997). Local influence assessment in the growth curve model with unstructured covariance. *Journal of Statistical Planning and Inference* **62** 263–278.

Pan, J., Fang, K. T. & von Rosen, D. (1998). On the posterior distribution of the covariance matrix of the growth curve model. *Statistics & Probability Letters* **38** 33–39.

Pan, J., Fang, K. T. & von Rosen, D. (1999). Bayesian local influence in the growth curve model with unstructured covariance. *Biometrical Journal* **41** 641–658.

Pan, J. & MacKenzie, G. (2003). On modelling mean–covariance structures in longitudinal studies. *Biometrika* **90** 239–244.

Potthoff, R. F. & Roy, S. N. (1964). A generalized multivariate analysis of variance model useful especially for growth curve problems *Biometrika* **51** 313–326.

Pourahmadi, M. (1999). Joint mean–covariance models with applications to longitudinal data: Unconstrained parameterisation. *Biometrika* **86** 677–90.

Pourahmadi, M. (2000). Maximum likelihood estimation of generalised linear models for multivariate normal covariance matrix. *Biometrika* **87** 425–435.

Rao, C. R. (1965). The theory of least squares when the parameters are stochastic and its application to the analysis of growth curves. *Biometrika* **52** 447–458.

Rao, C. R. (1966). Covariance adjustment and related problems in multivariate analysis. In *Multivariate Analysis* (P. Krishnaiah, ed.) 87–103. Academic Press, New York.

von Rosen, D. (1989). Maximum likelihood estimates in multivariate linear normal model. *Journal of Multivariate Analysis* **31** 187–200.

Ye, H. & Pan, J. (2004a). Modelling covariance structures in generalized estimating equations for longitudinal data. *Proceedings of the 19th International Workshop of Statistical Modelling* 244-248. Florence, Italy, July.

Ye, H. & Pan, J. (2004b). Nonparametrically modelling the mean and covariance structures for large longitudinal data. *Reprint.*

Wang, Y. & Carey, V. (2003). Working correlation structure misspecification, estimation and covariate design: implications for generalized estimating equations performance. *Biometrika* **90** 29–41.

On Kotz-Type Elliptical Distributions

Tõnu Kollo[1] and Anu Roos[2]

[1] University of Tartu, Tartu, Estonia (tonu.kollo@ut.ee)
[2] University of Tartu, Tartu, Estonia (anu.roos@ut.ee)

Summary. In the paper Kotz-type elliptical distributions are examined. Expressions of multivariate moments and cumulants in matrix form are given as well as kurtosis characteristics. It is proved that marginal distribution of a Kotz distribution is a mixture of a normal and a Kotz distribution. Also basic formulae for matrix Kotz distribution are given and a simulation rule based on the Metropolis-Hastings algorithm is presented.

Key words: Kotz-type distribution, marginal distribution, matrix Kotz distribution, multivariate kurtosis, moments and cumulants, simulation rule.

2000 Mathematics Subject Classification: 62H10, 60E05, 62E15

1 Introduction

Kotz-type distributions were introduced by Kotz (1975) in a special case and have become a topic of interest in applications in last ten years. The distributions present a wide 5-parameter class of elliptical distributions which contains the multivariate normal family. These five parameters include a location vector-parameter, a matrix scale (dispersion) parameter and three scalar parameters to give flexibility to the model. In Fang, Kotz & Ng (1990), Section 3.2 most of the results on Kotz-type distributions up to 1990 are collected. The characteristic function has been first derived in a special case by Iyengar & Tong (1989) and later obtained in the general case by Li (1993, 1994). The class of Kotz-type distributions was characterized by Liang & Bentler (1998). The relation of Kotz-type distributions with extreme value distributions was studied by Kotz & Nadarajah

Contemporary Multivariate Analysis and Experimental Design—In Honor Celebration of Professor Kai-Tai Fang's 65th birthday. Edited by Jianqing Fan and Gang Li. The World Scientific Publisher, 2005.

(2001). Moment expressions of elliptical distributions have got much attention. General expressions of moments and cumulants up to the order eight were presented in matrix form by Traat (1990). Moments of an elliptically distributed vector and its quadric forms are presented in Li (1990) while moment expressions for a matrix elliptical distribution can be found in Fang & Zhang (1990) and in a monograph by Gupta & Varga (1993). Expressions of the moments and cumulants of Kotz-type distributions are straightforwardly obtained from the corresponding general expressions for elliptical distributions using the characteristic generator. Estimation and inference problems have not got much attention, we point out here contributions of Zhao (1994) who considered estimation of the dispersion parameter and Fang & Li (1999) on Bayesian inference problems.

Several early applications are referred to in Koutras (1986). Recently the matrix Kotz distribution became a useful tool in parameter approximation in the Growth Curve model (Kollo, Roos & von Rosen (2004)). A more detailed presentation can be found in Kollo & von Rosen (2004). A review paper on this class of distributions has been written by Nadarajah (2003), where references to different recent applications can also be found.

We are going to concentrate on a special case which is of special interest in parameter approximation in the Growth Curve model. The obtained results and used procedures can be carried over to other fixed sets of parameters. The main text of the paper is divided into five sections. In Section 2 necessary notation and notions are introduced, Section 3 presents moments and cumulants of the distribution, also kurtosis characteristics are found. Marginal distributions are presented in Section 4. Section 5 deals with the matrix Kotz-type distributions and finally, in Section 6, a simulation rule for Kotz-type distributions is given.

2 Notation and notions

Kotz-type distributions form a subclass of elliptical (or elliptically contoured) distributions (see Fang, Kotz & Ng (1990)) or Anderson (2003). This class of distributions can be defined by the probability density function (pdf).

Definition 1 *Let $\boldsymbol{x} = (X_1, \dots, X_p)'$ be a random $p-$vector. The vector $\boldsymbol{x}$ has a Kotz-type distribution with the parameters $\boldsymbol{\mu}$, $\boldsymbol{V}$, N, s, r, if the density $f_{\boldsymbol{x}}(\boldsymbol{x})$ of $\boldsymbol{x}$ has a form:*

$$f_{\boldsymbol{x}}(\boldsymbol{x}) = C_p|\boldsymbol{V}|^{-\frac{1}{2}}[(\boldsymbol{x}-\boldsymbol{\mu})'\boldsymbol{V}^{-1}(\boldsymbol{x}-\boldsymbol{\mu})]^{N-1}\exp\{(-r[(\boldsymbol{x}-\boldsymbol{\mu})'\boldsymbol{V}^{-1}(\boldsymbol{x}-\boldsymbol{\mu})]^s)\},$$
$$r, s > 0,\ 2N + p > 2, \qquad (1)$$

where C_p is a normalizing constant:

$$C_p = \frac{s\Gamma(p/2)}{\pi^{p/2}\Gamma(2N+p-2/2s)} r^{(2N+p-2/2s)}. \qquad (2)$$

If we look the pdf as a univariate function of $(\boldsymbol{x}-\boldsymbol{\mu})'\boldsymbol{V}^{-1}(\boldsymbol{x}-\boldsymbol{\mu})$, we get a function $g(u)$, which is called a *density generator*:

$$g(u) = C_p u^{N-1} \exp(-ru^s), \quad r, s > 0,\ 2N + p > 2. \tag{3}$$

We write $\boldsymbol{x} \sim K_p(\boldsymbol{\mu}, \boldsymbol{V}, N, s, r)$ if $\boldsymbol{x}$ has pdf (1). When $N = 1$, $s = 1$ and $r = \frac{1}{2}$, we get the multivariate normal distribution.

The Kotz-type distribution $K_p(\boldsymbol{0}, \boldsymbol{I}_p, N, s, r)$ is called the *Kotz-type spherical distribution.* Let $\boldsymbol{y} \sim K_p(\boldsymbol{0}, \boldsymbol{I}_p, N, s, r)$, then it follows from the properties of density (1), that

$$\boldsymbol{x} = \boldsymbol{A}\boldsymbol{y} + \boldsymbol{\mu} \sim K_p(\boldsymbol{\mu}, \boldsymbol{V}, N, s, r), \text{ where } \boldsymbol{A}\boldsymbol{A}' = \boldsymbol{V}.$$

Next we describe shortly the influence of the parameters N, s and r to the shape of the density $f_{\boldsymbol{x}}(\boldsymbol{x})$ in the univariate case. For simplicity we take $\boldsymbol{\mu} = 0$. Parameter N determines the modality of the distribution: it is unimodal when $N \leq 1$ and bimodal when $N > 1$. The parameters s and r regulate dispersion of the distribution, s can be considered as a dispersion characteristic additional to $\boldsymbol{V}$ while r can be called the kurtosis parameter.

In the bivariate case meaning of the parameters N, s and r is the same as in the univariate case with just one difference that N regulates the volume of the "hole" at the origin of the coordinates. Shape of the distribution for two sets of parameters is shown in Figures 1 and 2.

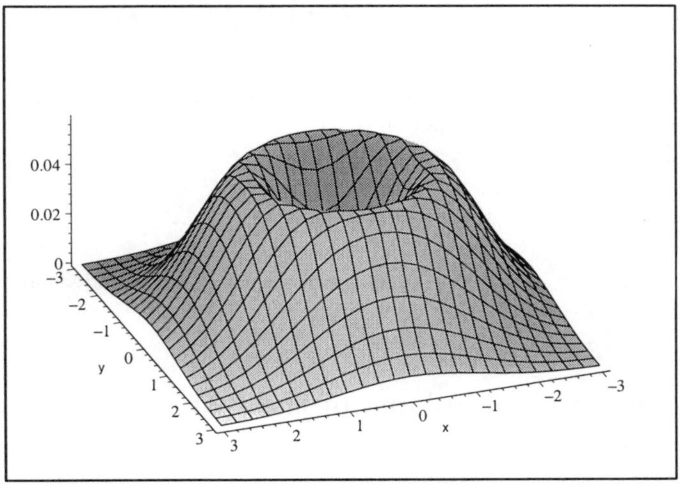

Fig. 1. Kotz distribution, parameters $N = 2$, $s = 1$, $r = \frac{1}{2}$

The characterization of a Kotz-type distribution through the characteristic function is more complicated than via the density function. In the case $s = 1$

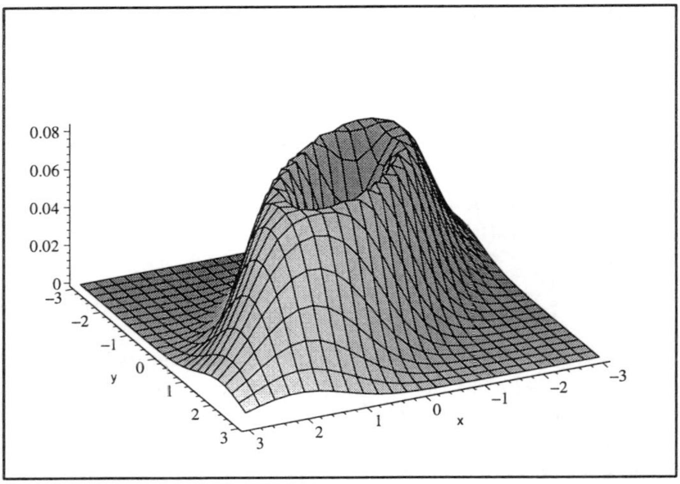

Fig. 2. Kotz distribution, parameters $N = 2$, $s = 1$, $r = \frac{1}{2}$, correlated components

the characteristic function $\varphi_{\boldsymbol{x}}(\boldsymbol{t})$ can be found from the book Fang, Kotz & Ng (1990), Section 3.2, the formula is quite complicated.

The expression simplifies considerably when $N = 2$, $r = \frac{1}{2}$:

$$\varphi_{\boldsymbol{x}}(\boldsymbol{t}) = \exp(i\boldsymbol{t}'\boldsymbol{\mu}) \exp\left(-\frac{\boldsymbol{t}'\boldsymbol{V}\boldsymbol{t}}{2}\right)\left(1 - \frac{\boldsymbol{t}'\boldsymbol{V}\boldsymbol{t}}{p}\right). \tag{4}$$

In this case (as when we considered the density) we may also look the characteristic function as a univariate function of $t'Vt$. Then we get a function $\phi(u)$ which is called the *characteristic generator*:

$$\phi(u) = \exp\left(-\frac{u}{2}\right)\left(1 - \frac{u}{p}\right). \tag{5}$$

We say that $\boldsymbol{x} \sim K_p(\boldsymbol{\mu}, \boldsymbol{V}, 2, 1, \frac{1}{2})$ has a *Kotz distribution* and denote it $\boldsymbol{x} \sim K_p(\boldsymbol{x}, \boldsymbol{V})$.

3 Moments and cumulants of the Kotz distribution

Consider the Kotz distribution, $\boldsymbol{x} \sim K_p(\boldsymbol{\mu}, \boldsymbol{V})$. Moments can be obtained by differentiating the characteristic function (4). Another way is to find the moments using general formulae for an elliptical distribution. We are going to use the second possibility and get the expressions straightforwardly from Kollo & von Rosen (2004). The first four moments are given in the next theorem.

Theorem 1. *Let random vector $\boldsymbol{x} \sim K_p(\boldsymbol{\mu}, \boldsymbol{V})$, then*

$$E(\boldsymbol{x}) = \boldsymbol{\mu}, \tag{6}$$

$$m_2(\boldsymbol{x}) = \left(1 + \frac{2}{p}\right)\boldsymbol{V} + \boldsymbol{\mu}\boldsymbol{\mu}', \tag{7}$$

$$m_3(\boldsymbol{x}) = \boldsymbol{\mu}(\boldsymbol{\mu}')^{\otimes 2} + \left(1 + \frac{2}{p}\right)(\boldsymbol{\mu}' \otimes \boldsymbol{V} + \boldsymbol{V} \otimes \boldsymbol{\mu}' + \boldsymbol{\mu}\mathrm{vec}'\boldsymbol{V}), \tag{8}$$

$$\begin{aligned} m_4(\boldsymbol{x}) = {} & \boldsymbol{\mu}(\boldsymbol{\mu}')^{\otimes 3} + \left(1 + \frac{2}{p}\right)\boldsymbol{\mu}(\mathrm{vec}\boldsymbol{V} \otimes \boldsymbol{\mu}')(\boldsymbol{I}_{p^3} + \boldsymbol{I}_p \otimes \boldsymbol{K}_{p,p}) \\ & + \left(1 + \frac{2}{p}\right)[(\boldsymbol{\mu}')^{\otimes 2} \otimes \boldsymbol{V} + \boldsymbol{\mu}' \otimes \boldsymbol{V} \otimes \boldsymbol{\mu}' + \boldsymbol{\mu}'\boldsymbol{\mu} \otimes \mathrm{vec}'\boldsymbol{V}) + \boldsymbol{V} \otimes (\boldsymbol{\mu}')^{\otimes 2}] \\ & + \left(1 + \frac{4}{p}\right)[(\boldsymbol{V} \otimes \mathrm{vec}'\boldsymbol{V}) + (\mathrm{vec}'\boldsymbol{V} \otimes \boldsymbol{V})(\boldsymbol{I}_{p^3} + \boldsymbol{I}_p \otimes \boldsymbol{K}_{p,p})]. \end{aligned} \tag{9}$$

In the theorem $\boldsymbol{K}_{p,p}$ denotes the *commutation matrix* (see Harville (1997), for example). The expressions of the dispersion matrix and the fourth central moment are given in the next theorem.

Theorem 2. *Let random vector $\boldsymbol{x} \sim K_p(\boldsymbol{\mu}, \boldsymbol{V})$, then*

$$D(\boldsymbol{x}) = \left(1 + \frac{2}{p}\right)\boldsymbol{V}, \tag{10}$$

$$\overline{m}_4(\boldsymbol{x}) = \left(1 + \frac{4}{p}\right)[(\boldsymbol{V} \otimes \mathrm{vec}'\boldsymbol{V}) + (\mathrm{vec}'\boldsymbol{V} \otimes \boldsymbol{V})(\boldsymbol{I}_{p^3} + \boldsymbol{I}_p \otimes \boldsymbol{K}_{p,p})]. \tag{11}$$

All odd central moments equal to zero.

The first cumulants are presented in the next theorem.

Theorem 3. *Let random vector $\boldsymbol{x} \sim K_p(\boldsymbol{\mu}, \boldsymbol{V})$, then*

$$c_1(\boldsymbol{x}) = \boldsymbol{\mu}, \tag{12}$$

$$c_2(\boldsymbol{x}) = D(\boldsymbol{x}) = \left(1 + \frac{2}{p}\right)\boldsymbol{V}, \tag{13}$$

$$\begin{aligned} c_4(\boldsymbol{x}) = {} & -\frac{4}{p^2}[(\boldsymbol{V} \otimes \mathrm{vec}'\boldsymbol{V}) \\ & +(\mathrm{vec}'\boldsymbol{V} \otimes \boldsymbol{V})(\boldsymbol{I}_{p^3} + \boldsymbol{I}_p \otimes \boldsymbol{K}_{p,p})]. \end{aligned} \tag{14}$$

All odd cumulants starting from $c_3(\boldsymbol{x})$ equal to zero.

We are also interested in kurtosis of the Kotz distribution. We use two measures for that. First we find the kurtosis parameter κ, which is defined for an elliptical distribution using the characteristic generator (Muirhead (1982), for example).

Definition 2 *Let a vector $\boldsymbol{x}$ be elliptically distributed with the characteristic generator $\phi(u)$, then*

$$\kappa = \frac{\phi''(0) - (\phi'(0))^2}{(\phi'(0))^2}. \tag{15}$$

is called the kurtosis parameter.

By straightforward calculations we get from (15) and (5) the value of the kurtosis parameter for the Kotz distribution:

$$\kappa = -\frac{4}{(p+2)^2}.$$

Another multivariate kurtosis measure was defined by Mardia (1970). We use the definition from Kollo & Srivastava (2000) to present Mardia's skewness and kurtosis measures through multivariate moments.

Definition 3 *Let $\mathbf{x}$ be a random p-vector with mean vector $\boldsymbol{\mu}$ and dispersion matrix $\boldsymbol{\Sigma}$ and $\boldsymbol{y} = \boldsymbol{\Sigma}^{-\frac{1}{2}}(\boldsymbol{x} - \boldsymbol{\mu})$. Let the third and fourth moment of vector $\boldsymbol{y}$ be $m_3(\boldsymbol{y})$ and $m_4(\boldsymbol{y})$, respectively, and the third and fourth cumulant $c_3(\boldsymbol{y})$ and $c_4(\boldsymbol{y})$, respectively. Then skewness $\beta_{1,p}$ and kurtosis $\beta_{2,p}$ of the vector $\mathbf{x}$ are defined by equalities:*

$$\beta_{1,p} = tr[c_3(\boldsymbol{y})^T c_3(\boldsymbol{y})] = tr[m_3(\boldsymbol{y})^T m_3(\boldsymbol{y})],$$
$$\beta_{2,p} = tr[m_4(\boldsymbol{y})] = tr[c_4(\boldsymbol{y})] + p^2 + 2p.$$

From Definition 3 we get an expression for the kurtosis of the Kotz distribution:

$$\beta_{2,p} = p^2 + 2p - \frac{4p}{2+p}.$$

Both measures of kurtosis, κ and $\beta_{2,p}$ show that the Kotz distribution has lighter tails than the normal distribution, because the same measures for normal distribution are $\kappa = 0$ and $\beta_{2,p} = p^2 + 2p$.

4 Marginal distribution

In this section we are going to derive marginal distribution of the Kotz distribution. It comes out that marginal distributions of the Kotz distribution do not follow the Kotz distribution of lower order and can be presented as a mixture of a normal and a Kotz distribution.

Theorem 4. *Let $\mathbf{x} \sim \mathbf{K_p}(\boldsymbol{\mu}, \boldsymbol{V})$ be a random p-vector. Let $\boldsymbol{x} = (\boldsymbol{x}_1, \boldsymbol{x}_2)'$, where $\boldsymbol{x}_1 : p_1 \times 1$ and $\boldsymbol{x}_2 : p_2 \times 1$, let $\boldsymbol{\mu} = (\boldsymbol{\mu}_1, \boldsymbol{\mu}_2)'$ be partitioned the same way and let*

$$\boldsymbol{V} = \begin{pmatrix} \boldsymbol{V}_{11} & \boldsymbol{V}_{12} \\ \boldsymbol{V}_{21} & \boldsymbol{V}_{22} \end{pmatrix}, \quad \textit{with dimensions} \quad \begin{pmatrix} p_1 \times p_1 & p_1 \times p_2 \\ p_2 \times p_1 & p_2 \times p_2 \end{pmatrix}.$$

Then the distribution of the vector $\boldsymbol{x}_1$ is a mixture of a Kotz distribution and a normal distribution with parameters $\boldsymbol{\mu}_1$ and $\boldsymbol{V}_{11}$, with weights $\frac{p_1}{p}$ and $\frac{p_2}{p}$, respectively.

Proof

Denote $\boldsymbol{z} = \boldsymbol{x} - \boldsymbol{\mu}$, $\boldsymbol{z}_1 = \boldsymbol{x}_1 - \boldsymbol{\mu}_1$ and $\boldsymbol{z}_2 = \boldsymbol{x}_2 - \boldsymbol{\mu}_2$.

We use a formula from Horn & Johnson (1989, p.31) for the inverse of a partitioned matrix:

$$\boldsymbol{V}^{-1} = \begin{pmatrix} \boldsymbol{V}_{11}^{-1} - \boldsymbol{V}_{11}^{-1}\boldsymbol{V}_{12}\boldsymbol{V}_{22\cdot1}^{-1}\boldsymbol{V}_{21}\boldsymbol{V}_{11}^{-1} & -\boldsymbol{V}_{11}^{-1}\boldsymbol{V}_{12}\boldsymbol{V}_{22\cdot1}^{-1} \\ -\boldsymbol{V}_{22\cdot1}^{-1}\boldsymbol{V}_{21}\boldsymbol{V}_{11}^{-1} & \boldsymbol{V}_{22\cdot1}^{-1} \end{pmatrix},$$

where $\boldsymbol{V}_{22\cdot1}$ is the Schur complement of $\boldsymbol{V}_{11}$. Calculate the quadric form $\boldsymbol{z}'\boldsymbol{V}^{-1}\boldsymbol{z}$:

$$\begin{aligned}\boldsymbol{z}'\boldsymbol{V}^{-1}\boldsymbol{z} &= \boldsymbol{z}_1'\boldsymbol{V}_{11}^{-1}\boldsymbol{z}_1 + \boldsymbol{z}_1'\boldsymbol{V}_{11}^{-1}\boldsymbol{V}_{12}\boldsymbol{V}_{22\cdot1}^{-1}\boldsymbol{V}_{21}\boldsymbol{V}_{11}^{-1}\boldsymbol{z}_1 \\ &\quad -\boldsymbol{z}_2'\boldsymbol{V}_{22\cdot1}^{-1}\boldsymbol{V}_{21}\boldsymbol{V}_{11}^{-1}\boldsymbol{z}_1 - \boldsymbol{z}_1'\boldsymbol{V}_{11}^{-1}\boldsymbol{V}_{12}\boldsymbol{V}_{22\cdot1}^{-1}\boldsymbol{z}_2 + \boldsymbol{z}_2'\boldsymbol{V}_{22\cdot1}^{-1}\boldsymbol{z}_2.\end{aligned}$$

Using the symmetry of matrix $\boldsymbol{V}$, we get

$$\begin{aligned}\boldsymbol{z}'\boldsymbol{V}^{-1}\boldsymbol{z} &= \boldsymbol{z}_1'\boldsymbol{V}_{11}^{-1}\boldsymbol{z}_1 + \boldsymbol{z}_1'\boldsymbol{V}_{11}^{-1}\boldsymbol{V}_{12}\boldsymbol{V}_{22\cdot1}^{-1}\boldsymbol{V}_{21}\boldsymbol{V}_{11}^{-1}\boldsymbol{z}_1 \\ &\quad -2\boldsymbol{z}_1'\boldsymbol{V}_{11}^{-1}\boldsymbol{V}_{12}\boldsymbol{V}_{22\cdot1}^{-1}\boldsymbol{z}_2 + \boldsymbol{z}_2'\boldsymbol{V}_{22\cdot1}^{-1}\boldsymbol{z}_2 \\ &= \boldsymbol{z}_1'\boldsymbol{V}_{11}^{-1}\boldsymbol{z}_1 + (\boldsymbol{z}_2'\boldsymbol{V}_{22\cdot1}^{-\frac{1}{2}} - \boldsymbol{z}_1'\boldsymbol{V}_{11}^{-1}\boldsymbol{V}_{12}\boldsymbol{V}_{22\cdot1}^{-\frac{1}{2}})(\boldsymbol{V}_{22\cdot1}^{-\frac{1}{2}}\boldsymbol{z}_2 - \boldsymbol{V}_{22\cdot1}^{-\frac{1}{2}}\boldsymbol{V}_{21}\boldsymbol{V}_{11}^{-1}\boldsymbol{z}_1) \\ &= \boldsymbol{z}_1'\boldsymbol{V}_{11}^{-1}\boldsymbol{z}_1 + (\boldsymbol{V}_{22\cdot1}^{-\frac{1}{2}}\boldsymbol{z}_2 - \boldsymbol{V}_{22\cdot1}^{-\frac{1}{2}}\boldsymbol{V}_{21}\boldsymbol{V}_{11}^{-1}\boldsymbol{z}_1)'(\boldsymbol{V}_{22\cdot1}^{-\frac{1}{2}}\boldsymbol{z}_2 - \boldsymbol{V}_{22\cdot1}^{-\frac{1}{2}}\boldsymbol{V}_{21}\boldsymbol{V}_{11}^{-1}\boldsymbol{z}_1).\end{aligned}$$

By a formula for determinant (Anderson (2003, p.637)) we get

$$|\boldsymbol{V}| = |\boldsymbol{V}_{11}||\boldsymbol{V}_{22\cdot1}|.$$

To get the distribution of $\boldsymbol{x}_1$ we have to integrate over $\boldsymbol{x}_2$. The integral can be found after the change of variables:

$$\begin{aligned}\boldsymbol{z}_1 &= \boldsymbol{x}_1 - \boldsymbol{\mu}_1, \\ \boldsymbol{z}_2 &= \boldsymbol{x}_2 - \boldsymbol{\mu}_2, \\ \boldsymbol{y} = \boldsymbol{z}_2 - \boldsymbol{V}_{21}\boldsymbol{V}_{11}^{-1}\boldsymbol{z}_1 \Rightarrow \boldsymbol{z}_2 &= \boldsymbol{y} - \boldsymbol{V}_{21}\boldsymbol{V}_{11}^{-1}\boldsymbol{z}_1 \Rightarrow d\boldsymbol{y} = d\boldsymbol{z}_2.\end{aligned}$$

The Jacobian of the transformation equals 1. By integration we get:

$$\begin{aligned}f_{Z_1}(\boldsymbol{z}_1) &= \frac{1}{(2\pi)^{\frac{p}{2}}p}|\boldsymbol{V}|^{-1}\exp(-\frac{1}{2}\boldsymbol{z}_1'\boldsymbol{V}_{11}^{-1}\boldsymbol{z}_1)\int_{\mathbb{R}^{p_2}}\exp(-\frac{1}{2}\boldsymbol{y}'\boldsymbol{V}_{22\cdot1}^{-1}\boldsymbol{y}) \\ &\quad \times(\boldsymbol{z}_1'\boldsymbol{V}_{11}^{-1}\boldsymbol{z}_1 + \boldsymbol{y}'\boldsymbol{V}_{22\cdot1}^{-1}\boldsymbol{y})d\boldsymbol{y}.\end{aligned}$$

Direct calculations yield

$$\begin{aligned}f_{Z_1}(\boldsymbol{z}_1) &= \frac{1}{(2\pi)^{\frac{p}{2}}p}|\boldsymbol{V}|^{-1}\exp(-\frac{1}{2}\boldsymbol{z}_1'\boldsymbol{V}_{11}^{-1}\boldsymbol{z}_1)\left(\boldsymbol{z}_1'\boldsymbol{V}_{11}^{-1}\boldsymbol{z}_1\int_{\mathbb{R}^{p_2}}\exp(-\frac{1}{2}\boldsymbol{y}'\boldsymbol{V}_{22\cdot1}^{-1}\boldsymbol{y})d\boldsymbol{y}\right. \\ &\quad \left.+\int_{\mathbb{R}^{p_2}}\boldsymbol{y}'\boldsymbol{V}_{22\cdot1}^{-1}\boldsymbol{y}\exp(-\frac{1}{2}\boldsymbol{y}'\boldsymbol{V}_{22\cdot1}^{-1}\boldsymbol{y})d\boldsymbol{y}\right) \\ &= \frac{1}{(2\pi)^{\frac{p}{2}}p}|\boldsymbol{V}|^{-1}\exp(-\frac{1}{2}\boldsymbol{z}_1'\boldsymbol{V}_{11}^{-1}\boldsymbol{z}_1)[(2\pi)^{\frac{p}{2}}\boldsymbol{z}_1'\boldsymbol{V}_{11}^{-1}\boldsymbol{z}_1|\boldsymbol{V}_{22\cdot1}| + (2\pi)^{\frac{p}{2}}p_2|\boldsymbol{V}_{22\cdot1}|] \\ &= \frac{1}{(2\pi)^{\frac{p_1}{2}}}|\boldsymbol{V}_{11}|^{-1}\exp(-\frac{1}{2}\boldsymbol{z}_1'\boldsymbol{V}_{11}^{-1}\boldsymbol{z}_1)\left(\frac{\boldsymbol{z}_1'\boldsymbol{V}_{11}^{-1}\boldsymbol{z}_1}{p} + \frac{p_2}{p}\right).\end{aligned}$$

Going back to the initial variables, we have the density of $\boldsymbol{x}_1$ of the form of a mixture of the Kotz distribution and the normal distribution with weights $\frac{p_1}{p}$ ja $\frac{p_2}{p}$, respectively.

QED

5 Matrix Kotz-type distribution

Let us define first a matrix Kotz-type spherical distribution.

Definition 4 *A random matrix* $\boldsymbol{X} : p \times n$ *has the matrix Kotz-type spherical distribution, if*

$$\mathrm{vec}\boldsymbol{X} \sim K_{pn}(\boldsymbol{0}, \boldsymbol{I}_{pn}).$$

For the vec-operator see Harville (1997), if necessary.

As in the multivariate case, we can introduce a matrix elliptical distribution via a linear transformation. Let $\boldsymbol{V} = \boldsymbol{\Delta}\boldsymbol{\Delta}'$ and $\boldsymbol{W} = \boldsymbol{\Gamma}\boldsymbol{\Gamma}'$ be two positive definite matrices with dimensions $p \times p$ and $n \times n$ respectively, where $\boldsymbol{\Delta}$ and $\boldsymbol{\Gamma}$ are square matrices of full rank.

Definition 5 *We say that* $\boldsymbol{Y} : p \times n$ *is matrix Kotz-type distributed, if*

$$\boldsymbol{Y} = \boldsymbol{M} + \boldsymbol{\Delta}\boldsymbol{X}\boldsymbol{\Gamma}',$$

where $\boldsymbol{M} : p \times n$ *is a constant matrix and the matrices* $\boldsymbol{\Delta}$ *and* $\boldsymbol{\Gamma}$ *are defined above,* $\boldsymbol{X}$ *has the matrix spherical Kotz-type distribution.*

From properties of the vec-operator:

$$\mathrm{vec}\boldsymbol{Y} = \mathrm{vec}\boldsymbol{M} + (\boldsymbol{\Gamma} \otimes \boldsymbol{\Delta})\mathrm{vec}\boldsymbol{X}$$

and when applying density expression (1) to vec$\boldsymbol{Y}$ we get:

$$f_{\boldsymbol{Y}}(\boldsymbol{Y}) = |\boldsymbol{V}|^{-\frac{n}{2}}|\boldsymbol{W}|^{-\frac{p}{2}}g(\mathrm{tr}\{\boldsymbol{V}^{-1}(\boldsymbol{Y}-\boldsymbol{M})\boldsymbol{W}^{-1}(\boldsymbol{Y}-\boldsymbol{M})'\}),$$

where $g(\cdot)$ is the density generating function of a Kotz-type distributions, given in (3).

The first two moments of the matrix Kotz distribution are given in the next theorem. Formulae for other special cases can be easily obtained using general moment expressions for elliptical distribution which are given in Kollo & von Rosen (2004).

Theorem 5. *Let a random matrix* $\boldsymbol{X} : p \times n$ *have the matrix Kotz distribution with parameters* $(N = 2,\ s = 1,\ r = \frac{1}{2},)$ $\boldsymbol{M}$, $\boldsymbol{V}$, $\boldsymbol{\Psi}$. *Then*

$$m_1(\boldsymbol{X}) = E\boldsymbol{X} = \boldsymbol{M}; \tag{16}$$

$$m_2(\boldsymbol{X}) = \left(1 + \frac{2}{pn}\right)(\boldsymbol{W} \otimes \boldsymbol{V}) + \mathrm{vec}\boldsymbol{M}\mathrm{vec}'\boldsymbol{M}. \tag{17}$$

The expressions of the dispersion matrix and the fourth central moment are given in next theorem.

Theorem 6. *Let a random matrix* $\boldsymbol{X} : p \times n$ *have the matrix Kotz distribution with parameters* ($N = 2$, $s = 1$, $r = \frac{1}{2}$,) $\boldsymbol{M}$, $\boldsymbol{V}$, $\boldsymbol{W}$. *Then all odd central moment equals to zero and*

$$\overline{m}_2(\boldsymbol{X}) = D\boldsymbol{X} = \left(1 + \frac{2}{pn}\right) \boldsymbol{W} \otimes \boldsymbol{V}; \tag{18}$$

$$\overline{m}_4(\boldsymbol{X}) = \boldsymbol{W} \otimes \boldsymbol{V} \otimes \mathrm{vec}'(\boldsymbol{W} \otimes \boldsymbol{V}) + (\mathrm{vec}'(\boldsymbol{W} \otimes \boldsymbol{V}) \otimes \boldsymbol{W} \otimes \boldsymbol{V})(I_{(pn)^3,(pn)^3} + I_{pn} \otimes K_{pn,pn}). \tag{19}$$

6 Simulation

We shall describe a simulation algorithm for the spherical Kotz-type distribution. A linear transformation to obtain elliptical Kotz-type distribution can be performed later.

One way to simulate a spherically distributed random vector $\boldsymbol{x}$ is to use the following stochastic representation.

Theorem 7. *Assume that a p-vector* $\boldsymbol{x}$ *is spherically distributed. Then the following stochastic representation holds*

$$\boldsymbol{x} \stackrel{d}{=} R\boldsymbol{u}, \tag{20}$$

where $\boldsymbol{u}$ *is uniformly distributed on the unit sphere,* $R \sim F(x)$ *is independent of* $\boldsymbol{u}$ *and* $F(x)$ *is a distribution function over* $[0, \infty)$.

For Kotz-type distribution the representation for R is given in Fang, Kotz & Ng (1990), Section 3.2. For the Kotz ditribution, which is of special interest for us, R^2 is Γ-distributed: $R^2 \sim \Gamma(n/2 + 1, 1/2)$.

The simulation rule for $\boldsymbol{u}$ is given in Fang & Wang (1994), Section 4.3. If the dimension of $\boldsymbol{x}$ is small, this rule is easy to use, but for larger dimensions the algorithm is rather complicated.

For the latter case we suggest another simulation rule, based on Metropolis-Hastings algorithm (see, for example, Gamerman (1997, p.172)).

When simulating random p-vector with pdf $f(\cdot)$ we have to set an arbitrary initial value $\boldsymbol{z}_0$ and choose a known and easy to simulate distribution with pdf $h(\cdot)$. This chosen distribution should be similar to the one being generated (from point of view of symmetry and shape, for example) and should have a larger variance than the pdf generated. First, take $i = 0$ and the algorithm is as follows:

1) $i = i + 1$;
2) simulate $\boldsymbol{y}_i$ from pdf $h(\cdot)$;

3) calculate the acceptance probability:

$$\alpha(\boldsymbol{z}_{i-1}, \boldsymbol{y}_i) = \min\left(1, \frac{f(\boldsymbol{y}_i)h(\boldsymbol{z}_{i-1})}{f(\boldsymbol{z}_{i-1})h(\boldsymbol{y}_i)}\right);$$

4) simulate u from $U(0,1)$;
5) if $u < \alpha$, then $\boldsymbol{z}_i = \boldsymbol{y}_i$, otherwise $\boldsymbol{z}_i = \boldsymbol{z}_{i-1}$.

Repeat 1)-5) until requested number of vectors is generated.

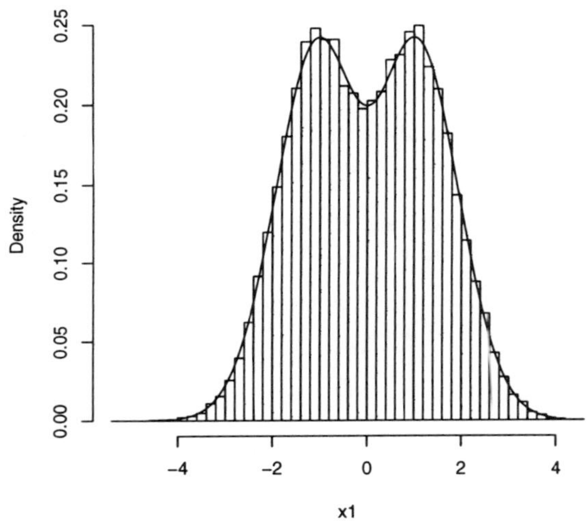

Fig. 3. The generated histogram of the first marginal and corresponding theoretical density

The suggestion for the choice of initial value $\boldsymbol{z}_0$ and pdf $h(\cdot)$ will be made in a special case.

Let us apply the simulation algorithm to the Kotz distribution with parameters $\boldsymbol{\mu} = \boldsymbol{0}$ and $\boldsymbol{V} = \boldsymbol{I}_2$.

We choose the easy-to-simulate distribution to be $N_p(0, 2\left(1 + \frac{2}{p}\right)\boldsymbol{I}_p)$, because it is easy to simulate, symmetric as the Kotz distribution, and its variance is twice the variance of the requested distribution.

To test the algorithm we generate 100 000 bivariate random vectors. We use two initial vectors: $(0,1,-0,1)$ ja $(25,-25)$, to test the convergence of the algorithm.

In Figure 3 the simulated marginal distribution is compared with the theoretical one, found in Theorem 4. The figure demonstrates good fit and tests validity of the algorithm.

Acknowledgements

Tõnu Kollo is thankful to Professor Kai-Tai Fang for inspiring discussions on various topics of multivariate statistics during the last decade. The authors are grateful to the Referees and Editors for their suggestions and comments which have improved the presentation of the material. We are also thankful to the Estonian Research Foundation for their support through Grant 5686. The research was also partly financed by the target-financed project 1776.

References

Anderson, T. W. (2003). *An Introduction to Multivariate Statistical Analysis.* Wiley, New York.

Fang, K. T. & Li, R. (1999). Bayesian statistical inference on elliptical matrix distributions. *J. Multivar. Anal.* **70** 66–87.

Fang, K. T., Kotz, S. & Ng, K. W. (1990). *Symmetric Multivariate and Related Distributions.* Chapman and Hall, London.

Fang, K. T. & Wang, Y. (1994). *Number-theoretic Methods in Statistics.* Chapman and Hall, London.

Fang, K. T. & Zhang, Y. T. (1990). *Generalized Multivariate Analysis.* Science Press/ Springer-Verlag, Beijing/ Berlin.

Gamerman, D. (1997). *Markov Chain Monte Carlo.* Chapman and Hall, London.

Gupta, A. K. & Varga, T. (1993). *Elliptically Contoured Models in Statistics.* Kluwer, Dordrecht.

Harville, D. A. (1997). *Matrix Algebra from a Statistician's Perspective.* Springer, New York.

Horn, R. A. & Johnson C. R. (1989). *Matrix Analysis.* Mir, Moscow, (in Russian).

Iyengar, S. & Tong, Y. L. (1989). Convexity properties of elliptically contoured distributions with applications. *Sankhya, Ser, A* **51** 13–29.

Kollo, T., Roos, A., von Rosen, D. (2004). Approximation of the Parameter Distributions of Growth Curve Model. *20th Nordic Conference on Mathematical Statistics NORDSTAT 2004 Abstracts*, University of Jyväskylä Finland, 1 p.

Kollo, T. & von Rosen, D. (2004). *Advanced Multivariate Statistics with Matrices.*, unpublished manuscript.

Kollo, T. & Srivastava, M. S. (2000). *A New Class of Skewed Multivarite Distributions.* Technical Report No. 2013. University of Toronto, Department of Statistics.

Kotz, S. (1975). Multivariate distributions at a cross-road. In: *Statistical Distributions in Scientific Work* **1** (G. P. Patil, S. Kotz, & J. K. Ord. D. Reidel Publishing Company eds.) 247–270. Dordrecht.

Kotz, S. & Nadarajah, S. (2001). Some extremal type elliptical distributions. *Statist. Probab. Lett.* **54** 171–182.

Koutras, M. (1986) On the generalized noncentral chi-squared distribution induced by an elliptical gamma law. *Biometrika* **73** 528–532.

Li, G. (1990). Moments of a random vector and its quadric forms. In *Statistical inference in Elliptical Contoured and Related Distributions* (K. T. Fang & T. W. Anderson eds.) 433-440. Allerton Press, New York.

Li, R. (1993). The characteristic functions of spherical matrix distribution. *Statist. Probab. Lett.* **17** 273–279.

Li, R. (1994). The characteristic functions of some subclasses of spherical distributions. *Chinese J. Appl. Probab. Statist.* **10** 290–296.

Liang, J. & Bentler, P. M. (1998). Characterizations of some subclasses of spherical distributions. *Statist. Probab. Lett.* **40** 155–164.

Mardia, K. V. (1970). Measures of multivariate skewness and kurtosis with applications. *Biometrika* **57** 519–530.

Muirhead, R. J. (1982). *Aspects of Multivariate Statistical Theory.* Wiley, New York.

Nadarajah, S. (2003). The Kotz-type distribution with applications. *Statistics* **37** 341–358.

Traat, I. (1990). Moments and cumulants of multivariate elliptical distribution with some applications. *Acta et Commentationes Universitatis Tartuensis* **893** 11–16.

Zhao, Y. (1994). Covariance matrices of quadratic forms in elliptical distributions. *Statist. Probab. Lett.* **21** 131–140.

Part II

Experimental Design

Optimized Two-Step Sequential U-Type Designs

Peter Winker

University of Erfurt, P.O. Box 900 221, D-99105 Erfurt, Germany
(Peter.Winker@uni-erfurt.de)

Summary. Uniform design has become a standard tool in experimental design over the last decade. Uniform design is particularly powerful when the specific form of the response to differing factor levels is unknown. Different criteria such as L_2-discrepancy and modifications thereof like the centered and wrap-around L_2-discrepancies are used to assess the uniformity of U-type designs. Recent advantages in the analysis of these criteria allow for efficient calculation and the derivation of lower bounds. In this contribution, these results are used to construct two-step sequential U-type designs. The construction is based on the optimization heuristic threshold accepting. The results are compared with theoretical lower bounds and ad hoc heuristics for the generation of sequential designs.

Key words: Discrepancy, threshold accepting, uniform design, sequential design

2000 Mathematics Subject Classification: 62K05, 62L05, 90C10

1 Introduction

Uniform design has become a standard tool in experimental design over the last decade. Uniform design is particularly powerful when the specific form of the response to differing factor levels is unknown. Consequently, applications include analysis of model robustness (Bates, Buck, Riccomagno & Wynn (1996)) and efficient numerical analysis (Fang & Wang (1994, Ch.5)), but are not restricted to these areas (Fang & Lin (2003)).

When the exact model is not known, uniform design might be used for a preliminary exploratory analysis.

Contemporary Multivariate Analysis and Experimental Design—In Honor Celebration of Professor Kai-Tai Fang's 65th birthday. Edited by Jianqing Fan and Gang Li. The World Scientific Publisher, 2005.

When no a priori information is available, a useful criterion for constructing experimental designs is uniformity. Consequently, in this case, the quality of an experimental design is expressed in terms of its discrepancy from uniformity. There exist several different definitions of discrepancy for this purpose including the classical "star discrepancy", the "L_p-discrepancy", and the "symmetric L_p-discrepancy". Recently, (Hickernell, (1998a, b)) suggested modified versions of the L_2-*discrepancy,* namely the "centered L_2-discrepancy" and the "wrap-around L_2-discrepancy". These measures are theoretically appealing and allow for efficient calculation and updating (Fang, Lu & Winker (2003)), while e.g. the calculation of the star discrepancy itself represents a complex optimization problem (Winker & Fang (1997)). Given a measure of discrepancy, the number of factors d, levels q and replications n, the goal consists in finding an experimental design minimizing the discrepancy.

While all measures of discrepancy are defined for sets of d-dimensional points in any super-rectangle Ω, after a linear transformation, we may assume $\Omega = [0,1]^d$. The problem of minimizing any measure of discrepancy on Ω is probably a NP hard problem. Only for the case $d = 1$, it is easy to obtain the optimal solution (Fang, Ma & Winker (2002)). In this case, if only q different levels are permitted for a set of $n = qk$ points on the unit interval, the minimum for the centered L_2-discrepancy is assumed if and only if the points take each of the equidistant values $\frac{2l-1}{2q}, l = 1, \ldots q$ equally often, i.e. k times. Consequently, the search for low discrepancy designs in higher dimensions is restricted to points lying on an equidistant grid on Ω. These designs are denoted as *U-type designs.* Fang, Ma & Winker (2002) analyze possible improvements of relaxing this constraint and find them to be rather small.

Finding good experimental designs or good U-type designs implies that with a given number of replications, a good approximation to the response surface can be obtained, or that for a required quality of the approximation, the number of experimental runs can be reduced. Unfortunately, as already mentioned, finding optimal U-type designs for a given measure of discrepancy is a complex integer programming problem. For this type of problems, the use of optimization heuristics appears to be an adequate way to obtain high quality results (Winker & Gilli (2004)). Winker & Fang (1997) provide the first implementation of the threshold accepting algorithm in this context. More details on their implementation and the heuristic can be found in Winker (2001).

So far, the problem of finding good experimental designs has been described for the one-step case. However, when experiments are time consuming and/or very costly, a sequential setup might become relevant. In practical applications, the idea is to use a low number of experiments in a first step in order to find out whether expected effects are likely to hold. Only if this first step provides a positive signal, additional experiments are run in order to obtain a better approximation. Obviously, for the final evaluation, the results of both steps are considered. Therefore, in this case, an optimal design has to take into account both steps. How should the sequential design be built? Three alternatives could be considered. First, use an optimal design for the first step and try to add additional runs in an optimal way afterwards. Second, use an optimal design for the total number of runs of both steps and try to select a subset for the first step in an optimal way. Third, apply explicit optimization to both steps of the sequential designs simultaneously. The latter approach will be followed in this paper. It leads to the additional question on how to measure and weigh the performance of both steps in the optimization algorithm.

This contribution presents a first application of the third approach to two- and three-level designs. The optimization heuristic uses the efficient updating procedure and the lower bounds introduced by Fang, Lu & Winker (2003). Section 2 provides a formal description of the sequential design problem. In section 3 the threshold accepting implementation is introduced. First results for the two- and three-level cases are reported in section 4. The main findings and an outlook on future research are provided in the concluding section 5.

2 The Sequential Design Problem

A two-step sequential U-type design $U(n_1, n_2, q^d)$ consists of a pair of q-level U-type designs $\{U(n_1, q^d), U(n_2, q^d)\}$, $n_1 < n_2$. Thereby, a U-type design $U(n, q^d)$ is given by a matrix of dimension $n \times d$ with each column having an equal number of entries $\frac{2l-1}{2q}$, $l = 1, \ldots, q$. By the linear transformation $\frac{2l-1}{2q} \to l$, a $U(n, q^d)$ can also be presented as a matrix of size $n \times d$, with each column having the same number of entries $1, 2, \ldots, q$. In this contribution, both representations are used without further specification. Let $\mathcal{U}(n, q^d)$ be the set of all q-level U-type designs with n runs (points) and d factors, and let $\mathcal{U}(n_1, n_2, q^d)$ stand for the set of all two-step sequential U-type designs with n_1 and n_2 runs, respectively.

The goal consists in generating two-step sequential U-type designs with a high degree of uniformity for both the first and the second step design. In the present application, uniformity is measured by means of the centered L_2-discrepancy (CL_2) for two-level designs and by the wrap-around L_2-discrepancy (WL_2) for three-level designs introduced by Hickernell (1998a, 1998b) . Besides attractive theoretical properties, these measures allow for a reexpression based on, respectively, the column balance and the row distance (Fang, Lu & Winker (2003)). Using these presentations of the two measures allows to derive lower bounds, which can be used as a benchmark, and to implement a fast local updating procedure within the local search heuristic threshold accepting (Winker (2001)), which is used for the generation of low discrepancy designs.

Once, the measure of uniformity is defined, the two-step sequential design problem is given by the following optimization problem:

$$min_{U=(U_1,U_2)\in\mathcal{U}(n_1,n_2,q^d)}\left[\omega_1 L_2(U_1) + \omega_2 L_2(U_2)\right], \tag{1}$$

where L_2 stands for CL_2 in the two-level case and for WL_2 in the three-level case. The ω_i define the weights of the two designs in the global measure of uniformity. For the present application, these weights are defined by the relative deviations from the best available lower bounds. For general applications, different choices of ω_i can be considered.

Given the definition of the problem as stated in (1), a solution appears to be straightforward given that $\mathcal{U}(n_1, n_2, q^d)$ is a finite set. However, the cardinality of $\mathcal{U}(n_1, n_2, q^d)$ precludes an enumeration approach for typical values of n_1, n_2 and q encountered in real applications. The scope of use of standard optimization tools like the Simplex method also appears to be limited to very small problem instances Fang, Ma & Winker (2002). Probably, the problem shares the NP hardness with the standard optimum design problem described above. Consequently, heuristic algorithms have to be considered.

In this contribution, two different approaches are considered. First, we start with an optimized design for the total number of runs n_2, which is obtained by the threshold accepting implementation. Then, we choose n_1 points out of the n_2 as the first step design. This selection can be done sequentially or simultaneously. For the present application, a full enumeration of all possible subsets, i.e. a simultaneous approach is used. Second, the optimization problem described in (1) can be tackled explicitly by heuristic optimization. In section 3 a threshold accepting implementation is described for this explicit approach. The results will be compared with the sequential method in section 4.

When a simultaneous optimization is performed, the choice of the weights ω_i in the objective function (1) becomes relevant. It might depend on the application considered, i.e. whether low discrepancy is considered to be more important for the first or for the second step of the analysis. In this paper, the weights are chosen relative to the value of the lower bound for the discrepancy of the two designs. Consequently, if lb_1 and lb_2 denote the theoretical lower bounds for the designs with n_1 and n_2 points, respectively, the weights are $\omega_1 = 1/lb_1$ and $\omega_2 = 1/lb_2$. In order to give the resulting measure an intuitive interpretation, we subtract the constant two. Thus, the objective function becomes

$$\left(\frac{1}{lb_1} L_2(U_1) - 1\right) + \left(\frac{1}{lb_2} L_2(U_2) - 1\right) \tag{2}$$

and can be interpreted as the sum of the relative deviations from the two lower bounds. However, the method described in the following section will also work for any other choice of the ω_i.

3 The Threshold Accepting Implementation

Threshold accepting is a modification of simulated annealing using a deterministic acceptance criterion instead of the probabilistic one in simulated annealing. Consequently, it belongs to the class of local search methods. Like most of the local search heuristics it allows uphill moves, i.e. accepts solutions which are worse than the previous one, in order to be able to escape local minima. A classification of optimization heuristics can be found in Winker & Gilli (2004), and a more detailed description of the threshold accepting algorithm is provided by Winker (2001). It has already been used repeatedly in the context of experimental design, e.g. to obtain lower bounds for the star-discrepancy (Winker & Fang (1997)), to generate low discrepancy U-type designs for the star-discrepancy (Winker & Fang (1997)), several modifications of the L_2-discrepancy (Fang, Lin, Winker & Zhang (2000)), for CL_2 (Fang, Ma & Winker (2002)), and for CL_2 and WL_2 (Fang, Lu & Winker (2003)). Algorithm 1 provides the pseudo-code for the threshold accepting implementation for the simultaneous optimization of the two-step sequential design problem.

Algorithm 1 Pseudo-code for Threshold Accepting

1: Initialize n_R, n_{S_r} and the sequence of thresholds τ_r, $r = 1, 2, \ldots, n_R$
2: Generate starting design $U^0 \in \mathcal{U}(n_1, n_2, q^d)$
3: **for** $r = 1$ to n_R **do**
4:     **for** $i = 1$ to n_{S_r} **do**
5:         Generate $U^1 \in \mathcal{N}(U^0)$ (neighbor of U^0)
6:         **if** $CD_2(U^1) < CD_2(U^0) \times (1 + \tau_r)$ **then**
7:             $U^0 = U^1$
8:         **end if**
9:     **end for**
10: **end for**

Thereby, CD_2 represents any of the weighted measures of discrepancy for a two-step designed as discussed before. Replacing $\mathcal{U}(n_1, n_2, q^d)$ by a single U-type designs with n_1 or n_2 runs, respectively, and an appropriate choice of CD_2 allows to use the same algorithm for the optimization of the single-step sequential design problems. In fact, this approach will be used in the sequential approach to obtain a low-discrepancy design with n_2 points.

Threshold accepting performs a refined local search on the set $\mathcal{U}(n_1, n_2, q^d)$. It starts with a randomly generated two-step design U^0 (2:) and continues by iterating local search steps. For each step, a two-step design U^1 has to be chosen in the neighborhood of the current design (5:). Then, the value of the objective function of both two-step designs is compared (6:). The new design is accepted if it is better than U^0, but also if it is not much worse. The extent of an accepted worsening is limited by the current value of the threshold sequence (τ_r), which decreases to zero during the course of iterations.

The performance of the threshold accepting implementation depends on a number of settings. In particular, the definition of neighborhoods for the choice of U^1, the sequence of threshold values τ_r and, finally, the total number of iterations are most relevant. In order to use a high number of iterations in given real time, we resort to a fast updating method described in Fang, Lu & Winker (2003) for the first time. Instead of recalculating $CD_2(U^1)$ completely, specific representations of CL_2 and WL_2 in terms of the Hamming distance of row pairs allow for a local updating, i.e. only a small number of calculations has to be performed when U^1 results from U^0 by a small perturbation. In this application, the meaning of small perturbation is defined by the exchange of few elements in a column between rows, thus keeping the characteristics of a U-type design. Given this framework, a number of up to 5 million replications could be used for the optimization.

The neighborhood structure used by the threshold accepting heuristic to generate search steps is also based on the Hamming distances of row pairs (for details see Fang, Lu & Winker (2003)). Consequently, while moving from one design to a neighboring one, the objective function can be updated at low computational cost. In fact, the speed up as compared to a full calculation of the discrepancy is of the order of $\frac{1}{n_2 d}$ (Fang, Lu & Winker (2003)). As an additional complication compared to the implementation in Fang, Lu & Winker (2003) for the one-step design we have to check whether a change affects both subdesigns or only the larger one. For a given U-type design U^0, a neighborhood $\mathcal{N}(U^0)$ is defined as a ε-sphere centered in U^0 using the metric derived from the Hamming distance, i.e. a U-type design is considered as being a neighbor of U^0 if and only if it differs in not more than ε entries from U^0. In order to obtain elements in $\mathcal{U}(n_1, n_2, q^d)$, changes to U^0 are restricted to exchanging elements within columns.

An alternative to this undirected definition of neighborhoods is provided by Fang, Tang & Yin (2004), who exploit specific features of uniform designs in order to define the neighborhoods. Using this information about the specific structure of the optimization problem might increase the efficiency of the threshold accepting implementation. It is left for future research to integrate their approach in the present context and to assess the gain in efficiency.

Finally, the threshold sequence τ_r, $r = 1, \ldots, n_R$ is obtained by a procedure first described in Winker (2001) which is basically data driven. The intuitive idea results from the observation that for a finite search space like $\mathcal{U}(n_1, n_2, q^d)$, the local changes of the objectivefunction can only take a finite number of values. Obviously, for the algorithm, only these values are relevant for the threshold sequence. Consequently, the threshold sequence is constructed as follows. First, a large number of two-step designs is generated at random. Then, for each of these random designs a neighbor is selected using the neighborhood definition introduced above. For each resulting pair of designs, the relative difference of the objective function values between the larger and the smaller value is calculated. Ordering these values provides an approximation to the distribution of local relative changes of the objective function. Consequently, a lower quantile of these sequence is employed as the threshold sequence.

In principle, the algorithm described in this section could be applied to any number of levels. Nevertheless, we start with implementations for two- and three-level designs as lower-bounds are available in Fang, Lu & Winker (2003) and previous results for one-step designs might serve as additional benchmarks.

4 Results for Two- and Three-level designs

The discrepancy values obtained by the optimization procedure for the two-step designs are difficult to evaluate without a relevant benchmark. For this purpose, different alternatives can be considered. First, the lower bounds for the individual designs with n_1 and n_2 runs, respectively, could be used. Given our choice of weights ω_i in (2), the value of this benchmark (BM_1) is zero by definition. Of course, in general, it will not be possible to reach these lower bounds, as the two-step design cannot generate the two sub-designs independently. Nevertheless, it might provide a useful benchmark. A second approach consists in starting the best result obtained for n_2 runs and to select the best subset of n_1 rows out of these n_2 runs. Obviously, this approach has a high computational complexity and will not be feasible for larger values of n_2. However, for most of the problem sizes considered in this paper, it is still possible to generate this best subset. The two-step design resulting from this sequential approach will serve as a second benchmark (BM_2). We did not implement the alternative strategy to start with a good design for n_1 runs and to add sequentially additional rows. However, previous experience with this type of sequential construction algorithms (Fang, Lin, Winker & Zhang (2000, p. 244)) indicates that we should not expect a high quality for the resulting design with n_2 runs, if $n_2 \gg n_1$.

We contrast the best results obtained by the simultaneous optimization approach described in the previous section with a maximum number of 1.000.000 iterations for the two-level case and 5.000.000 iterations for the three-level case with the two benchmark values BM_1 and BM_2. Given that by definition BM_1 has to be smaller or equal to the values obtained for the optimized two-step designs and BM_2, all values are expressed in relative deviations from BM_1.

The following subsections summarize our findings for the two- and three-level sequential designs, respectively. For the two-level designs, the centered L_2-discprepancy is used for the individual designs, while for the three-level case, the wrap-around L_2 discrepancy is employed.

4.1 Two-level sequential designs

The results for the two-level sequential designs are summarized in table 1. The measure of discrepancy employed in equation (1) is the centered L_2-discrepancy. The first three columns describe the dimension of the design: d is the number of factors (columns), and n_1 and n_2 are the number of runs (rows) for the first and second step designs, respectively. The first n_1 runs of the second step design are equal to the first step design.

The next four columns provide the results for the sequential approach described above. In a first step, the threshold accepting implementation is used to obtain a low-discrepancy design with n_2 runs. Then, by means of enumeration of all subsets of n_1 rows out of the n_2 the best first step design is identified. The column *enumeration* provides the number of first step designs considered in the selection process. The entries in the column n_1 correspond to the relative deviation of the best first step design found by the sequential approach relative to the maximum lower bound, while the entries in the column n_2 provide the relative deviation for the complete design with n_2 rows relative to the maximum lower bound for this dimension. Finally, the column *total* provides the sum of both relative deviations which is used as our objective criterion for the two-step design in this contribution. Obviously, the threshold accepting implementation run with up to 1.000.000 iterations provides optimal or near optimal designs for

Table 1. Deviation (in percent) from theoretical lower bound BM_1 for sequential designs ($U(n_1, n_2, 2^d)$)

dimension			sequential approach BM_2				TA optimized design		
d	n_1	n_2	enumeration	n_1	n_2	total	n_1	n_2	total
5	8	10	45	13.03	0.00	13.03	2.37	1.63	3.99
6	8	10	45	12.32	1.26	13.57	3.49	2.59	6.08
6	8	12	495	10.63	0.36	10.99	3.49	2.60	6.09
7	8	12	495	13.53	0.65	14.18	0.00	4.75	4.75
7	8	14	3003	19.85	0.56	20.41	0.00	1.79	1.79
8	10	12	66	6.55	1.01	7.56	3.11	2.57	5.68
8	10	14	1001	9.25	0.86	10.11	3.11	2.74	5.85
8	10	16	8008	9.25	0.93	10.18	3.11	4.77	7.88
9	10	12	66	8.14	1.42	9.56	5.50	3.63	9.13
9	10	14	1001	10.50	3.46	13.96	4.86	4.12	8.98
9	10	16	8008	8.47	4.40	12.87	5.39	5.60	10.99
9	10	18	43758	10.50	1.96	12.46	4.56	3.06	7.62
10	10	12	66	9.34	1.88	11.21	9.34	1.88	11.21
10	10	14	1001	7.76	5.38	13.13	5.97	5.38	11.35
10	10	16	8008	12.55	5.98	18.53	6.13	7.73	13.86
10	10	18	43758	8.94	2.95	11.89	5.97	4.23	10.20
11	10	12	66	7.51	0.00	7.51	7.51	0.00	7.51
11	10	14	1001	4.10	6.57	10.67	4.10	6.57	10.67
11	10	16	8008	10.23	7.52	17.76	4.33	9.08	13.41
11	10	18	43758	9.04	4.30	13.34	3.98	5.53	9.51
12	10	14	1001	10.09	6.66	16.75	2.23	7.55	9.78
12	10	16	8008	10.09	5.14	15.23	3.55	6.86	10.41
12	10	18	43758	6.58	5.83	12.41	2.23	6.91	9.14
13	10	14	1001	7.99	5.13	13.12	2.62	5.77	8.39
13	10	16	8008	7.99	3.52	11.51	2.73	5.98	8.71
13	10	18	43758	6.43	7.46	13.89	2.62	7.86	10.48
14	10	14	1001	6.21	4.00	10.21	2.07	4.87	6.94
14	10	16	8008	6.75	1.65	8.40	2.07	3.11	5.17
14	10	18	43758	7.01	7.93	14.94	2.57	8.24	10.82
15	10	14	1001	6.68	2.64	9.32	1.83	4.07	5.90
15	10	16	8008	12.41	0.00	12.41	1.83	4.69	6.52
15	10	18	43758	7.33	6.06	13.39	1.83	7.05	8.88
16	10	14	1001	7.00	2.27	9.27	1.57	4.20	5.77
16	10	18	43758	7.94	5.18	13.12	1.57	5.63	7.20
17	10	14	1001	7.70	1.82	9.52	0.74	3.32	4.06
17	10	18	43758	6.61	4.17	10.78	0.74	4.82	5.56
18	10	14	1001	5.14	1.81	6.95	0.00	3.18	3.18
18	10	18	43758	8.18	3.10	11.28	0.00	3.38	3.38

n_2 rows for many problem instances. However, even when performing a full enumeration of all subsets with n_1 rows, the resulting first step design is often far off the lower bound. Consequently, the total value of the objective criterion is quite high, ranging from about 8 to 20%.

Finally, the last three columns of table 1 provide the results for the simultaneous optimization approach. Again, the values in column n_1 are the relative deviations from the theoretical lower bound for the first n_1 rows, while the column n_2 provides the same information for the complete design. Obviously, the simultaneous optimization imposes some constraint on the complete design with n_2 rows. Consequently, it is not surprising that the values obtained for the full design are slightly larger, i.e. worse than for the sequential approach BM_2. However, the sub-designs with n_1 rows are much better than in the sequential approach. Therefore, the overall quality of the two-step designs resulting from the simultaneous optimization approach is much better than for the sequential approach. When interpreting the absolute values for the optimized two-step designs, it should be kept in mind that the benchmark BM_1 resulting from the two lower bounds for n_1 and n_2 rows, respectively, is only a hypothetical one. First, it cannot be guaranteed that for all n_1 and n_2 the relevant lower bound can be met. Second, the restrictions imposed by the two-step characteristic of the designs have to be taken into account. In general, they will not allow to reach the individual lower bounds.

4.2 Three-level sequential designs

The results for the three-level case are given in table 2. In contrast to the two-level case, the wrap-around L_2-discrepancy is used as a measure of uniformity in equation (1) as no lower bound was available for the three-level case and the centered discrepancy. However, current research provides such a lower bound (Fang, Tang & Yin (2004)). Consequently, future research might extend the analysis to the centered discrepancy for the three- and four-level case.

The presentation of the results follows the conventions for the two-level case. The first three columns describe the number of factors (d), and the number of rows for the first (n_1) and second (n_2) step of the two-step designs, respectively.

The following columns under the heading BM_2 summarize the findings for the sequential approach. As indicated by the first of these columns, the number of evaluations required for the full enumeration of all possible sub-designs of the n_2 runs increases rapidly. Consequently, in order to keep the computational time feasible, the sequential approach BM_2 is only performed when the number of enumeration does not exceed $5 \cdot 10^6$. For the other cases, only the value for the optimized design with n_2 rows is reported. The entries in the column n_1 correspond to the relative deviation of the best first step design found by the sequential approach relative to the maximum lower bound, while the entries in the column n_2 provide the relative deviation for the complete design with n_2 rows relative to the maximum lower bound for this dimension. The column *total* provides the sum of both relative deviations corresponding to equation (1). Again, the threshold accepting implementation performs well in generating optimal or near optimal designs for designs with n_2 rows while using up to 5.000.000 iterations. However, even when performing a full enumeration of all subsets with n_1 rows, the resulting design is often far off the lower bound. Consequently, the total value of the objective criterion is often quite high, for many instances exceeding 10%.

The results for the simultaneous optimization approach are presented in the last three columns. Due to the constraint imposed by the two-step design, the discrepancy for the second step design (n_2) is typically larger than in the case of the sequential approach. However, as for the two-level case, this disadvantage for the second step is overcompensated by the gain for the first step design (n_1). Consequently, the overall quality of the two-step designs as measured by (1) is much better for the simultaneous approach than for the sequential approach, which, in addition, becomes intractable for larger problem instances, in particular, when the increase from n_1 to n_2 is large.

5 Conclusion

This contribution extends previous analysis on the construction of low-discrepancy U-type designs by means of a threshold accepting implementation. Here, the feature of two-step designs is analyzed. Two alternatives for the construction of such designs are discussed and considered for the application to two- and three-level two-step designs.

It turns out that a suitable implementation of the threshold accepting heuristic is able to generate high-quality, i.e. low-discrepancy designs taking into account the constraints imposed by the two-step characteristic of the problem. In general, the results obtained with this simultaneous optimization approach are better than those resulting from a sequential procedure.

Future work will extend the analysis to more problem instances, e.g. designs with more than three levels. Furthermore, alternative objective functions or weighting scheme for the two steps will be considered. Finally, an additional benchmark might be generated from the independent optimization of designs for n_1 and n_2 rows, respectively.

References

Bates, R., Buck, R., Riccomagno, E. & Wynn, H. P. (1996). Experimental design and observations for large systems. *Journal of the Royal Statistical Society (B)* **57** 77–94.

Fang, K. T. & Lin, D. K. J., (2003). Uniform experimental designs and their applications in industry. In *Handbook of Statistics* (R. Khattree, & C. R. Rao, eds.) **22** 131–170. Elsevier, Amsterdam.

Fang, K. T., Lin, D. K. J., Winker, P. & Zhang, Y., (2000). Uniform Design: Theory and Application, *Technometrics* No. 3 **42** 237–248.

Fang, K. T., Lu, X. & Winker, P., (2003). Lower bounds for centered and wrap-around L_2-discrepancies and construction of uniform designs by Threshold Accepting. *Journal of Complexity* **19** 692–711.

Fang, K. T., Ma, C. X. & Winker, P. (2002). Centered L_2-discrepancy of Random Sampling and Latin hypercube Design, and Construction of Uniform Designs. *Mathematics of Computation* **71** 275–296.

Fang, K. T., Shiu, W. C., Pan, J. X. Uniform designs based on latin squares. Statistica Sinica **9** 905–912.

Fang, K. T., Tang, Y. & Yin, J.,(2004). Lower bounds for wrap-around L_2-discrepancy and constructions of symmetrical uniform designs. *Forthcomming.*

Fang, K. T. & Wang, Y., (1994). *Applications of Number Theoretic Methods in Statistics.* Chapman and Hall, London.

Hickernell, F. J., (1998a). Lattice rules: How well do they measure up. In *Random and Quasi-Random Point Sets* (P. Hellekalek & G. Larcher, eds.) 106–166. Springer, New York.

Hickernell, F. J., (1998b). A generalized discrepancy and quadratic error bound. *Mathematics of Computation* **67** 299–322.

Winker, P., (2001). *Optimization Heuristics in Econometrics.* Wiley, Chichester.

Winker, P. & Fang, K. T. (1997a). Application of Threshold Accepting to the Evaluation of the Discrepancy of a Set of Points. *SIAM Journal on Numerical Analysis* **34** 2028–2042.

Winker, P. & Fang, K. T., (1997b). Optimal U-type designs, In *Monte Carlo and Quasi-Monte Carlo Methods* (H. Niederreiter, P. Hellekalek, G. Larcher & P. Zinterhof, eds.) 436–488. Springer, New York.

Winker, P. & Gilli, M., (2004). Applications of optimization heuristics to estimation and modelling problems. *Computational Statistics and Data Analysis* **47** 211–223.

Table 2. Deviation (in percent) from theoretical lower bound BM_1 for sequential designs $(U(n_1, n_2, 3^d))$

dimension			sequential approach BM_2				TA optimized design		
d	n_1	n_2	enumeration	n_1	n_2	total	n_1	n_2	total
4	9	12	220	13.82	0.00	13.82	0.00	0.31	0.31
5	12	15	455	3.42	0.10	3.52	0.00	0.15	0.15
6	12	15	455	5.73	0.31	6.04	0.00	0.35	0.35
6	12	18	$1.86 \cdot 10^4$	1.70	0.23	1.93	0.46	1.17	1.63
7	12	15	455	8.64	0.56	9.20	0.14	2.04	2.18
7	12	18	$1.86 \cdot 10^4$	7.47	0.62	8.09	0.14	3.04	3.18
7	12	21	$2.94 \cdot 10^5$	5.86	1.23	7.09	0.14	1.43	1.57
11	12	15	455	7.82	1.50	9.32	0.78	3.46	4.24
11	12	18	$1.86 \cdot 10^4$	11.28	2.11	13.39	0.93	3.43	4.36
11	12	21	$2.94 \cdot 10^5$	10.39	2.86	13.24	0.93	4.01	4.94
11	12	24	$2.70 \cdot 10^6$	9.56	4.13	13.68	1.09	4.65	5.74
11	12	27	$1.74 \cdot 10^7$ *	n.a.	5.01	n.a.	0.78	5.73	6.51
11	12	30	$8.65 \cdot 10^7$ *	n.a.	3.81	n.a.	1.09	4.05	5.14
8	15	18	816	7.26	1.99	9.25	1.18	4.00	5.19
8	15	21	$5.43 \cdot 10^4$	5.06	1.82	6.88	1.18	2.15	3.33
8	15	24	$1.31 \cdot 10^6$	6.53	2.26	8.79	1.35	2.51	3.86
9	15	18	816	8.58	1.97	10.55	2.40	4.15	6.55
9	15	21	$5.43 \cdot 10^4$	7.45	2.43	9.88	2.43	2.92	5.35
9	15	24	$1.31 \cdot 10^6$	7.76	3.02	10.78	2.26	3.50	5.76
9	15	27	$1.74 \cdot 10^7$ *	n.a.	4.00	n.a.	2.90	4.72	7.62
10	15	18	816	6.89	2.35	9.24	2.01	3.70	5.72
10	15	21	$5.43 \cdot 10^4$	8.66	3.23	11.89	2.01	4.02	6.03
10	15	24	$1.31 \cdot 10^6$	7.29	3.84	11.14	2.01	4.35	6.37
10	15	27	$1.74 \cdot 10^7$ *	n.a.	5.36	n.a.	2.15	5.83	7.98
10	15	30	$1.55 \cdot 10^8$ *	n.a.	3.05	n.a.	2.01	3.30	5.32
12	15	21	$5.43 \cdot 10^4$	8.28	2.44	10.73	1.46	3.83	5.28
12	15	27	$1.74 \cdot 10^7$ *	n.a.	4.05	n.a.	1.73	4.83	6.57
12	15	30	$1.55 \cdot 10^8$ *	n.a.	4.61	n.a.	1.73	4.94	6.68
17	15	18	816	4.70	1.03	5.73	0.67	2.66	3.33
17	15	21	$5.43 \cdot 10^4$	6.73	1.30	8.03	0.72	2.71	3.44
17	15	24	$1.31 \cdot 10^6$	6.63	1.77	8.40	0.72	2.83	3.55
17	15	27	$1.74 \cdot 10^7$ *	n.a.	2.28	n.a.	0.78	3.23	4.01
17	15	30	$1.55 \cdot 10^8$ *	n.a.	2.71	n.a.	0.82	3.59	4.41

The entries in column *enumeration* marked with $*$ indicate that a full enumeration of all sub-desings with n_1 runs would require the evaluation of more than 5.000.000 designs. For these cases BM_2 is not calculated.

Granularity and Balance in Experimental Design

Fred J. Hickernell

Department of Mathematics, Hong Kong Baptist University, Kowloon Tong, Hong Kong SAR, China (fred@hkbu.edu.hk)

Summary. Granularity and perfect balance are defined and discussed for multiple factor designs. The granularity of a design is related to its discrepancy, an important concept in uniform experimental design. It indicates how fine a structure in the dependence of the response on the factors can be resolved. The balance of a design is similar to the resolution of fractional factorial designs, but it is defined for a much broader class of designs. The granularities and balance of various designs, including simple random designs, orthogonal arrays, digital nets, and integration lattices are compared. Two applications, the simple pendulum and blood glucose monitoring, are used to illustrate how granularity and balance can identify good designs.

Key words: Discrepancy, grid, integration lattice, orthogonal array, positive semi-definite kernel, simple random

2000 Mathematics Subject Classification: 11K38, 62K15, 62K20, 62K99

1 Introduction

My first recollection of performing a serious experiment was as a college freshman taking general physics. We were asked to determine the period of a simple pendulum as a function of its mass and length, and the amplitude of the motion. Before beginning the experiment we were taught how to make careful measurements and to record what we observed in a laboratory notebook. The instructor demonstrated the experimental apparatus, and guidance was given on how to write our lab reports. However, no instruction was given on how to *design* the experiment.

Contemporary Multivariate Analysis and Experimental Design—In Honor Celebration of Professor Kai-Tai Fang's 65th birthday. Edited by Jianqing Fan and Gang Li. The World Scientific Publisher, 2005.

The three parameters or factors to be varied were length of the pendulum, mass of the bob and amplitude of the motion. The single response was the period. The sources of experimental error included imprecision of the instruments and inexperience of the experimentalists. Some of us performed the experiment by varying one factor while keeping the rest fixed. Others tried a full factorial experiment, i.e., taking measurements of the period for all possible combinations of levels of the three factors, but his meant that we had to limit ourselves to only a few levels per factor. A few decades after that episode I have another experiment to perform — checking my blood glucose level. The blood glucose meter comes with a booklet listing seven different times per day: before breakfast, after breakfast, before lunch, after lunch, before dinner, after dinner, and before sleep. A person's blood glucose level normally peaks about 1–2 hours after eating. Individuals with diabetes should aim for a blood sugar level 2 hours after eating to be below 8.9 mmol/L (160 mg/dL), the level before breakfast, lunch, and supper to be 5.0–6.1 mmol/L (90–130 mg/dL), and the level before sleep to be 6.1–8.3 mmol/L (110–150 mg/dL) (Joslin Diabetes Center (2004)). Because my blood glucose level is moderately high, the doctor advised me to check it once per day. He also wants me to check it at different times of the day on different days to give an indication of the intra-daily fluctuations. However, no one has told me how to *design the experiment.*

Experiments require valuable resources: time, equipment and consumables. Measuring one's blood glucose level involves the discomfort of pricking a finger to take a blood sample. It is crucial that experimentalists be taught how to design their experiments to obtain the maximum amount of information given limited resources. This article discusses two important concepts for measuring the quality of an experimental design: *granularity* and *perfect balance.* These are formally defined in Definitions 1–3 and 5, but they are introduced briefly here. Granularity describes the effective number of levels per factor, or the fineness of scale over which the design allows you to detect fluctuations in the response. A design is perfectly balanced up to a certain granularity $(q_1, \ldots, q_s)$ if it looks like an evenly spaced $q_1 \times \cdots \times q_s$ grid.

It may be helpful to relate granularity and balance to other approaches to experimental design. Atkinson & Donev (1992) and Pukelsheim (1993) emphasize efficient estimation of the model parameters, assuming a known model for the response as a function of the factors. If the model is truly known then one only needs an experimental design with the granularity and balance dictated by the complexity of the model. However, when the model is not known and the experimentalist expects to infer the model from the data, it is advisable to make the granularity and balance as high as possible. Dey & Mukerjee (1999) and Hedayat, Sloane & Stufken (1999) assume that the model is unknown and try to maximize balance, which is a generalization of resolution or strength. However, such designs often presume a relatively small number of levels and so have a low granularity. On the other hand, digital nets (Niederreiter (1992), Larcher (1998), Niederreiter & Xing (2001)) and integration lattices (Niederreiter (1992) and Sloan & Joe (1994)) have higher granularity while not sacrificing much balance. These designs are more flexible in model selection and more robust against model misspecification than orthogonal designs.

Digital nets and integration lattices are examples of *uniform designs* (Fang (1980), Wang & Fang (1981), Fang & Wang (1994) and Fang, Lin, Winker & Zhang (2000)). They are constructed to minimize a model-independent criterion called the *discrepancy*. Granularity and balance as defined here are closely related to the discrepancy (see Remark 6) and the projection uniformity pattern (Hickernell & Liu (2002)).

2 Granularity for One Factor Designs

For simplicity consider the case of just one factor. This corresponds to fixing the mass and amplitude of the pendulum and studying how the period depends on its length. It also corresponds to measuring the blood glucose level several times during just one day or at the same times every day. Let x denote this factor, and let $\mathcal{X} \subseteq \mathbb{R}$ denote the set of possible values, i.e., the experimental domain. Although the definition of granularity is model-independent, a basic understanding of regression illuminates the qualities that make a good experimental design. Suppose that there are n runs or experiments to be performed. The set of different values of the factors, $P = \{x_1, \ldots, x_n\}$, is called the *experimental design*. Let y denote the value of the response, and suppose that a linear regression model describes the dependence of y on x:

$$y_i = \beta_1 g_1(x_i) + \cdots + \beta_p g_p(x_i) + \varepsilon_i. \tag{1}$$

The $n \times 1$ vector of response data is $\mathbf{y} := (y_1, \ldots, y_n)^T$, the $n \times p$ structure matrix is $\mathbf{G} := (g_j(x_i))_{i,j}$, and the $p \times 1$ unknown regression coefficient is $\boldsymbol{\beta} := (\beta_1, \ldots, \beta_p)^T$. The least squares estimate of the regression coefficient is $\hat{\boldsymbol{\beta}} := (\mathbf{G}^T\mathbf{G})^{-1}\mathbf{G}^T\mathbf{y}$, and the estimated model for the response is $\hat{y}(x) = \hat{\beta}_1 g_1(x) + \cdots + \hat{\beta}_p g_p(x)$. In this article all variables and functions are assumed to be real-valued, unless otherwise stated, but extensions to complex-valued functions are straight-forward. If the design P has just one level, i.e., $x_1 = \cdots = x_n$, then one can only estimate a model with a single term, i.e., $\hat{y}(x) = \hat{\beta}_1 g_1(x)$, since otherwise $\mathbf{G}$ is not of full rank and $\mathbf{G}^T\mathbf{G}$ is singular. Unless there is a priori knowledge to suggest otherwise, it is probably safest to assume an overall mean model, $\hat{y}(x) = \hat{\beta}_1$. If the design has p distinct levels, then one may fit a linear regression model with at most p terms.

To allow a wider class of models it is advisable for the design to have as many distinct levels as practical, the maximum being n. However, the number of distinct levels is not a sufficient criterion to describe the quality of a design. One may readily imagine a design where with n distinct, but nearly equal, levels. In this case, the matrix $\mathbf{G}$, while of full rank, would have large condition number, and imply a large $\text{cov}(\hat{\boldsymbol{\beta}})$. One would like a criterion for designs that indicates how finely structured or complex a model one may reasonably fit. One might call such a measure the *granularity*. Intuitively, a design with all points being evenly split among p well-spaced levels should have a granularity of p and be suitable for fitting models that have p terms. Here is a possible definition the granularity.

Definition 1. *Let $\mathcal{X} \subseteq \mathbb{R}$ be the experimental domain of the factor (independent variable). Let $K(x,w)$ be a symmetric, positive semi-definite function defined for all $x, w \in \mathcal{X}$. Let $P = \{x_1, \ldots, x_n\}$ be a subset of the points in $\mathcal{X}$, with the possibility that some points are repeated. The* granularity *of the design P is defined as*

$$\operatorname{gran}(P; K) = n \left\{ \sum_{i,k=1}^{n} K(x_i, x_k) \right\}^{-1/2}.$$

First, some examples are given to show how this definition matches one's intuition. Consider the case of $\mathcal{X} = [0,1)$ and

$$K_{\text{sh}}(x,w) = 6\,|x-w|\,(|x-w|-1)+1. \tag{2}$$

For this kernel designs of evenly spaced points with an arbitrary shift have a granularity of n:

$$P_{\text{sh}} = \{[(i-1)+\Delta]/n : i = 1, \ldots, n\}, \qquad \operatorname{gran}(P_{\text{sh}}; K_{\text{sh}}) = n, \tag{3}$$

where $\Delta \in [0,1)$ is arbitrary and $n \in \mathbb{N} := \{1, 2, \ldots\}$.

For the same experimental domain another useful kernel is

$$K_{\text{SC}}(x,w) = b^{1-2\xi(x,w)}, \qquad \text{where} \tag{4}$$
$$\xi(x,w) := \max\{m \in \mathbb{N} : \lfloor b^{m-1}x \rfloor = \lfloor b^{m-1}w \rfloor\}, \qquad \lfloor x \rfloor := x - (x \mod 1),$$

and $b \geq 2$ is a fixed integer. Jittered designs with $n = 1, b, b^2, \ldots$ runs have granularity n, i.e.,

$$P_{\text{SC}} = \{[(i-1)+\Delta_i]/n : i = 1, \ldots, n\}, \qquad \operatorname{gran}(P_{\text{SC}}; K_{\text{SC}}) = n, \tag{5}$$

where the Δ_i are arbitrary numbers in $[0,1)$.

For comparison consider a simple random design:

$$P_{\text{ran}} = \{x_1, \ldots, x_n : x_i \text{ are i.i.d. on } \mathcal{X} \text{ with distribution } F\}. \tag{6}$$

The granularity is also random and one may easily compute the inverse root mean square of the inverse granularity directly from Definition 1:

$$\begin{aligned}\left[E\{\operatorname{gran}(P_{\text{ran}}; K)\}^{-2}\right]^{-1/2} = \Bigg\{ & \frac{1}{n}\int_{\mathcal{X}} K(x,x)\,dF(x) \\ & + \left(1-\frac{1}{n}\right)\int_{\mathcal{X}^2} K(x,w)\,dF(x)\,dF(w)\Bigg\}^{-1/2}. \end{aligned} \tag{7}$$

The definition of granularity may be extended to arbitrary probability distributions, and even arbitrary signed measures F (Dudley (2002)), as follows:

$$\operatorname{gran}(F;K) := \left\{ \int_{\mathcal{X}^2} K(x,w)\, dF(x)\, dF(w) \right\}^{-1/2}. \tag{8}$$

Choosing F to be the empirical distribution function of a design P in (8) recovers the original definition of granularity. The granularity of a simple random design P_{ran} approaches $\operatorname{gran}(F;K)$ as $n \to \infty$. If F is the uniform distribution, it follows that

$$\operatorname{gran}(F;K) = \infty, \quad \left[E\{\operatorname{gran}(P_{\text{ran}};K)\}^{-2}\right]^{-1/2} = n^{1/2}, \quad K = K_{\text{sh}}, K_{\text{sc}}.$$

Thus, the granularity of a simple random design is generally less than n.

Consider this design with points restricted to the left half of $[0,1)$:

$$P_{\text{left}} = \{(i-1)/(2n) : i = 1, \ldots, n\}.$$

Figure 1 compares the granularities of various designs for the kernels K_{sh} and K_{sc}. The limiting values of the granularities of P_{left} are

$$\lim_{n\to\infty} \operatorname{gran}(P_{\text{left}}; K_{\text{sh}}) = \operatorname{gran}(F_{\text{left}}; K_{\text{sh}}) = 2,$$
$$\lim_{n\to\infty} \operatorname{gran}(P_{\text{left}}; K_{\text{sc}}) = \operatorname{gran}(F_{\text{left}}; K_{\text{sc}}) = \sqrt{2},$$

where F_{left} is the uniform distribution on $[0,1/2)$. Unlike the other designs considered so far the granularities of P_{left} have finite limits as $n \to \infty$ because part of the experimental domain (in this case $[1/2,1)$) is completely uncovered by experimental points.

Remark 1. The granularity is always positive, even for signed measures, because K is positive semi-definite.

Remark 2. The *ideal design*, F_{ideal}, may be defined as the probability distribution that maximizes the granularity. A necessary and sufficient condition for F_{ideal} is that

$$\int_{\mathcal{X}} K(x,w)\, dF_{\text{ideal}}(w) \text{ is constant for all } x \in \mathcal{X}. \tag{9}$$

To prove this suppose that F and F_{ideal} are probability distributions with $G = F - F_{\text{ideal}}$, so that G is a signed measure satisfying $\int_{\mathcal{X}} dG(x) = 0$. It follows that

$$\frac{1}{[\operatorname{gran}(F;K)]^2} = \frac{1}{[\operatorname{gran}(F_{\text{ideal}};K)]^2} + 2\int_{\mathcal{X}^2} K(x,w)\, dF_{\text{ideal}}(x)\, dG(w) + \frac{1}{[\operatorname{gran}(G;K)]^2}.$$

If F_{ideal} satisfies (9), then the second term on the right vanishes, and so $\operatorname{gran}(F;K) \le \operatorname{gran}(F_{\text{ideal}};K)$. If F_{ideal} does not satisfy (9), then there is some G which makes the second term, which is linear in G, negative and larger in magnitude than the the third term, which is quadratic in G. Thus, there exists an F for which $\operatorname{gran}(F;K) > \operatorname{gran}(F_{\text{ideal}};K)$.

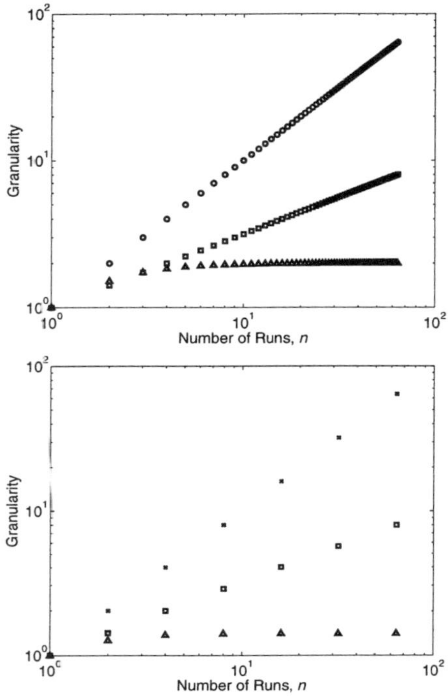

Fig. 1. Granularities defined by the kernels K_{sh} (left) and K_{SC} (right) and evaluated for P_{sh} (○), P_{SC} (×), P_{ran} (□) , and P_{left} (△).

Remark 3. Condition (9) is satisfied by the uniform distribution over the unit interval for the kernels K_{sh} and K_{SC}.

Remark 4. Definition 1 for the granularity has a functional analytic interpretation. Any symmetric, positive definite kernel K is the *reproducing kernel* for $\mathcal{H}$, some Hilbert space of functions (Aronszajn (1950) and Wahba (1990)). The inverse of the granularity is the largest possible average of function values over the design for a function with norm no larger than one (see the argument in Hickernell (2000) and elsewhere):

$$\mathrm{gran}(P;K) = \left\{ \sup_{\|g\|_{\mathcal{H}} \leq 1} \frac{1}{n} \sum_{i=1}^{n} g(x_i) \right\}^{-1} . \tag{10}$$

Remark 5. For a particular design the worst-case function that attains the above supremum is (Hickernell (1998, 2000))

$$g_{\mathrm{wor}}(x;P,K) = \frac{1}{n} \sum_{i=1}^{n} K(x,x_i).$$

The plot below shows $g_{\text{wor}}(x; P_{\text{sh}}, K_{\text{sh}})$ for $n = 8$.

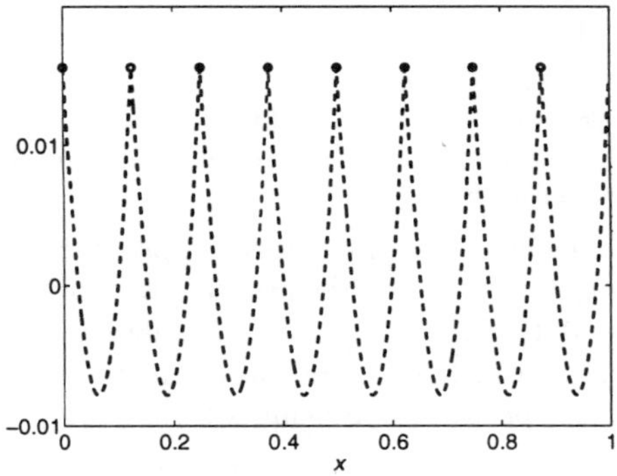

Remark 6. The granularity is related to the *discrepancy*, a crucial concept in the theory of uniform experimental designs. The discrepancy, $D(P; K, F)$, is the worst-case error for numerically approximating an integral of a function in $\mathcal{H}$ whose norm is no greater than one by the mean of function values over the design P (Niederreiter (1992), Hickernell (1998, 1999, 2000)), i.e.,

$$\begin{aligned} D(P; K, F) &:= \sup_{\|g\|_{\mathcal{H}} \le 1} \left| \int_{\mathcal{X}} g(x)\, dF(x) - \frac{1}{n} \sum_{i=1}^{n} g(x_i) \right| \\ &= \left\{ \int_{\mathcal{X}} K(x, w)\, dF(x)\, dF(w) - \frac{2}{n} \sum_{i=1}^{n} \int_{\mathcal{X}} K(x_i, w)\, dF(w) \right. \\ &\qquad \left. + \frac{1}{n^2} \sum_{i,k=1}^{n} K(x_i, x_k) \right\}^{1/2}. \end{aligned} \tag{11}$$

From this definition it follows that

$$\frac{1}{[\text{gran}(P; K)]^2} = \frac{1}{[\text{gran}(F_{\text{ideal}}; K)]^2} + [D(P; K, F_{\text{ideal}})]^2.$$

One may also define the discrepancy for an arbitrary signed measure, G, in which case the above formula holds with P replaced by G. For good designs the discrepancy tends to zero as $n \to \infty$, and the granularity tends to the granularity of F_{ideal}.

Remark 7. If a design P is composed of m copies of another design $\tilde{P}$ then their granularities are the same by Definition 1.

Remark 8. Although Definition 1 allows much freedom in the choice of the kernel, proper scaling is required for the granularity to match one's intuition. The kernel should be chosen such that $\text{gran}(F_{\text{ideal}}; K)$ equals the number of possible levels in the domain. For a continuous domain like $[0, 1)$, the number of levels is infinite, so one should ensure that $\mathrm{gran}(F_{\mathrm{ideal}}; K) = \infty$, a condition that holds for K_{sh} and K_{SC}. However, for $\mathcal{X} = \{1, \ldots, q\}$ one would want $\mathrm{gran}(F; K) = q$.

Remark 9. Another important constraint when choosing the kernel is that $\mathrm{gran}(P, K) = n$ for well-chosen designs. This is the case in (3) and (5). A well-chosen design may depend on the choice of kernel. For example, $\mathrm{gran}(P_{\mathrm{SC}}; K_{\mathrm{SC}}) = n$ for $n = b^m$, but $\mathrm{gran}(P_{\mathrm{SC}}; K_{\mathrm{sh}}) < n$ in general.

Remark 10. The discrepancy literature gives examples of many kernels, including K_{sh} and K_{SC}, but most of them are defined on the unit interval. Suppose that K is defined for $\mathcal{X} = [0, 1)$, and F_{ideal} is the ideal design. Let $\tilde{\mathcal{X}} \subseteq \mathbb{R}$ be another domain, and let $\tilde{F}$ be a probability distribution function on $\tilde{\mathcal{X}}$. Then define the following symmetric and positive semi-definite kernel:

$$\tilde{K}(x, w) = K(\tilde{F}(x), \tilde{F}(w)), \quad x, w \in \tilde{\mathcal{X}}.$$

It follows that $\tilde{F}_{\mathrm{ideal}}(x) := F_{\mathrm{ideal}}(\tilde{F}(x))$ is the ideal design for the granularity based on $\tilde{K}$ and the experimental domain $\tilde{\mathcal{X}}$. Furthermore, if $\tilde{P} \subseteq \tilde{\mathcal{X}}$ and $P := \{\tilde{F}(x) : x \in \tilde{P}\} \subseteq \mathcal{X}$, then these two designs have the same granularity, i.e., $\mathrm{gran}(\tilde{P}; \tilde{K}) = \mathrm{gran}(P; K)$. For example, if is $\tilde{\mathcal{X}} = [x_{\min}, x_{\max})$, then one may define $\tilde{K}$ as:

$$\tilde{K}(x, w) = K\left(\frac{x - x_{\min}}{x_{\max} - x_{\min}}, \frac{w - x_{\min}}{x_{\max} - x_{\min}}\right), \quad x, w \in [x_{\min}, x_{\max}).$$

and that $\tilde{F}_{\mathrm{ideal}}(x) := F_{\mathrm{ideal}}((x - x_{\min})/(x_{\max} - x_{\min}))$.

Remark 11. Relatively few kernels have been defined for discrete domains such as $\tilde{\mathcal{X}} = \{1, \ldots, q\}$ (Hickernell, Liu & Yam (2000) and Hickernell & Liu (2002)), however, one may do so using the previous remark. Given K defined for $[0, 1)$ and a q-run design $P = \{x_{(1)}, \ldots, x_{(q)}\}$ with $x_{(1)} \leq \cdots \leq x_{(q)}$ for which $\mathrm{gran}(P; K) = q$, define $\tilde{K}(i, k) := K(x_{(i)}, x_{(k)})$, $i, k \in \tilde{\mathcal{X}} = \{1, \ldots, q\}$. This kernel satisfies $\mathrm{gran}(F; \tilde{K}) = q$ if F is the uniform distribution over $\tilde{\mathcal{X}}$. For example, using the kernels K_{sh} and K_{SC} with $P = \{0, 1/q, \ldots, (q-1)/q\}$, one obtains

$$\tilde{K}_{\mathrm{sh}}(i, k) := \frac{6}{q}|i - k|\left(\frac{|i - k|}{q} - 1\right) + 1, \quad i, k \in \tilde{\mathcal{X}}$$

$$\tilde{K}_{\mathrm{SC}}(i, k) := b^{1 - 2\xi((i-1)/q, (k-1)/q)}, \quad i, k \in \tilde{\mathcal{X}}$$

where in the latter case q is assumed to be a power of b.

Remark 12. Whereas the granularity in Definition 1 is related to the discrepancy and theory of numerical integration, it may be possible to define the granularity based criteria for design quality arising in approximation theory or in optimal design theory. Discrepancy is known to be related to alphabetic optimality (Hickernell & Liu (2002)). Regardless of how granularity is defined, Remarks 1 and 7–9 should be preserved.

Remark 13. Although this section is restricted to one factor designs, many of the observations may be extended to multiple factor designs. Some of this is done in Section 4.

3 Balance for One Factor Designs

This section defines what it means for a one factor design to have perfect balance for a certain granularity, q. This is not the same as asking when a design P has $\mathrm{gran}(P; K) \geq q$.

The kernel introduced to define the granularity may be decomposed as follows (Wahba (1990)):

$$K(x, w) = \sum_{\nu=1}^{\infty} \lambda(\nu)\sigma(x, \nu)\bar{\sigma}(w, \nu) \tag{12}$$

where $^{-}$ denotes the complex conjugate. The (possibly complex) eigenfunctions $\sigma(x, \nu)$ and the eigenvalues $\lambda(\nu)$ satisfy

$$\int_{\mathcal{X}} K(x, w)\sigma(w, \nu)\, dF_{\mathrm{ideal}}(w) = \lambda(\nu)\sigma(x, \nu),$$

$$\int_{\mathcal{X}} \bar{\sigma}(x, \mu)\sigma(x, \nu)\, dF_{\mathrm{ideal}}(x) = \delta_{\mu,\nu}.$$

The eigenvalues are all non-negative because the kernel is positive semi-definite. Since F_{ideal} satisfies (9), the first eigenfunction may be labelled as $\sigma(x, 1) = 1$. The eigenfunctions with $\nu > 1$ satisfy $\int_{\mathcal{X}} \sigma(x, \nu)\, dF_{\mathrm{ideal}}(x) = 0$ and are ordered according to $\lambda(2) \geq \lambda(3) \geq \cdots$.

Based on the above eigenfunction decomposition one may write the granularity in terms of the sum of pieces:

$$\frac{1}{[\mathrm{gran}(P; K)]^2} = \sum_{\nu=1}^{\infty} \frac{1}{[\mathrm{gran}(P; K, \nu)]^2},$$

$$\mathrm{gran}(P; K, \nu) := \frac{n}{\sqrt{\lambda(\nu)}} \left| \sum_{i=1}^{n} \sigma(x_i, \nu) \right|^{-1}, \tag{13}$$

with the natural extension to signed measures, F. By (9) it follows that

$$\mathrm{gran}(F_{\mathrm{ideal}}; K, \nu) = \begin{cases} \mathrm{gran}(F_{\mathrm{ideal}}; K), & \nu = 1, \\ \infty, & \nu > 1. \end{cases} \tag{14}$$

A design with perfect balance up to a certain granularity is defined as one whose granularity pieces match those of F_{ideal}.

Definition 2. *Let the experimental domain, kernel and granularity be as in Definition 1, and let the pieces of the granularity be defined as in* (13). *A design P is* perfectly balanced *for granularity $q \in \mathbb{N}$ if* $\mathrm{gran}(P; K, 1) = \mathrm{gran}(F_{ideal}; K)$ *and* $\mathrm{gran}(P; K, \nu) = \infty$ *for $\nu = 2, \ldots, q$.*

Note that since $\sigma_1 = 1$ by convention, $\mathrm{gran}(P; K, 1) = \mathrm{gran}(F_{\mathrm{ideal}}; K)$ for any design P. Thus, any design is perfectly balanced for granularity one. For granularities $q > 1$ one only needs to check the condition $\mathrm{gran}(P; K, \nu) = \infty$ for $\nu = 2, \ldots, q$.

Some examples are given to illustrate this definition. The kernel K_{sh} defined in (2) may be decomposed in terms of trigonometric polynomials:

$$K_{\text{sh}}(x,w) = \frac{6}{\pi^2}\sum_{\nu=1}^{\infty}\frac{\cos(2\pi\nu(x-w))}{\nu^2},$$

where $\imath = \sqrt{-1}$. For an arbitrary n-run design, the pieces of the granularity are $\text{gran}(P; K_{\text{sh}}, 1) = \infty$,

$$\text{gran}(P; K_{\text{sh}}, 2\nu) = \text{gran}(P; K_{\text{sh}}, 2\nu+1) = \frac{n\pi\nu}{\sqrt{3}}\left|\sum_{i=1}^{n} e^{2\pi\imath\nu x_i}\right|^{-1}, \quad \nu \in \mathbb{N}.$$

In general a design is not perfectly balanced for granularities greater than one because even the condition $\text{gran}(P; K_{\text{sh}}, 2) = \infty$ is nontrivial. The structure of the shifted evenly spaced design, P_{sh}, fits the kernel K_{sh} so that $\text{gran}(P; K_{\text{sh}}, \nu) = \infty$ for all $\nu \geq 2$ satisfying $\nu \not\equiv 0, 1 \pmod{2n}$. This means that P_{sh} is perfectly balanced for granularity $2n-1$.

The kernel K_{SC} defined in (4) may be decomposed in terms of Walsh functions, which are piecewise constant functions. Let ${}_b0.x_{\cdot 1}x_{\cdot 2}\cdots$ denote the base b expansion of $x \in [0,1)$ and $(\cdots \nu_2\nu_1)_b$ denote the base b expansion of the non-negative integer ν, where $x_{\cdot l}$ and i_l are digits between 0 and $b-1$ inclusive. The Walsh functions, σ_{wa} are defined as

$$\sigma_{\text{wa}}(x, \nu+1) = \exp\left(\frac{2\pi\imath}{b}\sum_{l=1}^{\infty}\nu_l x_{\cdot l}\right), \quad \nu = 0, 1, \ldots.$$

The kernel K_{SC} may be written as

$$K_{\text{SC}}(x,w) = \sum_{\nu=1}^{\infty}\sigma_{\text{wa}}(x,\nu)\bar{\sigma}_{\text{wa}}(w,\nu)b^{1-2\lg(\nu-1)}, \tag{15}$$

where $\lg(\nu) := \max(0, \lfloor \log_b(\nu)\rfloor + 1)$. For an arbitrary n-run design, the pieces of the granularity are $\text{gran}(P; K_{\text{SC}}, 1) = \infty$,

$$\text{gran}(P; K_{\text{SC}}, \nu+1) = nb^{\lg(\nu)-1/2}\left|\sum_{i=1}^{n}\exp\left(\frac{2\pi\imath}{b}\sum_{l=1}^{\infty}\nu_l x_{il}\right)\right|^{-1}, \quad \nu \in \mathbb{N}.$$

Again a general design is typically not perfectly balanced for granularities ≥ 2 because even the condition $\text{gran}(P; K_{\text{sh}}, 2) = \infty$ is nontrivial. The structure of the jittered evenly spaced design, P_{SC}, fits the kernel K_{SC} so that $\text{gran}(P_{\text{SC}}; K_{\text{sh}}, \nu) = \infty$ for $\nu = 1, \ldots, n = 1, b, b^2, \ldots$. This means that P_{SC} is perfectly balanced for granularity $n = b^m$.

For simple random designs, P_{ran}, whose design points have distribution F over the domain $\mathcal{X}$ one may compute the inverse root mean square of the inverse granularity pieces as was done in (7):

$$\left[E\{\text{gran}(P_{\text{ran}};K,\nu)\}^{-2}\right]^{-1/2} = \left[\frac{\lambda(\nu)}{n}\int_{\mathcal{X}} |\sigma(x,\nu)|^2\, dF(x) \right.$$
$$\left. + \left(1-\frac{1}{n}\right)\{\text{gran}(F;K)\}^{-2}\right]^{-1/2}.$$

For $F = F_{\text{ideal}}$ this reduces to $\left[E\{\text{gran}(P_{\text{ran}};K,\nu)\}^{-2}\right]^{-1/2} = [n/\lambda(\nu)]^{1/2}$. Although the granularity of P_{ran} may be made arbitrary making n large enough, this design is never perfectly balanced for granularity ≥ 2 except in the trivial case of $0 = \lambda(2) = \lambda(3) = \cdots$. This is the case where the kernel K is constant and any design is perfectly balanced for all granularities.

It was noted that in the previous section that the design with evenly spaced points in the left of the unit interval has different limiting granularities for $n \to \infty$ for the kernels K_{sh} and K_{SC}. However, for both kernels P_{left} and F_{left} are perfectly balanced only for granularity 1 and no higher, even though $\text{gran}(F_{\text{left}};K_{\text{sh}}) = 2$. If a design, P is perfect balanced for granularity q it means that the eigenfunctions $\sigma(x,1) = 1, \sigma(x,2), \cdots, \sigma(x,q)$ have the same expectation with respect to the empirical distribution of the design as they do with respect to the ideal design, assuming that $\lambda(2) \geq \cdots \geq \lambda(q) > 0$. The eigenfunction 1 has expectation one, and the other $q-1$ eigenfunctions have expectation zero. If all pairwise products of the functions making up linear regression model (1), i.e., $g_j(x)g_l(x)$, $j,l = 1,\ldots,p$, may be written as a linear combinations of the first q eigenfunctions, then for designs, P, with perfect balance for granularity q have the same information matrix as F_{ideal}, i.e.,

$$\frac{1}{n}\mathbf{G}^T\mathbf{G} = \left(\int_{\mathcal{X}} g_j(x)g_l(x)\, dF_{\text{ideal}}(x)\right)_{j,l}.$$

Typically, the ideal design, F_{ideal}, is impossible to achieve with a limited number of runs. For example, F_{ideal} is the uniform distribution over the unit interval in some of the examples discussed above. However, a design with perfect balance for granularity q does just as well as the ideal design for certain classes of linear regression models, and this class grows with q. Returning to the example of the simple pendulum mentioned in the introduction, consider the period as a function of length only, with a fixed amplitude of 90°. For simplicity it is also assumed that air friction is negligible. Thus, a simulated experiment may be performed by solving the partial differential equation $\theta'' + (g/l)\sin(\theta) = 0$, where θ is the amplitude of the pendulum as a function of time, the g in this equation is the acceleration of gravity, and l is the length of the pendulum.

Consider a design of evenly spaced length values, $P = \{0.1\text{ m}, 0.2\text{ m}, \ldots,$ $0.9\text{ m}\}$. This 9 run, 9 level design has a granularity of 9 based on the kernels defined in Remark 11. Figure 2 shows the exact period as a function of l and some data simulated by adding random normal i.i.d. noise with a standard deviation of 1 sec to the exact values. Also shown is plot of the quadratic model in l fitted by linear regression. This model and all of its coefficients are significant, and the root mean square error of the fitted model with respect to the true period for length in the range of 0.1–0.9 m is 0.054 sec. When a cubic model is fit to this data, the cubic term is not significant.

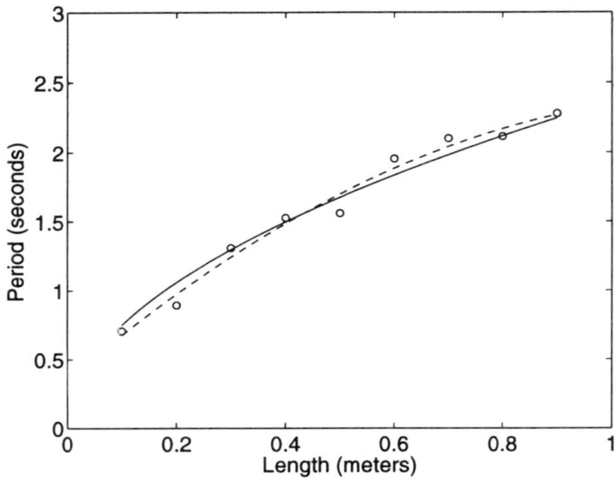

Fig. 2. Period of a pendulum as a function of length (solid), simulated experimental data (◯), and fitted value based on a quadratic model (dashed), where the amplitude is 90°.

An alternative to the above 9 level design is a 3 level optimal design for quadratic models: $\tilde{P} = \{0.1\text{ m}, 0.1\text{ m}, 0.1\text{ m},\ 0.5\text{ m}, 0.5\text{ m}, 0.5\text{ m}, 0.9\text{ m}, 0.9\text{ m}, 0.9\text{ m}\}$. This design has granularity 3. A quadratic model fit to simulated response data based on the design $\tilde{P}$ gives a somewhat better root mean square error for the fitted model of 0.047 sec. However, it is impossible to fit a cubic model using the design $\tilde{P}$ to check whether a cubic term is significant. If the noise is significantly smaller than was assumed here (unlikely for college students, but quite likely for professional engineers), then the regression model based on the design P is superior to that based on $\tilde{P}$ for two reasons. First of all, there is misspecification in the quadratic model, since the true period is proportional to $\sqrt{l}$. The higher granularity design P is more robust to this misspecification. Second of all, for smaller noise the design P allows one to fit a statistically significant regression model with higher degree polynomial terms that fits the true period better than the quadratic model.

4 Granularity for Multiple Factor Designs

This section and the next extend the concepts of granularity and perfect balance to the case of multiple ($s > 1$) factors or independent variables. The experimental domain for the jth factor is denoted $\mathcal{X}_j \subseteq \mathbb{R}$ and the full experimental domain is assumed to be the tensor product $\mathcal{X} = \mathcal{X}_1 \otimes \cdots \otimes \mathcal{X}_s \subseteq \mathbb{R}^s$. A point $\mathbf{x} = (x_1, \ldots, x_s) \in \mathcal{X}$ specifies the levels of the s different factors. For each factor there is assumed to be an associated symmetric, positive semi-definite kernel, $K_j(x_j, w_j)$, defined for all $x_j, w_j \in \mathcal{X}_j$. The full kernel is $K(\mathbf{x}, \mathbf{w}) = \prod_{j=1}^{s} K_j(x_j, w_j)$. For any set $u \subseteq \{1, \ldots, s\}$ of coordinate indices with cardinality $|u|$, let $\mathbf{x}_u$ denote the vector comprised of the x_j with $j \in u$, e.g., $\mathbf{x}_{\{1,2,4\}}$ is the 3-vector (x_1, x_2, x_4). Thus, $\mathbf{x}_u$ lies in $\mathcal{X}_u$ the projection of $\mathcal{X}$ into the coordinates indexed by u, i.e., $\mathcal{X}_{\{1,2,4\}} = \mathcal{X}_1 \otimes \mathcal{X}_2 \otimes \mathcal{X}_4$. Corresponding to $\mathcal{X}_u$ there is a kernel $K_u(\mathbf{x}_u, \mathbf{w}_u) = \prod_{j \in u} K_j(x_j, w_j)$. For any design P let P_u denote its projection into the coordinates indexed by u. In other words, P_u is the sub-design of P found by considering only the factors x_j with $j \in u$.

The definition of granularity for one factor designs in Definition 1 can be extended to multiple factor designs using the above notation. This is done below.

Definition 3. *For an experiment with s factors let $P = \{\mathbf{x}_1, \ldots, \mathbf{x}_n\}$ denote the design. The τ* factor granularity *of this design is defined for $\tau = 1, \ldots, s$ as*

$$\operatorname{gran}(P; K; \tau) = \left\{ \frac{1}{\binom{s}{\tau}} \sum_{|u|=\tau} \frac{1}{n^2} \sum_{i,k=1}^{n} K_u(\mathbf{x}_{iu}, \mathbf{x}_{ku}) \right\}^{-1/(2\tau)} \tag{16}$$

This definition reduces to Definition 1 for the case $\tau = s = 1$. It involves an average over all the kernels corresponding to τ coordinates. The τth root is taken so that the definition of granularity corresponds to one's intuition for grids.

Theorem 1. *Suppose that any projection, P_u, of the s factor design, P, into $|u| = \tau$ of its coordinates is a tensor product of one factor designs, each with granularity q. Then* $\operatorname{gran}(P; K; \tau) = q$.

Proof. When P_u is a tensor product of one factor designs, then the term $n^{-2} \sum_{i,k=1}^{n} K_u(\mathbf{x}_{iu}, \mathbf{x}_{ku})$ in (16) may be re-written as a product of τ factors, $n_j^{-2} \sum_{i,k=1}^{n_j} K_j(\mathbf{w}_{ij}, \mathbf{w}_{kj})$ for $j \in u$, where $\{w_{1j}, \ldots, w_{n_j j}\}$ is the one factor design and $n = \prod_{j \in u} n_j$. Each of these τ factors must be q^{-2}, since each one factor design has granularity q. This implies that

$$\operatorname{gran}(P; K; \tau) = \left\{ \frac{1}{\binom{s}{\tau}} \sum_{|u|=\tau} (q^{-2})^{\tau} \right\}^{-1/(2\tau)} = q,$$

as desired. □

This theorem may be applied to compute the granularities of orthogonal arrays and grids. These designs appear as tensor products of one factor designs whose points are evenly spread. Note that the definition of orthogonal arrays below makes assumptions about the granularity, which implies that the kernels must be chosen appropriately.

Definition 4. *An* orthogonal array *with $\mathbf{q} = (q_1, \ldots, q_s)$ levels and strength ρ is a design, P, all of whose ρ dimensional projections, P_u with $|u| = \rho$, are tensor products of one factor designs P_j, $j \in u$. Each P_j has q_j levels, granularity q_j, and is perfectly balanced for granularity q_j. A* grid *design is an orthogonal array of strength s.*

Corollary 1. *If P is an orthogonal array with $q = q_1 = \cdots = q_s$ levels of strength ρ, then* $\mathrm{gran}(P;K;\tau) = q$ *for $\tau = 1,\ldots,\rho$. Moreover, it follows that the number of levels is bounded above by $n^{1/\rho}$.*

Many of the observations about granularity for one factor designs can be extended to the multiple factor case. Some of these are summarized in the following theorem.

Theorem 2. *a. For arbitrary signed measures F defined on $\mathcal{X}$ with marginals F_u defined on $\mathcal{X}_u$ the definition of granularity may be extended as follows:*

$$\mathrm{gran}(F;K;\tau) = \left\{ \frac{1}{\binom{s}{\tau}} \sum_{|u|=\tau} \int_{\mathcal{X}} K_u(\mathbf{x},\mathbf{w})\, dF_u(\mathbf{x})\, dF_u(\mathbf{w}) \right\}^{-1/(2\tau)}$$

b. This granularity is positive for all designs or even signed measures.

c. The ideal design, *F_{ideal}, is defined as having independent marginals $F_{ideal,j}$ that satisfy* (9) *for experimental domains $\mathcal{X}_j$ and kernels K_j. This ideal design simultaneously maximizes all τ factor granularities.*

d. For the unit cube domain $\mathcal{X} = [0,1)^s$ and the kernels, K, defined as products of K_{sh} and/or K_{SC} the ideal design is the uniform distribution.

e. If a design P is composed of m copies of another design $\tilde{P}$ then their granularities are the same.

f. For $j = 1,\ldots,s$ let $\tilde{\mathcal{X}}_j \subseteq \mathbb{R}$ be domains, $\tilde{F}_j$ be probability distribution functions defined on $\tilde{\mathcal{X}}_j$, and $\tilde{K}_j(x_j,w_j) := K_j(\tilde{F}_j(x_j),\tilde{F}_j(w_j))$ for $x_j,w_j \in \tilde{\mathcal{X}}_j$ as in Remark 10. It follows that the ideal design for the granularity based on $\tilde{K}$ has independent marginals $\tilde{F}_{ideal,j}(x_j) := F_{ideal,j}(\tilde{F}_j(x_j))$ Furthermore, if $\tilde{P}$ is a design on $\tilde{\mathcal{X}}$, then the design $P := \{(\tilde{F}_1(x_1),\ldots,\tilde{F}_s(x_s)) : \mathbf{x} \in P\}$ has the same granularity, i.e., $\mathrm{gran}(\tilde{P};\tilde{K}) = \mathrm{gran}(P;K)$.

5 Balance for Multiple Factor Designs

Now the concept of balance is extended to multifactor designs. The kernels $K_j(x_j,w_j)$ are decomposed in terms of eigenvalues $\lambda_j(\nu)$ and eigenfunctions $\sigma_j(x_j,\nu)$ for $\nu = 1,2,\ldots$ as in (12). The τ factor granularity for $\tau = 1,\ldots,s$ may be written in terms of a sum of pieces:

$$\mathrm{gran}(P;K;\tau) = \left\{ \frac{1}{\binom{s}{\tau}} \sum_{|u|=\tau} \sum_{\boldsymbol{\nu}_u \in \mathbb{N}^u} \frac{1}{[\mathrm{gran}(P;K,\boldsymbol{\nu}_u)]^{2\tau}} \right\}^{-1/(2\tau)},$$

$$\mathrm{gran}(P;K,\boldsymbol{\nu}_u) := \left| \frac{1}{n} \sum_{i=1}^{n} \prod_{j\in u} \sqrt{\lambda_j(\nu_j)}\,\sigma_j(x_i,\nu_j) \right|^{-1/|u|}. \tag{17}$$

where $\mathbb{N} = \{1,2,3,\ldots\}$. Analogous to (14) it follows that $\mathrm{gran}(F_{\mathrm{ideal}};K,\boldsymbol{\nu}_u) = \infty$ for all $\boldsymbol{\nu}_u$ with at least one $\nu_j > 1$. A design with perfect balance up to a certain granularity is defined as one whose granularity pieces match those of F_{ideal}.

Definition 5. *For any* $\mathbf{q} \in \mathbb{N}^s$, *let* $u = \{j : q_j > 1\}$. *A design* P *is said to be perfectly balanced for the granularity* $\mathbf{q}$ *if* $\text{gran}(P; K, \boldsymbol{\nu}_u) = \text{gran}(F_{ideal}; K, \boldsymbol{\nu}_u)$ *for all* $\boldsymbol{\nu}_u \leq \mathbf{q}_u$.

Again any design is perfectly balanced for granularity $\mathbf{1}$. From this definition of perfect balance for multiple factor designs one may draw several conclusions based on the discussion in the previous two sections.

Theorem 3. *a. Suppose that the projection,* P_u, *of the* s *factor design,* P, *into the coordinates indexed by* $j \in u$ *is a tensor product of one factor designs* P_j *that are perfectly balanced for granularities* q_j. *Then* P *is perfectly balanced for granularity* $(\mathbf{q}_u, \mathbf{1})$, *the vector with elements* q_j *for* $j \in u$ *and* 1 *otherwise.*

b. If P *is an orthogonal array with* $\mathbf{q}$ *levels of strength* ρ, *then* P *is perfectly balanced for granularities* $(\mathbf{q}_u, \mathbf{1})$ *for all* $|u| \leq \rho$.

c. In the linear regression model $y_i = \beta_1 g_1(\mathbf{x}_i) + \cdots + \beta_p g_p(\mathbf{x}_i) + \varepsilon_i$, *suppose that all pairwise products* $g_k(\mathbf{x}) g_l(\mathbf{x})$ *may be written as a linear combination of the* $\prod_{j=1}^s \sigma(x_j; \nu_j)$ *for* $\boldsymbol{\nu} \leq \mathbf{q}$, *and that* $\lambda_j(q_j) > 0$ *for all* $q_j > 1$. *Then for all designs,* P, *with perfect balance for granularity* $\mathbf{q}$ *the information matrix for the design is the same as the information matrix* F_{ideal}.

Two important multiple factor experimental designs in the numerical integration literature are digital nets and integration lattices. Although they are typically used for large numbers of runs when evaluating integrals numerically, they have also been used with moderate numbers of runs for laboratory experiments (Fang & Wang (1994)). For digital nets the kernel K_{SC} is the most suitable for computing the granularity because it is based on Walsh functions, which are integrated well by nets (Larcher & Traunfellner (1994)). This is also a suitable kernel for orthogonal arrays, which are related to nets. For integration lattices, a generalization of grids, the kernel K_{sh} is the most suitable for computing the granularity because it is based on trigonometric polynomials, which are integrated well by integration lattices (Sloan & Joe (1994) and Hickernell (2000)). This kernel is also suitable for orthogonal arrays.

Table 1 gives an example of a 16 run, 8 level, 3 factor design that one might for the pendulum example from the introduction. A 16 level design is possible, but it was thought that 8 levels might be less cumbersome for the experimentalist. This design (Hickernell, Liu & Feng (2004)) comes from a digital net.

The 1, 2, and 3 factor granularities of the design in Table 1 are 8, 2.34, and 1.53, respectively, using the kernel K_{SC} with the variable transformation described in Remarks 10 and 11. This design is also perfectly balanced for granularities up to and including $(2,2,2)$, $(4,4,1)$, $(4,1,4)$, $(1,4,4)$, $(8,2,1)$, $(8,1,2)$, $(1,8,2)$, $(1,2,8)$. By contrast, a 2 level orthogonal array of strength 3 would have 1, 2, and 3 factor granularities all equal to 2 and only be balanced for granularities up to and including $(2,2,2)$. A 4 level orthogonal array of strength 2 would have 1, and 2 factor granularities all equal to 4, and be balanced for granularities up to and including $(4,4,1)$, $(4,1,4)$, $(1,4,4)$. The design in Table 1, therefore, is perfectly balanced for more different granularities than traditional orthogonal arrays. This makes it more flexible for fitting different kinds of models and more robust to model misspecification.

Table 1. High granularity design for studying the period of the pendulum

Run	Factors			Length	Amplitude	Mass
	1	2	3	(cm)	(degrees)	(kilograms)
1	1	1	1	20	10	0.04
2	1	5	5	20	50	0.12
3	2	3	7	30	30	0.16
4	2	7	3	30	70	0.08
5	3	1	3	40	10	0.08
6	3	5	7	40	50	0.16
7	4	3	5	50	30	0.12
8	4	7	1	50	70	0.04
9	5	2	6	60	20	0.14
10	5	6	2	60	60	0.06
11	6	4	4	70	40	0.10
12	6	8	8	70	80	0.18
13	7	2	8	80	20	0.18
14	7	6	4	80	60	0.10
15	8	4	2	90	40	0.06
16	8	8	6	90	80	0.14

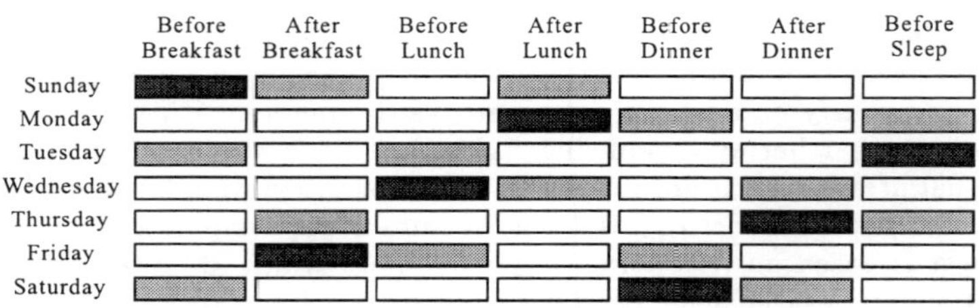

Fig. 3. High granularity design for monitoring blood glucose

The design in Figure 3 is suggested for monitoring blood glucose. There is a seven day cycle. For monitoring once a day one chooses the dark gray blocks, for monitoring twice a day one chooses the light gray blocks, and for monitoring four times a day one chooses the white blocks. Any other number of times per day may be monitored by choosing combinations of the three colors of blocks. The once per day design comes from a rank-1 integration lattice. The other colors of blocks are composed of shifted copies of this basic pattern. This design has several advantages. All times per day are covered every week for once per day monitoring, every 4 days for twice per day monitoring and every 3 days for four times per day monitoring. The designs for three and four times per day monitoring are BIB designs, where the the times of the day are treatments. In this case every pair of times per day is covered once and twice in a week, respectively, for for three and four times per day monitoring. There are other BIB designs, for example, one where once per day monitoring is done before breakfast, after breakfast, before lunch, etc. on successive days. However, this design has a lower 2 factor granularity using K_{sh} and the transformation in in Remarks 10 and 11 than the design in Figure 3. In many applications of BIB designs the ordering of the treatments is unimportant, but in cases where it is, such as this one, granularity may be a good way to distinguish among BIB designs.

6 Conclusion

At first glance an evenly spaced grid is a very appealing design because it seems to cover the experimental domain well. Unfortunately, given a limited budget of n runs and a significant number of factors, s, the number of levels per factor may be severely limited ($\leq n^{1/s}$). Orthogonal arrays with strength ρ improve upon a grid by requiring that the design only look like a grid when considering ρ or fewer factors. However, the number of levels per factor is still $\leq n^{1/\rho}$. Simple random designs allow one to have n levels per factor, but the points are no longer spread evenly. Digital nets and integration lattices improve upon both orthogonal arrays and simple random designs by maintaining an even spread of points but allowing more levels per factor. The concepts of granularity and perfect balance have been defined with grids in mind. Suppose that one only considers the factors $j \in u$ for some $u \subseteq \{1, \ldots, s\}$. An ideal n point grid for those $\tau = |u|$ factors would have q_j evenly spaced levels in the jth coordinate with $\prod_{j \in u} q_j = n$. Granularity and perfect balance have been defined so that a good design looks as much as possible like this ideal grid.

The definitions of granularity and perfect balance are related to other concepts in experimental design. Granularity is defined in terms of a symmetric, positive semi-definite kernel, as is the discrepancy. The advantage of granularity is that it has an intuitive meaning as the effective number of levels of a design. Granularity and perfect balance are related to resolution, strength, aberration and word length pattern, concepts that arise in orthogonal designs. However, these concepts usually assume a very special structure of the design, e.g., that it be a regular fractional factorial. Generalizations of these concepts have appeared in Deng & Tang (1999), Tang & Deng (1999), Fang & Mukerjee (2000), Ma & Fang (2001), Tang (2001), Xu & Wu (2001) and Hickernell & Liu (2002), and Xu (2003). The definitions here may be thought of as a further kind of generalization. They make only minimal assumptions about the structure of the design.

Acknowledgments

The author would like to congratulate Professor Kai-Tai Fang on his 65th birthday and thank Professor Fang for introducing him to the subject of experimental design. It has been a great pleasure to collaborate with Professor Fang for more than a dozen years and to learn so much from him. This research was supported in part by Hong Kong Research Grants Council grant HKBU/2007/03P.

References

Aronszajn, M. (1950). Theory of reproducing kernels. *Trans. Amer. Math. Soc.* **68** 337–404.

Atkinson, A. C. & Donev, A. N. (1992). *Optimum Experimental Designs.* Clarendon Press, Oxford.

Deng, L. Y. & Tang, B. (1999). Generalized resolution and minimum aberration criteria for Plackett-Burman and other nonregular factorial designs. *Statist. Sinica* **9** 1071–1082.

Dey, A. & Mukerjee, R. (1999). *Fractional Factorial Plans.* John Wiley & Sons, New York.

Dudley, R. M. (2002). *Real analysis and probability.* No. 74 *in* Cambridge Studies in Advanced Mathematics. Cambridge University Press, Cambridge.

Fang, K. T. (1980). Experimental design by uniform distribution. *Acta Math. Appl. Sinica* **3** 363–372.

Fang, K. T., Lin, D. K. J., Winker, P. & Zhang, Y. (2000). Uniform design: Theory and applications. *Technometrics* **42** 237–248.

Fang, K. T. & Mukerjee, R. (2000). A connection between uniformity and aberration in regular fractions of two-level factorials. *Biometrika* **87** 193–198.

Fang, K. T. & Wang, Y. (1994). *Number-Theoretic Methods in Statistics.* Chapman and Hall, New York.

Hedayat, A. S., Sloane, N. J. A. & Stufken, J. (1999). *Orthogonal Arrays: Theory and Applications.* Springer Series in Statistics, Springer-Verlag, New York.

Hickernell, F. J. (1998). A generalized discrepancy and quadrature error bound. *Math. Comp.* **67** 299–322.

Hickernell, F. J. (1999). Goodness-of-fit statistics, discrepancies and robust designs. *Statist. Probab. Lett.* **44** 73–78.

Hickernell, F. J. (2000). What affects the accuracy of quasi-Monte Carlo quadrature?, *in* H. Niederreiter & J. Spanier, eds, Monte Carlo and Quasi-Monte Carlo Methods 1998. Springer-Verlag, Berlin, pp. 16–55.

Hickernell, F. J., Liu, K. I. & Feng, K. (2004). Digital net designs. submitted for publication.

Hickernell, F. J. & Liu, M. Q. (2002). Uniform designs limit aliasing. *Biometrika* **89** 893–904.

Hickernell, F. J., Liu, M. Q. & Yam, Y. C. (2000). Discrepancy measures of uniformity. *J. Chinese Statistical Association* **38** 353–373.

Joslin Diabetes Center (2004). Goals for blood glucose control. `http://joslin.org/education/library/wbggoal.shtml`.

Larcher, G. (1998). Digital point sets: Analysis and applications. In *Random and Quasi-Random Point Sets* (P. Hellekalek & G. Larcher, eds.) of *Lecture Notes in Statistics* **138** 167–222. Springer-Verlag, New York.

Larcher, G. & Traunfellner, C. (1994). On the numerical integration of Walsh series by number-theoretic methods. *Math. Comp.* **63** 277–291.

Ma, C. X. & Fang, K. T. (2001). A note on generalized aberration in factorial designs. *Metrika* **53** 85–93.

Niederreiter, H. (1992). *Random Number Generation and Quasi-Monte Carlo Methods.* CBMS-NSF Regional Conference Series in Applied Mathematics, SIAM, Philadelphia.

Niederreiter, H. & Xing, C. (2001). *Rational Points on Curves over Finite Fields: Theory and Applications*, no. 285. In *London Mathematical Society Lecture Note Series.* Cambridge University Press, Cambridge.

Pukelsheim, F. (1993). *Optimal Design of Experiments.* John Wiley & Sons, New York.

Sloan, I. H. & Joe, S. (1994). *Lattice Methods for Multiple Integration.* Oxford University Press, Oxford.

Tang, B. (2001). Theory of J-characteristics for fractional factorial designs and projection justification of minimum G_2-aberration. *Biometrika* **88** 401–407.

Tang, B. & Deng, L. Y. (1999). Minimum G_2-aberration for nonregular fractional factorial designs. *Ann. Stat.* **27** 1914–1926.

Wahba, G. (1990). *Spline Models for Observational Data.* SIAM, Philadelphia.

Wang, Y. & Fang, K. T. (1981). A note on uniform distribution and experimental design. *Kexue Tongbao (Chinese Sci. Bull.)* **26** 485–489.

Xu, H. (2003). Minimum moment aberration for nonregular designs and supersaturated designs. *Statist. Sinica* **13** 691–708.

Xu, H. & Wu, C. F. J. (2001). Generalized minimum aberration for asymmetrical fractional factorial designs. *Ann. Stat.* **29** 549–560.

A New Class of Latin Hypercube for Computer Experiments

Scott D. Beattie[1] and Dennis K. J. Lin[2]

[1] Eli Lilly and Co. (scottbt@lilly.com)
[2] The Pennsylvania State University (DKL5@psu.edu)

Summary. Computer models can describe complicated physical phenomena. To use these models for scientific investigation, however, their generally long running times and mostly deterministic nature require a specially designed experiment. Standard factorial designs are inadequate; in the absence of one or more main effects, their replication cannot be used to estimate error but instead produces redundancy. A number of alternative designs have been proposed, but many can be burdensome computationally. This paper presents a class of Latin hypercube designs developed from the rotation of factorial designs. These rotated factorial designs are easy to construct and preserve many of the attractive properties of standard factorial designs: they have equally-spaced projections to univariate dimensions and yield uncorrelated regression effect estimates (orthogonality). They also rate comparably to maximin Latin hypercube designs by the minimum interpoint distance criterion used in the latter's construction.

Key words: Effect correlation, maximin distance, minimum interpoint distance, rotated factorial design

1 Introduction

Computer models are often used to describe complicated physical phenomena encountered in science and engineering. These phenomena are often governed by a set of equations, including linear, nonlinear, ordinary, and partial differential equations. The equations are often too difficult to be solved simultaneously by any person, but can be by a computer modeling program. These programs, due to the number and complexity of the equations, may have long running times, making their use difficult for comprehensive scientific investigation.

Contemporary Multivariate Analysis and Experimental Design—In Honor Celebration of Professor Kai-Tai Fang's 65th birthday. Edited by Jianqing Fan and Gang Li. The World Scientific Publisher, 2005.

The SOLA-PTS algorithm described in Daly & Torrey (1984), for example, has been developed at the Los Alamos National Laboratory for modeling the rapid cooling of a nuclear reactor wall as a result of cold water injected into the reactor's downcomer for containment during a nuclear accident. The authors' three-pronged goal is to study the response of the reactor, to study the turbulent mixture of the cold water and the warm fluid already in the downcomer, and to predict the onset and growth of cracks in the reactor wall as a result of the rapid cooling. This algorithm simultaneously solves eight partial differential equations with eight inputs and takes approximately 90 minutes on a Cray supercomputer to run. It solves a large number of differential equations, is very computationally expensive in running time, and has a "black box" quality – one does not know in advance which factors have large effects and one would like to examine the response over a wide range of input combinations. This algorithm is typical of computer models needing designed experiments.

One goal in this setting is to build an approximating program which, although not as precise as the computer model, would run fast enough to study the phenomenon in detail. Construction of an adequate approximating function (or program) to the computer model requires the selection of design points (a designed experiment) at which the computer model will be run to build an approximating function. Because the computer models are mostly deterministic, these computer experiments require special designs. In physical experiments, if certain factors have no effect on the response and are taken out of the approximation function (linear model), then the replicated design points in the reduced design space can be used to estimate the random error present in the system. However, with computer experiments, there is no random error – only lack of fit. Replication of classical factorial designs cannot be used to estimate this error, but instead produces redundancy. That is, they are hindered by their non-unique projections to lower dimensions.

This paper presents a new and simple strategy for designs for computer experiments, developed from the rotation of the standard factorial design to yield a Latin hypercube. Section 2 discusses a number of alternative designs that have been proposed. The following sections develop the rationale for these new designs, using the two-dimensional case for illustration (Section 3), and compare them to other previously proposed designs (Section 4). Section 5 shows the high-dimensional rotation theorems and the concluding remarks are given in Section 6.

2 Design Criteria and Related Work

Selection of an appropriate designed experiment depends to an extent on the experimental region, the model to be fit, and the method of analysis. This paper assumes the following: the experimental region is cuboidal (each factor is bound between values of interest), the true model is unknown to the experimenter and that he will approximate it by a polynomial of some degree *a priori* unknown to him, and the method of analysis will be ordinary least squares regression, although alternative methods are available (see Haaland, McMillan, Nychka & Welch (1994)).

In order to assess design criteria for computer experiments, it is valuable to study the progression of proposed designs. Koehler & Owen (1996) provide an overview of past and current approaches. The two main geometric designs are the standard (full or fractional) factorial designs and the Latin hypercube designs, but also include other traditional designs for physical experiments, such as central composite designs. Easterling (1989) points out that standard factorial designs have many attractive properties for physical experiments: balance (factor levels used an equal number of times), symmetry (permutation of design matrix columns yields same design), orthogonality (separability of main effects), collapsibility (projects to lower subspace as factorial design, sometimes redundantly), equally-spaced projections to each dimension, and straightforward measurability of main effects.

McKay, Beckman & Conover (1979) introduced the use of the Latin hypercube (LH) in computer experiments. A n-point LH design matrix is constructed by randomly permuting the integers $\{1, 2, \ldots, n\}$ for each factor and rescaling to the experimental region, so that the points project uniquely and equally-spaced to each dimension. The unique projections of LHs allow for great flexibility in model fitting. Box & Draper (1959) showed that when the true model is a polynomial of unknown degree, the *best* design (in the sense of various criteria discussed in their paper) places its points evenly spaced over the design region. Thus, equally-spaced projections are also of value. For these reasons, the LH has become the standard for computer experiments. However, random LHs are susceptible to high correlations between factors, even complete confounding, and to omitting regions of the design space.

Computer-generated designs include those of Sacks, Schiller & Welch (1989) and Sacks, Welch, Mitchell & Wynn (1989) that try to minimize the integrated mean square error (IMSE) of prediction when prediction errors are taken as a realization of a spatial stochastic process. Johnson, Moore & Ylvisaker (1990) proposed similar designs to minimize the correlations between observations when responses are taken as a realization of a spatial stochastic process. The latter authors' design D^* they call a maximin distance design if

$$\min_{x_1, x_2 \in D^*} d(x_1, x_2) = \max_{D} \min_{x_1, x_2 \in D} d(x_1, x_2), \tag{1}$$

where d is a distance measure and $\min_{x_1, x_2 \in D} d(x_1, x_2)$ is the minimum interpoint distance (MID) of design D; that is, its points are moved as far apart from one another as possible.

Attempts have been made to bridge the gap between geometric designs and computer-generated designs. Tang (1993) and Owen (1992) introduced orthogonal-array based LHs to guarantee coverage of all regions for every subset of r factors. Morris & Mitchell (1992) and Tang (1994) proposed LHs that attain the largest MID among all LHs, called maximin Latin hypercubes. Park (1994) tried to construct LHs that optimize the IMSE criterion. Owen (1994) attempted to control the correlations between design matrix columns of random LHs. These methods are a step forward in merging the good properties of Latin hypercubes with the optimization of computer-generated designs. However, being themselves computer-generated designs leaves many susceptible to the aforementioned problems.

With this in mind, we seek a new design for computer experiments with these properties: the unique and equally-spaced projections to each dimension and flexibility in model selection provided by Latin hypercube designs and the orthogonality and ease of construction provided by standard factorial designs. In addition, these new designs should perform reasonably well in terms of other criteria mentioned, such as MID, correlation, and coverage of the design space.

3 Rotated Factorial Designs in Two Dimensions

The strategy taken here is to modify the standard factorial design by rotation so as to yield a Latin hypercube. To see how this is done, first consider the standard 3^2 factorial design, represented by the 3×3 square of points and how it can be rotated to yield equally-spaced projections (see Figure 1). The key to finding all such rotations is in the relationship between points A-D. We focus on nontrivial angles between 0 and 45 degrees clockwise due to the symmetry of the rotation problem.

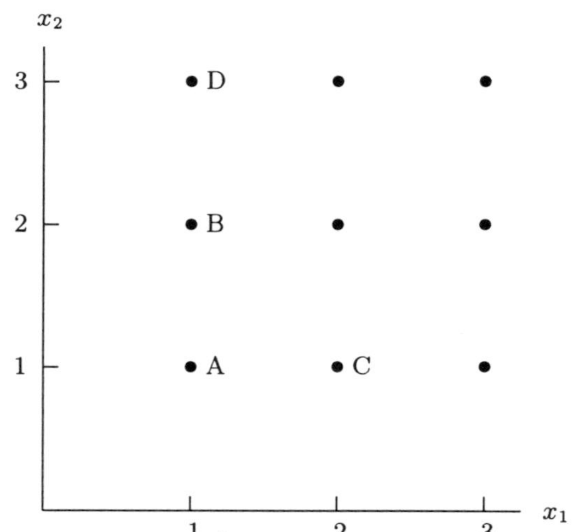

Fig. 1. Standard 3^2 factorial design before rotation

The matrix equation to rotate a set of points clockwise by an angle w about the origin is

$$[x_1\ x_2] \times \begin{bmatrix} \cos(w) & -\sin(w) \\ \sin(w) & \cos(w) \end{bmatrix}, \tag{2}$$

so that if (x_1, x_2) are the coordinates of a design point in the standard factorial design, then the rotation moves the point to $(x_1 \cos(w) + x_2 \sin(w), -x_1 \sin(w) + x_2 \cos(w))$.

Notice first that as the points are rotated clockwise about the origin that A will have the smallest x_1-coordinate for any angle between 0° and 45°. (A 45° rotation will place A directly on the x_1-axis and A is the closest point to the origin.) Also notice that the x_1-projections of points with the same initial x_1-coordinate (like A, B, and D) will be equally spaced, by $\sin(w)$, regardless of the rotation angle. Likewise, the x_1-projections of points with the same initial x_2-coordinate (like A and C) will be equally spaced, by $\cos(w)$, regardless of the rotation angle. It suffices to find all angles that make the x_1-projections of points A-D equally spaced. For the x_1-coordinates of A-D, see the table below.

point	x_1-coordinate
A	$\cos(w) + \sin(w)$
B	$\cos(w) + 2\sin(w)$
C	$2\cos(w) + \sin(w)$
D	$\cos(w) + 3\sin(w)$

Between 0° and 45°, $\sin(w) \leq \cos(w)$, so the point with the next smallest x_1-coordinate will always be B (although C will tie B when $w = 45°$) and the distance between the smallest two x_1-projections will always be $\sin(w)$. To achieve equally-spaced x_1-projections, the distance between all x_1-projections must equal $\sin(w)$. We've already seen that this is true when $w = 45°$ (equivalently, $\tan^{-1}(1)$) and both C and B have the second smallest x_1-coordinate (see Figures 2(b) and 2(c), for example).

Another possibility is that C will have the third smallest x_1-coordinate, and that the "x_1-distance" between B and C will be $\sin(w)$. However, the "x_1-distance" between B and D is always $\sin(w)$. In this case, C and D will have the same x_1-coordinate, hence

$$\cos(w) = 2\sin(w) \Longrightarrow w = \tan^{-1}(1/2).$$

Continuing in this manner, consider the case where C has the fourth smallest x_1-coordinate – after A, B, and D – and the "x_1-distance" between D and C is $\sin(w)$. Then

$$\cos(w) - 2\sin(w) = \sin(w) \Longrightarrow w = \tan^{-1}(1/3).$$

Point C cannot have the fifth smallest x_1-coordinate, so these three rotations are the only ones (again, among nontrivial angles between 0° and 45°) that yield equally-spaced x_1-projections from the 3^2 design. It is easily verified that these also yield equally-spaced x_2-projections.

Figure 2 displays the standard 3^2 factorial design, shown in open circles, and the designs that result from these rotations, shown in solid circles. Boxes are drawn around the rotated designs to identify the design regions. In practice, one would then scale this design (by subtraction and division) to the experimental region of interest. Along each axis, we have provided dot plots of the projections from which it is plain to see the equally-spaced property.

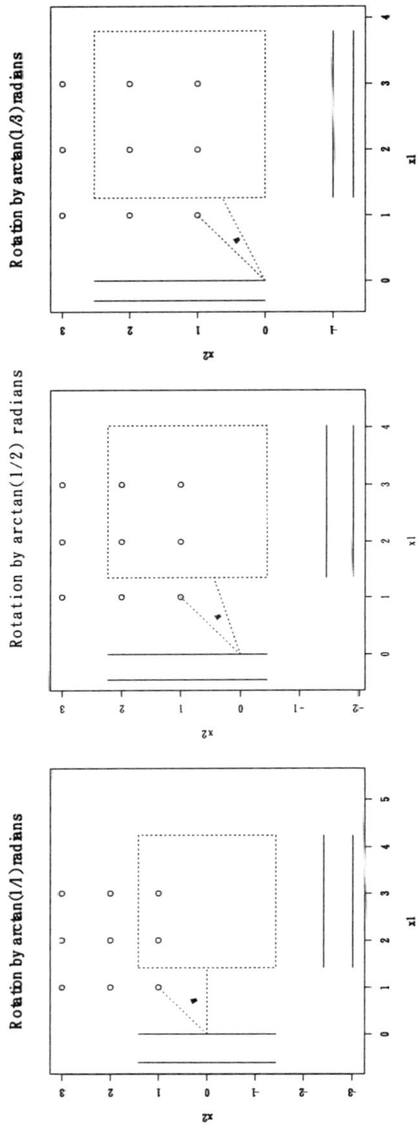

Fig. 2. Three rotations of a standard 3^2 factorial design:
(a) $w = \tan^{-1}(1)$, (b) $w = \tan^{-1}(1/2)$, (c) $w = \tan^{-1}(1/3)$

Following the argument above, a general result for factorial designs can be stated. (The proofs of Theorems 1 and 2 are straightforward and are thus omitted here.)

Theorem 1. *For nontrivial rotations between* $0°$ *and* $45°$, *a rotated standard* p^2 *factorial design will produce equally-spaced projections to each dimension if and only if the rotation angle is* $\tan^{-1}(1/k)$, *where* $k \in \{1, \ldots, p\}$.

Among the rotated standard p^2 factorial designs with equally-spaced projections, only those obtained from rotation angles of $\tan^{-1}(1/p)$ contain no redundant projections. Therefore, we define a *p^2-point rotated full factorial design* to be a rotated standard p^2 factorial design with unique, equally-spaced projections to each dimension (which is a Latin hypercube).

Theorem 2. *For a linear first-order regression model, any two-dimensional rotated factorial design has uncorrelated regression effects estimates.*

4 Two-Dimensional Subset Designs and Design Comparisons

Two-dimensional rotated full factorial designs can be easily modified to accomodate many design sizes other than p^2. After rotating the standard factorial design, remove the four most extreme points - two for each factor - to get a new design. This process can be repeated to get any design with the number of points equal to $p^2 - 4j$ for $j \in \{0, 1, \ldots, max(p-2, 0)\}$. When points are removed through this deletion process, the resulting design will no longer have the equally-spaced projection property, although it will have unique projections. We will refer to designs created by applying the deletion process to a rotated full factorial design as *Type U rotated factorial designs*, where U emphasizes these *unique* projections. Figure 3 shows the 12-point Type U rotated factorial design that is created by removing the four most extreme design points of the 16-point rotated full factorial design.

After the deletion process, these new designs can be given equally-spaced projections by adjusting the angle of rotation, although this may have the simultaneous effect of creating some redundant projections. We will refer to designs created by modifying the rotation angle of a Type U design to yield the greatest number of unique, equally-spaced projections as *Type E rotated factorial designs*, where E emphasizes the *equally-spaced* projections. Figure 4 shows the 12-point Type E rotated factorial design which has been given equally-spaced projections by adjusting the rotation angle to $tan^{-1}(2/3)$. Our preference is for Type E designs because of the equally-spaced projections, but others may choose Type U designs because of the unique projections. A complete illustration for the exact construction of $n = 16$, $n = 12$ Type U and Type E designs is given in the Appendix.

Table 1 presents the minimum interpoint distances calculated by scaling the designs to the unit square $[0, 1]^2$ and using Euclidean distance for these same

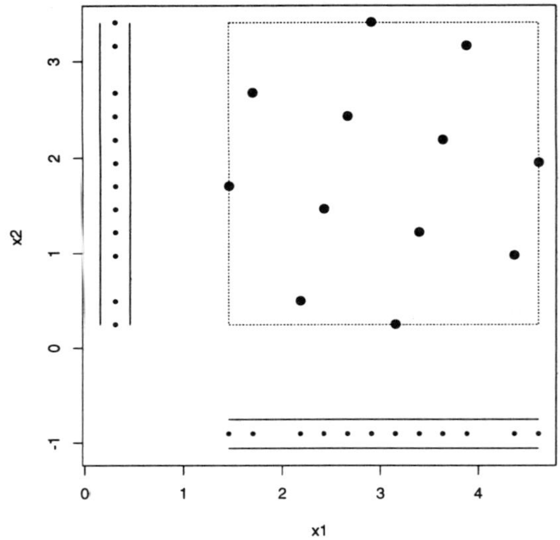

Fig. 3. 12-point $(4^2 - 4)$ Type U rotated factorial design

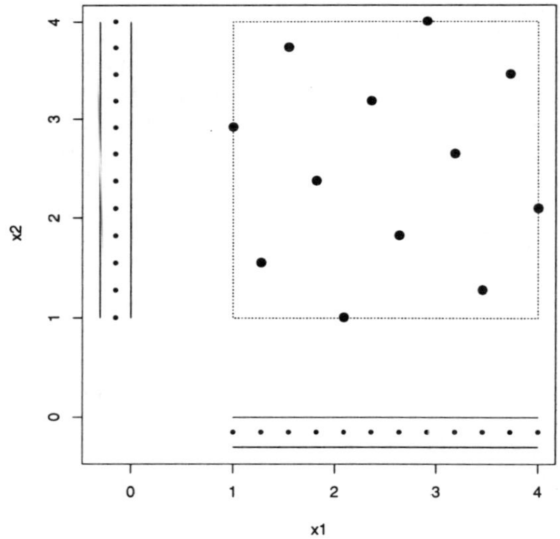

Fig. 4. 12-point $(4^2 - 4)$ Type E rotated factorial design

designs. Johnson, Moore & Ylvisaker (1990) gave ranges for the MIDs of maximin distance designs. These are listed merely as a reference for the other designs; no direct comparison will be made since maximin distance designs aren't necessarily appropriate for computer experiments (see, for example, Koehler & Owen (1996)). A few maximin distance designs were published in Johnson, Moore & Ylvisaker (1990) and in Koehler & Owen (1996) and the exact MIDs are listed for those designs.

Table 1. Minimum Interpoint Distance (MID) Comparisons for $d = 2$ Dimensional Designs

No. of Pts.	Maximin † Distance Design	Maximin ◇ Latin Hypercube	Rotated Factorial ◁ Design Type U	Rotated Factorial ◁ Design Type E
4	1.0000	.7454	.7454	.7454
5	.7071	.5590	.5270	.5590
8	.5000-1.0000	.4041	.3748	.4472
9	.5000	.3953	.3953	.3953
12	.3333-.5000	.3278	.3172	.3278
13	.3333-.5000	.3005	.2833	.3162
16	.3333	.2749	.2749	.2749
17	.2500-.3333	.2652	.2550	.2577
20	.2500-.3333	.2233	.2253	.2425

† Obtained via Johnson, Moore and Ylvisaker (1991).
◇ Obtained via Koehler and Owen (1996).
◁ Obtained by authors' algorithm.

In certain cases, the minimum interpoint distances of maximin LH and rotated factorial designs are equal – most notably when there are p^2 design points, but also when $n = 5, 12$. For $n = 8, 13, 20$, the MIDs are better for RFDs, while maximin LH designs are superior in the other listed case ($n = 17$). Maximin LH designs were constructed to have large MIDs while preserving the unique, equally-spaced projections of LHs, while RFDs were constructed to be LHs with a factorial design structure. The gains in MID from using maximin LH designs over rotated factorial designs, despite the significant increase in computer effort, are never very large when compared alongside maximin distance designs, the ideal according to minimum interpoint distance.

5 High-Dimensional Rotation Theory

Consider a standard full factorial design consisting of d factors, each with p levels. The goal is to rotate this design to convert it into a LH design, so that the p^d

points create unique and equally-spaced projections to each individual factor. For certain values of d (notably when d is a power of 2) such a rotation exists, but not for general d. The following proof proceeds in three parts: identification of the required form of the rotation matrix, construction of the power-of-2 rotation matrix, and failure of the transformation matrix to be a rotation matrix when d is not a power of two.

A p-level, d-factor standard full factorial design can be represented by a $p^d \times d$ matrix, D, with entries from $\{1, 2, \ldots, p\}$ and all p^d combinations represented.

$$D = \begin{bmatrix} 1\,1 & \cdots & 1 \cdots p\,p & \cdots & p \\ \vdots & & \vdots & & \vdots \\ 1\,1 \cdots 1 \cdots p\,p \cdots p & \cdots & 1\,1 \cdots 1 \cdots p\,p \cdots p \\ 1\,2 \cdots p \cdots 1\,2 \cdots p & \cdots & 1\,2 \cdots p \cdots 1\,2 \cdots p \end{bmatrix}^T$$

A rotation of this matrix is accomplished by post-multiplication by a $d \times d$ matrix R with the property that $R^T R = I_d$ where I_d is the $d \times d$ identity matrix. (In this section, we relax the definition of rotation to be a matrix R that satisfies $R^T R = kI_d$ for some scalar k, since the true rotation can be obtained as $(1/\sqrt{k})R$.) Let the multiplication matrix R have entries denoted as $r_{[i,j]}$, which is the entry from the ith row and jth column. Lemma 1 below will not be concerned with whether the multiplication matrix is indeed a rotation matrix, but with how such a matrix would yield unique and equally-spaced projections to each dimension.

Lemma 1. *The entries of each column of the transformation matrix R must be unique from the set $\{p^t | t = 0, 1, \ldots, d-1\}$ in order to yield unique and equally-spaced projections.*

The proof of Lemma 1 and all following lemmas and theorems are given in the Appendix.

The previous lemma shows that every column of the transformation matrix must be a permutation of the set $\{1, p, \ldots, p^{d-1}\}$ (allowing sign changes to elements and multiplication of entire columns by a constant). However, every rotation matrix R satisfies $R^T R = kI_d$, so that the sum of squares for all columns of R must be equal. Then, WLOG, every column of the transformation matrix must be a permutation of the set $\{1, p, \ldots, p^{d-1}\}$ (allowing only sign changes to elements).

It is obvious that the columns of the transformation matrix cannot be identical, for otherwise the columns of the transformed matrix would be identical. The following lemma shows that the ith entries for the d columns must be unique in magnitude in order for the transformation to be a rotation.

Lemma 2. *For a rotation matrix R, the ith entries of the d columns are unique in magnitude for all i.*

Lemmas 1 and 2 proved that all the rows and columns of the transformation matrix must be permutations of the set $\{1, p, \ldots, p^{d-1}\}$ (up to sign changes). However, this is not sufficient to guarantee that the matrix will also be a rotation.

Another requirement implied by the rotation condition $R^{\mathrm{T}}R = kI_d$ is that the columns of R must be orthogonal. Any matrix satisfying the requirements of the lemmas and this last condition will rotate factorial designs into Latin hypercubes. The remainder of this section shows how to create these matrices for d that are powers of two and illustrates why other choices of d, in general, have no such rotation matrix.

Let d be a power of 2. Let $c = \log_2 d$. Let

$$V_1 = [v_1 \quad v_2] = \begin{bmatrix} +1 & -p \\ +p & +1 \end{bmatrix}. \tag{3}$$

Now, for $c > 1$, let V_c be defined inductively from V_{c-1} as follows:

$$V_c = \begin{bmatrix} V_{c-1} & -(p^{2^{c-1}} V_{c-1})^* \\ p^{2^{c-1}} V_{c-1} & (V_{c-1})^* \end{bmatrix}, \tag{4}$$

where the operator $(\cdot)^*$ works on any matrix with an even number of rows by multiplying the entries in the top half of the matrix by -1 and leaving those in the bottom half unchanged.

Theorem 3. *The matrix V_c is a rotation of the d-factor ($d = 2^c$), p-level standard full factorial design which yields unique and equally-spaced projections to each dimension.*

Reviewing the two-dimensional result from section 3, when $d = 2 = 2^1$, equation (2) with $w = \tan^{-1}(1/p)$ can be re-expressed as

$$V_1 = \begin{bmatrix} \cos(\tan^{-1}(1/p)) & -\sin(\tan^{-1}(1/p)) \\ \sin(\tan^{-1}(1/p)) & \cos(\tan^{-1}(1/p)) \end{bmatrix} = \frac{1}{\sqrt{1+p^2}} \begin{bmatrix} +1 & -p \\ +p & +1 \end{bmatrix}, \tag{5}$$

which is the correctly scaled rotation matrix V_1 given in equation (3).

Other scaled rotation matrices for cases of interest ($d = 4, 8$ corresponding to $c = 2, 3$) are

$$V_2 = \sqrt{\frac{p^2-1}{p^8-1}} \begin{bmatrix} +1 & -p & +p^2 & -p^3 \\ +p & +1 & -p^3 & -p^2 \\ +p^2 & -p^3 & -1 & +p \\ +p^3 & +p^2 & +p & +1 \end{bmatrix} \tag{6}$$

and

$$V_3 = \sqrt{\frac{p^2-1}{p^{16}-1}} \begin{bmatrix} +1 & -p & +p^2 & -p^3 & +p^4 & -p^5 & +p^6 & -p^7 \\ +p & +1 & -p^3 & -p^2 & +p^5 & +p^4 & -p^7 & -p^6 \\ +p^2 & -p^3 & -1 & +p & -p^6 & +p^7 & +p^4 & -p^5 \\ +p^3 & +p^2 & +p & +1 & -p^7 & -p^6 & -p^5 & -p^4 \\ +p^4 & -p^5 & +p^6 & -p^7 & -1 & +p & -p^2 & +p^3 \\ +p^5 & +p^4 & -p^7 & -p^6 & -p & -1 & +p^3 & +p^2 \\ +p^6 & -p^7 & -p^4 & +p^5 & +p^2 & -p^3 & -1 & +p \\ +p^7 & +p^6 & +p^5 & +p^4 & +p^3 & +p^2 & +p & +1 \end{bmatrix}, \tag{7}$$

respectively.

The choice of rotation matrices for higher dimensions ($d > 2$) is not unique. Other inductive definitions for V_c in equation (4) are possible, namely

$$\begin{bmatrix} V_{c-1} & -p^{2^{c-1}} V_{c-1} \\ p^{2^{c-1}} V_{c-1} & V_{c-1} \end{bmatrix}. \tag{8}$$

However, the point is still clear, such rotations do exist.

Owen (1994) showed why orthogonality of design matrix columns is important in the estimation of Monte Carlo integrals and attempted to control the column correlations within Latin hypercubes. Theorem 4 will prove that all designs obtained by rotation of standard factorial designs, specifically rotated full factorial designs, will also be orthogonal. Let k be the sum of squares of the first column of X. As X is an orthogonal matrix with equal sum of squares for every column, $X^{\mathrm{T}}X = kI_d$. So $(XR)^{\mathrm{T}}(XR) = R^{\mathrm{T}}X^{\mathrm{T}}XR = R^{\mathrm{T}}kI_dR = kR^{\mathrm{T}}R = kI_d$, a diagonal matrix. Therefore, the rotated design matrix XR is an orthogonal design.

Theorem 4. *Let X be an orthogonal design matrix of n rows and d columns in which the sums of squares for columns are equal. Let R be a $d \times d$ rotation matrix. The design resulting from the matrix product XR is also an orthogonal design.*

Since computation of Monte Carlo integrals is, in effect, a computer experiment, it is beneficial for designs for computer experiments to have uncorrelated regression estimates of main effects. The following theorem shows this to be true for all designs obtained by rotation of standard full factorial designs, specifically rotated factorial designs.

Theorem 5. *Any p^d-point rotated factorial design has uncorrelated regression estimates of main effects.*

Recall that Johnson *et al.* (1990) introduced the use of minimum interpoint distance (MID) as an important design criterion (see equation (1)). It can be shown that the MID using Euclidean distance for a p^d-point rotated factorial design scaled to the unit hypercube, $[0,1]^d$, is $\sqrt{1 + p^2 + \cdots, +p^{d+1}}/(p-1) = \sqrt{(p^{2d}-1)/((p^2-1)(p-1)^2)}$. Additionally, it can be shown this is the maximal MID for $d = 2$. We are unable to obtain a formal proof for higher dimensions, however.

Table 2 lists the MIDs for several of the four-dimensional RFDs requiring fewer than 100 points and for the respective MmLH and MmU designs. Due to the computational requirements of obtaining designs from other methods, some were results not available (N/A). It is clear that the easily-constructed RFDs have similar (if not equal) MIDs to other computing-extensive constructed designs.

6 Concluding Remarks

This paper has presented a new class of experimental designs for computer experiments: the rotated factorial designs. Developed from a rotation of the standard

Table 2. MID Comparisons of Four-Dimensional MmLH, RFD, and MmU Designs

No. of Pts	Maximin Latin H-cube		Rotated Factorial Design Type U	Rotated Factorial Design Type E#	Maximin U Design	
8	0.9258	†	0.8692	0.7071 (3)	0.7954	◁
9	0.8101	†	0.5762	1.0000 (3)	0.6960	◁
10	0.7857	†	*	*	*	
11	0.7416	†	*	*	*	
12	0.7216	†	*	*	*	
16	0.6218	◇	0.6146	0.6146 (16)	0.5292	◁
24	0.5325	◇	0.3963	0.3963 (24)	N/A	
28	N/A		0.3951	0.4167 (7)	*	
36	N/A		0.3725	0.3725 (36)	N/A	
40	N/A		0.5192	0.5192 (40)	N/A	
41	0.4507	◇	0.5062	0.5062 (41)	*	
54	N/A		0.3641	0.3641 (54)	N/A	
67	N/A		0.3825	0.3825 (67)	*	
68	N/A		0.3751	0.3751 (68)	*	
81	N/A		0.3579	0.3579 (81)	N/A	

The number in parenthesis means the number of unique projected points.
* No design can be constructed as defined.
† Published in Morris & Mitchell (1992).
◇ Obtained via Morris & Mitchell (1992) algorithm by the author.
◁ Obtained by author's algorithm.

factorial design to produce a Latin hypercube, these designs have qualities that make them excellent candidates for use in today's computer experiments. The rotated full factorial designs possess the orthogonality of factorial designs and the unique and equally-spaced projections of Latin hypercubes, while maintaining a high spatial dispersion according to minimum interpoint distance. The Type U and E RFDs possess the orthogonality of factorial designs and either the unique or equally-spaced projections of Latin hypercubes, again while maintaining high spatial dispersion. All of the rotated factorial designs are extremely simple to construct, in contrast to the computer-intensive nature of most other recent designs, and perform well in terms of the minimum interpoint distance criterion used in the construction of a competing design. In terms of orthogonality, these RFDs perform even better. We have developed software to construct the rotated factorial designs presented in this paper. Users of S-Plus or C who are interested in obtaining this, please contact the authors.

Directions for future research in this area include finding alternative procedures for dimensions that are not powers of two, considering rotation of fractional factorial designs (or some other method to reduce the number of required points as d increases), and investigating the possibility of rotating mixed-level designs (perhaps as an alternative for the other dimensions). Johnson *et al.* (1990) also defined the index of a design - the number of pairs separated by the MID - as a second criterion to distinguish among several designs with identical MIDs. The performance of these designs may be investigated or modifications suggested, if and when this criterion becomes relevant. Some related recent work can be found in Bursztyn and Steinberg (2001, 2002).

Acknowledgements

This work was partially supported by the *National Science Foundation* via Grant DMS-9704711, *National Security Agency* via Grant MDA904-02-1-0054, and *National Science Council of ROC* via Contract NSC 87-2119-M-001-007. Professor Kai-Tai Fang has been a true leader in our society and has been a strong supporter of young fellows. His original work on uniform design had a significant impact on this work (see, for example, Fang and Lin (2003) and Fang, Lin, Winker and Zhang (2000)). It is our great privilege to contribute this work on computer experiments to this special issue in honor of his 65th birthday.

Appendices

6.1 A Sample Construction

(1) A 4^2=16-run rotated factorial design.
Start with a 4^2 standard factorial design.

$$\begin{bmatrix} 1\,1\,1\,1\,2\,2\,2\,2\,3\,3\,3\,3\,4\,4\,4\,4 \\ 1\,2\,3\,4\,1\,2\,3\,4\,1\,2\,3\,4\,1\,2\,3\,4 \end{bmatrix}^T$$

Rotate by $\tan^{-1}(1/4)$. This yields a 16-point rotated factorial design.

$$\begin{bmatrix}
1\cos(\tan^{-1}(1/4)) + 1\sin(\tan^{-1}(1/4)) & -1\sin(\tan^{-1}(1/4)) + 1\cos(\tan^{-1}(1/4)) \\
1\cos(\tan^{-1}(1/4)) + 2\sin(\tan^{-1}(1/4)) & -1\sin(\tan^{-1}(1/4)) + 2\cos(\tan^{-1}(1/4)) \\
1\cos(\tan^{-1}(1/4)) + 3\sin(\tan^{-1}(1/4)) & -1\sin(\tan^{-1}(1/4)) + 3\cos(\tan^{-1}(1/4)) \\
1\cos(\tan^{-1}(1/4)) + 4\sin(\tan^{-1}(1/4)) & -1\sin(\tan^{-1}(1/4)) + 4\cos(\tan^{-1}(1/4)) \\
2\cos(\tan^{-1}(1/4)) + 1\sin(\tan^{-1}(1/4)) & -2\sin(\tan^{-1}(1/4)) + 1\cos(\tan^{-1}(1/4)) \\
2\cos(\tan^{-1}(1/4)) + 2\sin(\tan^{-1}(1/4)) & -2\sin(\tan^{-1}(1/4)) + 2\cos(\tan^{-1}(1/4)) \\
2\cos(\tan^{-1}(1/4)) + 3\sin(\tan^{-1}(1/4)) & -2\sin(\tan^{-1}(1/4)) + 3\cos(\tan^{-1}(1/4)) \\
2\cos(\tan^{-1}(1/4)) + 4\sin(\tan^{-1}(1/4)) & -2\sin(\tan^{-1}(1/4)) + 4\cos(\tan^{-1}(1/4)) \\
3\cos(\tan^{-1}(1/4)) + 1\sin(\tan^{-1}(1/4)) & -3\sin(\tan^{-1}(1/4)) + 1\cos(\tan^{-1}(1/4)) \\
3\cos(\tan^{-1}(1/4)) + 2\sin(\tan^{-1}(1/4)) & -3\sin(\tan^{-1}(1/4)) + 2\cos(\tan^{-1}(1/4)) \\
3\cos(\tan^{-1}(1/4)) + 3\sin(\tan^{-1}(1/4)) & -3\sin(\tan^{-1}(1/4)) + 3\cos(\tan^{-1}(1/4)) \\
3\cos(\tan^{-1}(1/4)) + 4\sin(\tan^{-1}(1/4)) & -3\sin(\tan^{-1}(1/4)) + 4\cos(\tan^{-1}(1/4)) \\
4\cos(\tan^{-1}(1/4)) + 1\sin(\tan^{-1}(1/4)) & -4\sin(\tan^{-1}(1/4)) + 1\cos(\tan^{-1}(1/4)) \\
4\cos(\tan^{-1}(1/4)) + 2\sin(\tan^{-1}(1/4)) & -4\sin(\tan^{-1}(1/4)) + 2\cos(\tan^{-1}(1/4)) \\
4\cos(\tan^{-1}(1/4)) + 3\sin(\tan^{-1}(1/4)) & -4\sin(\tan^{-1}(1/4)) + 3\cos(\tan^{-1}(1/4)) \\
4\cos(\tan^{-1}(1/4)) + 4\sin(\tan^{-1}(1/4)) & -4\sin(\tan^{-1}(1/4)) + 4\cos(\tan^{-1}(1/4))
\end{bmatrix}$$

$$= \begin{bmatrix}
1.21 & 0.73 \\
1.46 & 1.70 \\
1.70 & 2.67 \\
1.94 & 3.64 \\
2.18 & 0.49 \\
2.43 & 1.46 \\
2.67 & 2.43 \\
2.91 & 3.40 \\
3.15 & 0.24 \\
3.40 & 1.21 \\
3.64 & 2.18 \\
3.88 & 3.15 \\
4.12 & 0.00 \\
4.37 & 0.97 \\
4.61 & 1.94 \\
4.85 & 2.91
\end{bmatrix}$$

This can be rescaled to be a 16-point Latin hypercube by multiplying by 15/3.64 then subtracting 3.99 from the first column and adding 1.00 to the second column.

$$\begin{bmatrix}
1 & 2 & 3 & 4 & 5 & 6 & 7 & 8 & 9 & 10 & 11 & 12 & 13 & 14 & 15 & 16 \\
4 & 8 & 12 & 16 & 3 & 7 & 11 & 15 & 2 & 6 & 10 & 14 & 1 & 5 & 9 & 13
\end{bmatrix}^T$$

(2) A 12-run Type U design.
To constuct a 12-point Type U design, remove the 4 most extreme design points (from the prescaled matrix): the 1st, 4th, 13th, and 16th.

(3) A 12-run Type E design.
To get a 12-point Type E rotated factorial design, adjust the rotation angle to $\tan^{-1}(2/3)$. Figuring out the correct rotation angle is easy. If the original design has p^2 points, then the angle is unadjusted if 0 points are removed and is adjusted to $\tan^{-1}(1/(p-1))$ if $\{2, 4, \ldots, 2p-2\}$ points are removed or to $\tan^{-1}(1/(p-2))$

if $\{2p, 2p+2, \ldots, 4p-8\}$ points are removed. However, there is one exception to this rule: if the new design has an even number of points which exceed a square by 3, then the angle is adjusted to $\tan^{-1}(2/(p-1))$. (Note that 12 is such a number, making the rotation angle $\tan^{-1}(2/3)$.)

$$\begin{bmatrix} 1\cos(\tan^{-1}(2/3)) + 2\sin(\tan^{-1}(2/3)) & -1\sin(\tan^{-1}(2/3)) + 2\cos(\tan^{-1}(2/3)) \\ 1\cos(\tan^{-1}(2/3)) + 3\sin(\tan^{-1}(2/3)) & -1\sin(\tan^{-1}(2/3)) + 3\cos(\tan^{-1}(2/3)) \\ 2\cos(\tan^{-1}(2/3)) + 1\sin(\tan^{-1}(2/3)) & -2\sin(\tan^{-1}(2/3)) + 1\cos(\tan^{-1}(2/3)) \\ 2\cos(\tan^{-1}(2/3)) + 2\sin(\tan^{-1}(2/3)) & -2\sin(\tan^{-1}(2/3)) + 2\cos(\tan^{-1}(2/3)) \\ 2\cos(\tan^{-1}(2/3)) + 3\sin(\tan^{-1}(2/3)) & -2\sin(\tan^{-1}(2/3)) + 3\cos(\tan^{-1}(2/3)) \\ 2\cos(\tan^{-1}(2/3)) + 4\sin(\tan^{-1}(2/3)) & -2\sin(\tan^{-1}(2/3)) + 4\cos(\tan^{-1}(2/3)) \\ 3\cos(\tan^{-1}(2/3)) + 1\sin(\tan^{-1}(2/3)) & -3\sin(\tan^{-1}(2/3)) + 1\cos(\tan^{-1}(2/3)) \\ 3\cos(\tan^{-1}(2/3)) + 2\sin(\tan^{-1}(2/3)) & -3\sin(\tan^{-1}(2/3)) + 2\cos(\tan^{-1}(2/3)) \\ 3\cos(\tan^{-1}(2/3)) + 3\sin(\tan^{-1}(2/3)) & -3\sin(\tan^{-1}(2/3)) + 3\cos(\tan^{-1}(2/3)) \\ 3\cos(\tan^{-1}(2/3)) + 4\sin(\tan^{-1}(2/3)) & -3\sin(\tan^{-1}(2/3)) + 4\cos(\tan^{-1}(2/3)) \\ 4\cos(\tan^{-1}(2/3)) + 2\sin(\tan^{-1}(2/3)) & -4\sin(\tan^{-1}(2/3)) + 2\cos(\tan^{-1}(2/3)) \\ 4\cos(\tan^{-1}(2/3)) + 3\sin(\tan^{-1}(2/3)) & -4\sin(\tan^{-1}(2/3)) + 3\cos(\tan^{-1}(2/3)) \end{bmatrix}$$

$$= \begin{bmatrix} 1.94 & 1.11 \\ 2.50 & 1.94 \\ 2.22 & -0.28 \\ 2.77 & 0.55 \\ 3.33 & 1.39 \\ 3.88 & 2.22 \\ 3.05 & -0.83 \\ 3.61 & 0.00 \\ 4.16 & 0.83 \\ 4.71 & 1.66 \\ 4.44 & -0.55 \\ 4.99 & 0.28 \end{bmatrix}$$

Once constructed, these designs can be rescaled to the experimental region. For example, to convert the 12-point Type E design matrix to LH notation, multiply by 11/3.05 then subtract 6.00 from the first column and add 3.99 to the second column.

$$\begin{bmatrix} 1 & 2 & 3 & 4 & 5 & 6 & 7 & 8 & 9 & 10 & 11 & 12 \\ 8 & 11 & 3 & 6 & 9 & 12 & 1 & 4 & 7 & 10 & 2 & 5 \end{bmatrix}^T$$

6.2 Proofs

Lemma 1: *The entries of each column of the transformation matrix R must be unique from the set $\{p^t | t = 0, 1, 1 \ldots, d-1\}$ in order to yield unique and equally-spaced projections.*

Proof: The multiplication of the factorial design by R yields a new $p^d \times d$ matrix X with entries labeled $x_{[i,j]}$:

$$X = D \times R$$

Note that the values of the jth column of X depend on the jth column of matrix R, but on none of its other columns. Without loss of generality (WLOG), consider only the first column of these matrices and examine how the choices of $r_{[1,1]}, \ldots, r_{[d,1]}$ affect the values of $x_{[1,1]}, \ldots x_{[p^d,1]}$.

The rows of the factorial design matrix can be arranged into p^{d-1} groups of rows where the rows in each group are identical in $d-1$ columns but unique in one column. They can be arranged, WLOG, as above with the first $d-1$ columns identical and the last column unique. Within each group the transformed coordinates differ only in respect to the value of $r_{[d,1]}$:

$$\begin{aligned}
x_{[(i_1-1)p^{d-1}+(i_2-1)p^{d-2}+\cdots+(i_{d-1}-1)p+1,1]} &= i_1 r_{[1,1]} + i_2 r_{[2,1]} + \cdots i_{d-1} r_{[d-1,1]} \\
&\quad +1 r_{[d,1]} \\
x_{[(i_1-1)p^{d-1}+(i_2-1)p^{d-2}+\cdots+(i_{d-1}-1)p+2,1]} &= i_1 r_{[1,1]} + i_2 r_{[2,1]} + \cdots i_{d-1} r_{[d-1,1]} \\
&\quad +2 r_{[d,1]} \\
&\;\;\vdots \\
x_{[(i_1-1)p^{d-1}+(i_2-1)p^{d-2}+\cdots+(i_{d-1}-1)p+p,1]} &\quad i_1 r_{[1,1]} + i_2 r_{[2,1]} + \cdots i_{d-1} r_{[d-1,1]} \\
&\quad +p r_{[d,1]}
\end{aligned}$$

where $i_1, \ldots, i_{d-1} \in \{1, \ldots, p\}$. For these points to be unique and equally-spaced requires only that $r_{[d,1]} \neq 0$. Let $r_{[d,1]} = 1$ (or -1), WLOG, so that the transformed points within any group differ by one unit and there are p^{d-1} such groups.

Now arrange the factorial design matrix into p^{d-2} groups of p^2 rows so that the rows within each group are identical in the first $d-2$ columns,subgrouped as before by the last column, and unique (by subgroups) in the $(d-1)$th column. For any group, examine the jth transformed point within each subgroup. Then their transformed coordinates differ only in respect to the value of $r_{[d-1,1]}$:

$$\begin{aligned}
x_{[(i_1-1)p^{d-1}+\cdots+(i_{d-2}-1)p^2+j,1]} &= i_1 r_{[1,1]} + \cdots i_{d-2} r_{[d-2,1]} + 1 r_{[d-1,1]} \\
&\quad +j r_{[d,1]} \\
x_{[(i_1-1)p^{d-1}+\cdots+(i_{d-2}-1)p^2+p+j,1]} &= i_1 r_{[1,1]} + \cdots i_{d-2} r_{[d-2,1]} + 2 r_{[d-1,1]} \\
&\quad +j r_{[d,1]} \\
&\;\;\vdots \\
x_{[(i_1-1)p^{d-1}+\cdots+(i_{d-2}-1)p^2+(p-1)p+j,1]} &= i_1 r_{[1,1]} + \cdots i_{d-2} r_{[d-2,1]} + p r_{[d-1,1]} \\
&\quad +j r_{[d,1]}
\end{aligned}$$

where $i_1, \ldots, i_{d-2}, j \in \{1, \ldots, p\}$. For these p points to be unique and equally-spaced requires only that $r_{[d-1,1]} \neq 0$. However, for all p^2 points within the group to be unique and equally-spaced requires that $r_{[d-1,1]} = \pm p$, since each of the p listed points represents one subgroup of p points differing by one unit. (Note that $r_{[d-1,1]} = \pm 1/p$ also satisfies the requirement; but then the entire matrix R could be multiplied by p to obtain $r_{[d-1,1]} = \pm 1$ and $r_{[d,1]} = \pm p$, essentially the same transformation.) This strategy yields p^{d-2} groups of p^2 points where the transformed points within any group differ by one unit.

Continuing in this manner, it is clear that to yield unique and equally-spaced points in the transformed space, the values of $r_{[1,1]}, \ldots, r_{[d,1]}$ must be unique from the set $\{1, p, \ldots, p^{d-1}\}$ (up to sign changes and multiplication by a constant). As the choice of columns to examine was arbitrary, so must the entries from each column of the transformation matrix be of this form.

□

Lemma 2: *For a rotation matrix R, the ith entries of the d columns are unique in magnitude for all i.*

Proof: Assume that R is a rotation matrix. Then $R^T R = kI_d$, which implies that $RR^T = kI_d$. This says that the sum of squares for all rows (in addition to columns) of R must be equal to

$$\sum_{j=0}^{d-1} p^{2j} = (p^{2d} - 1)/(p^2 - 1).$$

Suppose that one row has two (or more) entries with magnitude equal to p^{d-1}. Then its sum of squares is greater than

$$2p^{2(d-1)} > (2(p^2 - 1)/p^2)((p^{2d} - 1)/(p^2 - 1)).$$

Note that, since $p \geq 2$, we have $2(p^2 - 1)/p^2 > 1$. Thus its sum of squares is greater than $(p^{2d} - 1)/(p^2 - 1)$, and this row has greater sum of squares than is possible. Thus each row has exactly one entry with magnitude equal to p^{d-1}.

Now, suppose that one row has two (or more) entries with magnitude equal to p^{d-2}. Examining the sum of squares of that row shows that it too is larger than is possible. Therefore each row has exactly one entry with magnitude equal to p^{d-2}. Continuing in this manner, since d is finite, proves the lemma.

□

Theorem 1: *The matrix V_c is a rotation of the d-factor ($d = 2^c$), p-level standard full factorial design which yields unique and equally-spaced projections to each dimension.*

Proof: It suffices to show for each $c \geq 1$ that V_c is comprised of rows and columns of permutations of the set $\{1, p, p^2, \ldots, p^{d-1}\}$ (up to sign changes) and that the columns are all orthogonal. The proof proceeds by induction.

First, consider the simplest case where $d = 2$ and $c = 1$. Clearly V_1 meets these criteria and is therefore a rotation satisfying the projection criteria.

Suppose now that V_{c-1} is a rotation satisfying the projection criteria. If this implies that V_c is also such a rotation, the proof is completed.

Note these observations:

1. V_{c-1} is comprised of rows and columns of permutations of $\{1, p, p^2, \ldots, p^{2^{c-1}-1}\}$ (up to sign changes).

2. $V_{c-1}^{\mathrm{T}} V_{c-1} = k' I_{2^{c-1}}$, where $k' = 1 + p^2 + \ldots + p^{2^{c-1}}$.
3. The rows and columns of $p^{2^{c-1}} V_{c-1}$ are permutations (up to sign changes) of $\{p^{2^{c-1}}, p^{2^{c-1}+1}, \ldots, p^{2^c-1}\}$.
4. The operator $(\cdot)^*$ does not affect the magnitudes of entries in a matrix, only their signs.

From these 4 observations it follows that V_c is comprised of rows and columns of permutations of $\{1, p, p^2, \ldots, p^{2^c-1}\}$ (up to sign changes). All that remains to show is that the columns of V_c are orthogonal.

Recall that for an arbitrary matrix subdivided into 4 submatrices A $(n_1 \times p_1)$, B $(n_2 \times p_1)$, C $(n_1 \times p_2)$, and D $(n_2 \times p_2)$,

$$\begin{bmatrix} A & C \\ B & D \end{bmatrix}^{\mathrm{T}} \begin{bmatrix} A & C \\ B & D \end{bmatrix} = \begin{bmatrix} A^{\mathrm{T}} & B^{\mathrm{T}} \\ C^{\mathrm{T}} & D^{\mathrm{T}} \end{bmatrix} \begin{bmatrix} A & C \\ B & D \end{bmatrix} = \begin{bmatrix} A^{\mathrm{T}}A + B^{\mathrm{T}}B & A^{\mathrm{T}}C + B^{\mathrm{T}}D \\ C^{\mathrm{T}}A + D^{\mathrm{T}}B & C^{\mathrm{T}}C + D^{\mathrm{T}}D \end{bmatrix}. \tag{9}$$

Letting $A = V_{c-1}$, $B = p^{2^{c-1}} V_{c-1}$, $C = -(p^{2^{c-1}} V_{c-1})^*$, and $D = (V_{c-1})^*$, it suffices to show that $A^{\mathrm{T}}A + B^{\mathrm{T}}B = C^{\mathrm{T}}C + D^{\mathrm{T}}D = kI_{2^{c-1}}$ for some k and that $A^{\mathrm{T}}C + B^{\mathrm{T}}D = 0$.

First,

$$A^{\mathrm{T}}A = V_{c-1}^{\mathrm{T}} V_{c-1} = k' I_{2^{c-1}} \tag{10}$$

and

$$B^{\mathrm{T}}B = p^{2^{c-1}} V_{c-1}^{\mathrm{T}} p^{2^{c-1}} V_{c-1} = p^{2^c} V_{c-1}^{\mathrm{T}} V_{c-1} = p^{2^c} k' I_{2^{c-1}}. \tag{11}$$

Thus $A^{\mathrm{T}}A + B^{\mathrm{T}}B = (1 + p^{2^c}) k' I_{2^{c-1}}$. Now for simplicity let for any matrix M with even number of rows

$$(M)^* = \begin{bmatrix} -M^{(1)} \\ M^{(2)} \end{bmatrix}, \tag{12}$$

where $M^{(1)}$ and $M^{(2)}$ are the top and bottom half of M, respectively. Then

$$\begin{aligned} D^{\mathrm{T}}D &= (V_{c-1})^{*\mathrm{T}} (V_{c-1})^* \\ &= \begin{bmatrix} -V_{c-1}^{(1)} \\ V_{c-1}^{(2)} \end{bmatrix}^{\mathrm{T}} \begin{bmatrix} -V_{c-1}^{(1)} \\ V_{c-1}^{(2)} \end{bmatrix} \\ &= \begin{bmatrix} -V_{c-1}^{(1)\mathrm{T}} & V_{c-1}^{(2)\mathrm{T}} \end{bmatrix} \begin{bmatrix} -V_{c-1}^{(1)} \\ V_{c-1}^{(2)} \end{bmatrix} \\ &= V_{c-1}^{(1)\mathrm{T}} V_{c-1}^{(1)} + V_{c-1}^{(2)\mathrm{T}} V_{c-1}^{(2)} \\ &= \begin{bmatrix} V_{c-1}^{(1)\mathrm{T}} & V_{c-1}^{(2)\mathrm{T}} \end{bmatrix} \begin{bmatrix} V_{c-1}^{(1)} \\ V_{c-1}^{(2)} \end{bmatrix} \\ &= \begin{bmatrix} V_{c-1}^{(1)} \\ V_{c-1}^{(2)} \end{bmatrix}^{\mathrm{T}} \begin{bmatrix} V_{c-1}^{(1)} \\ V_{c-1}^{(2)} \end{bmatrix} \\ &= V_{c-1}^{\mathrm{T}} V_{c-1} \\ &= k' I_{2^{c-1}} \end{aligned} \tag{13}$$

and

$$\begin{aligned} C^{\mathrm{T}} C &= [-(p^{2^{c-1}} V_{c-1})^*]^{\mathrm{T}} [-(p^{2^{c-1}} V_{c-1})^*] \\ &= p^{2^c} [(V_{c-1})^*]^{\mathrm{T}} (V_{c-1})^* \\ &= p^{2^c} k' I_{2^{c-1}}. \end{aligned} \tag{14}$$

Thus $C^{\mathrm{T}} C + D^{\mathrm{T}} D = (1 + p^{2^c}) k' I_{2^{c-1}} = A^{\mathrm{T}} A + B^{\mathrm{T}} B$. Finally,

$$A^{\mathrm{T}} C = V_{c-1}^{\mathrm{T}} [-(p^{2^{c-1}} V_{c-1})^*] = -p^{2^{c-1}} V_{c-1}^{\mathrm{T}} (V_{c-1})^* \tag{15}$$

and

$$B^{\mathrm{T}} D = [p^{2^{c-1}} V_{c-1}]^{\mathrm{T}} (V_{c-1})^* = p^{2^{c-1}} V_{c-1}^{\mathrm{T}} (V_{c-1})^*. \tag{16}$$

Thus $A^{\mathrm{T}} C + B^{\mathrm{T}} D = 0$, and the columns of V_c are orthogonal. By the principle of mathematical induction, for all $c \geq 1$, V_c is a rotation of the d-factor ($d = 2^c$), p-level standard full factorial design which yields unique and equally-spaced projections to each dimension. That is, V_c turns standard factorial designs into Latin hypercubes.

Theorem 5: *Any p^d-point rotated factorial design has uncorrelated regression estimates of main effects.*

Proof: Let N be the model matrix with p^d rows and $d + 1$ columns: a first column (x_0) of 1s for an intercept and d centered and scaled columns ($x_1, \ldots, x_d$) representing the standard p^d full factorial design. The columns are centered so that $x_0^T x_i = \sum_{j=1}^{p^d} = 0$ for all $i = 1, \ldots, d$ and scaled so that $x_i^T x_i = \sum_{j=1}^{p^d} x_{ij}^2 = p^d$. Since each level of any one factor is used in combination with all other levels of any other factor, we have $x_0^T = 0$ for all $i \neq k$. That is the matrix $X^T X = p^d I_{d+1}$.

Let R be the $D \times d$ rotation matrix which transforms the factorial design into a rotated factorial design. Since the rotation matrix does not affect the intercept, the associated tranformation matrix on the model matrix X is the $(d+1) \times (d+1)$ matrix

$$R^* = \begin{bmatrix} 1 & 0 \\ 0 & R \end{bmatrix}.$$

Then

$$(XR^*)^T (XR^*) = p^d R^{*T} R^* = p^d \begin{bmatrix} 1 & 0 \\ 0 & I_d \end{bmatrix} p^d I_{d+1}.$$

It follows that the regression estimates of the main effects are uncorrelated.

References

Beattie, S. D. (1999). Contributions To the Design and Analysis of Experiments. PhD Thesis. The Pennsylvania State University, Department of Statistics.

Box, G. E. P. and Draper, N. R. (1959). A basis for the selection of a response surface design. *Journal of the American Statistical Association* **54** 622–654.

Bursztyn, D. and Steinberg D. M. (2001). Rotation designs for experiments in high bias situations. *Journal of Statistical Planning and Inference* **97** 399–414.

Bursztyn, D. and Steinberg D. M. (2002). Rotation designs: Orthogonal first-order designs with higher order projectivity. *Journal of Applied Stochastic Models in Business and Industry* **18** 197–206.

Crosier, R. B. (1993). Method for design rotation. Technical Report ERDEC-TR-099. Edgewood Research, Development & Engineering Center, U.S. Army Chemical and Biological Defense Agency, Aberdeen Proving Ground, Maryland 21010-5423.

Currin, C., Mitchell, T., Morris, M. & Ylvisaker (1991). Bayesian prediction of deterministic functions, with applications to the design and analysis of computer experiments. *Journal of the American Statistical Association* **86** 953–963.

Daly, B. J. & Torrey, M. D. (1984). SOLA-PTS: A transient, three-dimensional algorithm for fluid-thermal mixing and wall heat transfer in complex geometries. Technical Report LA-10132-MS. Los Alamos National Laboratory.

DeFeo, P. & Myers, R. H. (1992). A new look at experimental design robustness. *Biometrika* **79** 375–380.

Easterling, R. G. (1989). Comment: Design and analysis of computer experiments. *Statistical Science* **4** 425–427.

Fang, K. T. & Lin, D. K. J. (2003). Uniform Experimental Design and Its Applications in Industry. In *Handbook of Statistics in Industry* (C. R. Rao & R. Khattree, eds.) of *Handbook of Statistics in Industry* **22** Chapter 4 131–170. North Holland: New York.

Fang, K. T., Lin, D. K. J., Winker, P. & Zhang, Y. (2000). Uniform Design: Theory and Application. *Technometrics* **42** 237–248.

Haaland, P., McMillan, N., Nychka, D. & Welch, W. (1994). Analysis of space-filling designs. In *Computationally Intensive Statistical Methods* (J. Sall & A. Lehman, eds.) **26** of *Computing Science and Statistics* 111–120. Interface Foundation of North America. Proceedings of the 26th Symposium on the Interface.

Johnson, M. E., Moore, L. M. & Ylvisaker, D. (1990). Minimax and maximin distance designs. *Journal of Statistical Planning and Inference* **26** 131–148.

Koehler, J. R. & Owen, A. B. (1996). Computer experiments. In *Handbook of Statistics* (S. Ghosh & C. R. Rao, eds.) **13** Chapter 9 261–308. Elsevier Science B. V.

Liao, W. H. & Wang, K. W. (1996). Analysis and design of viscoelastic materials for active constrained layer damping treatments. In *Proceedings of SPIE Conference on Smart Structures and Materials* **2720** 212–223.

Lucas, J. M. (1996). Comments on computer experiments. *Journal of the American Statistical Association* **38** 197–199.

McKay, M. D., Beckman, R. J. & Conover, W. J. (1979). A comparison of three methods for selecting values of input variables in the analysis of output from a computer code. *Technometrics* **21** 239–245.

Morris, M. D. & Mitchell, T. J. (1992). Exploratory designs for computational experiments. Technical Report TM-12045. Oak Ridge National Laboratory.

Owen, A. B. (1992). Orthogonal arrays for computer experiments, integration and visualization. *Statistica Sinica* **2** 439–452.

Owen, A. B. (1994). Controlling correlations in Latin hypercube samples. *Journal of the American Statistical Association* **89** (428) 1517–1522.

Park, J. S. (1994). Optimal Latin-hypercube designs for computer experiments. *Journal of Statistical Planning and Inference* **39** 95–111.

Sacks, J., Schiller, S. B. & Welch, W. J. (1989). Designs for computer experiments. *Technometrics* **31** 41–47.

Sacks, J., Welch, W. J., Mitchell, T. J. & Wynn, H. P. (1989). Design and analysis of computer experiments. *Statistical Science* **4** 409–423.

Tang, B. (1993). Orthogonal array-based Latin hypercubes. *Journal of the American Statistical Association* **88** 1392–1397.

Tang, B. (1994). A theorem for selecting OA-based Latin hypercubes using a distance criterion. *Communications in Statistics: Theory and Methods* **23** 2047–2058.

Discrete Discrepancy and Its Application in Experimental Design

Min-Qian Liu[1], Hong Qin[2] and Min-Yu Xie[3]

[1] Department of Statistics, School of Mathematical Sciences and LPMC, Nankai University, Tianjin 300071, China (mqliu@nankai.edu.cn)
[2] Department of Statistics, Central China Normal University, Wuhan 430079, China (qinhong@mail.ccnu.edu.cn)
[3] Department of Statistics, Central China Normal University, Wuhan 430079, China (myxie@mail.ccnu.edu.cn,)

Summary. Discrepancy is a kind of important measure used in experimental design. Recently, a so-called discrete discrepancy has been applied to evaluate the uniformity of factorial designs. In this paper, we review some recent advances on application of the discrete discrepancy to several common experimental designs and summarize some important results.

Key words: Block design; discrepancy; factorial design; generalized minimum aberration; minimum moment aberration; orthogonality; uniformity; uniform design; supersaturated design.

2000 Mathematics Subject Classification: 62K15, 62K05, 62K99.

1 Introduction

Discrepancy has been employed to many fields of statistics, in particular, to experimental design. Based on discrepancy, Wang & Fang (1981) and Fang & Wang (1994) proposed a kind of novel experimental design, called uniform design, which favors a design with the smallest discrepancy value. In view of geometry, a uniform design spreads its experimental points uniformly over the experimental domain. Uniformity is an important concept related to uniform designs. Several important and popular measures of uniformity are discrepancies, such as the star discrepancy

Contemporary Multivariate Analysis and Experimental Design—In Honor Celebration of Professor Kai-Tai Fang's 65th birthday. Edited by Jianqing Fan and Gang Li. The World Scientific Publisher, 2005.

and the L_p-star discrepancy, etc, in the Quasi-Monte Carlo methods. The star discrepancy, introduced by Weyl (1916), measures the difference between the empirical distribution $F_n(\boldsymbol{x})$ of the set of design points, $P = \{\boldsymbol{z}_1, \ldots, \boldsymbol{z}_n\}$ in the unit hypercube $C^m = [0,1]^m$, and the uniform distribution $F_*(\boldsymbol{x})$ on C^m, and has been used in goodness-of-fit test named as the Kolmogorov-Smirov statistic. However, the star discrepancy is not easy to compute. The *L_p-star discrepancy*, viewed as an extension of the star discrepancy, has been widely used in Quasi-Monte Carlo methods. The set P is associated with an $n \times m$ matrix, $X_P = (x_{ki})$. It is well known that the L_p-star discrepancy is invariant to the permutation of rows and columns of X_P, but it is not invariant if the hypercube C^m is rotated by mapping x_{ki} to $1 - x_{ki}$. When n is small, the star discrepancy is not sensitive enough while the L_p-star discrepancy ignores differences between $F_n(\boldsymbol{x})$ and $F_*(\boldsymbol{x})$ in any low dimensional manifold. Unreasonable results of the L_2-star discrepancy may be easy found through many sets of points. Therefore, by using reproducing kernels in Hilbert space, Hickernell (1998a), Hickernell (1998b) proposed several modified versions of the L_p-star discrepancy, such as the centered L_p-discrepancy and the wrap-around L_p-discrepancy. These discrepancies can overcome the weakness of the L_p-star discrepancy mentioned above. In particular, when $p = 2$, analytical expressions of the centered L_2-discrepancy (CD, for short) and the wrap-around L_2-discrepancy (WD, for short) have also been obtained by Hickernell (1998a), Hickernell (1998b). The statistical justification for the CD/WD serving as a measure of uniformity for fractional factorial designs with two- or three-level has been interpreted by Fang & Mukerjee (2000), Fang, Lin, Winker & Zhang (2000), Ma & Fang (2001), Fang (2002), Fang & Ma (2002), Fang, Ma & Mukerjee (2002), Fang, Lin & Qin (2003), Ma, Fang & Lin (2003), Qin (2003), Fang & Qin (2004), Chatterjee, Fang & Qin (2004a) and Chatterjee, Fang & Qin (2004b).

Note that the above discrepancies are defined in a unite hypercube domain and used for measuring the uniformity of points corresponding to continuous variables. However, for factorial designs the number of possible levels for each factor may be restricted to a finite number. For example, a factor may have only two values (low and high) or three values (low, medium and high). In these situations it is reasonable to represent the experimental domain $\mathcal{X}$ as a discrete set, e.g., $\mathcal{X} = \{0, 1, \ldots, q_1 - 1\} \times \cdots \times \{0, 1, \ldots, q_m - 1\}$ for mixed levels. Liu & Hickernell (2002b) provided some justification for directly using the discrepancy defined on a discrete domain instead of on a continuous domain as a measure of uniformity of such design points. By using a reproducing kernel in Hilbert space, the so-called *discrete discrepancy* (DD, for short) was directly defined on such a discrete domain by Hickernell & Liu (2002), Liu & Hickernell (2002a), Liu (2002) and Fang, Lin & Liu (2003). Comparing with other discrepancies mentioned above, the DD not only enormously reduces the computational cost, particularly in constructing uniform designs, but also has itself statistical properties.

The main purpose of this paper is to review some recent developments on the application of the discrete discrepancy to experimental design.

2 Discrete discrepancy

We begin with a brief review of the discrete discrepancy. Let $\mathcal{X}$ be a measurable subset of $\boldsymbol{R}^m$. A kernel function $K(\boldsymbol{x},\boldsymbol{w})$ is any real-valued function defined on $\mathcal{X}\times\mathcal{X}$, and is symmetrical in its arguments and non-negative definite, i.e.,

$$K(\boldsymbol{x},\boldsymbol{w}) = K(\boldsymbol{w},\boldsymbol{x}), \text{ for any } \boldsymbol{x},\boldsymbol{w}\in\mathcal{X} \quad \text{and} \tag{1}$$

$$\sum_{i,j=1}^{n} a_i a_j K(\boldsymbol{x}^i,\boldsymbol{x}^j) \geq 0, \text{ for } a_i\in R,\ \boldsymbol{x}^i\in\mathcal{X},\ i=1,\ldots,n. \tag{2}$$

Let F_* denote the uniform distribution function on $\mathcal{X}$, $P=\{\boldsymbol{z}_1,\ldots,\boldsymbol{z}_n\}\subseteq\mathcal{X}$ be a set of design points and F_n denote the empirical distribution of P, where

$$F_n(\boldsymbol{x}) = \frac{1}{n}\sum_{\boldsymbol{z}\in P} 1_{\{\boldsymbol{z}\leq\boldsymbol{x}\}}.$$

Here $\boldsymbol{z}=(z_1,\ldots,z_m)\leq\boldsymbol{x}=(x_1,\ldots,x_m)$ means that $z_j\leq x_j$ for all j, 1_A is the indicator function of A. For a given kernel function $K(\boldsymbol{x},\boldsymbol{w})$, the discrepancy of P is defined as

$$\begin{aligned} D(P;K) &= \left\{\int_{\mathcal{X}^2} K(\boldsymbol{x},\boldsymbol{w})d[F_*(\boldsymbol{x})-F_n(\boldsymbol{x})]d[F_*(\boldsymbol{w})-F_n(\boldsymbol{w})]\right\}^{\frac{1}{2}} \\ &= \left\{\int_{\mathcal{X}^2} K(\boldsymbol{x},\boldsymbol{w})dF_*(\boldsymbol{x})dF_*(\boldsymbol{w}) - \frac{2}{n}\sum_{\boldsymbol{z}\in P}\int_{\mathcal{X}} K(\boldsymbol{x},\boldsymbol{z})dF_*(\boldsymbol{x}) \right. \\ &\quad \left. +\frac{1}{n^2}\sum_{\boldsymbol{z},\boldsymbol{z}'\in P} K(\boldsymbol{z},\boldsymbol{z}')\right\}^{\frac{1}{2}}. \end{aligned}$$

From the above definition, it is clear that the discrepancy measures how far apart the empirical distribution F_n is from the population distribution F_*. Consequently, for a fixed number of points, n, a design with low discrepancy is preferred. Several kernel functions were proposed and discussed by Hickernell (1998a,b; 2000).

Let d denote a factorial design with n runs and m factors, where the ith factor has q_i levels. The experimental domain $\mathcal{X}=\{0,1,\ldots,q_1-1\}\times\cdots\times\{0,1,\ldots,q_m-1\}$ is formed by all possible $\prod_{i=1}^m q_i$ level-combinations of the m factors, F_* is the discrete uniform distribution on $\mathcal{X}$. For notational convenience in this paper we define for given $a>0,\rho>1$,

$$\widetilde{K}(x_j,w_j) = \begin{cases} a\rho & \text{if } x_j = w_j, \\ a & \text{if } x_j \neq w_j, \end{cases} \quad \text{for } x_j,w_j\in\{0,1,\ldots,q_j-1\}, \tag{3}$$

and

$$K(\boldsymbol{x},\boldsymbol{w}) = \prod_{j=1}^{m}\widetilde{K}(x_j,w_j), \tag{4}$$

for any $\boldsymbol{x} = (x_1, \ldots, x_m)$ and $\boldsymbol{w} = (w_1, \ldots, w_m) \in \mathcal{X}$. Then $K(\boldsymbol{x}, \boldsymbol{w})$ is a kernel function. In particular, it satisfies conditions (1) and (2). The corresponding *discrete discrepancy*, denoted by $D(d; a, \rho)$, can be used for measuring the uniformity of design points over the domain $\mathcal{X}$ (Hickernell & Liu (2002); Liu & Hickernell (2002a)).

Consider the set, denoted by $\mathcal{D}(n; q_1 \cdots q_m)$, of asymmetrical factorials with n runs and m factors, where the ith factor has q_i levels, q_i is any positive integer ($\geq$ 2) and the n level-combinations are not necessarily distinct. If some q_i's are equal, we denote it by $\mathcal{D}(n; q_1^{s_1} \cdots q_r^{s_r})$, where $\sum_{i=1}^{r} s_i = m$. U-type designs play a key role in construction of uniform designs. A design d is called *U-type* if levels of each factor appear equally often (Fang, Lin, Winker & Zhang (2000)). Following Fang, Lin & Liu (2003) and Qin & Fang (2004), the squared DD-value, $(D(d; a, b))^2$, can be calculated as follows:

$$(D(d; a, b))^2 = -\prod_{i=1}^{m} \frac{(\rho + q_i - 1)a}{q_i} + \frac{(a\rho)^m}{n} + \frac{2a^m}{n^2} \sum_{k=1}^{n} \sum_{l=k+1}^{n} \rho^{\sigma_{kl}},$$

where σ_{ij} is the coincidence number between the ith and jth rows of d.

A lower bound of $D(d; a, b)$ over U-type designs in $\mathcal{D}(n; q_1 \cdots q_m)$ is given in the following theorem. A necessary and sufficient condition for a design reaching this lower bound is obtained also.

Theorem 1. *Let $d \in \mathcal{D}(n; q_1 \cdots q_m)$ be a U-type design. Then*

$$(D(d; a, \rho))^2 \geq L^A(d; a, \rho), \tag{5}$$

where

$$L^A(d; a, \rho) = -\prod_{i=1}^{m} \frac{(\rho + q_i - 1)a}{q_i} + \frac{(a\rho)^m}{n} + \frac{(n-1)[1 + (\rho - 1)(\sigma - \gamma)]a^m \rho^\gamma}{n},$$

$\sigma = \sum_{i=1}^{m}(n/q_i - 1)/(n-1)$ and γ is the integer part of σ. The lower bound of $L^A(d; a, \rho)$ can be achieved if and only if for any run d^k of d, among the $(n-1)$ values of σ_{kl} $(l \neq k)$, there are $(n-1)(\gamma + 1 - \sigma)$ with the value γ and $(n-1)(\sigma - \gamma)$ with the value $\gamma + 1$.

One lower bound of $D(d; a, \rho)$ for symmetrical design $d \in \mathcal{D}(n; q^m)$ can be obtained from Theorem 1. Recently, Qin & Li (2003) obtained the following lower bound of $D(d; a, \rho)$ for a design $d \in \mathcal{D}(n; q^m)$, which is sharper than the lower bound obtained from Theorem 1.

Theorem 2. *Let $d \in \mathcal{D}(n; q^m)$. Then*

$$(D(d; a, \rho))^2 \geq L^C(d; a, \rho),$$

where

$$L^C(d; a, \rho) = \frac{a^m}{n^2} \sum_{v=1}^{m} \binom{m}{v} (\rho - 1)^v R_{n,v,q} \left(1 - \frac{R_{n,v,q}}{q^v}\right),$$

$R_{n,v,q}$ is the residual of n (mod q^v).

Note that Theorems 1 and 2 hold for a wide range of DD measures in which the kernel satisfies (4), no matter what the values of a and ρ ($a > 0, \rho > 1$) are. The lower bound $L^A(d; a, \rho)$ or $L^C(d; a, \rho)$ can be used as a benchmark for searching uniform designs. A design $d \in \mathcal{D}(n; q_1 \cdots q_m)$ is called a *uniform design* under $D(d; a, \rho)$ if its DD value $D(d; a, \rho)$ achieves the minimum value over $\mathcal{D}(n; q_1 \cdots q_m)$. Based on Theorem 1 or 2, a design $d \in \mathcal{D}(n; q_1 \cdots q_m)$ or $\mathcal{D}(n; q^m)$ in which the squared DD-value equals the lower bound $L^A(d; a, \rho)$ or $L^C(d; a, \rho)$ is obvious a uniform design. In this paper, the uniformity criterion favors designs with the smallest discrete discrepancy.

3 Statistical inference for uniform designs measured by DD

3.1 Robustness of uniform designs measured by DD

At the initial stage of an experiment, it is often the case that a practitioner does not have enough information about models concerning the response and factors. Therefore, it is important to use a factorial design that is robust against the underlying model specifications. Since the uniform design spreads the design points evenly in the design space, it usually has robust performance with different modelling methods. Wiens (1991) gave two optimality properties of uniform designs. Hickernell (1999) and Yue & Hickernell (1999) proved that the uniform design is optimal and robust for approximate linear regression methods. Moreover, Xie & Fang (2000) proved that the uniform design is admissible and minimax under a certain sense in nonparametric regression model. Recently, Hickernell & Liu (2002) reported that although it is rare for a single design to be both maximally efficient and robust, uniform designs may limit the effects of aliasing to yield reasonable efficiency and robustness together.

3.2 Connections between DD and GMA/MMA

Minimum aberration (Fries & Hunter (1980); Franklin (1984)) and generalized minimum aberration (GMA, for short) (Tang & Deng (1999); Ma & Fang (2001); Xu & Wu (2001)) have become the popular and standard criteria for optimal factor assignment. Recently, Xu (2003) proposed the minimum moment aberration (MMA, for short) criterion to evaluate optimal factor assignment. Relationship between uniformity and aberration, which may raise the hope of improving the connection between uniform design theory and factorial design theory, has received a great deal of attention. The work of Fang & Mukerjee (2000) was a first attempt towards providing an analytic link between uniformity measured by CD and the word-length pattern of regular 2^{s-k} factorials. Fang & Ma (2002) and Fang, Ma & Mukerjee (2002) gave extensions of previous works for three- and higher-level factorials, respectively. For the discrete discrepancy, Qin & Fang (2004) obtained similar conclusions as follows.

Theorem 3. *Let $d \in \mathcal{D}(n; q_1 \cdots q_m)$. Then*

$$(D(d;a,\rho))^2 = \prod_{i=1}^{m} \frac{(\rho + q_i - 1)a}{q_i} \sum_{(j_1,\ldots,j_m)\in\mathcal{S}} \prod_{i=1}^{m} \left(\frac{\rho - 1}{\rho + q_i - 1}\right)^{j_i} \cdot C^*_{j_1\cdots j_m}(d),$$

where $C^*_{j_1\cdots j_m}(d)$'s are the MacWilliams transforms of the distance distribution of d, $\mathcal{S} = \{(j_1,\ldots,j_m) : 0 \le j_i \le 1, 1 \le i \le m, (j_1,\ldots,j_m) \ne (0,\ldots,0)\}$.

Corollary 1. *Let $d \in \mathcal{D}(n; q^m)$. Then*

$$(D(d;a,b))^2 = \left(\frac{(\rho + q - 1)a}{q}\right)^m \sum_{j=1}^{m} \left(\frac{\rho - 1}{\rho + q - 1}\right)^j A^{xw}_j(d),$$

where $A^{xw}_j(d) = \sum_{j_1+\cdots+j_m=j} C^*_{j_1\cdots j_m}(d)$, $(A^{xw}_1(d),\ldots,A^{xw}_m(d))$ is called the generalized word-length pattern by Xu & Wu (2001).

From Theorem 3 noting that the coefficient of $C^*_{j_1\cdots j_m}(d)$ in $(D(d;a,\rho))^2$ decreases exponentially with $(j_1,\ldots,j_m)$, we anticipate that factorials which keep $A^{xw}_{j_1+\cdots+j_m}(d)$ small for small values of $j_1 + \cdots + j_m$, that is those having less aberration, should behave well in terms of uniformity in the sense of keeping $(D(d;a,\rho))^2$ small. This shows that uniform designs under the DD and GMA designs are strongly related to each other, and provides a justification for the criterion of GMA by consideration of uniformity measured by the DD. Theorem 3 also shows us that the uniformity criterion does not completely agree with the GMA criterion. However, Qin & Fang (2004) indicated that for asymmetrical factorials, a special kind of uniform design has MMA, and uniform designs, MMA designs and GMA designs are equivalent in a special class of symmetrical factorials.

Recently, Hickernell & Liu (2002) defined a *projection discrepancy pattern* and proposed a *minimum projection uniformity* (MPU, for short) criterion in terms of this pattern, which considers the uniformity of low-dimensional projections of a design. Based on a specific kernel $K(\boldsymbol{x}, \boldsymbol{w})$ raised for asymmetrical factorial designs, the t-dimensional projection discrepancy $D_{(t)}(d;K)$ of a design $d = (d_{ij})$ is defined as

$$(D_{(t)}(d;K))^2 = \frac{1}{n^2} \sum_{|u|=t} \sum_{i,j=1}^{n} \prod_{l\in u} \left(-1 + q_l \delta_{d_{il}d_{jl}}\right), \tag{6}$$

where u is any subset of the set $\{1,\ldots,m\}$, $|u|$ denotes the cardinality of u, δ_{xw} denotes the Kronecker delta function, i.e., $\delta_{xw} = 1$ if $x = w$ and $\delta_{xw} = 0$ otherwise. The vector

$$\mathrm{PD}(d;K) = (D_{(1)}(d;K),\ldots,D_{(m)}(d;K))$$

is called the projection discrepancy pattern, and the MPU criterion is to sequentially minimize $D_{(t)}(d;K)$ for $t = 1,\ldots,m$. Based on (6), Hickernell & Liu (2002) showed that

Theorem 4. *Let $d \in \mathcal{D}(n; q_1 \cdots q_m)$, then $(D_{(t)}(d; K))^2 = A_t^{xw}(d)$, i.e. the MPU is equivalent to the GMA defined by Xu & Wu (2001). For the case of 2-level designs, the MPU is equivalent to the* minimum G_2-aberration *of Tang & Deng (1999).*

And their results show that the MPU criterion may be further generalized to cover designs that are not fractional factorials by using the discrepancy. It is also shown that minimum aberration designs and minimum discrepancy designs are equivalent in a certain limit.

3.3 Connection between DD and orthogonality

We know that strength is a good measure of orthogonality for factorial designs. Liu (2002) studied the connection between uniformity and strength. Taking $a = 1+\tau\beta$ and $a\rho = 1 + \beta$, where $\beta > 0$ and $-1/(q-1) < \tau < 1$ in (3), Liu (2002) obtained the following relation between discrepancies of an orthogonal array (Hedayat, Sloane & Stufken (1999)) on its low-dimensional projections and its strength.

Theorem 5. *Let $d = (d_{ij}) \in \mathcal{D}(n; q^m)$, then*
(i) *$D_{(t)}(d; a, \rho) = 0$ if and only if d is an $OA(n, m, q, t)$, where*

$$(D_{(t)}(d; a, \rho))^2 = -\left[\frac{\beta(1 + (q-1)\tau)}{q}\right]^t + \frac{\beta^t}{n^2} \sum_{|u|=t} \sum_{i,j=1}^{n} \prod_{l \in u} \tau^{1-\delta_{d_{il}d_{jl}}}.$$

(ii) *$D(d; a, \rho) = 0$ if and only if d is an $OA(n, m, q, m)$ (here n must be a multiple of q^m).*

Liu (2002) also showed that symmetrical saturated orthogonal arrays are the most uniform one among all the saturated factorial designs with the same parameters.

Recently, some new criteria, such as the B-criterion (Fang, Lu & Winker (2003)) and O-criterion (Fang, Ma & Mukerjee (2002)), have been utilized to measure and evaluate the orthogonality of factorial designs. These criteria can be viewed as extensions of the concept of strength in orthogonal array. For any t columns of $d \in \mathcal{D}(n; q^m)$, say $c_{l_1}, \ldots, c_{l_t}$, let $n_{\alpha_1 \cdots \alpha_t}^{(l_1 \cdots l_t)}$ be the number of runs in which $(c_{l_1}, \ldots, c_{l_t})$ takes the level-combination $(\alpha_1 \cdots \alpha_t)$, let

$$B_{l_1 \cdots l_t}(d) = \sum_{\alpha_1, \ldots, \alpha_t} \left(n_{\alpha_1 \cdots \alpha_t}^{(l_1 \cdots l_t)} - \frac{n}{q^t} \right)^2,$$

where the summation is taken over all possible level-combinations, and define

$$B_t(d) = \sum_{1 \le l_1 < \cdots < l_t \le m} B_{l_1 \cdots l_t}(d) \Big/ \binom{m}{t},$$

the *B-criterion* is to minimize $B_t(d)$ for $t = 1, \ldots, m$ sequentially. For symmetrical designs, Qin & Chen (2004) showed that B-criterion is equivalent to GMA. Qin & Li (2003) indicated that B-criterion and O-criterion are mutually equivalent, and gave the following connection between DD and B-criterion.

Theorem 6. *Let* $d \in \mathcal{D}(n; q^m)$. *Then*

$$(D(d; a, \rho))^2 = \frac{(a\rho)^m}{n^2} \sum_{v=1}^{m} \binom{m}{v} (\rho - 1)^v B_v(d).$$

3.4 Connection between DD and CD/WD

As mentioned in Section 1, usefulness of uniformity measured by the CD/WD in two- or three-level factorials has been discussed. The definitions and computation formulas for the CD and WD can refer to Hickernell (1998a) and Hickernell (1998b). For $d \in \mathcal{D}(n; q^m)$, its CD and WD are denoted by $CD(d)$ and $WD(d)$ respectively. Recently, Qin & Fang (2004) gave the following result, which connect the DD with the CD and WD.

Theorem 7. *For any design* $d \in \mathcal{D}(n; q^m)$, *we have the following equations:*
(i) when $q = 2$, $\rho = 5/4$ *and* $a = 1$,

$$(D(d; a, \rho))^2 = (CD(d))^2 + 2(35/32)^m - (13/12)^m - (9/8)^m;$$

(ii) when $q = 2$, $\rho = 6/5$ *and* $a = 5/4$,

$$(D(d; a, \rho))^2 = (WD(d))^2 + (4/3)^m - (11/8)^m;$$

(iii) when $q = 3$, $\rho = 27/23$ *and* $a = 23/18$,

$$(D(d; a, \rho))^2 = (WD(d))^2 + (4/3)^m - (73/54)^m.$$

It is well known that there is yet an open problem whether uniformity measured by the CD/WD may be utilized as a criterion for assessing factorials with high levels. However, the DD can be used to compare symmetrical and asymmetrical factorials with high levels. Hence, the DD can be regarded as a kind of generalization of the CD and WD. We strongly recommend to use the discrete discrepancy as a measure of uniformity for comparing fractional factorials in most cases.

3.5 Connection between DD and balance

Block design is an important kind of experimental design. Its basic ideas come from agricultural and biological experiments. But now the applications of these ideas are found in many areas of sciences and engineering. The most widely-used one is the balanced incomplete block (BIB, for short) design in which every pair of treatments occurs altogether in exact the same number of blocks. Another important one is the resolvable incomplete block (RIB, for short) design. For a thorough discussion of block designs, please refer to Dey (1986).

As we know the definitions in block designs reflect some "*balance*" among the treatments, the blocks, or the parallel classes. This kind of balance is easy to be accepted intuitively. While in existed works on block designs the criterion of balance is introduced from the estimation point of view. In fact the balance criterion can be regarded as a kind of *uniformity*. Recently, Liu & Chan (2004) and Liu & Fang (2004) studied the uniformity of block designs and obtained some satisfactory results. Liu & Chan (2004) used the DD measure to prove theoretically that BIB designs are the most uniform ones among all binary incomplete block designs. This is an important characteristic of BIB designs in terms of uniformity. While Liu & Fang (2004) obtained a sufficient and necessary condition under which a certain kind of RIB design is the most uniform one in the sense of the DD measure, and showed that this uniform design is connected. They also proposed a construction method for such designs via a kind of U-type designs. This method sets up an important bridge between this kind of RIB designs and U-type designs. All these results confirm our judgement that the "*balance*" criterion can be regarded as a kind of *uniformity*. Note that these results are obtained in the sense of the DD measure, but they also holds for any of the modified L_2-discrepancies proposed by Hickernell (1998a) and Hickernell (1998b).

4 Application of the DD in supersaturated designs

In the context of factorial designs, there has been recent interest in the study of the *supersaturated design* (SSD, for short). Whenever the run size of a design is insufficient for estimating all the main effects represented by the columns of the design matrix, the design is called supersaturated. In industrial statistics and other scientific experiments, especially in their preliminary stages, very often there are a large number of factors to be studied and the run size is limited because of cost. However, in many situations only a few factors are believed to have significant effects. Under this assumption of *effect sparsity* (Box & Meyer (1986)), SSDs can be used effectively, allowing the simultaneous identification of the active factors.

4.1 Connection between DD and $E(s^2)$ in 2-level SSDs

Most studies on SSDs have focused on the 2-level case. Booth & Cox (1962), in the first systematic construction of SSDs, proposed the $E(s^2)$ *criterion*, which is a measure of *non-orthogonality* under the assumption that *only two out of the* m factors are active. After Booth & Cox (1962), there was not much work on the subject of SSDs until Lin (1993). Other recent work focusing on constructions of $E(s^2)$-optimal SSDs includes, e.g. Liu & Zhang (2000), Butler, Mead, Eskridge & Gilmour (2001), Liu & Dean (2004) and the references therein.

Recently, Liu & Hickernell (2002a) showed that the $E(s^2)$ criterion shares the same optimal designs with the DD criterion. They constructed a DD, i.e. taking $a = 1 + \tau\beta$ and $a\rho = 1 + \beta$ ($\beta > 0, -1 \leq \tau < 1$) in (3), and showed that for 2-level factorial designs both $E(s^2)$ and the DD can be expressed in terms of the *Hamming distances* (or the *coincidence numbers*) between any two runs of the design. These expressions in terms of Hamming distances lead to a lower bound on $E(s^2)$ and the lower bound of (5) on DD for 2-level SSDs. It is interesting to note that if a design d can attain one of these lower bounds, then it attains both of them. In other words, an $E(s^2)$-optimal design is also uniform (minimal discrepancy) for the DD. They further showed that in what cases these lower bounds can be achieved, even though the DD is *not* equivalent to the $E(s^2)$ criterion.

Theorem 8. *Let d be a 2-level design with n runs and m factors, where each column has the same number of ± 1 elements. Suppose that $\tau\beta > -1$, and that $m = c(n-1) + e$ for $e = -1, 0$ or 1. Also, suppose that either a) n is a multiple of 4 and there exists an $n \times n$ Hadamard matrix, or b) c is even and there exists a $2n \times 2n$ Hadamard matrix. Then the lower bounds of $E(s^2)$ and DD can be attained.*

Moreover, the DD is a more general, and thus more flexible criterion than $E(s^2)$. For example, $E(s^2)$ ignores possible interactions of more than one factor. However, the DD includes interactions of all possible orders, and their importance may be increased or decreased by changing the value of β.

4.2 Connection between DD and $E(f_{NOD})$ in mixed-level SSDs

Two-level SSDs can be used for screening the factors in simple linear models. When the relationship between a set of factors and a response is nonlinear, or approximated by a polynomial response surface model, designs with multi-level and mixed-level are often required, e.g., to exploring nonlinear effects of the factors. Recently, Fang, Lin & Liu (2003) proposed a new criterion, called the $E(f_{NOD})$ *criterion*, for comparing SSDs. For a design $d \in \mathcal{D}(n; q_1 \cdots q_m)$, the criterion is defined as minimizing

$$E(f_{NOD}) = \sum_{1 \le i < j \le m} f_{NOD}^{ij} \Big/ \binom{m}{2},$$

where

$$f_{NOD}^{ij} = \sum_{u=1}^{q_i} \sum_{v=1}^{q_j} \left(n_{uv}^{(ij)} - \frac{n}{q_i q_j} \right)^2,$$

$n_{uv}^{(ij)}$ is the number of (u, v)-pairs in the ith and jth columns. Here, the subscript NOD stands for *non-orthogonality* of the design. Fang, Lin & Liu (2003) obtained a lower bound for $E(f_{NOD})$ which can serve as a benchmark of design optimality. They also studied the connection between DD and $E(f_{NOD})$. Fang, Ge, Liu & Qin (2004a) provided the following lower bound and the sufficient and necessary condition to achieve it for $E(f_{NOD})$, which includes the bound and condition of Fang, Lin & Liu (2003) as a special case.

Theorem 9. *Let $d \in \mathcal{D}(n; q_1 \cdots q_m)$ be a U-type design, then*

$$E(f_{NOD}) \ge \frac{n(n-1)}{m(m-1)} \left[(\gamma + 1 - \sigma)(\sigma - \gamma) + \sigma^2 \right] + C(n, q_1, \ldots, q_m), \qquad (7)$$

where $C(n, q_1, \ldots, q_m) = \frac{nm}{m-1} - \frac{1}{m(m-1)}\left(\sum_{i=1}^{m} \frac{n^2}{q_i} + \sum_{1\le i\neq j\le m} \frac{n^2}{q_i q_j}\right)$, σ, γ *and the sufficient and necessary condition for the lower bound to be achieved are the same as those of Theorem 1.*

Thus we conclude that

Theorem 10. *Let* $d \in \mathcal{D}(n; q_1 \cdots q_m)$ *be a U-type design, then* d *is a uniform design with its squared DD-value achieving the lower bound on the right hand side of (5) if and only if* d *is* $E(f_{NOD})$*-optimal with its* $E(f_{NOD})$ *achieving the lower bound on the right hand side of (7).*

Theorem 10 leads to a strong relation between $E(f_{NOD})$ optimality and uniformity measured by the DD of any SSD. The uniformity of $E(s^2)$- and aveχ^2-optimal (Yamada & Lin (1999)) SSDs can be obtained directly based on this theorem, as special cases of SSDs with equal-level factors.

4.3 Constructions of uniform SSDs measured by DD

To find uniform designs is an NP hard problem. There are several methods to construct uniform designs in literature, such as the good lattice method (Fang & Wang (1994)), Latin square method (Fang, Shiu & Pan (1999)) and optimization searching method (Fang, Ma & Winker (2002)). In these methods, computer algorithms play an important role to obtain uniform designs.

Recently, some combinatorial methods are introduced to construct uniform U-type designs in terms of DD as well as $E(f_{NOD})$. Note that in most cases, uniform U-type designs are supersaturated. So this kind of U-type designs are also called the *uniform SSDs*. Many infinite classes for the existence of uniform designs with the same Hamming distances between any distinct rows are also obtained simultaneously. These combinatorial approaches can be summarized as follows:

I. Constructing symmetrical uniform SSDs from

a. Resolvable balanced incomplete block designs, see Fang, Ge & Liu (2002b), Fang, Ge, Liu & Qin (2003);
b. Room squares, see Fang, Ge & Liu (2002a);
c. Resolvable packings and coverings, see Fang, Ge & Liu (2004) and Fang, Lu, Tang & Yin (2004);
d. Super-simple resolvable *t-designs*, see Fang, Ge, Liu & Qin (2004b).

II. Constructing asymmetrical uniform SSDs from

a. Resolvable group divisible designs, see Fang, Ge, Liu & Qin (2004a);
b. Latin squares, see Fang, Ge, Liu & Qin (2004a);
c. Resolvable partially pairwise balanced designs, see Fang, Tang & Yin (2004);
d. Other uniformly resolvable designs, see Fang, Ge, Liu & Qin (2004a).

In addition, Fang, Lin & Liu (2003) proposed a method by fractionalizing saturated orthogonal arrays for constructing asymmetrical uniform SSDs. The properties of the resulting uniform SSDs were also investigated in those papers.

5 Concluding remarks

Uniform experimental design has been widely used in many fields in the last two decades. Discrepancy is a measurement of the uniformity and is a criterion in experimental design. In this paper, we review the recent developments on the discrete discrepancy and summarize some important results. The uniformity of the common experimental designs, such as factorial design, orthogonal design, block design and supersaturated design, are also discussed in this paper. All these results show that orthogonality (non-orthogonality) and balance are strongly related to uniformity, and the discrete discrepancy plays an important role in evaluating such experimental designs.

Acknowledgements

The authors are grateful to Professors Kai-Tai Fang and Fred J. Hickernell for their helpful comments and suggestions. Thanks also go to the Editors and one anonymous Referee for their valuable comments. This work was partially supported by the National Natural Science Foundation of China (Grant Nos. 10171051, 10301015), the Outstanding Scholar Program of Nankai University, the Key Project of Chinese Ministry of Education, the Project-sponsored by SRF for ROCS (SEM) and the NSF of Hubei Province.

References

Booth, K. H. V. & Cox, D. R. (1962). Some systematic supersaturated designs. *Technometrics* **4** 489–495.

Box, G. E. P. & Meyer, R. D. (1986). An analysis for unreplicated fractional factorial *Technometrics* **28** 11–18.

Butler, N. A., Mead, R., Eskridge, K. M. & Gilmour, S. G. (2001). A general method of constructing $E(s^2)$-optimal supersaturated designs. *J. Roy. Stat. Soc. Ser. B* **63** 621–632.

Chatterjee, K., Fang, K. T. & Qin, H. (2004a). Uniformity in factorial designs with mixed levels.

Chatterjee, K., Fang, K. T. & Qin, H. (2004b). A lower bound for centered L_2-discrepancy on asymmetric factorials and its application. *Metrika.* To appear.

Dey, A. (1986). *Theory of Block Designs.* John Wiley, New York.

Fang, K. T. (2002). Some applications of Quasi-Monte Carlo methods in statistics. In *Monte Carlo and Quasi-Monte Carlo Methods 2000* (K. T. Fang, F. J. Hickernell & H. Niederreiter, eds.). Springer-Verlag, Berlin.

Fang, K. T., Ge, G. N. & Liu, M. Q. (2002a). Construction of $E(f_{NOD})$-optimal supersaturated designs via Room squares. *Calcutta Statist. Assoc. Bull.* **52** 71–84.

Fang, K. T., Ge, G. N. & Liu, M. Q. (2002b). Uniform supersaturated design and its construction. *Sci. China Ser. A* **45** 1080–1088.

Fang, K. T., Ge, G. N. & Liu, M. Q. (2004). Construction of optimal supersaturated designs by the packing method. *Sci. China Ser. A* **47** 128–143.

Fang, K. T., Ge, G. N., Liu, M. Q. & Qin, H. (2003). Construction of minimum generalized aberration designs. *Metrika* **57** 37–50.

Fang, K. T., Ge, G. N., Liu, M. Q. & Qin, H. (2004a). Combinatorial constructions for optimal supersaturated designs. *Discrete Math.* **279** 191–202.

Fang, K. T., Ge, G. N., Liu, M. Q. & Qin, H. (2004b). Construction of uniform designs via super-simple resolvable t-designs.

Fang, K. T., Lin, D. K. J. & Liu, M. Q. (2003). Optimal mixed-level supersaturated design. *Metrika* **58** 279–291.

Fang, K. T., Lin, D. K. J. & Qin, H. (2003). A note on optimal foldover design.. *Statist. Probab. Lett.* **62** 245–250.

Fang, K. T., Lin, D. K. J., Winker, P. & Zhang, Y. (2000). Uniform design: theory and applications. *Technometrics* **42** 237–248.

Fang, K. T., Lu, X., Tang, Y. & Yin, J. X. (2004). Constructions of uniform designs by using resolvable packings and coverings. *Discrete Math.* **274** 25–40.

Fang, K. T., Lu, X. & Winker, P. (2003). Lower bounds for centered and wrap-around L_2-discrepancies and construction of uniform designs by threshold accepting. *J. Complexity* **19** 692–711.

Fang, K. T. & Ma, C. X. (2002). Relationships between uniformity, aberration and correlation in regular fractions 3^{s-1}. In *Monte Carlo and Quasi-Monte Carlo Methods 2000* (K. T. Fang, F. J. Hickernell & H. Niederreiter, eds.). Springer-Verlag, Berlin.

Fang, K. T., Ma, C. X. and Mukerjee, R. (2002). Uniformity in fractional factorials. In *Monte Carlo and Quasi-Monte Carlo Methods 2000* (K. T. Fang, F. J. Hickernell & H. Niederreiter, eds.). Springer-Verlag, Berlin.

Fang, K. T., Ma, C. X. & Winker, P. (2002). Centered L_2-discrepancy of random sampling and Latin hypercube design, and construction of uniform design. *Math. Comp.* **71** 275–296.

Fang, K. T. & Mukerjee, R. (2000). A connection between uniformity and aberration in regular fractions of two-level factorials. *Biometrika* **87** 173–198.

Fang, K. T. & Qin, H. (2004). Uniform pattern and related criteria for two-level factorials. *Sci. China Ser. A* **34** 418–428.

Fang, K. T., Shiu, C. X. & Pan, J. (1999). Uniform designs based on Latin squares. *Statist. Sinica* **9** 905–912.

Fang, K. T., Tang, Y. & Yin, J. X. (2004). Resolvable partially pairwise balanced designs and their applications in computer experiments. Utiltas Math. in press.

Fang, K. T. & Wang, Y. (1994). *Number-Theoretic Methods in Statistics*. Chapman and Hall, London.

Franklin, M. F. (1984). Constructing tables of minimum aberration p^{n-m} designs. *Technometrics* **26** 225–232.

Fries, A. & Hunter, W. G. (1980). Minimum aberration 2^{k-p} designs. *Technometrics* **22** 601–608.

Hedayat, A. S., Sloane, N. J. A. & Stufken, J. (1999). *Orthogonal Arrays: Theory and Applications*. Springer, New York.

Hickernell, F. J. (1998a). A generalized discrepancy and quadrature error bound. *Math. Comp.* **67** 299–322.

Hickernell, F. J. (1998b). Lattice rules: how well do they measure up?. In *Random and Quasi-Random Point Sets* (P. Hellekalek & G. Larcher, eds.) **138** 109–166. Springer, New York.

Hickernell, F. J. (1999). Goodness-of-fit statistics, discrepancies and robust designs. *Statist. Probab. Lett.* **44** 73–78.

Hickernell, F. J. (2000). What affects the accuracy of Quasi-Monte Carlo quadrature? In *Monte-Carlo and Quasi-Monte Carlo Methods 1998 (H. Niederreiter & J. Spamer, eds.)* 16–55. Springer-Verlag, Berlin.

Hickernell, F. J. & Liu, M. Q. (2002). Uniform designs limit aliasing. *Biometrika* **89** 893–904.

Lin, D. K. J. (1993). A new class of supersaturated designs. *Technometrics* **35** 28–31.

Liu, M. Q. (2002). Using discrepancy to evaluate fractional factorial designs. In *Monte Carlo and Quasi-Monte Carlo Methods 2000* (K. T. Fang, F. J. Hickernell & H. Niederreiter, eds.) 357–368. Springer-Verlag, Berlin.

Liu, M. Q. & Chan, L. Y. (2004). Uniformity of incomplete block designs. *Int. J. Materials and Product Technology* **20** 143–149.

Liu, M. Q. & Fang, K. T. (2004). Some results on resolvable incomplete block designs. *Sci. China Ser. A.*. In press.

Liu, M. Q. & Hickernell, F. J. (2002a). $E(s^2)$-optimality and minimum discrepancy in 2-level supersaturated designs. *Statist. Sinica* **12** 931–939.

Liu, M. Q. & Hickernell, F. J. (2002b). The relationship between discrepancies defined on a domain and on its subset, Technical report MATH-345, Hong Kong Baptist University.

Liu, M. Q. & Zhang, R. C. (2000). Construction of $E(s^2)$ optimal supersaturated designs using cyclic BIBDs. *J. Statist. Plann. Inference* **91** 139–150.

Liu, Y. F. and Dean, A. (2004). k-circulant supersaturated designs. *Technometrics* **46** 32–43.

Ma, C. X. & Fang, K. T. (2001). A note on generalized aberration factorial designs. *Metrika* **53** 85–93.

Ma, C. X., Fang, K. T. & Lin, D. K. J. (2003). A note on uniformity and orthogonality. *J. Statist. Plann. Inference* **113** 323–334.

Qin, H. (2003). Characterization of generalized aberration of some designs in terms of their complementary designs. *J. Statist. Plann. Inference* **117** 141–151.

Qin, H. & Chen, Y. B. (2004). Some results on generalized minimum aberration for symmetrical fractional factorial designs. *Statist. Probab. Lett.* **66** 51–57.

Qin, H. & Fang, K. T. (2004). Discrete discrepancy in factorial designs. *Metrika* **60** 59–72.

Qin, H. & Li, D. (2003). Connection between uniformity and orthogonality for symmetrical factorial designs. Submitted.

Tang, B. & Deng, L. Y. (1999). Minimum G_2-aberration for nonregular fractional designs. *Ann. Statist.* **27** 1914–1926.

Wang, Y. & Fang, K. T. (1981). A note on uniform distribution and experimental design. *Chin. Sci. Bull.* **26** 485–489.

Weyl, H. (1916). Über die gleichverteilung der zahlen mod eins. *Math. Ann.* **77** 313–352.

Wiens, D. P. (1991). Designs for approximately linear regression: two optimality properties of uniform designs. *Statist. Probab. Lett.* **12** 217–221.

Xie, M. Y. & Fang, K. T. (2000). Admissibility and minimaxity of the uniform design in nonparametric regrssion model. *J. Statist. Plann. Inference* **83** 101–111.

Xu, H. (2003). Minimum moment aberration for nonregular designs and supersaturated designs. *Statist. Sinica* **13** 691–708.

Xu, H. & Wu, C. F. J. (2001). Generalized minimum aberration for asymmetrical fractional factorial designs. *Ann. Statist.* **29** 1066–1077.

Yamada, S. & Lin, D. K. J. (1999). Three-level supersaturated design. *Statist. Probab. Lett.* **45** 31–39.

Yue, R. X. & Hickernell, F. J. (1999). Robust designs for fitting linear models with misspecification. *Statist. Sinica* **6** 455–469.

Orthogonal Arrays of 2 and 3 levels for Lean Designs

Chang-Xing Ma[1] and Ling-Yau Chan[2]

[1] Department of Statistics, University of Florida, Gainesville, FL 32611, U.S.A. (cma@biostat.ufl.edu http://www.biostat.ufl.edu/~cma)
[2] Department of Industrial and Manufacturing Systems Engineering, The University of Hong Kong, Pokfulam Road, Hong Kong (plychan@hku.hk)

Summary. When an orthogonal array (OA) of n rows is used as the design matrix in an experiment, n is the number of runs. In an OA of q levels, n is an integer multiple of q^2. In an experiment, if the number of runs cannot be set exactly equal to the number of rows of an OA because of constraints in resources or other reasons, the experimenter may use a design matrix formed by omitting some rows of an OA. If such a design matrix is used, the number of observed response obtained may not be enough for estimation of all the effects corresponding to columns of the orthogonal array. A lean design is a design matrix formed by deleting some rows and columns of an OA, which still allows efficient estimation of the effects of the factors corresponding to the remaining columns of the OA. In this article, the authors discuss lean designs of 2 and 3 levels, and provide D-optimal OA's from which lean designs can be formed.

Key words: Design of experiments, orthogonal array, lean design, D-optimality

2000 Mathematics Subject Classification: 62K15, 62K05

1 Introduction

Orthogonal arrays (OAs) have been studied for several decades and have widely been used in design of experiments. Given positive integers $n, s, q_1, \ldots, q_s$, an OA$(n; q_1 \times \cdots \times q_s)$ of strength 2 is an $n \times s$ array of symbols such that (i) along each column, different symbols appear the same number of times, and (ii) in every

Contemporary Multivariate Analysis and Experimental Design—In Honor Celebration of Professor Kai-Tai Fang's 65th birthday. Edited by Jianqing Fan and Gang Li. The World Scientific Publisher, 2005.

$n \times 2$ matrix formed from any two columns, all possible pairs of symbols appear the same number times (Wu and Hamada (2000), Dey and Mukerjee (1999), Hedayat, Sloan and Stufken (1999)).

When an $OA(n; q^s)$ of n rows is used as the design matrix in an experiment, the number of runs will be n which is an integer multiple of q^2. In an industrial experiment, due to various constraints in the equipment used, the setup of the experiment and the availability of other resources, it is not always possible to perform the experiment with the numbers of runs exactly equal to a multiple of q^2. Thus the experimenter may be forced to conduct an experiment with the number of runs not equal to the n in the OA. If the number of runs of the experiment is less than the n in the $OA(n; q_1 \times \cdots \times q_s)$, from the results of the experiment it may not be possible to estimate all the s effects corresponding to the columns of the OA. Goh (1996) proposed to delete some rows and some columns from an OA to obtain a *lean design* to be used for such an experiment. In doing this, it is important that appropriate rows and columns are deleted from the original OA, for otherwise the resulting matrix obtained will be of low efficiency or even be rank deficient.

This article is concerned with construction of OA's for lean designs with 2 and 3 levels. In Section 2, the motivation for lean design will be illustrated with an example. Section 3 will be devoted to 2-level designs, and Section 4 will be devoted to 3-level designs. This article is concluded in Section 5.

Throughout this paper, we shall use $X(n)$ to denote an extended design matrix of n rows formed by adding column vectors of length n to an $OA(n; q^s)$. The forms of the column vectors to be added depend on the regression model considered, and in many cases the $n \times 1$ column vector of 1's, $\mathbf{1}_n$, is the first column of $X(n)$.

2 The motivation for lean designs

Consider the experiment performed according to an $OA(8; 2^7)$, quoted from Table 13.4 of Box, Hunter and Hunter (1978, Section 13.3). The layout and the results of the experiment are shown in Table 1, where the column of 1's beneath x_0 together with the 7 columns of the $OA(8; 2^7)$ beneath $x_1, ..., x_7$ form the extended design matrix $X(8)$, and the first column from the right hand side shows the observed responses y.

The data in Table 1 can be used to estimate all the eight coefficients $\beta_0, \beta_1, ..., \beta_7$ in theregression model

$$E[y] = \beta_0 + \beta_1 x_1 + \cdots + \beta_5 x_5 + \beta_6 x_6 + \beta_7 x_7. \tag{1}$$

Now, suppose that the resources available allow the experimenter to make only 6 runs in the experiment instead of 8, and as a result only Runs 1 to 6 were made. In this case, some conclusions can still be drawn from the observed data. If it is known that the factors corresponding to x_6 and x_7 do not have effect on y, the regression model becomes

$$E[y] = \beta_0 + \beta_1 x_1 + \cdots + \beta_5 x_5. \tag{2}$$

Table 1. An experiment performed according to an OA$(8; 2^7)$.

Run i	x_0	x_1	x_2	x_3	x_4	x_5	x_6	x_7	Observed response y
1	1	-1	-1	-1	1	1	1	-1	$y_1 = 68.4$
2	1	1	-1	-1	-1	-1	1	1	$y_2 = 77.7$
3	1	-1	1	-1	-1	1	-1	1	$y_3 = 66.4$
4	1	1	1	-1	1	-1	-1	-1	$y_4 = 81.0$
5	1	-1	-1	1	1	-1	-1	1	$y_5 = 78.6$
6	1	1	-1	1	-1	1	-1	-1	$y_6 = 41.2$
7	1	-1	1	1	-1	-1	1	-1	$y_7 = 68.7$
8	1	1	1	1	1	1	1	1	$y_8 = 38.7$

Using the six responses $y_1, ..., y_6$, all the six coefficients $\beta_0, ..., \beta_5$ in model (2) can be estimated.

The matrix $X(6)$ obtained by deleting the 7th row, 8th row, 7th column and 8th column from $X(8)$ is called a lean design with 6 runs. With this lean design, assuming that the effects of x_6 and x_7 are nil, the missing observations y_7 and y_8 can be estimated from the following two equations

$$\text{(effect of } x_6) = (y_1 + y_2 - y_3 - y_4 - y_5 - y_6 + \hat{y}_7 + \hat{y}_8)/8 = 0,$$
$$\text{(effect of } x_7) = (-y_1 + y_2 + y_3 - y_4 + y_5 - y_6 - \hat{y}_7 + \hat{y}_8)/8 = 0.$$

Substituting y_7 and y_8 by such estimates $\hat{y}_7$ and $\hat{y}_8$, all the coefficients $\beta_0, \beta_1, ..., \beta_6, \beta_7$ in (1) can be estimated. Using the property that $X'(8)X(8)$ is a diagonal matrix, it was proved (Chan, Ma and Goh (2002)) that the estimates of $\beta_0, \beta_1, ..., \beta_4, \beta_5$ obtained in this way are identical to those obtained from (2) without the observations y_7 and y_8.

On the other hand, out of the 8 runs, if Run 2 and Run 8 are not available, and if x_6 and x_7 have no effect on y, the coefficients $\beta_0, ..., \beta_5$ in model (2) cannot be estimated because the matrix obtained by deleting 2nd row, 8th row, 7th column and 8th column from $X(8)$ has rank < 6. Setting the effects of x_6 and x_7 zero, we have

$$\text{(effect of } x_6) = (y_1 + \hat{y}_2 - y_3 - y_4 - y_5 - y_6 + y_7 + \hat{y}_8)/8 = 0,$$
$$\text{(effect of } x_7) = (-y_1 + \hat{y}_2 + y_3 - y_4 + y_5 - y_6 - y_7 + \hat{y}_8)/8 = 0,$$

from which the missing observations y_2 and y_8 cannot be estimated. Therefore, the 6-run design obtained by deleting the 2nd row, 8th row and the last two columns of $X(8)$ is a poor design for estimation of the effects $x_0, x_1, ..., x_5$ in (2), and this design cannot be considered as a lean design.

3 Lean designs of 2 levels

From an OA$(n; 2^{n-1})$, L say, where n is an integer multiple of 4, lean designs with the number of runs less than n can be obtained by appropriately deleting rows and columns from $X(n) = [\mathbf{1}_n, L]$, so that an efficient $i{\times}i$ sub-matrix $X(i)$ $(i < n)$ will formed. If $X(i)$ is used as the extended design matrix of a lean design for estimation of coefficients in the model

$$y = \beta_0 + \beta_1 x_1 + \cdots + \beta_{i-1} x_{i-1} + \epsilon, \tag{3}$$

under the i.i.d. assumption on ϵ, the covariance matrix of the estimated $(\beta_0, \beta_1, ..., \beta_{i-1})'$ will be proportional to $(X'(i)X(i))^{-1}$. Using *D-optimality* as a measure for efficiency (Atkinson and Donev (1996, Chapters 9 and 10)), we look for $X(i)$ with large $\det(X'(i)X(i)) = |\det(X(i))|^2$.

Table 2. Orthogonal arrays A(n) for two-level lean designs

A(8) - OA(8; 2^7)	A(12) - OA(12; 2^{11})	A(16) - OA(16; 2^{15})
0 1 0 0 0 1 1	0 1 1 0 1 0 1 1 1 0 0	0 1 1 1 1 0 0 0 0 1 1 0 0 1 1
1 1 1 0 0 0 0	1 0 1 0 0 0 1 1 0 1 1	1 0 1 1 0 1 0 1 0 0 1 0 1 1 0
0 0 1 1 0 0 1	1 1 0 0 1 1 0 1 0 0 1	0 0 0 1 1 0 1 1 0 1 1 1 1 0 0
1 0 0 0 1 0 1	0 0 0 1 1 0 0 1 1 1 1	1 1 0 1 0 0 1 1 1 1 0 0 0 1 0
0 0 1 0 1 1 0	1 1 1 1 0 0 0 0 1 0 1	1 1 0 0 1 1 0 1 0 1 0 0 1 0 1
0 1 0 1 1 0 0	0 1 0 0 0 1 1 0 1 1 1	0 0 0 0 0 0 0 0 0 0 0 0 0 0 0
1 0 0 1 0 1 0	0 0 1 1 1 1 1 0 0 0 1	1 0 1 0 1 0 1 1 1 0 1 0 0 0 1
1 1 1 1 1 1 1	1 1 0 1 1 0 1 0 0 1 0	0 1 1 0 0 1 1 0 1 1 1 0 1 0 0
	0 0 0 0 0 0 0 0 0 0 0	1 0 1 0 1 1 1 0 0 1 0 1 0 1 0
	1 0 1 0 1 1 0 0 1 1 0	0 0 0 1 1 1 1 0 1 0 0 0 1 1 1
	1 0 0 1 0 1 1 1 1 0 0	1 0 1 1 0 0 0 0 1 1 0 1 1 0 1
	0 1 1 1 0 1 0 1 0 1 0	0 1 1 1 1 1 0 1 1 0 0 1 0 0 0
		1 1 0 1 0 1 1 0 0 0 1 1 0 0 1
		0 0 0 0 0 1 0 1 1 1 1 1 0 1 1
		0 1 1 0 0 0 1 1 0 0 0 1 1 1 1
		1 1 0 0 1 0 0 0 1 0 1 1 1 1 0

To construct a 2-level lean design of i runs, a given OA$(n, 2^s)$, say L, is used as an initial design, where $n \geq i$. The extended design matrix $X(n) = [\mathbf{1}_n, L]$ is transformed by exchange of rows and exchange of columns, so that the upper-left $i{\times}i$ sub-matrix $X(i)$ of the transformed matrix has the largest $|\det X(i)|$. In doing this, the matrix L, as a submatrix of $X(n)$, is also transformed. The transformed matrix of L, denoted by A(n), will be an optimal OA for a lean design of i runs. In order that one A(n) can be used for lean designs for several i's, we may maximize the quantity $|(\det X(n_1)) \times (\det X(n_2)) \times \cdots \times (\det X(n_j))|$ for some $n_1, n_2, ..., n_j$, where $n > n_1 > n_2 \cdots > n_j > 1$. Chan, Ma and Goh (2002) carried out this maximization using a numerical searching algorithm constructed based on "threshold accepting" (Winker and Fang (1997)). The optimal OA's A(8), A(12), A(16), A(20), A(24) and A(28) obtained in Chan, Ma and Goh (2002) are displayed in Table 2. The 0's in A(8), ..., A(20) represent -1.

The first column from the right hand side of Tables 3 (extracted from Chan, Ma, Goh (2002)) shows the maximum values of $|\det X(i)|$ obtained from numerical searching, where $X(i)$'s are the upper-left $i \times i$ sub-matrices of A(8), ..., A(28). Matrices $X(i)$'s with large values of $|\det X(i)|$ are preferred. Values beneath A(8),

Table 2. Orthogonal arrays A(n) for two-level lean designs (Continue)

A(20) - OA(20; 2^{19})	A(24) - OA(24; 2^{23})	A(28) - OA(28; 2^{27})
1001010001110110110	01001011101101100000111	010101011110101000001011000
0100111010100011110	00011100111010100110101	001111001100100010010100110
1101010110001000111	11010101011100110100010	001001111010010100010011111
1110100100101110100	10100101100101101110001	101101000110011111100001100
1110000011100101011	01110110000011100101011	000000000101111011011011111
0011110101101001010	11111000001001011100101	011100100000101101110010010
0111000011011011100	00110000111001111000110	110011100010100011010001001
0100011111111100000	10011010000110111010011	100110111000011010111001010
0000001101001111111	00000000000000000000000	111010011000110111001010100
1101101000011101010	00101111011000111001001	101110010111110000110010001
1010101111010000110	11100110110000010010111	011000111111011011010100000
1100110101010011001	10100010101110110101100	111010001101001100000001011
1001101011101010001	01101000010100101111110	111100100011110000001101110
1011011110000111000	11001101000011110011100	101001001001000001111111000
0001100110110101101	10000100011111001001111	000110010011000111000111010
0010110010011110011	01110111001110001010100	000011101111110101101000010
1010011000111001101	00111100101101010011010	100010100110001100011110100
0111001100110010011	10101011011011000110010	001111110001001001001000101
0111111001000100101	01001110110111011100000	111111111111101111111111111
0000000000000000000	10011111100000001101110	010110101100010001100111101
	11010010111001101011000	110001110101000010100010110
	11111001110110000001001	010011010001111100110101100
	00010011010101010111101	100000011010101001100100111
	01000001101010011111011	011011000010011010101110011
		100101101001111110000110001
		001000110100100110101101001
		110101010100010101011100011
		010100001011000110111000101

A(12), ... , A(28) in Table 3 are the ratios of $|\det X(i)/\max(\det(X(i))|$ for each $X(i)$. The maximum ratio, 1, is highlighted with square brackets.

If n is a multiple of 4, L is an OA($n; 2^s$) and $X(n) = [\mathbf{1}_n, L]$, it follows from the identity $X'(n)X(n) = nI_n$ (where I_n denotes the $n \times n$ identity matrix) and the Hadamard inequality (Searle (1982), pp.199–200)

$$\det(X'(n)X(n)) \leq \prod_{i=1}^{n} \|\mathbf{x}_i\|^2 \leq n^n,$$

where $\mathbf{x}_i$ denotes the ith column of $X(n)$, that the maximum attainable value of $|\det(X(n))|$ is $n^{n/2}$. For similar results and a general equivalence theorem involving determinants derived from orthogonal arrays of mixed levels, readers may refer to Chan, Fang and Mukerjee (2001). As far as the authors are aware, when L is an OA($n; 2^s$), $X(n) = [\mathbf{1}_n, L]$, $X(i)$ is an $i \times i$ sub-matrix of $X(n)$ and i is not a multiple of 4, no general results are available about the theoretical maximum of $|\det(X(i))|$.

Given the number of runs n, the most appropriate choice(s) of $X(n)$ for lean design are indicated by the [1]'s in Table 3. For example, if a lean design of $n = 6$ runs is needed, Table 3 shows that the maximum value of $|\det X(6)|$

Table 3. Values of $\max|\det X(i)|$ and $|\det X(i)|/\max|\det X(i)|$.

i	A(8)	A(12)	A(16)	A(20)	A(24)	A(28)	$\max\|\det X(i)\|$
2	[1]	[1]	[1]	[1]	0	0	2
3	[1]	[1]	[1]	[1]	[1]	0	2^2
4	[1]	[1]	[1]	0.50	0.50	0	2^4
5	0.67	[1]	0.67	0	[1]	0.33	3×2^4
6	0.80	[1]	0.80	0.40	[1]	0	5×2^5
7	0.89	[1]	0.89	0.22	0.89	0.67	$2^6 \times 3^2$
8	[1]	0.56	0.75	0	0.50	0.38	2^{12}
9		0.48	0.57	0	0.68	[1]	7×2^{11}
10		0.96	0.76	0.10	[1]	0.86	$3 \times 7 \times 2^{11}$
11		[1]	0.53	0	0.97	0.46	$2^{10} \times 3^5$
12		[1]	0.18	0	0.34	0.02	$2^{12} \times 3^6$
13			0.78	0.19	[1]	0.02	$3 \times 5 \times 11 \times 2^{15}$
14			[1]	0	0.88	0.02	2^{25}
15			[1]	0	0.53	0.02	2^{28}
16			[1]	0.12	0.18	0.001	2^{32}
17				[1]	0.80	0	$2^{16} \times 5^7$
18				[1]	0.60	0	$2^{17} \times 5^8$
19				[1]	0.43	0	$2^{18} \times 5^9$
20				[1]	0.09	0.01	$2^{20} \times 5^{10}$.
21					[1]	0.16	$2^{29} \times 3^9$
22					[1]	0.10	$2^{31} \times 3^{10}$
23					[1]	0.11	$2^{33} \times 3^{11}$
24					[1]	0.07	$2^{36} \times 3^{12}$
25						[1]	$2^{24} \times 7^{11}$
26						[1]	$2^{25} \times 7^{12}$
27						[1]	$2^{26} \times 7^{13}$
28						[1]	$2^{28} \times 7^{14}$

found from numerical searching is 5×2^5, and this maximum is attained if $X(6)$ is the upper-left 6×6 sub-matrices of the $X(12)$ generated from A(12) or the $X(24)$ generated from A(24). These two $X(6)$'s are therefore the most efficient extended design matrices for 2-level lean design of 6 runs. If a lean design of $n = 9$ runs is needed, Table 3 shows that the maximum value of $|\det X(9)|$ found from numerical searching is 7×2^{11}, and this maximum is attained if $X(9)$ is the upper-left 9×9 sub-matrix of the $X(28)$ generated from A(28). This $X(9)$ is therefore the most efficient extended design matrix for 2-level lean design of 9 runs found by the authors so far.

When the number of parameters to be estimated is less than the i in a regression equation, Chan, Ma and Goh (2002) indicated which matrices to select and which additional column(s) to be deleted.

4 Lean designs of 3 levels.

In this section, optimal 3-level orthogonal arrays for lean designs, denoted by A(9), A(18), A(27), A(36), B(9), B(18), B(27) and B(36), are constructed from $OA(9;3^4)$, $OA(18;3^7)$, $OA(27;3^{13})$, and $OA(36;3^{13})$. The upper-left square sub-matrices of these A(n)'s and B(n)'s will be the design matrices recommended for lean designs. We use $-1,0,1$ as the entries of $OA(n;3^s)$.

It is well-known that $f_1(x) = x$ and $f_2(x) = 3x^2 - 2$ form a system of orthogonal polynomials of degrees up to 2, where $x = -1,0,1$ (Pearson and Hartley (1966)). Given an $OA(n;3^s)$, say $L = [x_{ij}]$, we denote the jth column $\mathbf{x}_j = (x_{1j},...,x_{nj})'$ of L by $\mathbf{f}_1(\mathbf{x}_j) = (f_1(x_{1j}),...,f_1(x_{nj}))'$, define $\mathbf{f}_2(\mathbf{x}_j) = (f_2(x_{1j}),...,f_2(x_{nj}))'$, and define the $n \times (2s+1)$ extended design matrices

$$X_A(n) = [\mathbf{1}_n, \mathbf{f}_1(\mathbf{x}_1), \mathbf{f}_2(\mathbf{x}_1), ..., \mathbf{f}_1(\mathbf{x}_s), \mathbf{f}_2(\mathbf{x}_s)],$$

$$X_B(n) = [\mathbf{1}_n, \mathbf{f}_1(\mathbf{x}_1), ..., \mathbf{f}_1(\mathbf{x}_s), \mathbf{f}_2(\mathbf{x}_1), ..., \mathbf{f}_2(\mathbf{x}_s)].$$

Then both $X'_A(n)X_A(n)$ and $X'_B(n)X_B(n)$ will be $(2s+1)\times(2s+1)$ nonsingular diagonal matrices, which ensures that the coefficients β_0, β_i, β_{ii} $(i = 1,...,s)$ in the quadratic model

$$E[y] = \beta_0 + \beta_1 x_1 + \cdots + \beta_s x_s + \beta_{11}(3x_1^2 - 2) + \cdots + \beta_{ss}(3x_s^2 - 2) \qquad (4)$$

can be independently estimated (Walpole and Myers (1998, Section 12.7)). In a lean design, since the number of runs is less than n, it may not be possible to estimate all the coefficients β_i and β_{ii} in (4), and hence columns in $X_A(n)$ and $X_B(n)$ may have to be removed.

To construct a lean design of i runs, an $OA(n,3^s)$, say L, where $n > i$, is used as an initial design. An extended design matrix $X(n)$ is constructed from L. The matrix $X(n)$ is transformed by exchanging its rows and exchanging its columns, so that the upper-left $i\times i$ sub-matrix $X(i)$ of the transformed matrix has a maximum $|\det X'(i)X(i)|$. In doing this, the matrix L, as a submatrix of $X(n)$, is also transformed. The transformed matrix of L will be an optimal OA for a lean design of i runs. In order that one optimal OA can be used for lean designs for several i's, we may maximize the quantity $|(\det X(n_1)) \times (\det X(n_2)) \times \cdots \times (\det X(n_j))|$ for some $n_1, n_2, ..., n_j$, where $n > n_1 > n_2 \cdots > n_j > 1$. For this searching, the authors adopted a maximization algorithm constructed using threshold accepting. Since both linear and quadratic effects can be estimated from an OA of 3 levels, in searching for optimal OA's we need to consider the following two aspects.

(1) To construct lean designs, we may choose $X(n) = X_A(n)$, $X(n) = X_B(n)$, or other forms of $X(n)$. If we want to exclude the quadratic and linear effects of insignificant variables in the lean design and keep the linear and quadratic effects of significant variables, we will exchange rows and exchange columns of $X(n) = X_A(n)$ to produce optimal OA's, denoted by A(n), so that the upper-left sub-matrices of A(n) will contain columns $\mathbf{f}_1(\mathbf{x})$ and $\mathbf{f}_2(\mathbf{x})$ of the significant variables $\mathbf{x}$'s. The optimal OA's, A(n) $(n = 9, 18, 27, 36)$, are produced from numerical searching, and their transposes are displayed in

Table 4. Orthogonal arrays A(n) and B(n) for three-level lean designs

Transpose of A(9) - OA($9;3^4$)
1 1 3 3 2 2 1 3 2
3 2 2 1 1 3 1 3 2
1 3 1 3 1 3 2 2 2
2 3 1 2 3 1 1 3 2

Transpose of A(18) - OA($18;3^7$)
3 2 3 1 1 3 2 3 2 2 2 1 2 1 1 3 1 3
3 2 2 1 3 1 1 3 3 1 3 2 2 1 2 1 3 2
1 1 2 1 3 3 2 2 1 3 3 3 2 2 1 1 2 3
3 1 2 2 2 2 3 1 2 1 3 1 2 3 3 1 1 3
1 3 1 2 3 3 3 3 1 1 2 1 2 1 3 2 2 2
2 3 3 1 3 2 1 1 1 2 3 1 2 3 2 3 2 1
2 3 1 1 2 3 2 1 3 1 1 2 2 3 1 2 3 3

Transpose of A(27) - OA($27;3^{13}$)
322311132121112132331223323
121332123113232223311323112
331131223121321231323222131
213132132133231232121211233
313221311121233333312221212
231312311133322133132212122
333333321221122211123122211
112212332113323121322321331
223323111132322311321123231
211233323132113212332121123
133122211113111322333322223
123231232111323213133113222
232113232132231113313122312

Transpose of A(36) - OA($36;3^{13}$)
121323211331333212323222121332111123
233112213132121311223221133321123233
332311211213211213232122213333312213
313231212331222113121323122311232313
322122113311213331331321221223223113
131232112213331131133223212212133223
112133313233312322122121112123321323
213323112112122232332123231211311333
223332311223123221111322312131213133
231221313321131123312121321122332233
111111111122332333211322333232222323
322213312122213122213223333113131113
232123311313223133212211132312213321

Transpose of B(9) - OA($9;3^4$)
1 1 3 1 2 2 3 3 2
3 2 2 1 1 3 3 1 2
1 3 1 2 1 3 2 3 2
2 3 1 1 3 1 3 2 2

Transpose of B(18) - OA($18;3^7$)
3 2 1 3 1 3 2 2 2 1 1 3 2 3 3 2 1 1
1 3 2 3 1 2 3 1 2 2 1 2 2 1 3 1 3 3
3 2 3 1 1 3 3 1 1 2 2 1 2 2 2 3 3 1
2 1 3 3 1 1 2 3 1 3 2 2 2 1 3 3 1 2
1 1 2 2 2 3 2 1 3 1 3 1 2 2 3 3 1 3
1 3 3 2 1 2 1 2 1 1 2 3 2 3 1 3 2 3
3 3 3 3 2 1 1 1 3 1 3 2 2 1 2 2 2 1

Table 4. The upper-left sub-matrices of A(n) can be used as design matrices of efficient lean designs. For tidiness of presentation, the values $-1, 0, 1$ taken by $x_1, ..., x_s$ in A(n) (and also B(n) to be introduced below) in Table 4 are denoted by 1,2,3, respectively. On the other hand, we may also wish to take an alternative approach of including all linear effects before including any quadratic effects in the lean design. Following this approach, we choose $X(n) = X_B(n)$, exchange rows and exchangs columns of $X(n) = X_B(n)$ to produce optimal OA's, denoted by B(n), so that the upper-left sub-matrices of B(n) will include columns $\mathbf{f}_1(\mathbf{x})$ of all linear effects before any columns $\mathbf{f}_2(\mathbf{x})$ of quadratic effects are included. The optimal OA's, B(n) ($n = 9, 18, 27, 36$), are produced from numerical searching, and their transposes are displayed in

Table 4. Orthogonal arrays A(n) and B(n) for three-level lean designs (Continue)

Transpose of B(27) - OA(27; 3^{13})	Transpose of B(36) - OA(36; 3^{13})
333111123312232212313211223	311332112212222311331313112223122333
231131322123132123311123322	131332123131322223313221112132213213
123131331321223221213223131	113133223231211322221133132312123312
312131313222311322112323213	331133231323122212322132112311312123
212331233111232132223321311	122313311322322211233133121132321312
233311313231123322121212321	221231323123223131123333122211131213
311311331333332221222112212	233133212112313232133111122323211232
223231111132332111132222333	123332131223113123131212122121333322
131331212312323233122121123	132231323312131221232312132223313111
112231123333123212331322112	222231312231132132311122112332331331
331231132231211313233122221	212332112323331333222231132111221131
133211233123311132232213122	313231331331313311112321122232222122
323331221213111331321221232	322133231112231113213223132133132221

Table 4. The upper-left sub-matrices of B(n) can be used as design matrices of efficient lean designs.

(2) Since entries of $X(n)$, where $X(n) = X_A(n)$ or $= X_B(n)$, are $-1, 0, 1$, and $f_1(x) = x$ and $f_2(x) = 3x^2 - 2$, it is clear that the sums of squares of entries along columns $\mathbf{f}_1(\cdot)$ in $X(n)$ is $2n/3$ and those along columns $\mathbf{f}_2(\cdot)$ have a considerably large value $2n$. Thus the value of $\det(X'(i)X(i))$ will be dominated by the columns $\mathbf{f}_2(\cdot)$, which will create instability during the numerical searching process for the maximum of $|(\det(X(n_1)) \times \det(X(n_2)) \times \cdots \times \det(X(n_j))|$. To overcome this problem, we may normalize all columns in $X(n)$ before applying the searching algorithm. In doing this we need only to put a factor $3^{-1/2}$ to $f_2(x) = 3x^2 - 2$, since $\sum_x (3^{-1/2} f_2(x))^2 = \sum_x (f_1(x))^2 = 2$, where the summation $\sum_x$ is taken for $x = -1, 0, 1$. Alternatively, we may also normalize the sums of squares of columns of the sub-matrix $X(i)$ obtained from $X(n)$ at each step in the numerical searching process, and in doing this we need to divide each entry in column $(f_\nu(x_{1j}), ..., f_\nu(x_{ij}))'$ in $X(i)$ by $(f_\nu^2(x_{1j}) + \cdots + f_\nu^2(x_{ij}))^{1/2}$ ($\nu = 1, 2$). In this article, we have chosen to normalize all columns of $X(n)$ before the searching process, since this approach produces more stable results. We have chosen $n_1 = 1$, ... , $n_j = 5$ so that $|\det(X(n)) \times \cdots \det(X(5))|$ is maximized.

For each of the extended design matrices $X_A(n)$ and $X_B(n)$ obtained from A(n) and B(n), respectively, its D-efficiency value defined in Wu and Hamada (2000) and Ma, Fang and Erkki (2000) are calculated, and the results are shown in Table 5. Given the number of runs i, the most appropriate choice(s) of $X_A(i)$ or $X_B(i)$ for lean design are the one(s) with the highest D-efficiency value. The highest D-efficiency values are high-lighted with square brackets in Table 5. The value of D-efficiency is 1 if and only if the design is an orthogonal array. Table 5 shows that the D-efficiency values for the lean designs constructed from A(n) and B(n) are rather high, larger than 0.7 and in many cases larger than 0.8. Figure 1

Table 5. D-efficiency of lean designs derived from A(n) and B(n)

run	A(9)	A(18)	A(27)	A(36)	B(9)	B(18)	B(27)	B(36)
5	**0.755**	**0.755**	**0.755**	**0.755**	0.805	0.932	**0.959**	0.941
6	**0.909**	0.827	**0.909**	0.881	0.860	**0.937**	0.813	0.891
7	0.814	0.874	0.814	**0.877**	0.899	**0.901**	0.785	0.883
8	**0.924**	0.858	0.923	0.865	**0.924**	0.877	0.820	0.841
9	**1.000**	0.815	0.917	0.880	**1.000**	0.841	0.841	0.828
10		0.822	0.822	**0.823**		0.815	**0.858**	0.791
11		0.809	0.783	**0.869**		0.834	**0.861**	0.808
12		0.800	0.784	**0.812**		0.840	**0.876**	0.759
13		**0.797**	0.730	0.772		0.823	**0.878**	0.763
14		**0.800**	0.756	0.769		0.820	**0.967**	0.764
15		**0.842**	0.750	0.764		0.841	**0.900**	0.746
16		**0.888**	0.773	0.775		**0.887**	0.864	0.745
17		**0.940**	0.783	0.747		**0.940**	0.873	0.750
18		**1.000**	0.800	0.760		**1.000**	0.863	0.755
19			**0.767**	0.743			**0.869**	0.765
20			**0.809**	0.753			**0.843**	0.774
21			**0.775**	0.754			**0.846**	0.763
22			**0.834**	0.773			**0.856**	0.750
23			**0.802**	0.795			**0.874**	0.760
24			**0.876**	0.855			**0.893**	0.769
25			**0.868**	0.808			**0.903**	0.813
26			**0.938**	0.805			**0.938**	0.777
27			**1.000**	0.795			**1.000**	0.795
28				**0.819**				**0.819**
29				**0.844**				**0.845**
30				**0.872**				**0.872**
31				**0.892**				**0.892**
32				**0.913**				**0.913**
33				**0.936**				**0.936**
34				**0.956**				**0.956**
35				**0.977**				**0.977**
36				**1.000**				**1.000**

shows the comparison of the high D-efficiency values of these lean designs, with the low D-efficiency values of those designs obtained by merely choosing the upper-left $i \times i$ submatrices of the extended design matrices formed from the initial designs L without exchanging any rows or exchanging any columns.

Given the number of runs i, the most efficient matrix (or matrices) $X(i)$ for lean design are indicated in Table 5. The larger the D-efficiency, the more efficient the design in terms of *D-optimality*. Consider an example in which a 3-level lean design of $i = 6$ runs is needed, and less significant variables are removed before the more significant ones. Table 5 shows that the largest D-efficiency 0.909 is attained when the upper-left 6×6 extended design matrices $X_A(6)$ derived from A(9) and A(27), respectively. The extended matrices $X_A(6)$ derived from A(9) and A(27) are given by

[$X_A(6)$ derived from A(9)]

x_0	x_1	$(3x_1^2-2)$	x_2	$(3x_2^2-2)$	x_3
1	-1	1	1	1	-1
1	-1	1	0	-2	1
1	1	1	0	-2	-1
1	1	1	-1	1	1
1	0	-2	-1	1	-1
1	0	-2	1	1	1

[$X_A(6)$ derived from A(27)]

x_0	x_1	$(3x_1^2-2)$	x_2	$(3x_2^2-2)$	x_3
1	1	1	-1	1	1
1	0	-2	0	-2	1
1	0	-2	-1	1	-1
1	1	1	1	1	-1
1	-1	1	1	1	1
1	-1	1	0	-2	-1

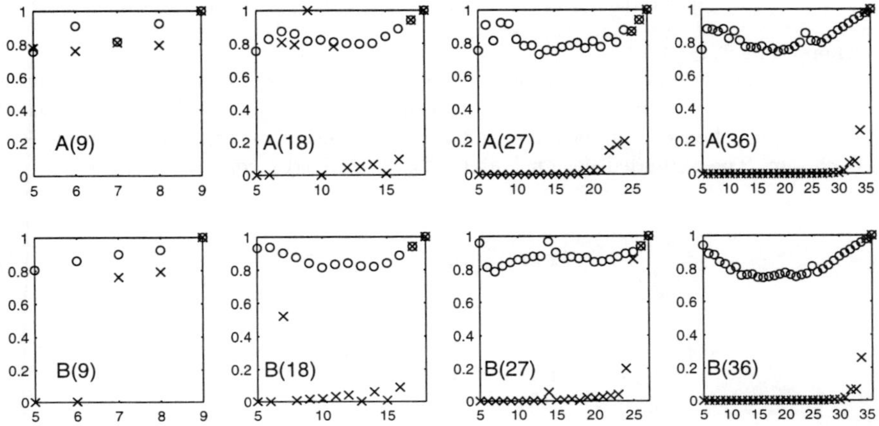

Fig. 1. D-efficiency vs Runs: Comparison of D-efficiency of the extended design matrices derived from the optimal OA's, A(n) and B(n) (circles), and the initial OA's used in the construction (crosses).

These designs can be used for estimation of all the coefficients β's in the model

$$E[y] = \beta_0 x_0 + \beta_1 x_1 + \beta_{11}(3x_1^2 - 2) + \beta_2 x_2 + \beta_{22}(3x_2^2 - 2) + \beta_3 x_3,$$

where $x_0 \equiv 1$.

Consider another example in which a lean design of $n = 7$ runs is needed and all insignificant quadratic effects are removed before linear effects. Table 5 shows that the largest D-efficiency 0.901 is attained when the upper-left 7×7 extended design matrix derived from B(18) is chosen. The extended design matrix $X_B(7)$ derived from B(18) is given by

[$X_B(7)$ derived from B(18)]

x_0	x_1	x_2	x_3	x_4	x_5	x_6
1	1	-1	1	0	-1	-1
1	0	1	0	-1	-1	1
1	-1	0	1	1	0	1
1	1	1	-1	1	0	0
1	-1	-1	-1	-1	0	-1
1	1	0	1	-1	1	0
1	0	1	1	0	0	-1

This design can be used for estimation of all the coefficients β's in the model

$$E[y] = \beta_0 x_0 + \beta_1 x_1 + \beta_2 x_2 + \beta_3 x_3 + \beta_4 x_4 + \beta_5 x_5 + \beta_6 x_6,$$

where $x_0 \equiv 1$.

If a lean design of $n = 8$ runs is needed and all insignificant quadratic effects are removed before linear effects. Table 5 shows that the largest D-efficiency 0.924 is attained when the upper-left 8×8 extended design matrix derived from B(9) is chosen. The extended design matrix $X_B(8)$ derived form B(9) is given by

[$X_B(8)$ derived from B(9)]

x_0	x_1	x_2	x_3	x_4	$(3x_1^2-2)$	$(3x_2^2-2)$	$(3x_3^2-2)$
1	-1	1	-1	0	1	1	1
1	-1	0	1	1	1	-2	1
1	1	0	-1	-1	1	-2	1
1	-1	-1	0	-1	1	1	-2
1	0	-1	-1	1	-2	1	1
1	0	1	1	-1	-2	1	1
1	1	1	0	1	1	1	-2
1	1	-1	1	0	1	1	1

This design can be used for estimation of all the coefficients β's in the model

$$E[y] = \beta_0 x_0 + \beta_1 x_1 + \beta_2 x_2 + \beta_3 x_3 + \beta_4 x_4 + \beta_{11}(3x_1^2 - 2) + \beta_{22}(3x_2^2 - 2) + \beta_{33}(3x_3^2 - 2),$$

where $x_0 \equiv 1$.

5 Conclusion

This article is concerned with forming 2-level and 3-level optimal OA's for lean designs for experiments. The motivation for doing this is that the number of runs allowed in an actual experiment may not be equal to the number of rows specified in an OA. The principle of construction of optimal OA's is to exchange rows and exchange columns of known OA's, so that the upper-left sub-matrices of the extended design matrices generated from the optimal OA's will have large values of determinant. For 3-level designs, two types of matrices A(n) and B(n) are constructed in this article, which deal with the two cases when we wish to remove both the quadratic and linear effects of insignificant variables, and when we wish to remove all insignificant quadratic effects before removing linear effects. Many other scenarios are not included – for example, when the effects $x_1, x_1^2, x_2, x_3, x_4^2$ are to be included and all other effects are insignificant. For such cases, individual optimal OA's can be constructed from numerical sercahing by first defining specific extended design matrices from the initial OA.

References

Atkinson, A. C. & Donev, A. N. (1996). *Optimum Experimental Designs.* Clarendon Press, Oxford, U.K.

Box, G. E. P., Hunter, W. G. & Hunter, J. S. (1978). *Statistics for Experiments – An Introduction to Design, Data Analysis, and Model Building.* John Wiley, New York, U.S.A.

Chan, L. Y., Fang, K. T. & Mukerjee, R. (2001). A characterization for orthogonal arrays of strength two via a regression model. *Statistics and Probability Letters* **54** 189–192.

Chan, L. Y., Ma, C. X. & Goh, T. N. (2002). Orthogonal Arrays for Experiments with Lean Designs. *Journal of Quality Technology* **35** No. 12 123–138.

Dey, A. & Mukerjee, R. (1999). *Fractional Factorial Plans.* Wiley, New York, NY.

Goh, T. N. (1996). Economical experimentation via 'lean design'. *Quality and Reliability Engineering International* **12** 383–388.

Hedayat, A. S., Sloan, N. J. A. & Stufken, J. (1999). *Orthogonal Arrays – Theory and Applications.* Springer, New York, U.S.A.

Ma, C. X., Fang, K. T. & Erkki, L. (2000). A new approach in constructing orthogonal and nearly orthogonal arrays. *Metrika* **50** 255-268.

Pearson, E. S., Hartley, H. O. (eds.) (1966). *Biometrika Tables for Statisticians.* Biometrika Trustees, University Press, Cambridge, U.K.

Searle, S. R. (1982). *Matrix Algebra Useful For Statistics.* Wiley, New York, U.S.A.

Walpole, R. E. & Myers, R. H. (1998). *Probability and Statistics for Engineers and Scientists.* Prentice Hall, New Jersey, U.S.A.

Winker, P. & Fang, K. T. (1997). Application of threshold accepting to the evaluation of the discrepancy of a set of points. *SIAM Journal on Numerical Analysis* **34** 2028–2042.

Wu, C. F. J. & Hamada, M. (2000). *Experiments: planning, analysis, and parameter design optimization.* John Wiley & Sons, New York.

Quasi-Random Sampling for Estimation of Integrals of Random Fields

Yingcai Su

Department of Mathematics, Southwest Missouri State University
(yis780f@smsu.edu)

Summary. The problem of interest is to estimate an integral (or the total) of a random field from observations at a finite number of sampling points. The quality of an estimator is measured by mean square error (MSE). Sampling points are appropriately selected so that the resulting MSE is as close to zero as possible. In this article, we begin by introducing the basic elements of estimation of integrals of random fields, including a review of the number-theoretic method. We then show how the number-theoretic method is related to the design of sampling points along with the sample mean estimator.

Key words: Random field, integral of random field, spatial average, sampling design, extreme discrepancy, mean square discrepancy, spatial statistics.

2000 Mathematics Subject Classification: 60G60, 62K99, 62M30, 65C05

1 Introduction

Consider a random field $X(\mathbf{t})$ on the d-dimensional unit cube $C^d = [0,1]^d$ with mean $f(\mathbf{t}) = EX(\mathbf{t})$ and covariance $R(\mathbf{t},\mathbf{s}) = E[(X(\mathbf{t}) - f(\mathbf{t}))(X(\mathbf{s}) - f(\mathbf{s}))]$. The random field is sampled at a number of n sampling points $T_n = \{\mathbf{t}_{i,n}\}_{i=1}^{n} \subset C^d$. Denote the observations of X at T_n by $X'_{T_n} = (X(\mathbf{t}_{1,n}), \ldots, X(\mathbf{t}_{n,n}))$. Here, observations are not only coordinate-referenced (time-referenced when $d = 1$ and geo-referenced when $d = 2, 3$) but also correlated. On the basis of observations X_{T_n}, the aim is to estimate the integral

Contemporary Multivariate Analysis and Experimental Design—In Honor Celebration of Professor Kai-Tai Fang's 65th birthday. Edited by Jianqing Fan and Gang Li. The World Scientific Publisher, 2005.

$$I(X) = \int_{C^d} X(\mathbf{t})d\mathbf{t}$$

of X over C^d. The performance of estimation is measured by mean square error (MSE). Sampling points T_n are designed in such a way that the MSE is as close to zero as possible.

The integral of a random field over a region is an important summarizing quantity of the random field, which represents the spatial total of the random field in the region. For instance, it represents the total area precipitation in hydrology, the total amount of acid concentration in the environmental science and the total reserve of a certain metal deposit within a region in the mining industry. The integral divided by the area or the volume of the region is the spatial average of the random field. Other types of statistical inferences concerning a random field can be found in Bras &Rodriguez-Iturbe (1985), Cressie (1993), Christakos (1992) and Journel & Huijbregts (1978). One can find discussions of various technical aspects of estimation of integrals of random processes and random fields in Cambanis (1985), Matérn (1986), Stein (1999) and Ylvisaker (1975).

It is a common practice to use a linear estimator, a linear combination of observations, each term of which is a product of a coefficient and an observation. Linear estimators range from the optimal linear estimator to the sample mean estimator, in accordance with available statistical features of the random field. The optimal linear estimator requires the mean f to be known up to a regression form and the precise knowledge of the covariance R. The sample mean estimator is the simplest linear estimator and relatively nonparametric in the sense that its coefficients do not depend on the mean and the covariance.

Denote by $MSE(T_n)$ the mean square error of an estimator constructed from observations X_{T_n}. A minimizer T_n^o of $MSE(T_n)$ constitutes an optimal sampling design, namely, $MSE(T_n^o) = \inf_{T_n} MSE(T_n)$, where the infinimum is taken over all possible sampling designs of size n. A sampling design T_n^* is asymptotically optimal if and only if

$$MSE(T_n^*) \asymp MSE(T_n^o),$$

where and throughout, $a_n \asymp b_n$ means that $a_n/\ b_n \to 1$, as $n \to \infty$. That is, an asymptotically optimal design T_n^* minimizes the MSE as the sample size tends to infinity.

A widely discussed sampling method in one-dimension is the regular sampling, which takes the percentiles of a properly chosen density as sampling points. Regular sampling designs are asymptotically optimal, as shown in Sacks & Ylvisaker (1966). However, regular sampling designs do not have a direct extension to higher ($d \geq 2$) dimensions. The simplest sampling design in higher dimensions is the product design which selects each coordinate of sampling points separately, but it has a rather poor convergent rate, as indicated by the work in Ylvisaker (1975). For prediction of isotropic random fields, a particular piecewise regular sampling design is used in Su (1997), in which the region C^2 is tessellated into piecewise regular hexagons or rhombi and the centers of these hexagons or rhombi are taken to be a design of sampling points. The sampling design constructed this way is asymptotically optimal for a nonlinear type of predictor. It is generally not an easy task to characterize optimal or asymptotically optimal sampling designs in higher dimensions, which has baffled researchers for years.

The number-theoretic method is a rule to generate uniformly distributed point sets on C^d. The quality of uniformity is measured by extreme discrepancy or mean square discrepancy. The number-theoretic method is extensively used in Fang & Wang (1994) to tackle a variety of statistical computation and inference problems, including the design of experiments. In this article, the approach in Fang & Wang (1994) is extended to the problem of estimation of integrals of random fields. Our discussions are also motivated by the work in Woźniakowski (1991), though it is in a different setup. Hereinafter, a design of sampling points produced by the number-theoretic method is called a quasi-random sampling design.

2 Quasi-Random Sampling Designs

There are two kinds of point sets that are constructed in order to minimize the discrepancy: finite point sets and infinite point sequences. A finite point set is of single-stage in nature while an infinite point sequence has the stepwise feature. A finite point set is of the form $T_n = \{\mathbf{t}_{i,n}\}_{i=1}^n$ while an infinite point sequence $T_n = \{\mathbf{t}_i\}_{i=1}^n$ can be viewed as the set of the first n points truncated from an infinite sequence $\{\mathbf{t}_i\}_{i=1}^\infty$.

For a point set T_n and $\mathbf{u} \in C^d$, denote by $T_n(\mathbf{u})$ the number of points in T_n that fall in the box $[\mathbf{0}, \mathbf{u}] = [0, u_1] \times \ldots \times [0, u_d]$. Then $T_n(\mathbf{u})/n$ is the proportion of points of T_n that are in $[\mathbf{0}, \mathbf{u}]$. The discrepancy function $\mu_{T_n}(\mathbf{u})$ defined on C^d by

$$\mu_{T_n}(\mathbf{u}) = \frac{T_n(\mathbf{u})}{n} - \prod_{\tau=1}^{d} u_\tau, \ \mathbf{u} \in C^d, \tag{1}$$

measures the local discrepancy at $\mathbf{u}$ of uniformity of the distribution of T_n. The overall degree of discrepancy of uniformity of T_n is quantified either by the extreme discrepancy $D(T_n) = \sup_{\mathbf{u} \in C^d} |\mu_{T_n}(\mathbf{u})|$ or by the rooted mean square discrepancy $L_2(T_n, dF) = \{\int_{C^d} \mu_{T_n}^2(\mathbf{u}) dF(\mathbf{u})\}^{1/2}$, where F is a probability distribution function on C^d. It is trivial to see that $L_2(T_n, dF) \leq D(T_n)$. Thus an upper bound for $D(T_n)$ is also for $L_2(T_n, dF)$ and a lower bound for $L_2(T_n, dF)$ also holds for $D(T_n)$. A thorough survey of facts concerning extreme discrepancy and mean square discrepancy is given in Fang & Wang (1994), Niederreiter (1978) and Niederreiter (1992).

The convergent rate of $\inf_{T_n} D(T_n)$ as well as constructions of a point set to achieve such a rate remains undecided. But it is widely believed that the convergent rate of $\inf_{T_n} D(T_n)$ is in the order of $(\log n)^{(d-1)}/n$ for finite point sets, while in the order of $(\log n)^d/n$ for infinite point sequences. Halton sequences and Hammersley point sets are such point sets that achieve the conjectured convergent rates, respectively, as shown in Niederreiter (1992).

2.1 Halton Sequences

Let $p \geq 2$ be an integer, and then every natural number i can be uniquely expanded into a digit representation in base p: $i = \sum_{j=0}^{M_i} a_j(i)p^j$, where the integers $0 \leq a_j(i) < p$ and $M_i < \infty$. The radical-inverse function $\sigma_p(\cdot)$ in base p is defined on all natural numbers by

$$\sigma_p(i) = \sum_{j=0}^{M_i} a_j(i)p^{-j-1}.$$

Now, let $p_1, \ldots, p_d$ be d different prime numbers, the set of points $\{\mathbf{h}_i\}_{i=1}^{\infty}$ defined by

$$\mathbf{h}_i = (\sigma_{p_1}(i), \ldots, \sigma_{p_d}(i))$$

is called a Halton sequence of points in the bases $p_1, \ldots, p_d$. The extreme discrepancy of a Halton sequence is in the order of $(\log n)^d/n$.

2.2 Hammersley Point Sets

Let $p_1, \ldots, p_{d-1}$ be $d-1$ different prime numbers, the set of points $\{\mathbf{h}_i\}_{i=1}^{n}$ defined by

$$\mathbf{h}_i = ((i-1)/n, \sigma_{p_1}(i), \ldots, \sigma_{p_{d-1}}(i))$$

is called a Hammersley point set in the bases $p_1, \cdots, p_{d-1}$. Hammersley point sets and Halton sequences differ only in the first component of points. However, Hammersley point sets are finite point sets. The extreme discrepancy of Hammersley point sets is in the order of $(\log n)^{d-1}/n$.

2.3 Modified Hammersley Point Sets

A modified version of Hammersley point sets is given in Fang & Wang (1994), in which the first components in a set of n points are $\{(2i-1)/(2n)\}_{i=1}^{n}$. The modified Hammersley point sets $\{\mathbf{h}_i\}_{i=1}^{n}$ take the form

$$\mathbf{h}_i = ((2i-1)/(2n), \sigma_{p_1}(i), \ldots, \sigma_{p_{d-1}}(i)).$$

Asymptotic properties of the modified Hammersley point sets are not available. However in one-dimension, the point set $\{(2i-1)/(2n)\}_{i=1}^{n}$ minimizes both extreme discrepancy and mean square discrepancy, as shown in Niederreiter (1992).

It is called a median sampling design in Cambanis (1985), and it outperforms other periodic sampling designs for a variety of inference problems concerning a time series.

For finite point sets, the results in Roth (1980) together with the continuity of $L_2(T_n, dF)$ in T_n suggest that there is a constant $c(d, F)$ such that

$$\inf_{T_n} L_2^2(T_n, dF) \asymp c(d, F)[(\log n)^{d-1}/n^2], \tag{2}$$

where $c(d, F)$ is a constant that depends on the dimension d and possibly also on F. The optimal constant $c(d, F)$ is known only in one dimension: $c(1, F) = 1/12$. If the unconfirmed rate $(\log n)^{(d-1)}/n$ of $D(T_n)$ is true for finite point sets, then the limiting equation (2) implies that $D(T_n)$ has not only a larger magnitude but also a slower convergent rate than $L_2(T_n, dF)$ (by a factor of $(\log n)^{(d-1)/2}$).

Next we analyze the MSE of estimation of integrals of random fields with the quasi-random sampling design, which is done for the case where the mean f is known or equivalently equals to zero and for the case where f is unknown, respectively.

3 Random Fields with Mean Zero

The sample mean estimator is $\bar{I}(X, T_n) = \sum_{i=1}^{n} X(\mathbf{t}_{i,n}) \,/\, n = \mathbf{1}_n' \, X_{T_n} \,/\, n$, where $\mathbf{1}_n' = (1, \ldots, 1)$ is the $1 \times n$ unit vector, and its MSE is $MSE(T_n) = E[\bar{I}(X, T_n) - I(X)]^2$.

Hereinafter, g stands for a function on C^d defined by $g(\mathbf{t}) = \int_{C^d} R(\mathbf{t}, \mathbf{s}) d\mathbf{s}$ and $\sigma_R^2 = \int_{C^d} g(\mathbf{t}) d\mathbf{t} = \int_{C^d} \int_{C^d} R(\mathbf{t}, \mathbf{s}) d\mathbf{t} d\mathbf{s}$ for the integrated value of g over C^d. When $f(\mathbf{t}) \equiv 0$, the mean square error of $\bar{I}(X, T_n)$ is equal to the variance of $\bar{I}(X, T_n)$, which can be written as

$$Var(T_n) = \sigma_R^2 - 2\mathbf{1}_n' g_{T_n}/n + \mathbf{1}_n' R_{T_n} \mathbf{1}_n / n^2, \tag{3}$$

where $g_{T_n}' = (g(\mathbf{t}_{1,n}), \ldots, g(\mathbf{t}_{n,n}))$ is the vector of values of the function g at T_n and $R_{T_n} = (R(\mathbf{t}_{i,n}, \mathbf{t}_{j,n}))_{n \times n}$ is the variance-covariance matrix of X_{T_n}.

We first establish a connection between $Var(T_n)$ and discrepancy of T_n. To do so, we need to introduce Riemann - Stieltjes integrals of functions on $C^d \times C^d$ with respect to R.

Let $\Gamma_m = \{\Gamma_{1,m}, \ldots, \Gamma_{m,m}\}$ be a partition of C^d into a number of $m = \prod_{\tau=1}^{d} m_\tau$ d-dimensional parallelepipeds (parallel to the axes): $\Gamma_{i,m} = [a_{i_1}^1, a_{i_1+1}^1] \times \cdots \times [a_{i_d}^d, a_{i_d+1}^d]$, where $0 \le i_\tau \le m_\tau - 1$, $\tau = 1, \ldots, d$ and $0 = a_0^\tau < a_1^\tau < \ldots < a_{m_\tau}^\tau = 1$. Let $\Gamma_{m'} = \{\Gamma_{1,m'}, \ldots, \Gamma_{m',m'}\}$ be another partition of C^d into a number of $m' = \prod_{\tau=1}^{d} m_\tau'$ parallelepipeds (parallel to the axes): $\Gamma_{j,m'} = [b_{j_1}^1, b_{j_1+1}^1] \times \cdots \times [b_{j_d}^d, b_{j_d+1}^d]$, where $0 \le j_\tau \le m_\tau' - 1$, $\tau = 1, \ldots, d$ and $0 = b_0^\tau < b_1^\tau < \ldots < b_{m_\tau'}^\tau = 1$. Denote by

$$\Delta_{i,j}(R) = \sum (-1)^\theta R((t_1, \ldots, t_d), (s_1, \ldots, s_d))$$

the assigned value of R to the parallelepiped $\Gamma_{i,m} \times \Gamma_{j,m'}$, where the sum is taken over all $t_\tau = a^\tau_{i_\tau}$ or $a^\tau_{i_\tau+1}$, $s_\tau = b^\tau_{j_\tau}$ or $b^\tau_{j_\tau+1}$ and θ is the sum of the number of t_τ which are $a^\tau_{j_\tau+1}$ and the number of s_τ which are $b^\tau_{j_\tau+1}$. When $d = 1$, dropping the superscript 1 from a^1_i and b^1_j, $\Delta_{i,j}(R)$ becomes

$$\Delta_{i,j}(R) = R(a_{i+1}, b_{j+1}) - R(a_{i+1}, b_j) - R(a_i, b_{j+1}) + R(a_i, b_j).$$

The total variation of R on $C^d \times C^d$, denoted by $V(R; C^d \times C^d)$, is defined by

$$V(R; C^d \times C^d) = \sup \sum_{i,j} |\Delta_{i,j}(R)|,$$

where the supreme is taken over all m and m' and all possible parallelepiped subdivisions Γ_m and $\Gamma_{m'}$. The function R is of bounded variation on $C^d \times C^d$ if and only if $V(R; C^d \times C^d) < \infty$.

For a function f on $C^d \times C^d$, $\mathbf{s}^*_i \in \Gamma_{i,m}$ and $\mathbf{s}^*_j \in \Gamma_{j,m'}$, form a sum $\sum_{i,j} f(\mathbf{s}^*_i, \mathbf{s}^*_j)\Delta_{i,j}(R)$. The Riemann - Stieltjes integral of f on $C^d \times C^d$ with respect to R is defined as

$$\int_{C^d} \int_{C^d} f(\mathbf{t}, \mathbf{s}) d^2 R(\mathbf{t}, \mathbf{s}) = \lim_{\delta_{m,m'} \to 0} \sum_{i,j} f(\mathbf{s}^*_i, \mathbf{s}^*_j)\Delta_{i,j}(R),$$

where $\delta_{m,m'}$ is the width of the longest side among all parallelepipeds in Γ_m and $\Gamma_{m'}$. It is straightforward to see that

$$\left| \int_{C^d} \int_{C^d} f(\mathbf{t}, \mathbf{s}) d^2 R(\mathbf{t}, \mathbf{s}) \right| \leq \sup_{(\mathbf{u},\mathbf{v}) \in C^d \times C^d} |f(\mathbf{u}, \mathbf{v})| \, V(R; C^d \times C^d). \tag{4}$$

Let $B^1_k = \{\mathbf{t} \in C^d : \mathbf{t} = (1, \ldots, t_k, \ldots, 1)\}$, $1 \leq k \leq d$ be the $\binom{d}{1}$ one-dimensional face-regions of C^d, $B^2_{k,\ell} = \{\mathbf{t} \in C^d : \mathbf{t} = (1, \ldots, t_k, \ldots, t_\ell, \ldots, 1)\}$, $1 \leq k \neq \ell \leq d$ be the $\binom{d}{2}$ two-dimensional face-regions of C^d, ..., and $B^{d-1}_{2,\ldots,d} = \{\mathbf{t} \in C^d : \mathbf{t} = (1, t_2, \ldots, t_d)\}$,..., $B^{d-1}_{1,\ldots,d-1} = \{\mathbf{t} \in C^d : \mathbf{t} = (t_1, \ldots, t_{d-1}, 1)\}$ be the $\binom{d}{d-1}$ $(d-1)$-dimensional face-regions of C^d.

Denote by $\mathcal{B}$ the collection of all these $\binom{d}{1} + \ldots + \binom{d}{d-1} = 2^d - 2$ face-regions of C^d. For $B', B'' \in \mathcal{B}$, the total variation $V(R; B' \times B'')$ of the restriction of R to $B' \times B''$ and the Riemann - Stieltjes integral of a function f on $B' \times B''$, with respect to the restriction of R, are defined similarly. We will make reference to the projection, denoted by T'_n, of T_n onto B', and also to the discrepancy function $\mu_{T'_n}(\cdot)$ of the point set T'_n on B', which is defined in a similar manner to μ as in (1). The projection of a Halton sequence (a Hammersley point set) onto $B' \in \mathcal{B}$ is still a Halton sequence (a Hammersley point set) in B'. It is well known that marginal distributions of an uniformly distributed random point set are still uniformly distributed. Thus, it is reasonable to assume that a similar property still holds for a low discrepancy point set, that is, the projection T'_n of a low discrepancy point set T_n onto B' shall have a low discrepancy as well. Now we state the following results.

Theorem 1. *Suppose that R is of bounded variation on $C^d \times C^d$ and $I(X)$ is estimated by $\bar{I}(X, T_n)$, then*

$$\begin{aligned} Var(T_n) = & \int_{C^d}\int_{C^d} \mu_{T_n}(\mathbf{t})\mu_{T_n}(\mathbf{s})\, d^2R(\mathbf{t},\mathbf{s}) \\ & + \sum_{B\in\mathcal{B}} \int_B\int_B \mu_{T_n'}(\mathbf{a})\mu_{T_n'}(\mathbf{b})\, d^2R(\mathbf{a},\mathbf{b}) \\ & + \sum_{B'\neq B''} (-1)^{d-\theta} \int_{B'}\int_{B''} \mu_{T_n'}(\mathbf{a})\mu_{T_n''}(\mathbf{b})\, d^2R(\mathbf{a},\mathbf{b}), \end{aligned} \tag{5}$$

where T_n' and T_n'' are the respective projections of T_n onto B' and B'', $\mathbf{a} = (a_1, \ldots, a_d)$ and $\mathbf{b} = (b_1, \ldots, b_d)$ with $a_i = t_i$ or 1 and $b_i = s_i$ or 1, θ is the number of a_i and b_i which are 1 and the second summation is over all $B' \neq B'' \in \mathcal{B}$.

The proof of the theorem is given in the section of **Proofs**. With the help of (4), an estimate on the variance $Var(T_n)$ can be immediately obtained from the theorem.

Corollary 1. *For the same setup as in the theorem,*

$$\begin{aligned} Var(T_n) \leq & V(R; C^d \times C^d)D^2(T_n) + \sum_{B\in\mathcal{B}} V(R; B\times B)D^2(T_n') \\ & + \sum_{B'\neq B''} V(R; B'\times B'')D(T_n')D(T_n''). \end{aligned}$$

It follows from the corollary that if T_n is a Halton sequence,

$$Var(T_n) \leq \sum_{k=1}^{d} c_k[(\log n)^k/n]^2 + \sum_{k\neq\ell} c_{k,\ell}[(\log n)^{k+\ell}/n^2],$$

where $c_k, c_{k,\ell}$ are constants, and thus $Var(T_n)$ is of the order $[(\log n)^d\ /n]^2$. Similarly, for a Hammersley point set T_n, $Var(T_n)$ is of the order $[(\log n)^{d-1}\ /n]^2$. In one-dimension, the equality (5) becomes

$$Var(T_n) = \int_0^1\int_0^1 \mu_{T_n}(t)\mu_{T_n}(s)d^2R(t,s).$$

When $d = 2$, it is

$$Var(T_n) = \int_{C^2}\int_{C^2} \mu_{T_n}(\mathbf{t})\mu_{T_n}(\mathbf{s})d^2R(\mathbf{t},\mathbf{s}) \tag{6}$$

$$+\int_{C^2}\mu_{T_n'}(t_1)\mu_{T_n'}(t_2)d^2R(\mathbf{t},(1,1))+\int_{C^2}\mu_{T_n''}(s_1)\mu_{T_n''}(s_2)d^2R((1,1),\mathbf{s})$$
$$+2\int_{C^2}\mu_{T_n'}(t_1)\mu_{T_n''}(s_2)d^2R((t_1,1),(1,s_2))-2\int_{C^2}\int_0^1$$
$$\mu_{T_n}(\mathbf{t})\mu_{T_n''}(s_2)d^2R(\mathbf{t},(1,s_2))-2\int_0^1\int_{C^2}\mu_{T_n'}(t_1)\mu_{T_n}(\mathbf{s})d^2R((t_1,1),\mathbf{s}),$$

where T_n' and T_n'' are the projections of T_n onto the horizontal axis and the vertical axis, respectively.

As an example, consider a random field with mean zero and a particular covariance

$$R_d(\mathbf{t},\mathbf{s})=\int_0^{\min(t_1,s_1)}\cdots\int_0^{\min(t_d,s_d)}\phi(\mathbf{u})d\mathbf{u},$$

where ϕ is a positive continuous function on C^d. It is elementary calculus to verify that when $d=1$, the variance of the sample mean is

$$Var(T_n)=\int_0^1\mu_{T_n}^2(u)\phi(u)du\ ,$$

and the set of points $T_n=\{(2i-1)/(2n)\}_{i=1}^n$ minimizes $Var(T_n)$. When $d=2$

$$Var(T_n)=\int_{C^2}\left[\mu_{T_n'}(u_1)+\mu_{T_n''}(u_2)-\mu_{T_n}(\mathbf{u})\right]^2\phi(\mathbf{u})d\mathbf{u}.$$

Substituting $T_n=\{\mathbf{t}_{i,n}\}_{i=1}^n$ by $S_n=\{\mathbf{s}_{i,n}\}_{i=1}^n$ with $\mathbf{s}_{i,n}=\mathbf{1}-\mathbf{t}_{i,n}$, then it is verified that $Var(S_n)=\int_{C^2}\mu_{S_n}^2(\mathbf{u})\phi(\mathbf{1}-\mathbf{u})d\mathbf{u}$. It just takes some extra but similar efforts to verify that the same equation holds in every d-dimension for covariance R_d, namely,

$$Var(S_n)=\int_{C^d}\mu_{S_n}^2(\mathbf{u})\phi(\mathbf{1}-\mathbf{u})d\mathbf{u},\tag{7}$$

which together with (2) implies that $\inf_{T_n}Var(T_n)\asymp c(d,\phi)[(\log n)^{d-1}/n^2]$, where $c(d,\phi)$ is a constant depending on d and ϕ.

4 Random Fields with Unknown Mean

Here the mean function $f(\mathbf{t})=EX(\mathbf{t})$ is unknown but it is assumed to have parallel smoothness to functions $R(\mathbf{t},\cdot)$, $\mathbf{t}\in C^d$, that is, it belongs to the reproducing kernel Hilbert space generated by the covariance R.

Write the random field X as

$$X(\mathbf{t})=f(\mathbf{t})+\xi(\mathbf{t}),\ \mathbf{t}\in C^d,$$

where $\xi(\mathbf{t})$ are random errors with mean zero and covariance $R(\mathbf{t},\mathbf{s}) = E[\xi(\mathbf{t})\xi(\mathbf{s})]$. In addition to the notations g_{T_n} and R_{T_n} introduced in the preceding section, put $f'_{T_n} = (f(\mathbf{t}_{1,n}), \dots, f(\mathbf{t}_{n,n}))$ for the values of f at T_n. Then the mean square error can be written as

$$MSE(T_n) = Bias^2(T_n) + Var(T_n),$$

where $Var(T_n)$ is as in (3), $Bias(T_n) = \mathbf{1}'_n f_{T_n}/n - I(f)$ and $I(f) = \int_{C^d} f(\mathbf{t})d\mathbf{t}$.

Let H(ξ) be the linear space of the random field ξ, which is the linear span of the random variables $\xi(\mathbf{t})$, $\mathbf{t} \in C^d$. That is, each random variable in H(ξ) is either a finite linear combination of random variables $\xi(\mathbf{t}), \mathbf{t} \in C^d$ or a mean square limit of a sequence of such finite linear combinations. The reproducing kernel Hilbert space, denoted by H(R), is the Hilbert space of all functions $h : C^d \mapsto (-\infty, \infty)$ of the form $h(\mathbf{t}) = E[\xi(\mathbf{t})\eta], \mathbf{t} \in C^d$, $\eta \in$ H(ξ). The linear space H(ξ) and the reproducing kernel Hilbert space H(R) are isomorphic under the correspondence $\xi(\mathbf{t}) \longleftrightarrow R(\mathbf{t}, \cdot)$. Thus the inner product in H(R) is provided by $< h_1, h_2 >_R = E[\eta_1\eta_2]$, where $h_i(\mathbf{t}) = E[\xi(\mathbf{t})\eta_i]$, $\eta_i \in$ H(ξ), $i = 1, 2$, and the norm of h_i in H(R) is $\|h_i\|_R = E^{1/2}\eta_i^2$. More properties concerning a reproducing kernel Hilbert space can be found in Aronszajn (1950) and Parzen (1967).

In particular the mean square integral $I(\xi) = \int_{C^d} \xi(\mathbf{t})d\mathbf{t}$ belongs to H(ξ). The function $g(\mathbf{t}) = \int_{C^d} R(\mathbf{t},\mathbf{s})d\mathbf{s} = E[\xi(\mathbf{t})I(\xi)]$ is a member of H(R) and $\|g\|_R^2 = EI^2(\xi) = \sigma_R^2$.

There is a connection between the problem of estimating the mean square integral $I(\xi)$ and the problem of approximating a definite integral of functions in H(R). For a fixed T_n, suppose that $I(\xi)$ is estimated by $\mathbf{1}'_n\xi_{T_n}/n$, where $\xi'_{T_n} = (\xi(\mathbf{t}_{1,n}), \dots, \xi(\mathbf{t}_{n,n}))$, and the definite integral $I(h) = \int_{C^d} h(\mathbf{t})d\mathbf{t}$, $h \in$ H(R), is approximated by the quadrature formula $\mathbf{1}'_n h_{T_n}/n$, where $h'_{T_n} = (h(\mathbf{t}_{1,n}), \dots, h(\mathbf{t}_{n,n}))$, then the variance of $\mathbf{1}'_n\xi_{T_n}/n$ is

$$Var(T_n) = \sup_{\|h\|_R \le 1} [\mathbf{1}'_n h_{T_n}/n - I(h)]^2,$$

where the supreme is taken over all $h \in$ H(R) with $\|h\|_R \le 1$. Since the mean f is in H(R), the function $h^* = f/\|f\|_R \in$ H(R) and $\|h^*\|_R = 1$. The squared error of approximation of the definite integral $I(h^*)$ by use of a quadrature formula $\mathbf{1}'_n h^*_{T_n}/n$ with sampling points (or nodes) T_n is bounded above by $Var(T_n)$. Thus

$$Bias^2(T_n) = \|f\|_R^2 \, [\mathbf{1}'_n h^*_{T_n}/n - I(h^*)]^2 \le \|f\|_R^2 \, Var(T_n),$$

which yields the following proposition.

Proposition 1. *When the mean f is in H(R) and the integral $I(X)$ is estimated by the sample mean estimator $\bar{I}(X, T_n)$, the following inequality holds*

$$MSE(T_n) \le (\|f\|_R^2 + 1) \, Var(T_n),$$

where $\|f\|_R$ is the norm of f in H(R) and $Var(T_n)$ is as in (3).

It should be pointed out that the inequality as in the proposition is true for any linear estimator. But as far as this article concerns, it is stated just for the sample mean estimator. The variance of $\mathbf{1}_n'\xi_{T_n}/n$ in estimation of $I(\xi)$ is equal to the supreme of the squared error of approximation of integrals of all functions in H(R) while the squared bias is the squared error of approximation of the integral of one particular function f in H(R). Thus when the mean function is unknown but belongs to the reproducing kernel Hilbert space of R, it is reasonable to ignore the mean (take it to be zero) and just search for T_n to minimize $Var(T_n)$. The following corollary follows from **Theorem** 1 and the proposition.

Corollary 2. *Under the setup as in the theorem, if the mean f of X is unknown but belongs to H(R) and $I(X)$ is estimated by $\bar{I}(X,T_n)$, the following inequality holds*

$$MSE(T_n) \le (\|f\|_R^2 + 1)\left\{\int_{C^d}\int_{C^d} \mu_{T_n}(\mathbf{t})\mu_{T_n}(\mathbf{s})\ d^2R(\mathbf{t},\mathbf{s})\right.$$
$$+\sum_{B\in\mathcal{B}}\int_B\int_B \mu_{T_n'}(\mathbf{a})\mu_{T_n'}(\mathbf{b})\ d^2R(\mathbf{a},\mathbf{b})$$
$$\left.+\sum_{B'\neq B''}(-1)^{d-\theta}\int_{B'}\int_{B''} \mu_{T_n'}(\mathbf{a})\mu_{T_n''}(\mathbf{b})\ d^2R(\mathbf{a},\mathbf{b})\right\},$$

where $\|f\|_R$ is the norm of f in H(R).

5 Proofs

5.1 Proof of Theorem 1

For simplicity, we prove **Theorem** 1 in $d = 2$ dimension. The proofs of cases with $d \ge 3$ can be done similarly.

Denote the projection of T_n onto the x-axis by $T_n' = \{t_i\}_{i=1}^m$, where $0 \le t_1 < \ldots < t_m \le 1$, and onto the y-axis by T_n''. Write T_n as $T_n = \cup_{i=1}^m\{(t_i, s_{i,j})\}_{j=1}^{n_i}$, where n_i is the number of points in T_n with the x-coordinate t_i. Clearly, $n_1 + \ldots + n_m = n$. Then the sample mean estimator can be written as $\bar{I}(X,T_n) = \sum_{i=1}^m\sum_{j=1}^{n_i}X(t_i,s_{i,j})/n$.

Note that $I(X) = \int_0^1\int_0^1 X(t,s)\ dtds$. By use of the technique of integration by parts first with respect to s then to t, we obtain

$$\int_0^1\int_0^1 X(t,s)dtds = X(1,1) - \int_0^1 t\ d_tX(t,1)$$
$$-\int_0^1 s\ d_sX(1,s) + \int_0^1\int_0^1 t\ s\ d^2X(t,s), \qquad (8)$$

where the differentials d_t and d_s are with respect to t and s, respectively, and the last term is defined as the mean square Riemann - Stieltjes integral with respect to X.

For $i = 1, \ldots, m$, put $T_{i,n} = \{s_{i,j}\}_{j=1}^{n_i}$. It is elementary calculus to check that for a function f of bounded variation on $[0, 1]$

$$\sum_{j=1}^{n_i} f(s_{i,j}) = n_i[f(1) - \int_0^1 p_{T_{i,n}}(t)\, df(t)],$$

where $p_{T_{i,n}}(t)$ is the proportion of points in $T_{i,n}$ that falls in the interval $[0, t]$. For each $i = 1, \ldots, m$, substituting f by $X(t_i, \cdot)$ in the equation above, it follows that

$$\sum_{j=1}^{n_i} X(t_i, s_{i,j}) = n_i[X(t_i, 1) - \int_0^1 p_{T_{i,n}}(s)\, d_s X(t_i, s)].$$

Observe that for each $s \in [0, 1]$, $X(t_i, s) = X(1, s) - \int_{t_i}^1 d_t X(t, s)$, which implies

$$X(t_i, 1) = X(1, 1) - \int_{t_i}^1 dX(t, 1),\ d_s X(t_i, s) = d_s X(1, s) - \int_{t_i}^1 d^2 X(t, s).$$

Thus

$$\sum_{j=1}^{n_i} X(t_i, s_{i,j}) = n_i[X(1, 1) - \int_{t_i}^1 dX(t, 1) - \int_0^1 p_{T_{i,n}}(s) d_s X(1, s)$$
$$+ \int_{t_i}^1 \int_0^1 p_{T_{i,n}}(s) d^2 X(t, s),$$

and

$$\bar{I}(X, T_n) = X(1, 1) - \sum_{i=1}^m \int_{t_i}^1 n_i\, dX(t, 1)/n$$
$$- \int_0^1 \sum_{i=1}^m n_i p_{T_{i,n}}(s)\, d_s X(1, s)/n + \sum_{i=1}^m \int_{t_i}^1 \int_0^1 n_i p_{T_{i,n}}(s) d^2 X(t, s)/n$$
$$= X(1, 1) - \int_0^1 p_{T_n'}(t) dX(t, 1) - \int_0^1 p_{T_n''}(s) dX(1, s) + \int_{C^2} p_{T_n}(t, s) d^2 X(t, s),$$

which together with (8) produces

$$I(X) - \bar{I}(X, T_n) = -\int_0^1 \mu_{T_n'}(t) dX(t, 1) - \int_0^1 \mu_{T_n''}(s) dX(1, s)$$
$$+ \int_{C^2} \mu_{T_n}(t, s) d^2 X(t, s), \tag{9}$$

Squaring both sides of (9) and then taking expectation will produce the equation (5). This completes the proof of **Theorem** 1.

Acknowledgement

The author would like to thank the referee for valuable suggestions which have led to an improved version of the article.

References

Aronszajn, N. (1950). Theory of reproducing kernels. *Trans. Amer. Math. Soc.* **68** 337–404.

Bras, R. L. & Rodriguez-Iturbe, I. (1993). *Random functions and hydrology.* Dover, New York.

Cambanis, S. (1985). Sampling designs for time series. In *Handbook of Statistics* (E. J. Hannan, P. R. Krishnaiah & M. M. Rao, eds.) **5** 337–362.

Cressie, N. (1993). *Statistics for spatial data.* Wiley, New York.

Christakos, G. (1992). *Random fields models in earth sciences*, Academic Press, San Diego, CA.

Fang, K. T. & Wang, Y. (1994). *Number-theoretic methods in statistics.* Monographs on Statistics and Applied Probability **51**, Chapman&Hall.

Journel, A. G. & Huijbregts, C. J. (1978). *Mining geostatistics.* Academic Press, New York.

Matérn, B. (1986). *Spatial variation.* Lecture Notes in Statistics **36**. Springer-Verlag, Berlin.

Niederreiter, H. (1978). Quasi-Monte Carlo methods and pseudo-random numbers. *Bull. Amer. Math. Soc.* **84** 957–1041.

Niederreiter, H. (1992). *Random number generation and Quasi-Monte Carlo methods.* CBMS-NSF SIAM **63**.

Parzen, E. (1967). *Time series analysis papers.* Holden-Day, San Francisco, CA.

Roth, K. F. (1980). On irregularities of distribution IV. *Acta Arithmetica* **37** 67–75.

Sacks, J. & Ylvisaker, D. (1966). Designs for regression problems with correlated errors. *Ann. Math. Statist* **37** 66–89.

Stein, M. J. (1999). *Interpolation of spatial data, some theory for kriging.* Springer Series in Statistics, Springer.

Su, Y. (1997). Estimation of random fields by piecewise constant estimators. *Stochastic Processes and Their Applications* **71** 145–163.

Woźniakowski, H. (1991). Average case complexity of multivariate integration. *Bull. Amer. Math. Soc.* **24** 185–194.

Ylvisaker, D. (1975). Designs on random fields. In *A Survey of Statistical Designs and Linear Models* (J. N. Srivastava, ed.) 595–607. North-Holland Amersterdam.

Uniform Sampling and Saddlepoint Approximation for Probabilistic Sensitivity Analysis in Engineering Design

Agus Sudjianto[1], Xiaoping Du[2] and Wei Chen[3]

[1] Senior Vice President, Risk Management Quality & Productivity, Bank of America, Charlotte, NC 28255-0001, U.S.A. (agus.sudjianto@bankofamerica.com)
[2] Assistant Professor, Department of Mechanical and Aerospace Engineering, University of Missouri–Rolla, Rolla, MO 65409-0050, U.S.A. (dux@umr.edu),
[3] Associate Professor, Department of Mechanical Engineering, Northwestern University, Evanston, IL 60208-3111, U.S.A. (weichen@northwestern.edu)

Summary. Sensitivity analysis plays an important role to help engineers gain knowledge of complex model behaviors and make informed decisions regarding where to spend engineering effort. In design under uncertainty, probabilistic sensitivity analysis (PSA) is performed to quantify the impact of uncertainties in random variables on the uncertainty in model outputs. One of the most challenging issues for PSA is the intensive computational demand for assessing the impact of probabilistic variations. An efficient approach to PSA is presented in this article. Our approach employs the Kolmogorov-Smirnov (KS) distance to quantify the importance of input variables. The saddlepoint approximation approach is introduced to improve the efficiency of generating cumulative distribution functions (CDFs) required for the evaluation of the KS distance. To further improve efficiency, optimized uniform samples are used to replace the direct Monte Carlo simulations for determining the cumulant generating function (CGF) in saddlepoint approximation. Efficient construction of a uniform design necessary to generate the "best" samples in a multidimensional space is presented. Our approach is illustrated with a structural design problem. It has the potential to be the most beneficial for high dimensional engineering design problems that involve expensive computer simulations.

Contemporary Multivariate Analysis and Experimental Design—In Honor Celebration of Professor Kai-Tai Fang's 65th birthday. Edited by Jianqing Fan and Gang Li. The World Scientific Publisher, 2005.

Key words: Probabilistic sensitivity analysis, Kolmogorov-Smirnov (KS) distance, saddlepoint approximation, latin hypercube sampling, monte carlo, low discrepancy, uniform design.

1 Introduction

In the past few years, with the advance of computing technologies and numerical approaches, scientific and engineering disciplines have experienced tremendous growth in the use of sophisticated computer models to assist scientific investigation and engineering analysis and design. Engineers and scientists make use of the models to perform various tasks and decision-making by interrogating the models to predict behaviors of systems under different input variable settings. The typical input-output relationship represented by a computer model is expressed as follows.

$$y = f(\mathbf{x}), \tag{1}$$

where $\mathbf{x} \in R^d$ are input variables, y is an output or response variable representing product or system performance, and $f(\cdot)$ is the relationship function between inputs and the output. In complex engineering applications, $f(\cdot)$ typically does not have an analytic formula. In product development such as automobile, sophisticated engineering computer models are eminent. These models are important for many reasons such as to guide significant upfront design decision making prior to the availability of physical prototypes, to substitute physical testing that can be too expensive, time consuming, harmful, or even, in some situations, prohibitive, and to gain insights into certain phenomena which may be lacking from physical experiments due to measurement system limitations or their practicality.

As engineering design process becomes more complex because of ever increasing customer expectation toward product quality, deductive approach in design using physics-based models alone is inadequate, and variability information must be integrated into the decision process. The seminal work by Taguchi (1993) has been very influential in introducing the concept of *robust design* for which a product or system must be designed by choosing the right setting of "control" variables (variables that engineers choose to control) such that the "ideal function" is insensitive to noise factors (i.e., variations due to piece-to-piece, degradation over time, environment, load, system interactions). Complementary to the robust design view, the product or system must be designed with high reliability (i.e., low probability of failure). The later view is the subject in reliability-based design discipline (Du, Sudjianto & Chen (2004)). The needs to address both robustness and reliability design necessitate the integration of probabilistic analysis with deterministic computer models. In this framework, the inputs to the computer models are treated as random variables with assumed distributions.

The interest in the probabilistic analysis approach is to understand the probabilistic characteristics (e.g., mean, μ_y, standard deviation, σ_y, or probability distribution, p_y) of the response variable, y, due to the stochastic nature of input variables, $\mathbf{x}$. Unfortunately, in most practical situations, the above needs are not easy to meet because stochastic information of input variables is often imprecise, and acquiring such information can be a very expensive proposition. To remedy this problem, the sensitivity analysis approach is employed with intention to rank order or to assess the importance of random input variables among each other. Through this analysis - though the stochastic information of input variable may be imprecise, and thus the distribution of the response variability may not be fully trustworthy - one can still acquire useful information for engineering decision making such as to focus the effort to reduce the variation due to important variables, to gather more precise stochastic information for the important variables, or to eliminate insignificant variables thus to simplify further analysis.

Because the inputs to computer models can be numerous, probabilistic sensitivity analysis involves various integral analyses in a high dimensional space. Unfortunately, computer models in engineering such as computational fluid dynamics and finite element models are usually computationally intensive. Thus, exercising the model by means of Monte Carlo simulation is not practical and often prohibitive. Therefore, a computationally more efficient technique that requires much fewer number of samples than that of Monte Carlo technique is needed. To this end, we present an approximation approach using the Saddlepoint Approximation in combination with uniform design to reduce the sample size yet maintaining a reasonable accuracy. This paper has the following flow. Section 2 introduces the concepts of probabilistic design and sensitivity analysis. The saddlepoint approximation technique required to calculate the sensitivity analysis is discussed in Section 3. Efficient construction of uniform design necessary to generate the "best" samples to calculate the saddlepoint approximation is presented in Section 4. Section 5 illustrates the use of our method for an engineering application. Finally, the conclusion is presented in Section 6.

2 Probabilistic Design and Sensitivity Analysis

In probabilistic design, the effects of input variability on product performance need to be addressed through rigorous variability analysis to prevent or to reduce the probability of failure occurrence or performance variation that leads to quality losses. The major task of probabilistic analysis is to obtain the probability distribution of the performance function (response) y given the distributions of the vector of random input variables $\mathbf{x}$. For a given performance target requirement, $y \leq Y$, the probability of the performance to meet the requirement can be calculated by a multi-dimensional integral,

$$P(y \leq Y) = \int_{f(\mathbf{x}) \leq Y} \cdots \int p(\mathbf{x})\, d\mathbf{x} \quad , \tag{2}$$

where $p(\mathbf{x})$ is the joint probability density function of random variables $\mathbf{x}$. The equality at the integration boundary $f(\mathbf{x}) = Y$ is called the *limit-state*, separating between "acceptable" and "unacceptable" (or safe and failure) regions of input variable space. Obviously, due to the multi-dimensional integration and the non-linear limit-state, the solution to Eq. (2) is analytically or numerically difficult to obtain; thus, Monte Carlo integration technique often becomes the method of choice. The solution is given by

$$P(y \leq Y) = \frac{1}{n} \sum_{i=1}^{n} I(y_i \leq Y) \quad \text{where} \begin{cases} 1 & \text{for } y_i \leq Y \\ 0 & \text{otherwise} \end{cases} , \tag{3}$$

where y_i is obtained by evaluating $f(\mathbf{x}_i)$ where $\mathbf{x}_i = \{x_i^1, x_i^2, \ldots, x_i^d\}$ are independent and identical distributions (i.i.d) random samples, and d is the dimension or the number of input random variables.

In robust design (see Chen, Allen, Mistree & Tsui (1996)), engineers would like to minimize the variation of y which can be represented by a dispersion measure such as standard deviation, σ_y, or quantile difference (Du, Sudjianto & Chen (2004)), $\delta_y = y^{1-\alpha} - y^{\alpha}$, where α is a prespecified quantile (see Figure 1) with $y^{\alpha} = P^{-1}(\alpha)$ and $y^{1-\alpha} = P^{-1}(1-\alpha)$. The quantile difference measure may be more prefereable than the standard deviation when the distribution of y consists of signficant higher moments (e.g., skewness and kurtosis).

The ability to calculate the distribution of y is needed for another important focus of probabilistic design: probabilistic sensitivity analysis (PSA) to quantify the impact of variability of input random variables on the variability of a model output. Results from PSA can be crucial to assist engineering design decisions, such as to help reduce the dimension of a design problem by identifying the probabilistically insignificant factors; to check the validity of a model structure and the assumptions made on the probability distributions of random inputs; to obtain insights about the design space and the probabilistic behavior of a model response; and to investigate the potential improvement on a probabilistic response by reducing the uncertainty in random inputs (Saltelli, Chan & Scott (2000)). If the interest is to study the effects of input variance on the output variance, variance-based method can be used to quantify the importance of input variables to an output (Saltelli, Chan & Scott (2000)). The variance-based method, however, will not be sufficient when the problem involves performance distributions with higher moments (Liu, Chen & Sudjianto (2004)). In this situation, sensitivity analysis must include complete stochastic information of the distribution. Considering this need, in the following discussion, we employ Kolmogorov-Smirnov (KS) distance to quantify the importance of an input variable.

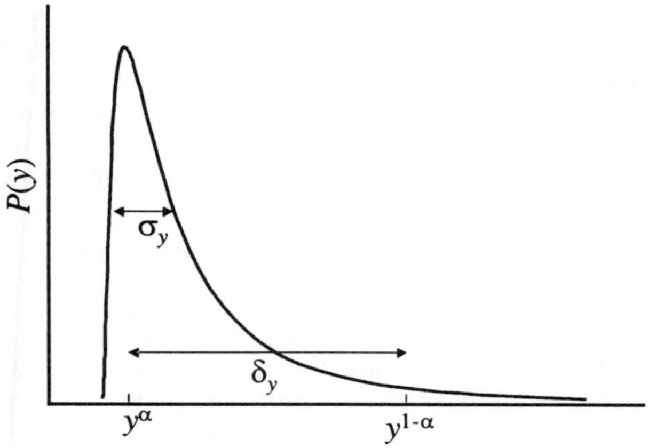

Fig. 1. Measure of response variability.

The KS distance, d_{KS}, measures the difference between two cumulative distribution functions (CDF), P_1 and P_2, as follows,

$$d_{KS}(P_1, P_2) = \sup_y |P_1(y) - P_2(y)|, \quad y \in R. \tag{4}$$

That is, the KS distance measures the maximum discrepancy between two distributions. In the context of sensitivity analysis, the KS distance can be used to quantify the main and total effects (i.e., the effect of a variable including all its interaction terms). The main effect of the jth variable, x^j, can be calculated as follows,

$$d^j_{KS}(P_0, P_j) = \sup_y \left| P_0(y|\mathbf{x}) - P_j\left(y\left|x^j\right.\right)\right|, \quad y \in R, \tag{5}$$

where $P_0(y|\mathbf{x})$ is the CDF of y by including all variability of input variables where $P_j\left(y\left|x^j\right.\right)$is the CDF of y including only the variability of the x^j and setting the rest of variables to constant values (e.g., their mean values). The smaller the value of d^j_{KS}, the closer $P_j\left(y\left|x^j\right.\right)$ is to $P_0(y|\mathbf{x})$, and the more dominant the variability of x^j is to define P_0; therefore, the smaller the value of d^j_{KS} is, the more important the variable is to the distribution of response variable. The total effect of x^j including the effect of its interactions with other variables can be calculated as,

$$d^{\sim j}_{KS}(P_0, P_{\sim j}) = \sup_y |P_0(y|\mathbf{x}) - P_{\sim j}(y|\mathbf{x}_{\sim j})|, \quad y \in R, \tag{6}$$

where $\mathbf{x}_{\sim j}$ is the set of all variables excluding the variability of x^j (i.e., setting x^j to a constant value such as its mean) and correspondingly $P_{\sim j}\left(y\left|\mathbf{x}_{\sim j}\right.\right)$ is the CDF of y by excluding the variability of x^j. Thus, when x^j is the dominant variable then by excluding it, the discrepancy between the two distributions will be larger. In this case, the larger the $d_{KS}^{\sim j}$ value, the more important x^j is. Based on the KS distance of the total effect, the importance of x^j can be ranked according to their order of importance.

As discussed above, probabilistic analysis including the sensitivity analysis requires numerous evaluations of $f(\mathbf{x})$ to calculate the integral in (2). Note, however, $f(\mathbf{x})$ is represented by a complex computer model with nonlinear behavior and expensive to compute. Because the evaluation of $f(\mathbf{x})$ is expensive and $P(y \leq Y)$ is typically very large for a highly reliable product, Monte Carlo integration may not be a practical alternative. This computation difficulty has led to the development of various approximation methods using the linearization of limit state (see Du, Sudjianto & Chen (2004)) which in some situations may not lead to satisfactory results. In Section 3, we propose an alternative method to the above problem.

3 Saddlepoint Approximation for Probabilistic Analysis and Sensitivity Analysis

3.1 Review of Saddlepoint Approximation

The Saddlepoint Approximation was introduced for approximating the probability density function (PDF) by Daniels (1954). Since then, the research and applications of Saddlepoint Approximations have vastly increased (e.g., Jensen (1995)).

Given a random variable y variable with a density function $p(y)$, then the characteristic function of y is

$$\xi\left(t\right) = \int_{-\infty}^{+\infty} e^{ity} p\left(y\right)\, dy, \tag{7}$$

where $i = \sqrt{-1}$. The cumulant generating function (CGF) $K(t)$ of y is defined as

$$K\left(t\right) = \log\left[\xi\left(t\right)\right], \tag{8}$$

where *log* is the natural logarithm. The PDF of y can be restored from $K(t)$ by the inverse Fourier transformation,

$$p(y) = \frac{1}{2\pi}\int_{-\infty}^{+\infty} e^{-ity}\xi(t)\, dt = \frac{1}{2\pi}\int_{-i\infty}^{+i\infty} e^{[K(t)-ty]}\, dt. \tag{9}$$

The key idea of obtaining the PDF of y is to accurately approximate the above integral through the concept of *saddlepoint* approximation. Simple formulae to calculate the PDF and CDF have been derived; consequently, their use is fairly straightforward. Daniels (1954) used the exponential power series expansion to estimate the integral in Eq. (9) as

$$p(y) = \left\{\frac{1}{2\pi K''(t_s)}\right\}^{\frac{1}{2}} e^{[K(t_s)-t_s y]}, \tag{10}$$

where $K''(\cdot)$ is the second derivative of the CGF, and t_s is the *saddlepoint*, which is the solution to the equation at the point of interest, Y.

$$K'(t) = Y, \tag{11}$$

where $K'(\cdot)$ is the first derivate of the CGF. Lugannani and Rice (1980) provided a very concise formula to approximate CDF,

$$P(y \le Y) = \Phi(w) + \phi(w)\left(\frac{1}{w} - \frac{1}{v}\right), \tag{12}$$

where $\Phi(\cdot)$ and $\phi(\cdot)$ are CDF and PDF of the standard normal distribution, respectively,

$$w = sgn(t_s)\ \{2[t_s Y - K(t_s)]\}^{1/2} \tag{13}$$

and

$$v = t_s\left\{2K''(t_s)\right\}^{1/2}, \tag{14}$$

where $sgn(t_s)$= +1, -1, or 0, depending on whether t_s is positive, negative or zero.

The Saddlepoint Approximation has several excellent features: (1) It yields extremely accurate probability estimation, especially in the tail area of a distribution; (2) it requires only the process of finding one saddlepoint without any integration; and (3) it provides estimations of the PDF and CDF simultaneously. In the following subsections, we will discuss how to combine Saddlepoint Approximations with simulation samples to conduct probabilistic sensitivity analysis.

3.2 Estimation of CDF by Saddlepoint Approximations

As discussed in the preceding section, the use of Saddlepoint Approximation rests on the ability to estimate the CGF of a general performance function $y = f(\mathbf{x})$. In some situations, a proper approximation can be developed through a linearization process to approximate the CGF (Du, Sudjianto & Chen (2004)). However, in general, the empirical estimation of CGF using sample dataset may be necessary as follows.

- Generate n samples for the d input random variables, $\mathbf{X} = \{x_i^j\}, i = 1, 2, \ldots, n; j = 1, 2, \ldots, d$. Various sampling techniques are available for this purpose such as Monte Carlo random sample, Quasi Monte Carlo, lattice points, Latin Hypercube Sampling (LHS), and uniform design (Fang and Wang, 1994; Owen, 1997; Fang et al., 2000). The choice of sampling technique is crucial in probabilistic engineering design to achieve high accuracy of cumulant estimation while employing only limited sample size because of computationally expensive engineering models. In Section 4, we will discuss this step with a greater depth.
- Acquire outputs by applying the sample dataset to the engineering computer model. For the ith sample, $\mathbf{x}_i = (x_i^1, x_i^2, \cdots, x_i^d)$, the output of computer model is $y_i = f(\mathbf{x}_i)$.
- Estimate cumulants of the response variable, y, based on the sample output. The first four cumulants are

$$\left.\begin{aligned} \kappa_1 &= \frac{s_1}{n} \\ \kappa_2 &= \frac{ns_2 - s_1^2}{n(n-1)} \\ \kappa_3 &= \frac{2s_1^3 - 3ns_1s_2 + n^2 s_3}{n(n-1)(n-2)} \\ \kappa_4 &= \frac{-6s_1^4 + 12ns_1^2s_2 - 3n(n-1)s_2^2 - 4n(n+1)s_1s_3 + n^2(n+1)s_4}{n(n-1)(n-2)(n-3)} \end{aligned}\right\}, \tag{15}$$

where s_r, $r = 1, 2, 3, 4$, are the rth power sum from the sample of output

$$s_r = \sum_{i=1}^{n} y_i^r. \tag{16}$$

The empirical CGF is calculated based on series expansions of powers of t

$$K(t) = \log \xi(t) = \sum_{j=1}^{\infty} \kappa_j \frac{t^j}{j!}. \tag{17}$$

- Calculate the saddlepoint solution. Since the empirical CGF in Eq. (17) is in a polynomial form, its first and second derivatives can be derived analytically.

If the higher order terms (i.e., after the fourth cumulant) in Eq. (17) are omitted, Eq. (11) is expressed as

$$K'(t) = \kappa_1 + \sum_{j=2}^{4} \kappa_j \frac{t^{j-1}}{(j-1)!} = Y. \tag{18}$$

Solving the above equation, we get the saddlepoint t_s.Then the PDF and CDF can be calculated using Eqs. (10) and (12),respectively.

4 Efficient Construction of Uniform Samples

The efficiency of the saddlepoint approximation approach (Section 3) can be improved by a proper choice of sampling technique. In the following discussion we review available sampling techniques and their computational efficiency. In particular, we employ a combination of Latin Hypercube Sampling (LHS) and low discrepancy samples known as uniform design (Fang, Lin, Winker & Zhang (2000)) by optimizing the uniformity of samples in multidimensional space. Because the optimization process to attain maximum uniformity is a difficult combinatorial problem, we present an efficient heuristic algorithm as a solution alternative. Additionally, we also propose an algorithm to efficiently calculate the uniformity criterion.

4.1 Latin Hypercube and Low Discrepancy Sampling

The probabilistic sensitivity calculation by means of saddlepoint approximation requires generation of i.i.d samples, $\mathbf{X} = \{x_i^j\}, i = 1, 2, \ldots, n; j = 1, 2, \ldots, d$, where each variable x^j follows a cumulative distribution G_j. These samples can be generated using Quantile-Quantile transformation from i.i.d samples, $\mathbf{U} = \{u_i^j\}$ uniformly distributed on $[\mathbf{0}, \mathbf{1})^d$,

$$x_i^j = G_j^{-1}\left(u_i^j\right). \tag{19}$$

Writing the multiple integration required to calculate probabilistic sensitivity in a canonical form yields,

$$I(f) = \int_{[\mathbf{0},\mathbf{1})^d} f(\mathbf{x})\, d\mathbf{x}. \tag{20}$$

The sample mean approximation to $I(f)$ is given by

$$\hat{I}(f, D) = \frac{1}{n} \sum_{i=1}^{n} f(\mathbf{x}_i). \tag{21}$$

When Monte Carlo random sampling is employed, it is known that $I - \hat{I}$ is unbiased (i.e., mean zero) and has asymptotic variance σ^2/n, therefore the random sampling errors are $O(n^{-\frac{1}{2}})$. McKay, Conover & Beckman (1979) introduced the use of Latin Hypercube Sampling (LHS) for computer experiments where the samples are

$$x_i^j = \frac{\pi_j(i) - u_i^j}{n}, \tag{22}$$

or its median version

$$x_i^j = \frac{\pi_j(i) - 0.5}{n}. \tag{23}$$

LHS stratifies each variable individually into equal intervals. Owen (1997) showed that for finite samples, LHS is never worse than Monte Carlo random samples,

$$var_{LHS}\left(\hat{I}\right) \leq \frac{\sigma^2}{n-1}. \tag{24}$$

Koksma-Hlawka inequality gives an upper bound for the approximation error

$$\left|I(f) - \hat{I}(f, \mathbf{X})\right| \leq D(\mathbf{X})\, V_{HK}(f), \tag{25}$$

where $V_{HK}(f)$ is the variation of f in the sense of Hardy and Krause (Niederreiter (1992)) and $D(\mathbf{X})$ is a measure of discrepancy or nonuniformity of the samples. The measure of discrepancy that has been widely used in the Quasi Monte Carlo theory and uniform design is the star L_p-discrepancy (Fang & Wang (1994)), i.e.,

$$D(\mathbf{X}) = \left\{ \int_{C^d} \left| \frac{N(\mathbf{X}, [\mathbf{0}, \mathbf{x}))}{n} - Vol([\mathbf{0}, \mathbf{x})) \right|^p d\mathbf{x} \right\}^{\frac{1}{p}}. \tag{26}$$

where $[\mathbf{0}, \mathbf{x})$ is the interval $[0, x_1) \times \cdots \times [0, x_d)$, $N(\mathbf{X},[\mathbf{0},\mathbf{x}))$ is the number of samples lies in $[\mathbf{0}, \mathbf{x})$, and Vol($[\mathbf{0}, \mathbf{x})$) is the volume of $[\mathbf{0}, \mathbf{x})$. Instead of using random samples, Quasi Monte Carlo uses deterministic sequence to generate samples in such a way to minimize sample cluster and gaps appear in random sampling; thus, minimizing the discrepancy. Nieiderreiter (1992) provided the asymptotic convergence rate of Quasi Monte Carlo as $O(n^{-1} \log(n)^{d-1})$. This bound, however quickly becomes larger than that of simple Monte Carlo when n is fixed and

d is large, which is the case for engineering design applications involving expensive to run computer models. For example, for $n = 500$, the Quasi Monte Carlo bound is greater than that of Monte Carlo for $d = 5$. The empirical experience, on the other hand, suggests that, in practice, the Quasi Monte Carlo provides better accuracy than that of simple Monte Carlo. Recently, Papageorgiou (2003) showed that the Quasi Monte Carlo samples converge at the rate of not worse than$O(n^{-1-\frac{p}{2}\log(n)})$ with $p \geq 0$ for integration over all reals. That is, independent of the dimensionality of a problem, the Quasi Monte Carlo sampling is significantly better than random sampling. Hickernell (1998) explained the weakness of the L_p-discrepancy and proposed several alternatives, among which the centered L_2-discrepancy (CL_2) is the most attractive (Fang, Lin, Winker & Zhang (2000)).

$$\begin{aligned} CL_2(\mathbf{X})^2 = {} & \left(\tfrac{13}{12}\right)^2 - \tfrac{2}{n}\sum_{i=1}^{n}\prod_{k=1}^{m}\left(1 + \tfrac{1}{2}\left|x_{ik} - 0.5\right| - \tfrac{1}{2}\left|x_{ik} - 0.5\right|^2\right) \\ & + \tfrac{1}{n^2}\sum_{i=1}^{n}\sum_{j=1}^{n}\prod_{k=1}^{m}\left(1 + \tfrac{1}{2}\left|x_{ik} - 0.5\right| + \tfrac{1}{2}\left|x_{jk} - 0.5\right| - \tfrac{1}{2}\left|x_{ik} - x_{jk}\right|\right) \end{aligned} . \quad (27)$$

A sample set, $\mathbf{X}$, is called uniform if it has the minimum $CL_2(\mathbf{X})^2$. For the one-dimensional case, Fang, Ma & Winker (2002) showed that the sample set with equidistant stratification (i.e., LHS) has the lowest discrepancy. For higher dimensional case, Fang, Ma & Winker (2002) showed that LHS has better expected value of $CL_2(\mathbf{X})^2$compared to that of Monte Carlo random samples:

$$E\left(CL_2\left(\mathbf{X}_{MC}\right)\right)^2 - E\left(CL_2\left(\mathbf{X}_{LHS}\right)\right)^2 = \left(\frac{13}{12}\right)^{d-1}\frac{d}{6n}\left(1 - \frac{2d+11}{26n}\right) + O\left(n^{-3}\right). \quad (28)$$

The advantage of LHS is more dramatic when the sample size is small and the dimension is large as shown in the figure below.

Combining the advantageous features of Quasi Monte Carlo and stratified sampling such as LHS, in this paper, we pursue to optimize LHS with respect to CL_2-discrepancy. That is, to arrive to optimal samples with LHS-type stratification which are uniformly distributed in the entire space and not only in the one-dimensional projection: the feature of *uniform design* (Fang, Lin, Winker & Zhang (2000)). When sample size is large, one may employ lattice sampling or digital nets to generate the best LHS design (Fang & Wang (1994)). However, when samples are expensive such as in our applications for computationally expensive computer models, the aforementioned approaches may produce unsatisfactory results. For example, two-dimensional projection plots of sample sets generated using Halton sequence and uniform design for $n = 16$ and $d = 5$ are shown in Figure 3 below. Considering that the uniformity in lower dimensions (especially 2 or 3) is usually of the particular interest in engineering designs where *hierarchy principle* is often applicable (i.e., the lower order interaction effects are usually more important than the higher order interaction effects), the sample

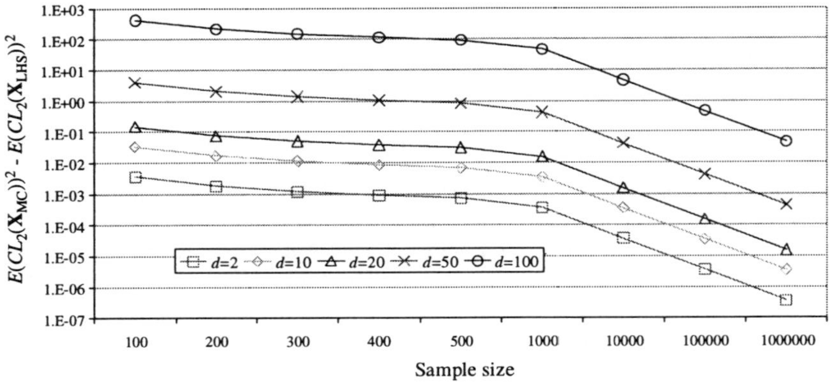

Fig. 2. Difference between the expected value of CL_2-discrepancy of monte carlo random sampling and LHS.

set generated using digital nets is inferior compared to that of uniform design. Motivated by overcoming this problem, we employ an optimization approach to search for LHS that minimizes CL_2-discrepancy. Searching the optimal uniform designs, however, is a difficult optimization problem to solve. Several heuristic combinatorial optimization approaches have been proposed. The computational cost of the existing algorithms, e.g., the simulated annealing (SA) algorithm used by Morris & Mitchell (1995), the *columnwise-pairwise* (CP) algorithm by Ye, Li & Sudjianto (2000), and the threshold accepting (TA) algorithm adopted by Fang, Ma & Winker (2002) for constructing optimal LHD, is generally high. (Ye, Li & Sudjianto 2000) reported that generating an optimal 25×4 LHSs using CP could take several hours on a Sun SPARC 20 workstation. For a design as large as 100×10, the computational cost could be formidable. Motivated by reducing this computational cost, an efficient algorithm for constructing optimal experimental designs is developed and introduced in this section. This new algorithm significantly improves the computational efficiency as it cuts the computation time from hours to minutes and seconds. There are two major ideas behind this algorithm (Jin, Chen & Sudjianto (2004)). One is on the use of an efficient global optimal search algorithm, named as the enhanced stochastic evolutionary (ESE) algorithm. The other is on the use of efficient methods for evaluating optimality criteria. Some details of the algorithm and results from comparative studies are provided in the following subsections.

4.2 Algorithm for Optimizing Uniformity

The strategy to construct uniform LHS is summarized as follows,

1. Start from a randomly chosen LHS, $\mathbf{X}_0$;

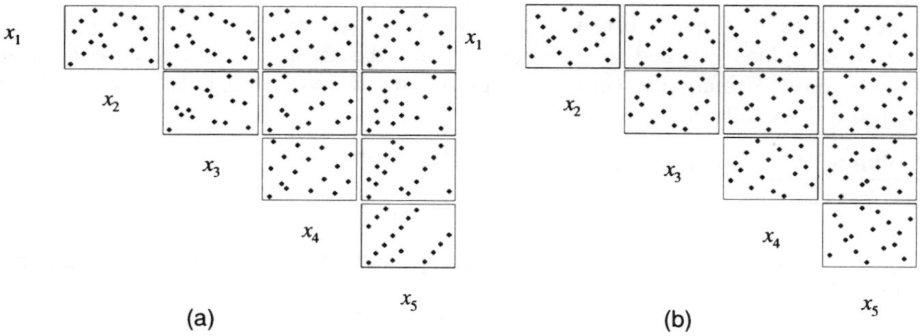

Fig. 3. Two-dimensional projections of samples for n = 16 and d = 5 generated using (a) Halton sequence and (b) Uniform design.

2. Construct a new design (or a set of new designs) through *columnwise* operations on the current design;
3. Compute optimality criterion (e.g., the centered L_2 discrepancy criterion) value of the new design and decide whether to replace the current design with the new one.
4. Repeat steps 2 and 3 until a stopping criterion is met.

The *columnwise element-exchange* operations are used in the step 2 of the search to maintain the structure property of LHS. The element-exchange within a column interchanges two distinct elements in a column and guarantees to retain the LHS property. As shown in Figure 4 for a 5×4 LHS, after the element-exchange, the balance property of the 2nd column is retained, and the sample is still a LHS after the exchange.

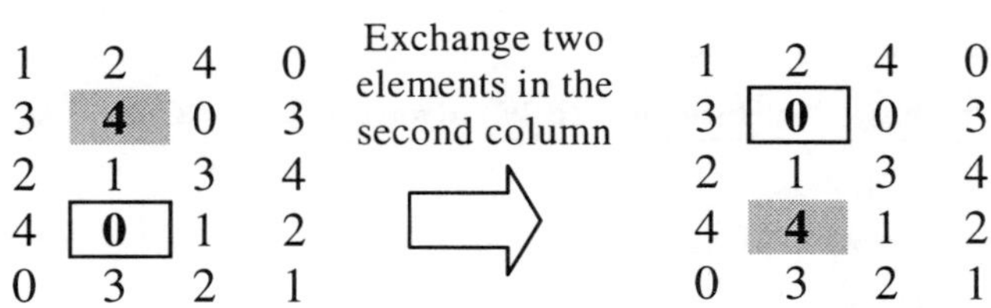

Fig. 4. Element-exchange in a 5×4 LHS.

In step 3 of the search process, a new sample set, $\mathbf{X}^{new}$, replaces the incumbent, $\mathbf{X}$, if it leads to an improvement in terms of the criterion, i.e., $CL_2(\mathbf{X}^{new}) <$

$CL_2(\mathbf{X})$. Otherwise, it will replace $\mathbf{X}$ with probability of $p(T,\ CL_2(\mathbf{X}^{new}),\ CL_2(\mathbf{X}))$ where T is a "threshold" of acceptance parameter. Several search algorithms have been applied to construct optimal design in the context of computer experiments. Principally, they differ in the strategy of threshold acceptance of $p(.)$ and T as follow:

Column Pair-wise (CP) algorithm (Li & Wu (1997))

$$p\left(T, CL_2\left(\mathbf{X}^{\text{new}}\right), CL_2\left(\mathbf{X}\right)\right) = \begin{cases} 1 & \text{for } CL_2\left(\mathbf{X}^{\text{new}}\right) - CL_2\left(\mathbf{X}\right) < T \text{ where } T = 0 \\ 0 & \text{otherwise} \end{cases} \tag{29}$$

Threshold Acceptance (TA) algorithm (Winker & Fang (1997))

$$p\left(T, CL_2\left(\mathbf{X}^{\text{new}}\right), CL_2\left(\mathbf{X}\right)\right) = \begin{cases} 1 & \text{for } CL_2\left(\mathbf{X}^{\text{new}}\right) - CL_2\left(\mathbf{X}\right) < T^k \\ 0 & \text{otherwise} \end{cases}, \tag{30}$$

where T^k is a "threshold" parameter, initially set to $T^0 = T$ and will be monotonically reduced by some schedule $T^k = \alpha T$ where α $(0 \leq \alpha < 1)$ is a constant.

Simulated Annealing (SA) algorithm(Morris & Mitchell (1995))

$$p\left(T, CL_2\left(\mathbf{X}^{\text{new}}\right), CL_2\left(\mathbf{X}\right)\right) = \exp\left\{-\frac{1}{T^k}\left[CL_2\left(\mathbf{X}^{\text{new}}\right) - CL_2\left(\mathbf{X}\right)\right]\right\}, \tag{31}$$

where T^k is also known as "temperature" parameter analogous to the physical process of annealing of solids which initially set to $T^0 = T$ and will be monotonically reduced by some *cooling* schedule$T^k = \alpha T$, where α $(0 \leq \alpha < 1)$ is a constant called *cooling* factor.

Enhanced Stochastic Evolutionary (ESE) algorithm(Jin, Chen & Sudjianto (2004))

$$p\left(T, CL_2\left(\mathbf{X}^{\text{new}}\right), CL_2\left(\mathbf{X}\right)\right) = \begin{cases} 1 & \text{for } CL_2\left(\mathbf{X}^{\text{new}}\right) < CL_2\left(\mathbf{X}\right) \\ 1 - \frac{1}{T^k}\left[CL_2\left(\mathbf{X}^{\text{new}}\right) - CL_2\left(\mathbf{X}\right)\right] & \text{for } CL_2\left(\mathbf{X}^{\text{new}}\right) - CL_2\left(\mathbf{X}\right) < T^k \\ 0 & \text{otherwise} \end{cases}, \tag{32}$$

where T^k is a "threshold" parameter, initially set to $T^0 = T$ and will be reduced or increased by some schedule $T^k = \alpha_1 T$ and $T^k = T/\alpha_2$ where α_j $(0 \le \alpha_j < 1,\ j = 1{,}2)$ are a chosen set of constants. The scheduling of the threshold value (reduced or increased) is adaptively determined by the history of the search results. Among the above strategies, the enhanced stochastic evolutionary (ESE) algorithm is the algorithm we recommend. It is adapted from the stochastic evolutionary (SE) algorithm (Saab & Rao (1991)). The algorithm uses a sophisticated combination of warming schedule and cooling schedule to control the threshold so that the algorithm can be self-adjusted during the search process. Details of the algorithm implementation can be found in (Jin, Chen & Sudjianto (2004)).

4.3 Efficient Optimality Criterion Calculation

Let $\mathbf{Z} = \{z_i^j\}$be the centered design matrix of $\mathbf{X}$, i.e., $z_i^j = x_i^j - 0.5$. Let $\mathbf{C} = [c_{ij}]_{n\times n}$be a symmetric matrix, whose elements are:

$$c_{ij} = \begin{cases} \frac{1}{n^2}\prod_{k=1}^{d}\frac{1}{2}(2 + |z_i^k| + |z_j^k - |z_i^k - z_j^k|) & \text{if } i \neq j, \\ \frac{1}{n^2}\prod_{k=1}^{d}(1 + |z_i^k|) - \frac{2}{n}\prod_{k=1}^{d}(1 + \frac{1}{2}|z_i^k| - \frac{1}{2}{z_i^k}^2) & \text{otherwise.} \end{cases} \tag{33}$$

Let

$$g_i = \prod_{k=1}^{d}(1 + |z_i^k|) \tag{34}$$

and

$$h_i = \prod_{k=1}^{d}(1 + \frac{1}{2}|z_i^k| - \frac{1}{2}{z_i^k}^2) \tag{35}$$

$$= \prod_{k=1}^{d}\frac{1}{2}(1 + |z_i^k|)(2 - |z_i^k|), \tag{36}$$

then,

$$c_{ii} = g_i/n^2 - 2h_i/n. \tag{37}$$

It can be proved easily that

$$CL_2(\mathbf{X})^2 = \left(\frac{13}{12}\right)^2 + \sum_{i=1}^{n}\sum_{j=1}^{n} c_{ij} \quad . \tag{38}$$

From Eq. (33)- Eq. (35), the computational complexity to calculate the $\mathbf{C}$ matrix (and thus, CL_2) is $O(dn^2)$. Note, however, that each updating operation using columnwise element-exchanges for generating a new sample set, only involves two elements in the sample matrix. That is, with the element exchange operation, $x_{i_1}^k \leftrightarrow x_{i_2}^k$, only elements in i_1 and i_2 rows and i_1 and i_2 columns of $\mathbf{C}$ are changed. Considering this situation, we seek a more efficient CL_2 calculation after an element exchange without recalculating the entire $\mathbf{C}$ matrix. For any $1 \le j \le n$ and $j \ne i_1, i_2$, let

$$\gamma(i_1, i_2, k, j) = (2 + |z_{i_2}^k| + |z_j^k| - \left|z_{i_2}^k - z_j^k\right|)\Big/(2 + |z_{i_1}^k| + |z_j^k| - \left|z_{i_1}^k - z_j^k\right|), \tag{39}$$

then,

$$c'_{i_1 j} = c'_{j i_1} = \gamma(i_1, i_2, k, j) c_{i_1 j}, \tag{40}$$

and

$$c'_{i_2 j} = c'_{j i_2} = c_{i_2 j}/\gamma(i_1, i_2, k, j). \tag{41}$$

Let $\alpha(i_1, i_2, k) = (1 + |z_{i_2}^k|)/(1 + |z_{i_1}^k|)$and $\beta(i_1, i_2, k) = (2 - |z_{i_2}^k|)/(2 - |z_{i_1}^k|)$, then:

$$c'_{i_1 i_1} = \alpha(i_1, i_2, k) g_{i_1}/n^2 - 2\alpha(i_1, i_2, k)\beta(i_1, i_2, k) h_{i_1}/n, \tag{42}$$

and

$$c'_{i_2 i_2} = g_{i_2}/[n^2 \alpha(i_1, i_2, k)] - 2h_{i_2}/[n\alpha(i_1, i_2, k)\beta(i_1, i_2, k)]. \tag{43}$$

The new CL_2 can be computed by:

$$\begin{aligned}(CL_2^2)' &= CL_2^2 + c'_{i_1 i_1} - c_{i_1 i_1} + c'_{i_2 i_2} - c_{i_2 i_2} \\ &+2 \times \sum_{1 \le j \le n, j \ne i_1, i_2}^{n} (c'_{i_1 j} - c_{i_1 j} + c'_{i_2 j} - c_{i_2 j}).\end{aligned} \tag{44}$$

Now, the computational complexity of calculating CL_2 after an element exchange operation becomes $O(n)$, which is much less than $O(dn^2)$. This efficient updating calculation enables us to search larger size optimal samples.

4.4 Example and Verifications

The search algorithm above can be used for optimizing various classes of designs of experiments, including but not limited to LHS, general balanced designs, Orthogonal Array with various optimization criteria other than Eqn.(27) (see Jin,

Chen & Sudjianto (2004)). Here we provide one example of optimal LHS based on the CL_2 criterion. As shown in Figure 5, before optimization, the initial LHS is a random LHS sample with good one-dimensional projective property but not so good space-filling property. After optimization, the projective property is maintained while the space filling property is much improved.

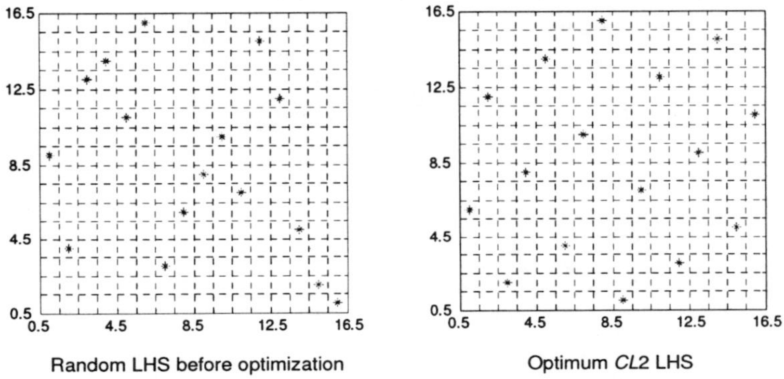

Fig. 5. LHS sample before and after optimization using CL_2 Criterion.

In Jin, Chen & Sudjianto (2004), the new algorithm is compared to existing techniques and found to be much more efficient in terms of the computation time, the number of exchanges needed for generating new designs, and the achieved optimality criteria. Specifically, it has cut the computation time from hours to minutes and seconds, which makes the just-in-time generation of relatively large size optimal samples possible. For the problems tested, we find that with the same number of exchanges, the optimal designs generated by ESE are generally better than those generated by Simulated Annealing (SA) and the columnwise-pairwise (CP) algorithm. To obtain a design statistically significantly better than those generated by SA and CP, ESE needs far less number of exchanges (typically around $1/6 \sim 1/2$ of exchanges needed by SA or CP for small-sized designs and $1/33 \sim 1/4$ of exchanges needed by CP for large-sized designs).

Through our comparative studies (Jin, Chen & Sudjianto (2004)), it was discovered that the CL_2 criterion is much more efficient to evaluate than other optimality criteria such as MAXIMIN distance criterion (Morris & Mitchell (1995)) and the entropy criterion (Ye, Li & Sudjianto (2000)). For the problems tested, the computing time for the MAXIMIN criterion is $2.3 \sim 3.0$ times as much as that for the CL_2 criterion. The larger the size of an experimental design, the more computational savings the method will make. For example, for 100x10 LHS, our new method for evaluating CL_2criteria only requires 1/82.1 of the computation effort compared to re-evaluating the whole matrix.

As the global optimal samples may never be known, one way to access how good the optimal designs are by estimating the probability of a randomly generated LHS to be better than that of optimal samples, $P(CL_2(\mathbf{X}_{random}) \leq CL_2(\mathbf{X}_{opt}))$. For the purpose of example, we generated 2×10^7 sets of 50×5 (n=50, d=5) LHS samples and calculate their CL_2 values. Figure 6 shows the empirical CDF of CL_2 values of random 50×5 LHSs. As we are only interested in the left tails of CDF curves (i.e., small CL_2 values), the right part of CDF curves have been truncated in the figure.

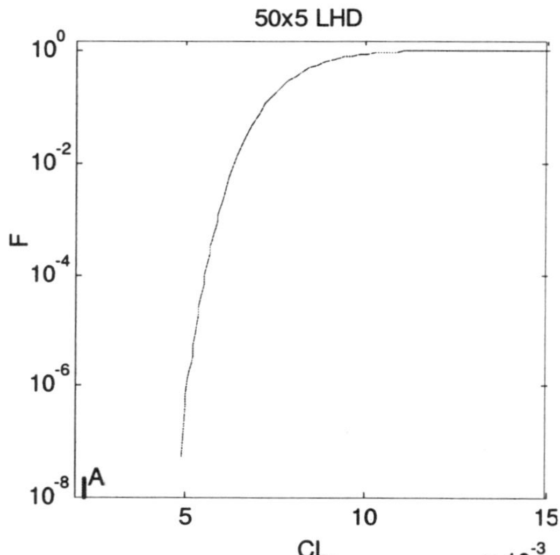

Fig. 6. Empirical Cumulative Distribution of CL_2 values of random 50×5 LHSs.

In this case, fitting a line through the points at the tail region, we estimated that $P(CL_2(\mathbf{X}_{random}) \leq CL_2(\mathbf{X}_{opt})) \approx 10^{-19}$ where $CL_2(\mathbf{X}_{opt}) = 0.002249$. Similar observations were obtained for 100×10 LHSs. These indicate that the optimal designs constructed by ESE generally have significantly lower CL_2 values (better uniformity) than randomly generated LHSs.

5 Application

To illustrate the application, we present a simple example where the engineering model has an analytic form and thus computationally cheap. Typical real world engineering models do not have analytic forms and computationally much more expensive (see for example Ejakov, Sudjianto & Pieprzak (2004)). Nevertheless, this example sufficiently demonstrates the use and the effectiveness of our proposed method. A composite beam with Young's modulus E_w and A mm wide by B mm deep by L mm long, has an aluminum plate with Young's modulus E_a and a net section C mm wide by D mm high securely fastened to its bottom face, as shown in Fig. 7. Six external vertical forces, P_1, P_2, P_3, P_4, P_5 and P_6 are applied at six different locations along the beam, L_1, L_2, L_3, L_4, L_5, and L_6. The allowable tensile stress is S. In this problem, there are twenty random variables as follows (Details of these random variables are given in Table 1),

$$\mathbf{X} = [X_1, \cdots, X_{20}]^T$$
$$= [A, B, C, D, L_1, L_2, L_3, L_4, L_5, L_6, L, P_1, P_2, P_3, P_4, P_5, P_6, E_a, E_w, S]^T$$

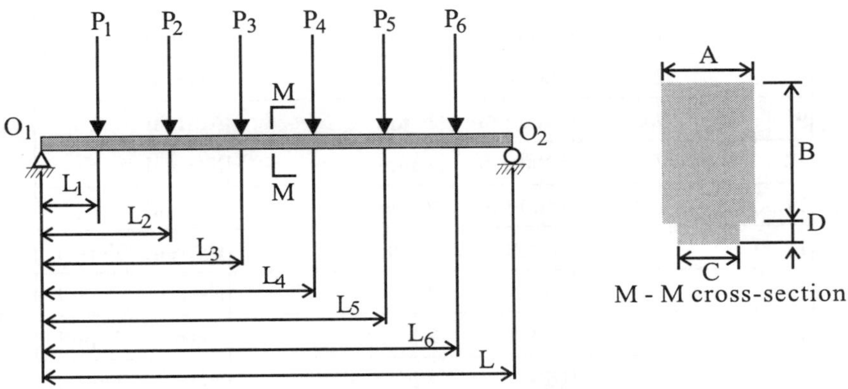

Fig. 7. A Composite Beam with 20 random variables.

The maximum stress occurs in the middle cross-section M-M and is given by

$$\sigma = \frac{\left[\frac{\sum_{i=1}^{6} P_i(L-L_i)}{L} L_3 - P_1(L_2 - L_1) - P_2(L_3 - L_2)\right]\left[\frac{0.5AB^2 + \frac{E_a}{E_w}DC(B+D)}{AB + \frac{E_a}{E_w}DC}\right]}{\frac{1}{12}AB^3 + AB\left\{\left[\frac{0.5AB^2 + \frac{E_a}{E_w}DC(B+D)}{AB + \frac{E_a}{E_w}DC}\right] - 0.5B\right\}^2 + \frac{1}{12}\frac{E_a}{E_w}CD^3 + \frac{E_a}{E_w}CD\left\{0.5D + B - \left[\frac{0.5AB^2 + \frac{E_a}{E_w}DC(B+D)}{AB + \frac{E_a}{E_w}DC}\right]\right\}^2}. \tag{45}$$

The response model is defined as the difference between the stress σ and the allowable stress (strength) S as below,

$$y = g(x) = S - \sigma. \tag{46}$$

The probability of failure p_f is defined by the probability of the strength less than the stress, i.e.,

Table 1. Random Variables of the Beam Reliability Problem

Variable No.	Variable	Mean value	Standard deviation	Distribution type
1	A	100 mm	0.2mm	Normal
2	B	200 mm	0.2 mm	Normal
3	C	80 mm	0.2 mm	Normal
4	D	20 mm	0.2 mm	Normal
5	L_1	200 mm	1 mm	Normal
6	L_2	400 mm	1 mm	Normal
7	L_3	600 mm	1 mm	Normal
8	L_4	800 mm	1 mm	Normal
9	L_5	1000 mm	1 mm	Normal
10	L_6	1200 mm	1 mm	Normal
11	L	1400 mm	2 mm	Normal
12	P_1	20 kN	4 kN	Extreme Type 1
13	P_2	20 kN	4 kN	Extreme Type 1
14	P_3	15 kN	2 kN	Extreme Type 1
15	P_4	15 kN	2 kN	Extreme Type 1
16	P_5	15 kN	2 kN	Extreme Type 1
17	P_6	15 kN	2 kN	Extreme Type 1
18	E_a	70 GPa	7Gpa	Normal
19	E_w	8.75 GPa	1 Gpa	Normal
20	S	27MPa	2.78 MPa	Normal

$$p_f = \Pr(S - \sigma < 0). \tag{47}$$

Since there is no analytic solution for (44), we employed relatively large size Monte Carlo samples (n = 1,000,000) as a reference for comparison. From this Monte Carlo sample set, we found that $p_f = 2.45\times10^{-4}$. The saddlepoint approximation with an optimal LHS of n = 500 produces $p_f = 2.397 \times10^{-4}$. Accordingly, the probabilistic sensitivity (i.e., total sensitivity) is obtained with optimized LHS of n =500 using Eq. (6) The results are summarized in Figure 8.

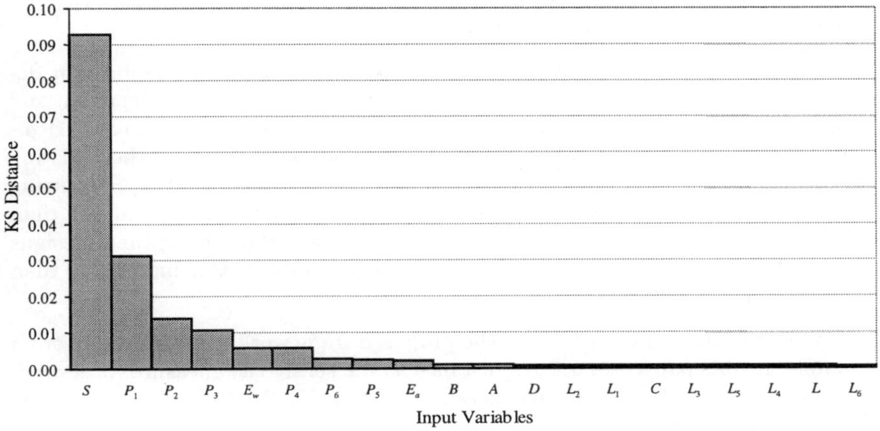

Fig. 8. KS distance and ranking of input random variables.

From the chart, it is noted that the most important variable is the material strength, S (about 50% of the output variation is due to this variable). This is in agreement with an observation of Eq. (43) and the variability information in Table 1. The other important variables include P_1, P_2, P_3, E_w, and P_4.

6 Conclusion

A comprehensive uniform sample-based approach to probabilistic analysis and sensitivity analysis is presented in this work. The efficiency of the probabilistic sensitivity analysis is enhanced from several aspects.

First, the saddlepoint approximation approach is used to improve the efficiency as well as the accuracy for probabilistic analysis when generating the whole cumulative density functions to evaluate the importance of random input variables. The accuracy is maintained because the saddlepoint approximation yields extremely accurate probability estimation, especially in the tail area of a distribution. The approach is also efficient as it requires only the process of finding one saddlepoint without any integration and provides estimations of PDF and CDF simultaneously.

Second, for implementing the saddlepoint approximation approach, the uniform samples are used to replace the intensive Monte Carlo simulations. A combination of Latin Hypercube Sampling (LHS) and low discrepancy criterion known as uniform design is employed by optimizing the uniformity of samples in multidimensional space. Given that for finite samples LHS is never worse than Monte

Carlo random samples, the advantages of using optimal LHS is more dramatic when sample size is small and the dimensionality is large.

Third, an efficient algorithm is developed for constructing the uniform designs. The new algorithm employs both an enhanced global search algorithm and a method for efficient evaluation of the uniformity criterion. The proposed algorithm to calculate CL_2 criterion cuts the computation time of other existing algorithms in this area from hours to minutes and seconds. That is, optimizing uniform designs using the CL_2 criterion is much more efficient to evaluate than other optimality criteria. The statistical tests indicate that the optimal designs constructed by our proposed algorithm have significantly better uniformity than randomly generated LHSs.

We demonstrated the utility of the proposed framework using an engineering example. In this paper, we use KS distance as the sensitivity measure. If desired, one may employ other sensitivity measures such as Kullback-Leibler divergence. This is subject to future research.

References

Chen, W., Allen, J. K., Mistree, F. & Tsui, K. L. (1996). A Procedure for Robust Design: Minimizing Variations Cause by Noise Factors and Control Factors. *ASME Journal of Mechanical Design* **118** No. 4 478–485.

Daniels, H. E., (1954). Saddlepoint Approximations in Statistics. *Annals of Mathematical Statistics* **25** 631–650.

Du, X. & Sudjianto A. (2004). First-Order Saddlepoint Approximation for Reliability Analysis. *AIAA Journal* **42** No. 6 1199–1207.

Du, X., Sudjianto, A. & Chen, W. (2004). An Integrated Framework for Optimization Under Uncertainty Using Inverse Reliability Strategy. *ASME Journal of Mechanical Design* **126** 1–9.

Ejakov, M., Sudjianto, A. & Pieprzak, J. (2004). Robustness and Performance Optimization of Engine Bearing System Using Computer Model and Surrogate Noise, DETC2004-57327. *Proceedings of DETC'04: ASME 2004 Design Engineering Technical Conferences and Computers and Information in Engineering Conference.* September 28 – October 3, 2004 in Salt Lake City, Utah.

Fang, K. T., Lin, D. K. J., Winker, P. & Zhang, Y. (2000). Uniform Design: Theory and Application. *Technometrics* **42** No. 3 237–248

Fang, K. T. & Wang, Y. (1994). *Number Theoretic Method in Statistics.* Chapman and Hall, London.

Fang, K. T., Ma, C. X. & Winker, P. (2002). Centered L_2-discrepancy of Random Sampling and Latin hypercube Design, and Construction of Uniform Designs. *Mathematics of Computation* **71** 275–296.

Hickernell, F. J., (1998). A Generalized Discrepancy and Quadrature Error Bound. *Mathematics of Computation* **67** 299–322.

Jensen L. J. (1995). *Saddlepoint Approximations.* Clarendon Press, Oxford.

Jin, R. Chen, W., & Sudjianto, A. (2004). An Efficient Algorithm for Constructing Optimal Design of Computer Experiments. to appear in *Journal of Statistical Planning and Inference.*

Li, W. W. & Wu, C. F. J. (1997). Columnwise-pairwise Algorithms with Applications to the Construction of Supersaturated Designs. *Technometrics* **39** 171–179.

Liu, H., Chen, W., & Sudjianto, A. (2004). Relative Entropy Based Method for Global and Regional Sensitivity Analysis in Probabilistic Design. *DETC2004/DAC-57500, Proceedings of DETC'04: ASME 2004 Design Engineering Technical Conferences and Computers and Information in Engineering Conference*, September 28 – October 3, 2004 in Salt Lake City, Utah.

Lugannani, R. and Rice, S. O. (1980). Saddlepoint Approximation for the Distribution of the Sum of Independent Random Variables. *Advances in Applied Probability* **12** 475–490.

McKay, M. D., Conover, W. J., & Beckman, R. J. (1979). A Comparison of Three Methods for Selecting Values of Input Variables in the Analysis of Output from a Computer Code. *Technometrics* No. 21 **2** 239–245.

Morris, M. D. & Mitchell, T. J. (1995). Exploratory Designs for Computational Experiments. *Journal of Statistical Planning and Inference* **43** 381–402.

Niederreiter, H. (1992). *Random Number Generation and Quasi Monte Carlo Methods.* SIAM, Philadelphia.

Owen, A. B. (1997). Monte Carlo Variance of Scrambled Equidistribution Quadrature. *SIAM Journal of Numerical Analysis* **34** No. 5 1884–1910.

Papageorgiou, A. (2003). Sufficient Conditions for Fast Quasi-Monte Carlo Convergence. *Journal of Complexity* **19** No.3 332–351.

Saab, Y. G. & Rao, Y. B. (1991). Combinatorial Optimization by Stochastic Evolution. *IEEE Transactions on Computer-aided Design* **10** No. 4 525–535.

Saltelli, A., Chan, K., & Scott, E. M. (2000). *Sensitivity Analysis.* John Wiley & Sons, Chichester, New York.

Taguchi, G. (1993). *Taguchi on Robust Technology Development: Bringing Quality Engineering Upstream.* ASME Press, New York.

Winker, P. & Fang, K. T. (1997). Application of Threshold Accepting to the Evaluation of the Discrepancy of a Set of Points. *SIAM Journal on Numerical Analysis* **34** 2028–2042.

Ye, K. Q., Li, W. & Sudjianto, A. (2000). Algorithmic Construction of Optimal Symmetric Latin Hypercube Designs. *Journal of Statistical Planning and Inference* **90** 145–159.

Schur-Convex Discrimination of Designs Using Power and Exponential Kernels

Aijun Zhang

Department of Statistics, University of Michigan, Ann Arbor, MI 48109, U.S.A. (ajzhang@umich.edu)

Summary. This paper elaborates the kernel selection problem in the majorization framework by Zhang, Fang, Li & Sudjianto (2004) for experimental designs. For designs with qualitative factors, the row-wise coincidence distribution and its raw, central and factorial moments are studied. Under the effects hierarchy principle, two protocols are recommended to employ power and exponential kernels, which are shown equivalent to some classical criteria for fractional factorial designs and uniform designs, respectively. In addition, an extension of majorization framework is given to uniform designs with quantitative factors under wrap-around discrepancy criterion.

Key words: discrepancy, fractional factorial design, majorization, minimum aberration, orthogonal array, Schur-convex, uniform design.

2000 Mathematics Subject Classification: 62K15, 26B25, 05B15, 94B05

1 Background

Experimental designs have drawn much attention recently for process improvement in industrial and computer experiments; see e.g. Wu & Hamada (2000) and Santner, Williams & Notz (2003). Two major types of popular designs are studied here: (I) fractional factorial design (FFD), either regular or non-regular, assessed by minimum abberation criterion for qualitative factors; (II) uniform

Contemporary Multivariate Analysis and Experimental Design—In Honor Celebration of Professor Kai-Tai Fang's 65th birthday. Edited by Jianqing Fan and Gang Li. The World Scientific Publisher, 2005.

design (UD), of space-filling type, assessed by categorial discrepancy criterion for qualitative factors and by wrap-around discrepancy for quantitative factors.

Consider in general the experiments with s factors each having q levels. Let $\{L_1, \ldots, L_q\}$ denote either categorial levels for qualitative factors or scale levels for quantitative factors. The lattice space $\mathcal{L}(q^s)$ is defined by the s-fold Cartesian product of $\{L_1, \ldots, L_q\}$, by taking each factor as a coordinate. An experimental design with design points from $\mathcal{L}(q^s)$ is termed as *lattice design* in a unified manner, which covers both FFD and UD. Let $\mathscr{U}(n, q^s)$ be the set of balanced designs such that the projection of design points onto each coordinate scatter uniformly, which correspond to the U-type designs in the UD context. Recently, Zhang, Fang, Li & Sudjianto (2004) derived a general two-stage framework for balanced designs with qualitative factors, by the *majorization* technique and *Schur-convex* kernel functions. The monograph Marshall & Olkin (1979) provides a comprehensive text of majorization and Schur-convex functions. See Shaked (1985), Cheng, Steinberg & Sun (1999), Fang & Zhang (2004), Zhang (2004) for the references of using majorization technique in experimental designs.

For each candidate design $\mathbf{X} = (x_{ij})_{n\times s}$ in a given interesting pool $\mathscr{D}(n, q^s) \subseteq \mathscr{U}(n, q^s)$, firstly compute the m-long $(m \equiv n(n-1)/2)$ vector $\boldsymbol{\beta}(\mathbf{X}) = [\beta(\mathbf{x}_i, \mathbf{x}_k)]$ for $i = 1, \ldots, k-1$ and $k = 2, \ldots, n$, where $\beta(\mathbf{x}_i, \mathbf{x}_k)$ counts the number of coincidences of the (i, k)-th runs. Such $\boldsymbol{\beta}(\mathbf{X}) = (\beta_1, \ldots, \beta_m)'$ is called an *informative pairwise coincidence* (IPC) vector. The sum of $\boldsymbol{\beta}(\mathbf{X})$ is a constant $\frac{1}{2}ns(\frac{n}{q}-1)$, so the average $\bar{\beta} = \frac{s(n-q)}{q(n-1)}$ holds for all $\mathbf{X} \in \mathscr{U}(n, q^s)$. For uniqueness, such constant $\bar{\beta}$ is called IPC-mean. Schur-Psi criteria are defined by Schur-convex functions of IPC-vectors, in particular the following *separable-convex* functions,

$$\Psi(\mathbf{X}; \sigma) = \sum_{r=1}^{m} \sigma(\beta_r) \quad \text{w.r.t. convex kernel } \sigma \text{ on non-negative } \mathbb{R}_+. \tag{1}$$

The two-stage investigation scheme of *majorization framework* by Zhang, Fang, Li & Sudjianto (2004) is summarized briefly as follows,

- **Beta-stage:** stringent majorization check.
  Let $\boldsymbol{\beta}(\mathbf{X})$ be sorted as $\beta_{[1]} \le \ldots \le \beta_{[m]}$. The champion of majorant design $\mathbf{X}^*$ is characterized by $\boldsymbol{\beta}(\mathbf{X}^*) \preceq \boldsymbol{\beta}(\mathbf{X})$ for all $\mathbf{X} \in \mathscr{D}(n, q^s)$, i.e.

$$\sum_{r=1}^{k} \beta_{[r]}(\mathbf{X}^*) \ge \sum_{r=1}^{k} \beta_{[r]}(\mathbf{X}), \quad \text{for } k = 1, 2, \cdots, m-1. \tag{2}$$

  Such $\mathbf{X}^*$ is universally optimum and absolutely recommended for experimentation, while any design on the right-hand side of (2) is inadmissible and prohibitive. If there exists no majorant design, take all the admissible candidates to Psi-stage for further comparison.
- **Psi-stage:** flexible Schur-convex discrimination.
  Select an *a priori* convex kernel $\sigma : \mathbb{R}_+ \to \mathbb{R}$ to define a specific Schur-Psi criterion by (1). Rank-order admissible designs by $\Psi(\mathbf{X}; \sigma)$ and determine

Schur-σ optimum one(s). The lower bound of $\Psi(\mathbf{X};\sigma)$ is given by Lemma 1 due to Zhang, Fang, Li & Sudjianto (2004). Follow-up kernels can be selected to further discriminate candidates of the same rank.

Lemma 1 (Zhang, Fang, Li & Sudjianto (2004)). *A balanced lattice design is majorant if and only if it is Schur-optimum w.r.t. every convex kernel function. For all $\mathbf{X} \in \mathscr{U}(n, q^s)$ under Schur-Psi criterion, we have*

$$\Psi(\mathbf{X};\sigma) \geq m(1-f)\sigma(\theta) + mf\sigma(\theta+1)$$

in which θ is the integral part of IPC-mean $\bar{\beta}$ and $f = \bar{\beta} - \theta$. The lower bound can be achieved by weak equidistant designs whose $\beta(\mathbf{x}_i, \mathbf{x}_k)$ for all $1 \leq i < k \leq n$ differ at most 1.

It seems however a bit ambiguous in choosing the Schur-convex kernels at Psi-stage. This paper elaborates the kernel selection problem for specific discrimination. Section 2 studies the pairwise coincidence measurements of qualitative factorials in order to guide lines of selecting Schur-convex kernels. It is shown that the coincidence moments are able to characterize the minimum aberration criterion. Power kernels are suggested in Section 3 to assess FFDs. Exponential kernels are suggested in Section 4 to assess UDs. In Section 5, the majorization framework is extended to UDs with quantitative factors under wrap-around discrepancy. Finally, Section 6 gives concluding remarks and acknowledgements.

Throughout the paper, for integers $j \geq k \geq 1$, $S(j,k)$ denotes the Stirling number of the second kind, i.e., the number of partitions of j elements into k non-empty sets; $C(x,j)$ denotes the factorial function $x(x-1)\cdots(x-j+1)$ for $x \geq j$. The combinatorial function $\binom{x}{j} = C(x,j)/j!$, $\binom{x}{0} = 1$ and $\binom{x}{j} = 0$ if $x < j$. Besides, the parameters θ and f, respectively for integral part and fractional part of the IPC-mean $\bar{\beta} \equiv \frac{s(n-q)}{q(n-1)}$, are used globally in the paper.

2 Coincidence Distribution and Moments

Given an arbitrary design $\mathbf{X} \in \mathscr{U}(n, q^s)$ and coincidence measure $\beta(\cdot,\cdot)$ between design points, let

$$P_j(\mathbf{x}) := \text{Prob}\big(\beta(\mathbf{x}, \mathbf{w}) = j : \mathbf{w} \in \mathbf{X}, \mathbf{w} \neq \mathbf{x}\big)$$

for $\mathbf{x} \in \mathbf{X}$ and for $j = 0, 1, \ldots, s$. Define the coincidence distribution $(P_0, P_1, \ldots, P_s)$ with components evaluated by

$$P_j(\mathbf{X}) = \frac{1}{n} \sum_{k=1}^{n} P_j(\mathbf{x}_k), \quad \text{for } j = 0, 1, \ldots, s.$$

It is clear that the probability that $\beta(\mathbf{x}_i, \mathbf{x}_k) = j$ for $i \neq k$ is $P_j(\mathbf{X})$ and $\sum_{j=0}^{s} P_j(\mathbf{X}) = 1$. The coincidence distribution has a tight link with Hamming distance distribution $(E_0, E_1, \ldots, E_s)$ in algebraic coding theory

$$\begin{cases} E_0(\mathbf{X}) = 1 + (n-1)P_s(\mathbf{X}), \\ E_j(\mathbf{X}) = (n-1)P_{s-j}(\mathbf{X}), \quad j = 1, \ldots, s. \end{cases} \tag{3}$$

2.1 Coincidence moments

In $\mathbf{X}(n, q^s)$, there are $n(n-1)/2$ random variable of $\beta(\mathbf{x}_i, \mathbf{x}_k)$ for all $i < k$. Consider the following three types of statistical moments,

$$\begin{cases} \kappa_j(\mathbf{X}) := \frac{2}{n(n-1)} \sum\limits_{i<k} \left[\beta(\mathbf{x}_i, \mathbf{x}_k)\right]^j & \text{(raw moments)} \\ \\ \mu_j(\mathbf{X}) := \frac{2}{n(n-1)} \sum\limits_{i<k} \left[\beta(\mathbf{x}_i, \mathbf{x}_k) - \bar{\beta}\right]^j & \text{(central moments)} \\ \\ c_j(\mathbf{X}) := \frac{2}{n(n-1)} \sum\limits_{i<k} C\big(\beta(\mathbf{x}_i, \mathbf{x}_k), j\big) & \text{(factorial moments)} \end{cases} \tag{4}$$

for $j = 2, 3 \ldots$. The corresponding first-order moments all coincide with the constant mean $\bar{\beta} = \frac{s(n-q)}{q(n-1)}$ for balanced designs of $\mathscr{U}(n, q^s)$. In terms of IPC-vector $\boldsymbol{\beta}(\mathbf{X})$ introduced in the previous section,

$$\kappa_j(\mathbf{X}) = \frac{1}{m}\sum_{r=1}^{m} \beta_r^j, \quad \mu_j(\mathbf{X}) = \frac{1}{m}\sum_{r=1}^{m} (\beta_r - \bar{\beta})^j \text{ and } c_j(\mathbf{X}) = \frac{1}{m}\sum_{r=1}^{m} C(\beta_r, j)$$

or equivalently $\sum_{k=0}^{s} k^j P_k(\mathbf{X}), \sum_{k=0}^{s} (k - \bar{\beta})^j P_k(\mathbf{X})$ and $\sum_{k=0}^{s} C(j,k) P_k(\mathbf{X})$ in terms of coincidence distribution, respectively. Each type of moment statistics completely characterize the nature of coincidence distribution.

For integer-valued random variable $\beta(\mathbf{x}_i, \mathbf{x}_k) = 0, 1, 2, \ldots$, it is natural to consider the probability generating function of power series in nonnegative y,

$$\mathscr{E}_{\mathbf{X}}(y) = \frac{2}{n(n-1)} \sum_{i<k} y^{\beta(\mathbf{x}_i, \mathbf{x}_k)} = \sum_{j=0}^{s} y^j P_j(\mathbf{X}),$$

similar to the enumerator of distance distribution that underlies MacWilliams transform in coding theory. Among (4), the factorial moments $c_j(\mathbf{X})$ are tied firmly with $\mathscr{E}_{\mathbf{X}}(y)$. Both the coincidence distribution and factorial moments can be evaluated from the derivatives of $\mathscr{E}_{\mathbf{X}}(y)$ at $y = 0$ and $y = 1$, respectively, i.e.,

$$P_j(\mathbf{X}) = \mathscr{E}_{\mathbf{X}}^{(j)}(0)/j! \quad \text{and} \quad c_j(\mathbf{X}) = \mathscr{E}_{\mathbf{X}}^{(j)}(1), \quad \text{for } j = 1, \ldots, s.$$

Factorial moments are important in the study of orthogonal arrays, see e.g. Bose & Bush (1952) and Dey & Mukerjee (1999, Chap5). For orthogonal array $\mathrm{OA}(n, s, q, t)$, denote the constants

$$c_j(\mathrm{OA}) = \frac{n/q^j - 1}{n-1} C(s, j), \quad \text{for } j = 1, \ldots, s. \tag{5}$$

By noting that in Theorem 2 of Zhang, Fang, Li & Sudjianto (2004) the Schur-combinatorial criteria

$$\Psi_C(\mathbf{X}; j) = 2m\big(c_j(\mathbf{X}) - c_j(\mathrm{OA})\big)/j!$$

that are all non-negative, the following lemma is readily obtained. The necessity was obtained early by Bose & Bush (1952); however, the sufficiency had been an open problem, see e.g. Seiden & Zemach (1966) for their doubt.

Lemma 2 (Bose & Bush (1952); Zhang, Fang, Li & Sudjianto (2004)). *The factorial moments $c_j(\mathbf{X})$ of any design $\mathbf{X}(n, q^s)$ are bounded from below by the constants $c_j(\mathrm{OA})$ for $j = 1, \ldots, s$. They can be achieved simultaneously for $j = 1, \ldots, t$ if and only if the design is an orthogonal array $\mathrm{OA}(n, s, q, t)$.*

2.2 Minimum aberration

Resolution and minimum aberration are important criteria for factorial designs based on the *word-length pattern* $(A_1, \ldots, A_s)$, which was originally defined by Fries & Hunter (1980) from defining contrast subgroup for regular designs. For general designs $\mathbf{X}(n, q^s)$, the word-length pattern has been shown by Ma & Fang (2001) and Xu & Wu (2001) independently to have a natural link with *MacWilliams transform* of distance distribution in coding theory,

$$A_j(\mathbf{X}) = \frac{1}{n} \sum_{k=0}^{s} \mathscr{P}_j(k; s, q) E_k(\mathbf{X}), \quad \text{for } j = 1, \ldots, s, \tag{6}$$

or $A_j(\mathbf{X}) = \frac{n-1}{n} \sum_{k=0}^{s} \mathscr{P}(s-k; s, q) P_k(\mathbf{X}) + \frac{(q-1)^j}{n} \binom{s}{j}$ by (3), where $\mathscr{P}_j(x; s, q) = \sum_{w=0}^{j} (-1)^w (q-1)^{j-w} \binom{x}{w} \binom{s-x}{j-w}$ are *Krawtchouk polynomials*; see MacWilliams & Sloane (1977, Chap5) and Hedayat, Sloane & Stufken (1999, Chap4) for details. The *resolution* of design $\mathbf{X}$ is defined to be the smallest index R such that $A_R(\mathbf{X}) > 0$ and $A_j(\mathbf{X}) = 0$ for all $j < R$. Orthogonal designs have resolution at least 3. Conversely, the resolution-R designs have orthogonal strength $R - 1$. For comparing two designs, $\mathbf{X}_1$ is said to have less aberration than $\mathbf{X}_2$ if there exists an index j such that $A_j(\mathbf{X}_1) < A_j(\mathbf{X}_2)$ and $A_k(\mathbf{X}_1) = A_k(\mathbf{X}_2)$ for all $k < j$. The *minimum aberration* design $\mathbf{X}^*$ is obtained if it has less aberration than all other competing designs.

Instead of using complicated Krawtchouk polynomials, Zhang, Fang, Li & Sudjianto (2004) introduced a *deviation pattern* $(B_1, \ldots, B_s)$ using Schur-convex function of separable combinatorial form based on pairwise coincidence measurements

$$B_j(\mathbf{X}) = \left\{ \sum_{k=j}^{s} \binom{k}{j} P_k(\mathbf{X}) - \binom{s}{j}\left(\frac{n^2}{q^j} - n\right)\right\}^{\frac{1}{2}} \equiv \alpha\sqrt{c_j(\mathbf{X}) - c_j(\mathrm{OA})} \qquad (7)$$

for $j = 1, \ldots, s$, where the constant $\alpha = \sqrt{n(n-1)q^{-j}/j!}$. The deviation pattern is easy to analyze from the majorization perspective. Its benchmark can be derived by Lemma 1 and can be achieved by weak equidistant designs; see Zhang, Fang, Li & Sudjianto (2004) for details. It was also shown linearly related to word-length pattern,

$$B_j^2(\mathbf{X}) = \frac{n^2}{q^{2j}} \sum_{k=1}^{j} \binom{s-k}{j-k} A_k(\mathbf{X}), \quad \text{for } j = 1, \ldots, s. \qquad (8)$$

Both patterns are equivalent under the sequential minimization procedure from $j = 1$ to s. So, the deviation pattern is able to characterize minimum aberration.

The sequential minimization of either A_j's or B_j's in the spirit of the minimum aberration criterion is supported by the *effects hierarchy principle*:

1) factorial effects of lower orders are more likely to be important; and
2) factorial effects of the same order are equally likely to be important.

Under this principle and the sequential minimization procedure, the moments in (4) are equivalent to word-length pattern. Therefore, all three moments are able to characterize minimum aberration.

Theorem 1. *For any design $\mathbf{X} \in \mathscr{U}(n, q^s)$, their moments in (4) have the following relationships with word-length pattern*

$$c_j(\mathbf{X}) = \kappa_j(\mathbf{X}) + T_1 = \mu_j(\mathbf{X}) + T_2 = \alpha A_j(\mathbf{X}) + T_3, \quad j = 2, \ldots, s$$

where the constant $\alpha = j!nq^{-j}/(n-1) > 0$, T_1, T_2 and T_3 involve only lower-order moments. Therefore, the sequential minimization of these moments from low to high orders are equivalent to the minimum aberration criterion.

2.3 Guidelines for selecting kernels

Let us continue discussing the illustrative example in Zhang, Fang, Li & Sudjianto (2004), with the present focus placed to the Psi-stage of Schur-convex discrimination for admissible designs. Recall the experimental scenario of selecting 4-factor sub-designs from $\mathbf{X}(27, 3^8)$, tabulated in Table 1, as well as the demonstrative sub-designs $\mathbf{X}_1$ to $\mathbf{X}_4$. Both $\mathbf{X}_3$ and $\mathbf{X}_4$ are inadmissible at Beta-stage due to

$$\{\beta(\mathbf{X}_1), \beta(\mathbf{X}_2)\} \prec \beta(\mathbf{X}_3) \prec \beta(\mathbf{X}_4),$$

Table 1. Full U-type design $\mathbf{X}(27, 3^8)$ for sub-design selection, obtained from the UD-web http://www.math.hkbu.edu.hk/UniformDesign/, constructed by Fang, Ma and Winker (2002)

A	B	C	D	E	F	G	H
1	1	2	1	1	0	1	2
2	2	2	1	2	0	2	0
0	0	1	2	1	0	2	2
2	0	1	0	1	1	2	0
0	0	1	1	2	0	0	1
0	2	1	2	2	2	1	1
0	2	0	1	0	1	2	2
1	2	1	0	0	0	0	0
2	1	0	2	2	1	0	1
1	0	0	1	0	2	0	2
0	2	2	0	2	1	0	2
2	1	1	2	0	1	0	2
2	2	0	2	1	0	1	2
1	0	0	2	2	1	2	0
0	1	0	0	2	0	1	0
2	0	2	1	2	2	1	2
0	1	2	0	0	2	2	1
2	1	1	1	0	2	1	0
0	1	0	1	1	2	0	0
1	1	1	0	2	2	2	2
1	2	2	2	1	2	0	0
1	2	1	1	1	1	1	1
2	2	0	0	1	2	2	1
1	1	2	2	0	0	2	1
1	0	0	0	0	0	1	1
0	0	2	2	0	1	1	0
2	0	2	0	1	1	0	1

In the space $\mathscr{D}(27, 3^4)$ of 70 sub-designs, we choose

$$\mathbf{X}_1 = \{A, C, G, H\},$$

$$\mathbf{X}_2 = \{B, C, G, H\},$$

$$\mathbf{X}_3 = \{A, B, D, F\},$$

$$\mathbf{X}_4 = \{A, D, E, F\},$$

for demonstration, same as Zhang, Fang, Li & Sudjianto (2004).

so we mainly concentrate on $\mathbf{X}_1, \mathbf{X}_2$ and other admissible designs.

It is flexible to choose a kernel for defining Schur-Psi criterion (1), as long as the univariate kernel is a convex function. The three toy kernels used by Zhang, Fang, Li & Sudjianto (2004) are of the following variance, power and exponential forms,

$$\sigma_1(\beta) = \frac{1}{m}(\beta - \bar{\beta})^2, \quad \sigma_2(\beta) = \beta^\pi \quad \text{and} \quad \sigma_3(\beta) = \left(\frac{1+\sqrt{5}}{2}\right)^\beta,$$

which result in the consistent rank-order for the corresponding Schur-Psi criteria defined by (1),

$$\Psi(\mathbf{X}_1; \sigma_{\{1,2,3\}}) \le \Psi(\mathbf{X}_2; \sigma_{\{1,2,3\}}) < \Psi(\mathbf{X}_3; \sigma_{\{1,2,3\}}) < \Psi(\mathbf{X}_2; \sigma_{\{1,2,3\}}).$$

To study the kernel selection problem, it is natural to ask

1. can we choose any other convex kernel, e.g. another toy kernel of the negative entropy form (modified to make $\beta(\mathbf{x}_i, \mathbf{x}_k) + 1 > 0$ for log function)

$$\sigma_4(\beta) = (\beta + 1)\log(\beta + 1) \ ?$$

2. how should we select follow-up kernel(s) to make further discrimination if there are many designs ranked No.1 under the *a priori* kernel?

For the first question, we get a reverse ordering of $\Psi(\mathbf{X}_1; \sigma_4) > \Psi(\mathbf{X}_2; \sigma_4)$ which contradicts with the first three kernels. Further, we claim that the negative entropy kernel $\sigma_4(\beta)$ is inadequate for the following reason. Both $\mathbf{X}_1$ and $\mathbf{X}_2$ are orthogonal designs of resolution 3, so their 1st and 2nd-order raw moments are constant by Theorem 1. By Taylor expansion of $\sigma_4(\beta)$ on β, we have

$$\begin{aligned}\Psi(\mathbf{X}; \sigma_4) &= \sum_{r=1}^{m} \left(\beta_r + \frac{1}{2}\beta_r^2 - \frac{1}{6}\beta_r^3 + \frac{1}{12}\beta_r^4 - \cdots\right) \\ &= m\bar{\beta} + \frac{m}{2}\kappa_2(\mathbf{X}) - \frac{m}{6}\kappa_3(\mathbf{X}) + \frac{m}{12}\kappa_4(\mathbf{X}) - \cdots \\ &= \text{fixed term} - \frac{m}{6}\kappa_3(\mathbf{X}) + \text{higher-order term},\end{aligned}$$

which tends to maximize the 3rd order raw moment and therefore destroy the 3-factor orthogonality. So, the answer to the first question is negative. Instead, the convex kernels should be carefully selected; the negative entropy kernel is a counter example for orthogonal designs.

Under the variance kernel $\sigma_1(\beta)$, the Schur-Psi criterion is equivalent to the 2nd-order central moment, i.e. $\Psi(\mathbf{X}; \sigma_1) = m\mu_2(\mathbf{X})$. It is fixed for all orthogonal designs by the necessary part of Lemma 2,

$$\Psi(\mathbf{X}; \sigma_1) = 2c_2(\text{OA}) + c_1(\text{OA}) - c_1^2(\text{OA}) \equiv \frac{ns(q-1)}{q^2(n-1)^2}(n - 1 - s(q-1)).$$

(Remarks: On the other hand, $\Psi(\mathbf{X};\sigma) \geq f(1-f)$ for any designs according to Lemma 1. Thus, we have a necessary condition for parameters (n, s, q) for the existence of $\text{OA}(n, s, q, 2)$, which is well known as Bose-Bush approach studied by Mukerjee & Wu (1995).) For the above example, $\Psi(\text{OA}(27, 4, 3, 2); \sigma_1)$ is evaluated by 0.6391, which is the minimum possible value for the candidate designs in $\mathscr{D}(27, 3^4)$. Conversely, any sub-design $\mathbf{X}(27, 3^4)$ with $\Psi(\mathbf{X};\sigma_1) = 0.6391$ is an orthogonal design, by the sufficient part of Lemma 2. That is to say, the Schur-variance criteria $\Psi(\mathbf{X};\sigma_1)$ can be used to screen out orthogonal candidates from the design pool, and it has successfully picked out 8 orthogonal sub-designs, including $\mathbf{X}_1$ and $\mathbf{X}_2$, as listed in Table 2.

Table 2. Ranked orthogonal designs (under multiple criteria) from $\mathscr{D}(27, 3^4)$, where the designs at the same row block have the same IPC-vector and rank, the boldfaced numbers in parentheses denote the specific ranks among all 70 candidates

Rank	Factors	$\Psi(\mathbf{X};\sigma_1)$	$\Psi(\mathbf{X};\sigma_2)$	$\Psi(\mathbf{X};\sigma_3)$	GWP-aberration	WL$_2$-discrepancy
1($\mathbf{X}_1$)	A, C, G, H	0.6391	1658.7(**2**)	683.4(**1**)	$(0, 0, \frac{10}{9}, \frac{8}{9})$ (**1**)	0.42416(**1**)
2	C, E, F, H	0.6391	1691.0(**5**)	684.4(**3**)	$(0, 0, \frac{38}{27}, \frac{16}{27})$(**2**)	0.42434(**2**)
3	A, B, C, G	0.6391	1707.1(**16**)	684.8(**6**)	$(0, 0, \frac{14}{9}, \frac{4}{9})$ (**3**)	0.42443(**3**)
3	A, B, C, D	0.6391	1707.1(**16**)	684.8(**6**)	$(0, 0, \frac{14}{9}, \frac{4}{9})$ (**3**)	0.42443(**3**)
5	C, E, G, H	0.6391	1707.8(**18**)	685.0(**9**)	$(0, 0, \frac{14}{9}, \frac{2}{3})$ (**5**)	0.42444(**5**)
5	A, B, G, H	0.6391	1707.8(**18**)	685.0(**9**)	$(0, 0, \frac{14}{9}, \frac{2}{3})$ (**5**)	0.42444(**5**)
7($\mathbf{X}_2$)	B, C, G, H	0.6391	1724.5(**30**)	685.6(**21**)	$(0, 0, \frac{46}{27}, \frac{20}{27})$(**7**)	0.42454(**7**)
8	A, B, C, H	0.6391	1732.0(**35**)	685.7(**25**)	$(0, 0, \frac{16}{9}, \frac{4}{9})$ (**8**)	0.42458(**8**)

Let us now try to answer the second question. Suppose the above variance kernel is selected *a priori*, under which the 8 orthogonal sub-designs listed in Table 2 are ranked equally the best candidates. Our focus is to make further discrimination by selecting certain follow-up kernels. We claim that a follow-up kernel is supposed to have higher functional order. Compared to the 2nd-order σ_1, the orders $O(\sigma_4) < 2 < O(\sigma_2) < O(\sigma_4)$, which implies that both π-th power and golden-ratio exponential toy kernels are feasible choices of follow-up kernels. The ranked order is consistent among 8 orthogonal candidates, using $\Psi(\mathbf{X};\sigma_1)$ plus either $\Psi(\mathbf{X};\sigma_2)$ or $\Psi(\mathbf{X};\sigma_3)$. Further, it is consistent with classical criteria, e.g. minimum aberration for orthogonal designs based on word-length pattern (GWP) and uniformity measure for uniform designs based on wrap-around discrepancy (WL$_2$). As indicated by Zhang, Fang, Li & Sudjianto (2004), the sub-design $\mathbf{X}_1$

is both a minimum aberration design and a uniform design in the complete pool $\mathscr{D}(27, 3^4)$ of 70 competing designs.

Formally, let us consider the linear model of the following ANOVA decomposition form under design $\mathbf{X}(n, q^s)$,

$$y(\mathbf{x}) = \sum_{u\in\mathscr{G}} \chi_u(\mathbf{x})\theta_u + \varepsilon = \sum_{wt(u)=0}^{s} \chi_u(\mathbf{x})\theta_u + \varepsilon, \tag{9}$$

where the group $\mathscr{G} = \{0, 1, \ldots, q-1\}^s$, ε is the random error and $wt(u)$ counts the number of nonzero elements. For $u \in \mathscr{G}$, θ_u and χ_u are factorial effects and orthonormal contrast coefficients, respectively Xu & Wu (2001). In (9), factorial effects are decomposed into 1) the grand mean $\theta_\emptyset$ and main effects θ_u with $wt(u) = 1$; and 2) j-factor interaction effects θ_u with $wt(u) = 2, \ldots, s$. Due to scarce resources, the effects hierarchy principle is commonly relied on. Experimental designers usually work on the *main effects model* $\mathbf{y} = \mathbf{G}\boldsymbol{\theta}+\varepsilon$, where $\boldsymbol{\theta}$ contains the grand mean and main effects, $\mathbf{y}$ and $\mathbf{G}$ are the corresponding $n\times 1$ response vector and $n\times(s+1)$ regression matrix under design $\mathbf{X}(n, q^s)$. The least squares estimate is given by $\widehat{\boldsymbol{\theta}} = (\mathbf{G}^H\mathbf{G})^{-1}\mathbf{G}^H\mathbf{y}$, where $\mathbf{G}^H$ denotes the complex conjugate of $\mathbf{G}$. It is known that an orthogonal design is universally optimal in the sense of making small the covariance matrix of $\widehat{\boldsymbol{\theta}}$. However, bias of $\widehat{\boldsymbol{\theta}}$ may be caused in the presence of interactions effects that are non-negligible. For $j = 2, \ldots, s$, denote the $n \times (q-1)^j\binom{s}{j}$ matrix $\mathbf{H}_j = [\chi_u(\mathbf{x}_i)]$ for $i = 1, \ldots, n$ and all $u \in \mathscr{G}$ with $wt(u) = j$. The alias matrix of all j-factor interactions on $\widehat{\boldsymbol{\theta}}$ is given by $\mathbf{C}_j = (\mathbf{G}^H\mathbf{G})^{-1}\mathbf{G}^H\mathbf{H}_j$.

Our objective is to minimize the moments of coincidence distribution in a certain sense. For the main effects model, it is known that a minimum aberration design guards against aliasing from higher order interactions, since

$$\|\mathbf{C}_j\|_F^2 = (s-j+1)(q-1)A_{j-1} + j(q-2)A_j + (j+1)A_{j+1}, \quad j = 2, \ldots, s$$

due to Xu & Wu (2001), where $A_{s+1} \equiv 0$ and the Frobenius norm $\|\mathbf{C}_j\|_F^2 = \text{trace}(\mathbf{C}_j^H\mathbf{C}_j)$ is used as a measure of aliasing effect caused by all j-factor interactions. By Theorem 1, it is equivalent to sequentially minimize power, central or factorial moments. Without loss of generality, we take the raw moments and

(I) minimize $\kappa_2(\mathbf{X}), \kappa_3(\mathbf{X}), \ldots$ one by one sequentially.

This minimum aberration approach follows *strictly* the effects hierarchy principle, but it is somewhat extreme. Consider the situation where 3-factor interactions are potential. Given two orthogonal designs $\mathbf{X}_1$ with word-length pattern $(0, 0, 0, A_4, \ldots)$ and $\mathbf{X}_2$ with $(0, 0, A_3', 0, \ldots)$ where A_4 is small and A_3' is large, the minimum aberration criterion contradicts the real need. It is logical to minimize some weighted sum of raw moments at different orders. By the effects hierarchy principle, more weights should be placed on the lower orders, so we should

(II) minimize $\sum_{j=2}^{\infty} \gamma_j\kappa_j(\mathbf{X})$, where the constants $\gamma_j > \gamma_k$ if $j < k$.

The above two approaches constitute the guidelines for the kernel selection procedure at Psi-stage of majorization framework. In what follows, we shall show that they correspond to the selection of power and exponential kernels for fractional factorial designs and uniform designs, respectively.

3 Power Kernels for Fractional Factorial Designs

For an FFD $\mathbf{X} \in \mathscr{U}(n, q^s)$ that has resolution at least 2, consider the Schur-Psi criteria (1) w.r.t. the *power kernels* $\sigma(\beta) = \beta^t$ with orders $t \geq 2$. We restrict the orders to be integers, rather than 3.14159... for the toy power kernel σ_2 discussed in Section 2.3. It is clear that Schur-power criteria are equivalent to the raw moments, i.e.

$$\Psi(\mathbf{X}; \beta^j) = \sum_{r=1}^{m} \beta_r{}^j \equiv m\kappa_j(\mathbf{X}), \quad j = 2, 3, \ldots \tag{10}$$

The sequential minimization of $\Psi(\mathbf{X}; \beta^2), \Psi(\mathbf{X}; \beta^3), \ldots$ characterizes the minimum aberration, according to Theorem 1. Formally,

Protocol 1 (Power kernels) *At Psi-stage of majorization framework, we suggest to select the 2nd-order power kernel $\sigma(\beta) = \beta^2$ a priori; then increase the power order one by one if necessary.*

For the example discussed in the previous section, the $\Psi(\mathbf{X}; \beta^2)$ is equivalent to $\Psi(\mathbf{X}; \sigma_1)$ w.r.t. the variance kernel, and therefore able to screen out the 8 orthogonal sub-designs. The $\Psi(\mathbf{X}; \beta^3)$ is used to assess these 8 resulting designs, but it can not discriminate rank-3 and 5 designs (whose 3rd GWP components are identical). Then, the $\Psi(\mathbf{X}; \beta^4)$ is needed for the further discrimination.

By Lemma 1, we have the lower bounds for Schur-power criteria (10), which can serve as the benchmark of Schur-optimum designs.

Theorem 2. *For $\mathbf{X} \in \mathscr{U}(n, q^s)$ and $j = 2, \ldots, s$,*

$$\Psi(\mathbf{X}; \beta^j) \geq m(1-f)\theta^j + mf(\theta+1)^j$$

where the lower bound can be attained by weak equidistant designs.

We note that Xu (2003) studied intensively the raw moments $\kappa_2(\mathbf{X}), \kappa_3(\mathbf{X}), \ldots$ for factorial designs. In Xu (2003), each raw moment $\kappa_j(\mathbf{X})$ is expressed as an affine transform of word-length $A_k(\mathbf{X})$ for $k \leq j$; similar to Theorem 1. Then, the lower bounds of raw moments can be obtained by the non-negativeness of word-length pattern. The transform involves the generalized Pless power moment identities in coding theory and is complicated in general. Xu (2003) only considered the lower bounds in terms of orthogonal arrays only up to strength 3. In this paper, we suggest to study the behavior of Schur-power criteria (10) through factorial moments $c_j(\mathbf{X})$ in (4) to derive their explicit lower bounds.

Theorem 3. *Schur-power criteria can be expressed through raw moments by*

$$\Psi(\mathbf{X};\beta^j) = m\sum_{k=1}^{j} S(j,k)c_k(\mathbf{X}) \geq m\sum_{k=1}^{j} S(j,k)c_k(\text{OA})$$

for $j = 2, 3, \ldots$, *where* $S(j,k)$ *are Stirling numbers of the second kind. The lower bounds can be achieved simultaneously for* $j = 2, \ldots, t$ *if and only if the design* $\mathbf{X}(n, q^s)$ *is an orthogonal array* $\text{OA}(n, s, q, t)$.

By the triangle of Stirling numbers of the second kind, with properties $S(1,1) = 1$ and the recurrence relation $S(j,k) = S(j-1,k-1) + kS(j-1,k)$ in which $S(j,k) \equiv 0$ if $k = 0$ or $k > j$, we have that $\Psi(\mathbf{X};\beta) = m\bar{\beta}$,

$$\begin{aligned}
\Psi(\mathbf{X};\beta^2) &= m\big(c_1(\mathbf{X}) + c_2(\mathbf{X})\big)\\
\Psi(\mathbf{X};\beta^3) &= m\big(c_1(\mathbf{X}) + 3c_2(\mathbf{X}) + c_3(\mathbf{X})\big)\\
\Psi(\mathbf{X};\beta^4) &= m\big(c_1(\mathbf{X}) + 7c_2(\mathbf{X}) + 6c_3(\mathbf{X}) + c_4(\mathbf{X})\big)\\
\Psi(\mathbf{X};\beta^5) &= m\big(c_1(\mathbf{X}) + 15c_2(\mathbf{X}) + 25c_3(\mathbf{X}) + 10c_4(\mathbf{X}) + c_5(\mathbf{X})\big)
\end{aligned}$$

and so forth. Their lower bounds can be written down similarly in terms of the constants $c_j(\text{OA})$ defined in (5). This generalizes the results by Xu (2003).

4 Exponential Kernels for Uniform Designs

For a U-type design $\mathbf{X} \in \mathscr{U}(n, q^s)$, consider the Schur-Psi criterion w.r.t. *exponential kernel* $\sigma(\beta) = \rho^\beta$, i.e.

$$\Psi(\mathbf{X};\rho^\beta) := \sum_{r=1}^{m} \rho^{\beta_r}, \quad (\rho > 1). \tag{11}$$

By the Taylor expansion of the exponential function ρ^β on β, we have

$$\begin{aligned}
\Psi_E(\mathbf{X};\rho) &= m(1 + \bar{\beta}\log\rho) + \sum_{j=2}^{\infty}\frac{(\log\rho)^j}{j!}\sum_{r=1}^{m}\beta_r^j\\
&= m(1 + \bar{\beta}\log\rho) + \frac{(\log\rho)^2}{2}\kappa_2(\mathbf{X}) + \frac{(\log\rho)^3}{6}\kappa_3(\mathbf{X}) + \cdots
\end{aligned}$$

where $\kappa_2(\mathbf{X}), \kappa_3(\mathbf{X}), \ldots$ are raw moments (4) of pairwise coincidences.

Schur-exponential criterion (11) corresponds to the approach (II) justified in Section 2.3. The total weight assigned to raw moments is evaluated by

$$\sum_{j=2}^{\infty}\frac{(\log\rho)^j}{j!} = \rho - 1 - \log\rho,$$

of which the proportions $\frac{(\log\rho)^j}{j!}$ divided by the total $(\rho - 1 - \log\rho)$ are assigned to $\kappa_j(\mathbf{X})$ for $j = 2, 3, \ldots$. The following facts can be easily verified.

a) The weight for $\kappa_j(\mathbf{X})$ is decreasing as j increases, provided that $\rho < e^3$;
b) The weight for $\kappa_2(\mathbf{X})$ is decreasing as ρ increases. This weight is large when ρ is close to 1, e.g., it is as large as 94.7% when $\rho = 1.174$;
c) The weights for higher-order $\kappa_j(\mathbf{X})$ $(j \geq 3)$ is increasing of the base ρ.

For an illustration, Figure 1 plots both the individual and cumulative weights assigned to raw moments $\kappa_2(\mathbf{X})$ to $\kappa_6(\mathbf{X})$, provided different choices of $\rho > 1$.

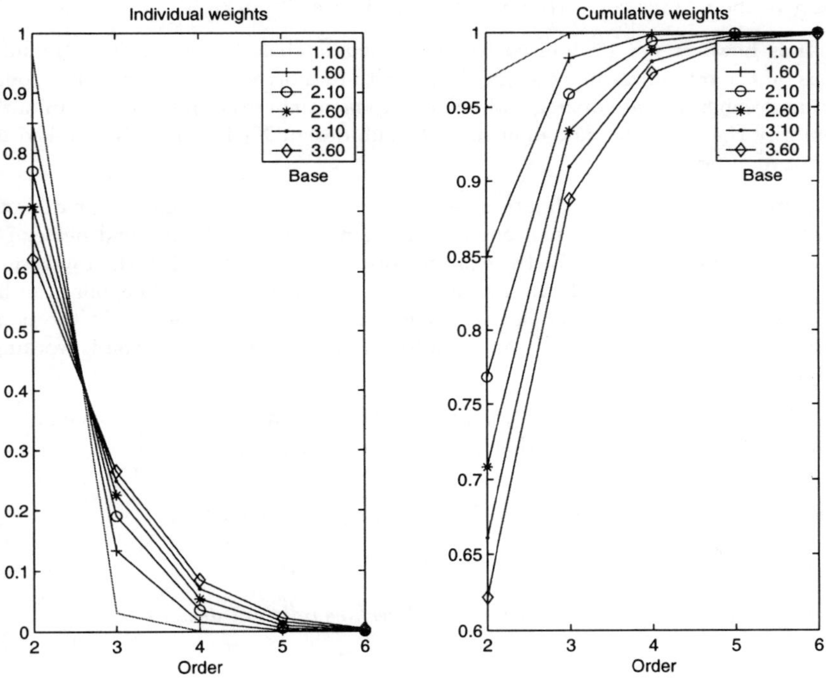

Fig. 1. Weights assigned to interactions effects by Schur-exponential criteria.

The discrepancy is the main criterion for uniform designs (Fang & Wang (1994)). For designs with qualitative factors, Zhang, Fang, Li & Sudjianto (2004) studied the categorial discrepancy, which is tightly related with Schur-exponential criterion by

$$D(\mathbf{X}; a \vee b) = \frac{1}{n}\sqrt{2\Psi(\mathbf{X}; \rho^{\beta}) + n(1+a)^s - n^2(1+\mu)^s},$$

where a, b are certain parameters for defining discrepancy by reproducing kernels, $\rho = (1+a)/(1+b) > 1$ and $\mu = (a+(q-1)b)/q$; see also Hickernell & Liu (2002). So Schur-exponential criterion is equivalent to categorical discrepancy, which leads to the following way of assessing U-type designs with qualitative factors.

Protocol 2 (Exponential kernels) *At Psi-stage of majorization framework, we suggest to select the exponential kernel $\sigma(\beta) = \rho^\beta$ with ρ slightly greater than* 1 *(say,* 1.174*); if necessary, increase the base for further discrimination.*

The Schur-exponential criterion with pilot choice of base 1.174 *de facto* approximates the wrap-around L_2-discrepancy (up to a scaling factor and an offset) for assessing 3-level uniform designs, as discussed by Zhang, Fang, Li & Sudjianto (2004). Shown in Table 2, it can successfully discriminate all 8 orthogonal designs and give their ranks correctly among 70 candidates in $\mathscr{D}(27, 3^4)$.

It is however not prohibited to choose smaller bases *a priori*, if we mainly focus on the minimization of $\kappa_2(\mathbf{X})$ in order to find orthogonal designs. For competing designs that are orthogonal, Schur-exponential criterion turns to minimize primarily the 3rd-order raw moment. The sub-design $\mathbf{X}_1$ in Table 2 is ranked in such a manner.

Exponential kernels with larger bases can be selected to assess U-type designs when the higher-order interactions effects are potential, but they had better be used as follow-up kernels. For the example discussed in Section 2.3, the toy kernel σ_3 based on golden ratio 1.618... seems a bit over, since many orthogonal candidates are ranked lower than the non-orthogonal ones. Nevertheless, in the pool of orthogonal sub-designs, the golden ratio Schur-exponential criterion is working properly.

The lower bound of Schur-exponential criterion can be otained immediately from Lemma 1, by observing that (11) is Schur-convex of $\beta(\mathbf{X})$.

Theorem 4. *For a U-type design $\mathbf{X} \in \mathscr{U}(n, q^s)$ and any $\rho > 1$,*

$$\Psi(\mathbf{X}; \rho^\beta) \geq m(1 - f + \rho f)\rho^\theta,$$

where the lower bound can be attained by weak equidistant designs.

5 An Extension to Quantitative Designs

All the above texts are based on the qualitative designs with categorical levels. In this section, we consider an extension of the majorization framework to quantitative designs, in particular the UDs under the wrap-around L_2-discrepancy (WL$_2$) criterion.

Without loss of generality, let the domain of quantitative factors be the hypercube $[0, 1)^s$, as denoted by C^s. As suggested by Fang & Wang (1994), C^s can be approximated by the following lattice

$$\mathcal{L}(q^s) = \{L_1, \ldots, L_q\}^s \quad \text{where} \quad L_k = (k - 0.5)/q, \quad k = 1, \ldots, q.$$

and for a U-type design $\mathbf{X}(n, q^s)$, each column vector is an permutation of

$$\Big\{\underbrace{\frac{1}{2q},\ldots,\frac{1}{2q}}_{n/q},\underbrace{\frac{3}{2q},\ldots,\frac{3}{2q}}_{n/q},\ldots,\underbrace{\frac{2q-1}{2q},\ldots,\frac{2q-1}{2q}}_{n/q}\Big\}. \tag{12}$$

The uniform designs can be assessed by different versions of discrepancy criteria, among which we consider the popular wrap-around one by Hickernell (1998)

$$\text{WL}_2(\mathbf{X}) = \Big\{ -\Big(\frac{4}{3}\Big)^s + \frac{1}{n^2}\sum_{i,k=1}^{n}\prod_{j=1}^{s}\Big[\frac{3}{2} - |x_{ij} - x_{kj}|(1 - |x_{ij} - x_{kj}|)\Big]\Big\}^{\frac{1}{2}},$$

which is important in Quasi-Monte Carlo methods for periodic integrands, response surface fitting by periodic functions and otherwise.

Let $\alpha(x, y) = \frac{3}{2} - |x - y|(1 - |x - y|)$. Similar to the IPC-vector introduced in Section 1, define for $\mathbf{X}(n, q^s)$ the informative vector $\boldsymbol{\beta}(\mathbf{X}) = [\beta(\mathbf{x}_i, \mathbf{x}_k)]$ with length $n(n-1)/2$, where

$$\beta(\mathbf{x}_i, \mathbf{x}_k) = \sum_{j=1}^{s} \ln \alpha(x_{ij}, x_{kj})$$

for $i = 1, \ldots, k-1$ and $k = 2, \ldots, n$. In terms of $\boldsymbol{\beta}(\mathbf{X})$ and separable exponential functions, define

$$\Psi(\mathbf{X}; \rho^\beta) = \sum_{k=2}^{n}\sum_{i=1}^{k-1} e^{\beta(\mathbf{x}_i, \mathbf{x}_k)}$$

which gives an extension of Schur-exponential criteria (11) for uniform designs, from qualitative factors to quantitative factors. Clearly,

$$\text{WL}_2^2(\mathbf{X}) = -\Big(\frac{4}{3}\Big)^s + \frac{1}{n}\Big(\frac{3}{2}\Big)^s + \frac{2}{n^2}\Psi(\mathbf{X}; e^\beta). \tag{13}$$

It is easy to check from (12) that the sum of $\boldsymbol{\beta}(\mathbf{X})$ is fixed for any designs $\mathbf{X} \in \mathscr{U}(n, q^s)$, as given concretely by

$$\begin{aligned} S = \sum_{k=2}^{n}\sum_{i=1}^{k-1}\beta(\mathbf{x}_i, \mathbf{x}_k) &= \frac{1}{2}\sum_{j=1}^{s}\Big[\sum_{i,k=1}^{n}\ln\alpha(x_{ij}, x_{kj}) - \sum_{i=1}^{n}\ln\alpha(x_{ij}, x_{ij})\Big] \\ &= \frac{ns}{2q}\sum\nolimits_{a,b=1}^{q}\ln\alpha(L_a, L_b) - \frac{ns}{2}\ln\frac{3}{2}. \end{aligned}$$

Similar to Lemma 1, we can derive the lower bounds for $\Psi(\mathbf{X}; e^\beta)$,

$$\Psi(\mathbf{X}; e^\beta) \geq \frac{n(n-1)}{2} e^{\frac{2S}{n(n-1)}}$$

with equality at the condition that $\beta(\mathbf{x}_i, \mathbf{x}_k) = \frac{2S}{n(n-1)}$ for all $i \neq k$. Then, by (13), we can derive the lower bounds for $\text{WL}_2(\mathbf{X})$, as stated in Theorem 5.

Theorem 5. *For U-type design* $\mathbf{X} \in \mathscr{U}(n, q^s)$, *its squared wrap-around* L_2-*discrepancy has a lower bound*

$$\mathrm{WL}_2^2(\mathbf{X}) \geq -\left(\frac{4}{3}\right)^s + \frac{1}{n}\left(\frac{3}{2}\right)^s + \frac{n-1}{n}\left\{\frac{2}{3}\left[\prod_{a,b=1}^{q} \alpha(L_a, L_b)\right]^{\frac{1}{q}}\right\}^{\frac{s}{n-1}}$$

where $L_a = (a - 0.5)/q$ *for* $a = 1, \ldots, q$. *The lower bound can be achieved when* $\prod_{j=1}^{s} \alpha(x_{ij}, x_{kj})$ *is a constant for all* $i \neq k$.

6 Concluding Remarks and Acknowledgments

The paper discusses Schur-convex kernel selection problem in the majorization framework for experimental designs, from a genuine perspective under the effects hierarchy principle. It is suggested to use power and exponential kernels, with respect to which Schur-optimum designs are justified to coincide with minimum aberration fractional factorial designs and minimum categorical discrepancy uniform designs, respectively. Thus, by Zhang, Fang, Li & Sudjianto (2004) and the current work, a unified approach is provided to assess fractional factorial designs and uniform designs, as well as their benchmarks (i.e. lower bounds of Schur-Psi criteria w.r.t. different convex Psi-kernel functions).

While using power kernels, there are two possible extensions, namely decimal power-orders (e.g. 2.5 or irrational π) and polynomial functions (e.g. $ax^3 + bx^2 + cx$), as long as they are convex on $\mathbb{R}_+$. We find that the decimal-order power kernels have stronger discrimination ability. For the example discussed in Section 2.3, $\Psi(\mathbf{X}; \beta^{2.5})$ can discriminate rank-3 and 5 designs that are indistinguishable under $\Psi(\mathbf{X}; \beta^3)$. Polynomial kernels can be viewed as a linear combination of individual power functions, similar to the weighted sum approach used by exponential kernel selection; but they make non difference in the sequential minimization procedure. The reformulation of minimum aberration for two-level designs by Butler (2003) can be also covered by Schur-power criteria, by observing that the so-called "confounding between runs" in Butler (2003) has a natural link with coincidences

$$T_{ik} = 2\beta(\mathbf{x}_i, \mathbf{x}_k) - s, \quad i, k = 1, \ldots, n.$$

Lower bounds for Schur-power criteria, as presented in Theorem 2, can be improved when the corresponding orthogonal arrays do not exist. Xu (2003) made an effort to consider weak orthogonal strength and derived explicitly the improved bounds up to order 3, using majorization (though not declared) in integer-valued frequency of level-combinations. By Theorem 3 and Lemma 2, the explicit lower bounds can be written down explicitly up to any order. Besides, for orthogonal designs with resolution 3 or higher, additional constraints can be set onto the pairwise coincidences, e.g. the conditions by Lemma 2 implied from orthogonal strength. Then, the lower bounds in Theorem 2 can be improved by constrained integer programming.

Uniform design by Fang and Wang (Fang (1980), Fang & Wang (1994), Fang (2004)) provides an alternative approach in terms of minimizing raw moments of coincidence distribution, by using exponential kernels to define Schur-Psi criterion as elaborated in Section 4. For the extension to quantitative designs in Section 5, an advanced pairwise measure is used to define the informative vector in the majorization framework, then majorization technique is used to derive Theorem 5. An important property that allows for such extension is the shift-invariance of the reproducing kernel that defines wrap-around discrepancy.

For the construction of Schur-optimum designs under majorization framework, Zhang, Fang, Li & Sudjianto (2004) provided a Robin-hood swapping algorithm that determines the local search direction of columnwise-pairwise algorithm (Li & Wu (1997)), in order to make pairwise coincidences spread as equally as possible. Stochastic optimization methods are crucial in the global search, e.g. the threshold accepting heuristic used by Fang, Lu & Winker (2003) to construct uniform designs.

Lastly, I would like to congratulate Professor Kai-Tai Fang on his 65th birthday and wish to express my deep gratitude to Professor Fang for not only enlightening me about the research of experimental designs and statistics, but also his inspiring guidance and consistent encouragement during my study in the Department of Mathematics, Hong Kong Baptist University (1999 - 2004). I would also thank Professor Fred Hickernell, Professor Runze Li and Professor Rahul Mukerjee for their long-term encouragement and support.

Appendices

Proof of Theorem 1: The relationships among the raw, central and factorial moments at the same order are clear by their definitions (4). By (7) and (8), we have that

$$c_j(\mathbf{X}) = \frac{j!n}{(n-1)q^j}\sum_{k=1}^{j}\binom{s-k}{j-k}A_k(\mathbf{X}) + c_j(\text{OA}) = \frac{j!n}{(n-1)q^j}A_j(\mathbf{X}) + T_3$$

which proves the theorem. □

Proof of Theorem 3: The lower bound is clear from Lemma 2, and we only need to prove that Stirling numbers of the second kind connect the raw and factorial moments by $\kappa_j(\mathbf{X}) = \sum_{k=1}^{j} S(j,k)c_k(\mathbf{X})$. If suffices to show that for any integer $\beta \geq 1$ and $j = 1, 2, \ldots$,

$$\beta^j = \sum_{k=1}^{j} S(j,k)C(\beta,k).$$

At $j = 1$, the identity holds obviously. Suppose it is true for $j = 1, \ldots, J$, then

$$\begin{aligned}\beta^{J+1} &= \beta \sum_{k=1}^{J} S(J,k)C(\beta,k) = \sum_{k=1}^{J} S(J,k)C(\beta,k)(\beta - k + k) \\ &= \sum_{k=1}^{J} S(J,k)C(\beta,k+1) + k\sum_{k=1}^{J} S(J,k)C(\beta,k) \\ &= C(\beta, J+1) + \sum_{k=1}^{J}[S(J,k-1) + kS(J,k)]C(\beta,k).\end{aligned}$$

By the recurrent property $S(J+1,k) = S(J,k-1)+kS(J,k)$ and $S(J+1,J+1) = 1$, we have that $\beta^{J+1} = \sum_{k=1}^{J+1} S(J+1,k)C(\beta,k)$. By mathematical induction, the theorem is established. □

References

Bose, R.C. & Bush, K.A. (1952). Orthogonal arrays of strength two and tree. *Ann. Math. Statist.* **23** 508–524.

Butler, N. A. (2003). Minimum aberration construction results for nonregular two-level fractional factorial designs. *Biometrika* **90** 891–898.

Cheng, C. S., Steinberg, D. M. & Sun, D. X. (1999). Minimum aberration and model robustness for two-level fractional factorial designs. *J. R. Statist. Soc. B* **61** 85–93.

Dey, A. & Mukerjee, R. (1999). *Fractional Factorial Plans.* Wiley, New York.

Fang, K. T. (1980). The uniform design: application of number–theoretic methods in experimental design. *Acta Math. Appl. Sinica.* **3** 363–372.

Fang, K. T. (2004). Uniform designs. *Encylopedia of Statistics*, to appear.

Fang, K. T., Lu, X. & Winker, P. (2003). Lower bounds for centered and wrap-around L_2-discrepancies and construction of uniform designs by threshold accepting. *J. Complexity* **19** 692–711.

Fang, K. T. & Wang, Y. (1994). *Number-Theoretic Methods in Statistics.* Chapman and Hall, London.

Fang, K. T. & Zhang, A. (2004). Minimum aberration majorization in non-isomorphic saturated designs. *J. Statist. Plan. Inference* **126** 337–346.

Fries, A. & Hunter, W. G. (1980). Minimum aberration 2^{k-p} designs. *Technometrics* **22** 601–608.

Hedayat, A. S., Sloane, N. J. A. & Stufken, J. (1999). *Orthogonal Arrays. Theory and Applications.* Springer-Verlag.

Hickernell, F. J. (1998). A generalized discrepancy and quadrature error bound. *Math. Comp.* **67** 299–322.

Hickernell, F. J. & Liu, M. Q. (2002). Uniform designs limit aliasing. *Biometrika* **89** 893–904.

Li, W. W. & Wu, C. F. J. (1977). Columnwise-pairwise algorithms with application to the construction of supersaturated designs. *Technometrics* **53** 171–179.

Ma, C. X. & Fang, K. T. (2001). A note on generalized aberration in factorial designs. *Metrika* **53** 85–93.

MacWilliams, F. J. & Sloane, N. J. A. (1977). *The Theory of Error-Correcting Codes.* North Holl and, Amsterdam.

Marshall, A. W. & Olkin, I. (1979). *Inequalities. Theory of Majorization and Its Applications.* Academic Press, New York.

Mukerjee, R. & Wu, C. F. J. (1995). On the existence of saturated & nearly saturated asymmetrical orthogonal arrays. *Ann. Statist.* **23** 2102–2115.

Santner, T. J., Williams, B. J. & Notz, W. I. (2003). *The Design and Analysis of Computer Experiments.* Springer-Verlag.

Seiden, E. & Zemach, R. (1966). On orthogonal arrays. *Ann. Math. Statist.* **37** 1355–1370.

Shaked, M. (1985). Majorization and Schur-convexity. In *Encyclopedia of Statistical Sciences* (S. Knotz, N. L. Johnson & C. B. Read, eds.) **5** 182–189. Wiley, New York.

Wu, C. F. J. & Hamada, M. (2000). *Experiments. Planning, Analysis, and Parameter Design Optimization.* Wiley, New York

Xu, H. (2003). Minimum moment aberration for nonregular designs and supersaturated designs. *Statist. Sinica* **13** 691–708

Xu, H. & Wu, C. F. J. (2001). Generalized minimum aberration for asymmetrical fractional factorial designs. *Ann. Statist.* **29** 1066–77.

Zhang, A. (2004). Majorization methodology for experimental designs. M.Phil. Thesis, Hong Kong Baptist University

Zhang, A., Fang, K. T., Li, R. & Sudjianto, A. (2004). Majorization framework for balanced lattice designs. *Ann. Statist.* tentatively accpeted.

Part III

Advances in Biostatistics

An Overview on Variable Selection for Survival Analysis

Jianqing Fan[1], Gang Li[2], and Runze Li[3]

[1] Department of Operation Research and Financial Engineering, Princeton University, NJ 08544, U.S.A. (jqfan@princeton.edu)
[2] Department of Biostatistics, University of California Los Angeles, CA 90095-1772, U.S.A. (vli@ucla.edu)
[3] Department of Statistics, The Pennsylvania State University, University Park, PA16802-2111, U.S.A. (rli@stat.psu.edu)

Summary. Variable selection is fundamental in high-dimensional statistical modeling. Many authors have proposed various variable selection criteria and procedures for linear regression models (Miller (2002)). Variable selection for survival data analysis poses many challenges because of complicated data structure, and therefore receives much attention in the recent literature. In this article, we will review various existing variable selection procedures for survival analysis. We further propose a unified framework for variable selection in survival analysis via a nonconcave penalized likelihood approach. The nonconcave penalized likelihood approach distinguishes from the traditional variable selection procedures in that it deletes the non-significant covariates by estimating their coefficients as zero. With a proper choice of the penalty function and the regularization parameter, we demonstrate the resulting estimate possesses an oracle property, namely, it performs as well as if the true submodel were known in advance. We further illustrate the methodology by a real data example.

Key words: Accelerated life model, Cox's model, Cox's frailty model, marginal hazards model, variable selection.

2000 Mathematics Subject Classification: 62N01

Contemporary Multivariate Analysis and Experimental Design—In Honor Celebration of Professor Kai-Tai Fang's 65th birthday. Edited by Jianqing Fan and Gang Li. The World Scientific Publisher, 2005.

1 Introduction

Variable selection is vital to survival analysis. In practice, many covariates are often available as potential risk factors. At the initial stage of modeling, data analysts usually introduce a large number of predictors. To enhance model predictability and interpretation, a parsimonous model is always desirable. Thus, selecting significant variables plays crucial roles in model building and is very challenging in the presence of a large number of predictors. Let us first review recent developments of model selection and variable selection for survival data analysis.

Bayesian model selection procedures have been proposed for survival analysis. Faraggi & Simon (1997) and Faraggi (1998) extended the ideas of Lindley (1968) to Cox's proportional hazard models with right censored survival data. To avoid specifying a prior on the baseline hazard function, they use the partial likelihood as the basis for their proposed Bayesian variable selection procedures rather than the full likelihood. Thus, their method indeed is not a proper Bayesian method. Ibrahim, Chen & MacEachern (1999) proposed a full Bayesian variable selection procedure for the Cox model by specifying a nonparametric prior for the baseline function and a parametric prior for the regression coefficients. To implement their methodology, Markov chain Monte Carlo (MCMC) was proposed to compute the posterior model probabilities. Ibrahim and his co-authors (Ibrahim & Chen (2000)), Ibrahim, Chen & Sinha (2001), Sinha, Chen & Ghosh (1999) further proposed several Bayesian model assessment criteria. Giudici, Mezzetti & Muliere (2003) proposed a Bayesian nonparametric approach to selecting significant variables in survival analysis based on mixtures of products of Dirichlet process priors. Bayesian variable selection procedures are simple in concept, but hard to implement in high-dimensional modeling due to computational demand for calculating posterior model probabilities.

Most variable selection criteria are closely related to penalized least squares and penalized likelihood. Some traditional variable selection criteria, such as Akaike information criterion (AIC, Akaike (1974)) and Bayesian information criterion (BIC, Schwarz (1978)) can be easily extended to survival analysis. Volinsky & Raftery (2000) extended the BIC to the Cox model. They propose a modification of the penalty term in the BIC so that it is defined in terms of the number of uncensored events instead of the number of observations. Traditional variable selection procedures require subset selection, such as stepwise deletion and the best subset selection. While they are practically useful, subset selection procedures ignore stochastic errors inherited at the stage of variable selections. Hence, their theoretic properties are somewhat difficult to understand. Furthermore, the best subset selection suffers from several drawbacks, the most severe of which is its lack of stability (Breiman (1996)). To retain virtues of the subset selection and to avoid the unstability of the subset selection, Tibshirani (1996) proposed the LASSO variable selection procedures for linear regression models and generalized linear models. The LASSO procedure was further extended to the Cox model in Tibshirani (1997). In an attempt to automatically and simultaneously select variables, Fan & Li (2001) proposed nonconcave penalized approaches for linear regression, robust linear models and generalized linear models, and suggested the use of smoothly clipped absolute deviation (SCAD) penalty. For simplicity of presentation, we will refer the procedures related to the SCAD penalized likelihood as SCAD. The SCAD is a useful amelioration of LASSO. Fan & Li (2001) demonstrated the SCAD possesses an oracle property, namely, the resulting estimate can correctly identify the true model as if it were known in advance, while the LASSO does not possess this oracle property. Fan & Li (2002) derived a nonconcave penalized partial likelihood for the Cox model and the Cox frailty model, and further illustrate the oracle property of their proposed procedures. In this paper, we aim to provide a unified framework of variable selection for various survival models, including parametric models and the Cox model for univariate survival data, and the Cox frailty model and the marginal hazard model for multivariate failure time.

The paper is organized as follows. In Section 2, we briefly introduce penalized likelihood approaches and extend the nonconcave penalized likelihood approach to parametric models in survival analysis. We derive a penalized partial likelihood procedure for the Cox model using Breslow's "least informative" nonparametric modeling for the cumulative baseline hazard function in Section 3. We extend nonconcave penalized likelihood variable selection procedures to multivariate survival data in Section 4. We deal with some practical implementation issues in Section 5. A real data example in Section 6 is used to illustrate the nonconcave penalized likelihood approach.

2 Nonconcave penalized likelihood approach

2.1 Penalized least squares and penalized likelihood

Most variable selection procedures are related to *penalized least squares*. Suppose that we have the $(d+1)$-dimensional random sample $(\mathbf{x}_i, y_i)$, $i = 1, \cdots, n$, from a population $(\mathbf{x}, y)$, where $\mathbf{x}$ is a d-dimensional random vector, and y is a continuous random variable. Consider a linear regression model

$$y_i = \mathbf{x}_i^T \boldsymbol{\beta} + \varepsilon_i,$$

where $\boldsymbol{\beta}$ is unknown regression coefficients, and ε_i is random error with mean zero and variance σ^2. Define a penalized least squares as

$$Q(\boldsymbol{\beta}) = \frac{1}{2}\sum_{i=1}^{n}(y_i - \mathbf{x}_i^T\boldsymbol{\beta})^2 + n\sum_{j=1}^{d} p_{\lambda_{jn}}(|\beta_j|), \tag{1}$$

where $p_{\lambda_{jn}}(\cdot)$ is a given nonnegative penalty function, and λ_{jn}s are regularization parameters, which may depend on n and can be chosen by a data-driven criterion, such as cross-validation (CV) and generalized cross-validation (GCV, Craven & Wahba (1979)). Minimizing (1) yields a penalized least squares estimator. It is worth to note that the penalty functions $p_{\lambda_{jn}}(\cdot)$ in (4) are not necessarily the same for all j. For example, one may wish to keep important predictors in a parametric model and hence not be willing to penalize their corresponding parameters. For simplicity of presentation, we will assume that the penalty functions for all coefficients are the same, denoted by $p_{\lambda_n}(\cdot)$. Extensions to the case with different thresholding functions do not involve any extra difficulties.

Many variable selection criteria can be derived from the above penalized least squares. Take the penalty function to be the L_0 *penalty*, namely, $p_{\lambda_n}(|\beta|) = \frac{1}{2}\lambda_n^2 I(|\beta| \neq 0)$, where $I(\cdot)$ is the indicator function. Note that $\sum_{j=1}^{d} I(|\beta_j| \neq 0)$ equals the number of nonzero regression coefficients in the model. Hence many popular variable selection criteria can be derived from (1) with the L_0 penalty by choosing different values of λ_n. For instance, the C_p (Mallows (1973)), AIC (Akaike (1974)), and BIC (Schwarz (1978)) correspond to $\lambda_n = \sqrt{2}(\sigma/\sqrt{n})$, $\sqrt{2}(\sigma/\sqrt{n})$ and $\sqrt{\log n}(\sigma/\sqrt{n})$, respectively, although these criteria were motivated from different principles.

Since the L_0 penalty is discontinuous, it requires an exhaustive search over all possible subsets of predictors to find the solution. That is, the algorithm must find the best subset of J predictors for each J in $1, ..., d$, and then choose J to optimize (1). This approach is very expensive in computational cost. Furthermore, the best subset selection suffers from other drawbacks, the most severe of which is its lack of stability as analyzed, for instance, by Breiman (1996).

To avoid the drawbacks of the best subset selection, expensive computational cost and the lack of stability, Tibshirani (1996) proposed the LASSO, which can be viewed as the solution of (1) with the L_1 *penalty*, defined by $p_{\lambda_n}(|\beta|) = \lambda_n|\beta|$. He further demonstrated that LASSO retains the virtues of both best subset selection and ridge regression. Frank & Friedman (1993) considered the L_q penalty, $p_{\lambda_n}(|\beta|) = \lambda_n|\beta|^q$, $(0 < q < 1)$, which yields a "bridge regression". The nonnegative garrote (Breiman (1995)) is in the same spirit as bridge regression. Efron, Hastie, Johnstone & Tibshirani (2004) further provides deep insights into procedures of the LASSO and the least angle regression. The issue of selection penalty function has been studied in depth by various authors, for instance, Antoniadis & Fan (2001). Fan & Li (2001) suggested the use of the smoothly clipped absolute deviation (SCAD) penalty, defined by

$$p'_{\lambda_n}(\beta) = \lambda_n\{I(\beta \leq \lambda_n) + \frac{(a\lambda_n - \beta)_+}{(a-1)\lambda_n} I(\beta > \lambda_n)\} \text{ for some } a > 2 \quad \text{and } \beta > 0,$$

with $p_{\lambda_n}(0) = 0$. This penalty function involves two unknown parameters λ_n and a. Justifying from a Bayesian statistical point of view, Fan & Li (2001) suggested using $a = 3.7$. The Bayes risk cannot be reduced much with other choices of a, and simultaneous data-driven selection of a and λ_n does not have any significant improvements from our experience. Figure 1 depicts the plots of the SCAD, $L_{0.5}$ and L_1 penalty functions.

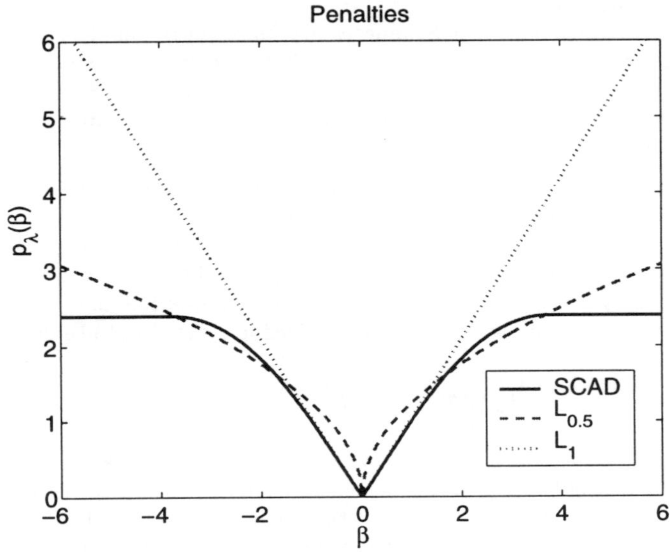

Fig. 1. *Plot of Penalty Functions*

As shown in Figure 1, the three penalty functions all are singular at the origin. This is a necessary condition for sparsity in variable selection: the resulting estimator automatically sets some small coefficients to be zero (Antoniadis & Fan (2001)). Furthermore, the SCAD and $L_{0.5}$ penalties are nonconvex over $(0, +\infty)$ in order to reduce estimation bias. We refer to penalized least squares with the nonconvex penalties over $(0, \infty)$ as *nonconvex penalized least squares* in order to distinguish from the L_2 penalty, which yields a ridge regression. The SCAD is an improvement over the L_0-penalty in two aspects: saving computational cost and resulting in a continuous solution to avoid unnecessary modeling variation. Furthermore, the SCAD improves bridge regression by reducing modeling variation in model prediction. Although similar in spirit to the L_1-penalty, the SCAD may also improve the L_1-penalty by avoiding excessive estimation bias because the solution of the L_1-penalty could shrink all regression coefficients by a constant, for instance, the soft thresholding rule (Donoho & Johnstone (1994) and Tibshirani (1996)).

Antoniadis & Fan (2001) and Fan & Li (2001) discussed extensively the choice of the penalty functions. They gave necessary conditions for the penalty function such that penalized least squares estimators to possess the following three desired properties. (i) **Sparsity**: The coefficients of insignificant variables should be estimated as zero. This achieves the purpose of the variable selection. (ii) **Continuity**: The estimated coefficients should be continuous in data to enhance the model stability. This avoids unnecessary variation in the prediction. (iii) **Unbi-**

asedness: When the true coefficients are large, they should be estimated asymptotically unbiasedly. This avoids unnecessary biases in the model selection steps. Antoniadis & Fan (2001) and Fan & Li (2001) gave several useful penalty functions that possess these three conditions. This includes the SCAD. Of course, the class of penalty functions satisfied the aforementioned three properties are infinitely many.

The discussion so far has assumed that y is continuous. When the response Y is discrete, such as binary output and count data, generalized linear models (McCullagh & Nelder (1989) may be used to fit the data. The penalized least squares approach can be adopted to this setting. Conditioning on $\mathbf{x}_i$, suppose that y_i has a density $f_i\{g(\mathbf{x}_i^T\boldsymbol{\beta}), y_i\}$, where g is a known link function. Let $\ell_i = \log f_i$ denote the conditional log-likelihood of y_i. Define a penalized likelihood as

$$\sum_{i=1}^{n} \ell_i(g(\mathbf{x}_i^T\boldsymbol{\beta}), y_i) - n\sum_{j=1}^{d} p_{\lambda_n}(|\beta_j|). \tag{2}$$

Maximizing the penalized likelihood results in a penalized likelihood estimator. The penalized likelihood with a nonconvex penalty over $(0, +\infty)$ is referred to as *nonconcave penalized likelihood*. For certain penalties, such as the SCAD, the selected model based on the nonconcave penalized likelihood satisfies $p_{\lambda_n}(|\beta_j|) = 0$ for certain β_j's. Therefore, model estimation is performed at the same time as model selection. Because the nonconcave penalized likelihood selects variables and estimates parameters simultaneously, this allows us to establish the sampling properties of the resulting estimators. Under certain regularity conditions, Fan & Li (2001) demonstrated how the rates of convergence for the penalized likelihood estimators depend on the regularization parameter λ_n. They further showed that the penalized likelihood estimators perform as well as the oracle procedure in terms of selecting the correct model, when the regularization parameter is appropriately chosen. In practice, a data-driven approach to selecting the regularization parameter is recommended. In Section 5, we present a data-driven method for choosing λ_n using the generalized cross-validation. The optimization of the nonconcave penalized likelihood can be accomplished by the modified Newton-Raphson algorithm with local quadratic approximations (LQA) to the penalty function (Fan & Li (2001)). The local quadratic approximation algorithm is also given in Section 5.

2.2 Parametric models in survival data analysis

The penalized likelihood approach can be directly applied for parametric models in survival analysis. Let T, C and $\mathbf{x}$ be respectively the survival time, the censoring time and their associated covariates. Correspondingly, let $Z = \min\{T, C\}$ be the observed time and $\delta = I(T \le C)$ be the censoring indicator. It is assumed that T and C are conditionally independent given $\mathbf{x}$ and that the censoring mechanism is noninformative. When the observed data $\{(\mathbf{x}_i, Z_i, \delta_i) : i = 1, \cdots, n\}$ is an independently and identically distributed random sample from a certain population

$(\mathbf{x}, Z, \delta)$, a complete likelihood of the data is given by

$$L = \prod_u f(Z_i|\mathbf{x}_i) \prod_c \overline{F}(Z_i|\mathbf{x}_i) = \prod_u h(Z_i|\mathbf{x}_i) \prod_{i=1}^n \overline{F}(Z_i|\mathbf{x}_i), \tag{3}$$

where the subscripts c and u denote the product of the censored and uncensored data respectively, and $f(t|\mathbf{x})$, $\overline{F}(t|\mathbf{x})$ and $h(t|\mathbf{x})$ are the conditional density function, the conditional survival function and the conditional hazard function of T given $\mathbf{x}$. Statistical inference in this paper will be based on the likelihood function (3).

In the reminder of this section, we illustrate how to extend the penalized likelihood approach for parametric survival models. Here we focus on accelerated life models, which is one of the most popular parametric life models (Badgonavičius & Nikulin (2002)). The proposed procedure is ready for applying to other parametric models. The accelerated life models use a linear regression model to fit $\log(T)$, the natural logarithm of T. In other words, the accelerated life models consider

$$\log(T) = \mu + \mathbf{x}^T \boldsymbol{\beta} + \varepsilon. \tag{4}$$

Different choices for the error distribution of ε yields different regression models. Let $\overline{F}_0(t)$ denote the survival function of T when $\mathbf{x} = 0$, i.e., $\overline{F}_0(t)$ is the survival function of $\exp(\mu + \varepsilon)$. Then

$$\overline{F}(t|\mathbf{x}) = P\{T > t|\mathbf{x}\} = \overline{F}_0\{t \exp(-\mathbf{x}^T \boldsymbol{\beta})\}.$$

Furthermore, with $h_0(\cdot)$ being the hazard risk of $\overline{F}_0(\cdot)$,

$$h(t|\mathbf{x}) = h_0\{t \exp(-\mathbf{x}^T \boldsymbol{\beta})\} \exp(-\mathbf{x}^T \boldsymbol{\beta}).$$

Using (3), the log-likelihood of the observed data $\{(\mathbf{x}_i, Z_i, \delta_i) : i = 1, \cdots, n\}$ is

$$\ell_a(\boldsymbol{\beta}, \boldsymbol{\theta}) = \sum_u \left(-\mathbf{x}_i^T \boldsymbol{\beta} + \log[h_0\{Z_i \exp(-\mathbf{x}_i^T \boldsymbol{\beta})\}]\right) + \sum_{i=1}^n \log[\overline{F}_0\{Z_i \exp(-\mathbf{x}_i^T \boldsymbol{\beta})\}], \tag{5}$$

where $\boldsymbol{\theta}$ consists of the unknown parameter involved in the distribution of $\mu + \varepsilon$. Thus, a penalized likelihood for the accelerated life model is

$$\ell_a(\boldsymbol{\beta}, \boldsymbol{\theta}) - n \sum_{j=1}^d p_{\lambda_n}(|\beta_j|). \tag{6}$$

Maximizing (6) yields a penalized likelihood estimator for $\boldsymbol{\beta}$ and $\boldsymbol{\theta}$. With a proper choice of p_{λ_n}, many of estimated coefficients will be zero and hence their corresponding variables do not appear in the model. This achieves the objectives of variable selection.

Example 1. Let the error distribution in (4) be $N(0, \sigma^2)$. This yields a log-normal regression model. Then the survival function of T when $\mathbf{x} = 0$ is

$$\overline{F}_0(t) = 1 - \Phi\left(\frac{\log(t) - \mu}{\sigma}\right),$$

where $\Phi(\cdot)$ is the cumulative distribution of the standard normal distribution. Furthermore, the hazard function when $\mathbf{x} = 0$ is

$$h_0(t) = \frac{\exp\{-(\log(t) - \mu)^2/(2\sigma^2)\}}{t\sqrt{2\pi}\sigma\left\{1 - \Phi\left(\frac{\log(t)-\mu}{\sigma}\right)\right\}}.$$

Plugging $\overline{F}_0(t)$ and $h_0(t)$ into (5), we can derive a closed form for the log-likelihood function. In this example, $\boldsymbol{\theta} = (\mu, \sigma^2)^T$.

Example 2. In this example, we consider the error distribution in (4) to be an extreme value distribution with the following density function

$$f(\varepsilon) = \alpha \exp\{\alpha\varepsilon - \exp(\alpha\varepsilon)\}.$$

The regression model (4) becomes a Weibull regression. By some straightforward calculation, we have

$$\overline{F}_0(t) = \exp(-\nu t^\alpha), \quad \text{and} \quad h_0(t) = \alpha\nu t^{\alpha-1},$$

where $\nu = \exp(-\alpha\mu)$. In this example,

$$h(t|\mathbf{x}) = \alpha\nu t^{\alpha-1}\exp(-\alpha\mathbf{x}^T\boldsymbol{\beta})$$

which is a proportional hazard model. Substituting $\overline{F}_0(t)$ and $h_0(t)$ into (5), the log-likelihood function of the collected data is

$$\ell_a(\boldsymbol{\beta}, \boldsymbol{\theta}) = \sum_u \{-\alpha\mathbf{x}_i^T\boldsymbol{\beta} + \log(\alpha\nu) + (\alpha - 1)\log(Z_i)\} - \nu\sum_{i=1}^n Z_i^\alpha \exp(-\alpha\mathbf{x}_i^T\boldsymbol{\beta}).$$

where $\boldsymbol{\theta} = (\nu, \alpha)^T$. Maximizing

$$\ell_a(\boldsymbol{\beta}, \boldsymbol{\theta}) - n\sum_{j=1}^d p_{\lambda_n}(|\beta_j|).$$

yields a penalized maximum likelihood estimate for $(\boldsymbol{\beta}, \boldsymbol{\theta})$.

Example 3. Take the error distribution in (4) to be logistic distribution with density

$$f(\varepsilon) = \frac{\exp(\alpha\varepsilon)}{\sigma\{1 + \exp(\alpha\varepsilon)\}^2}.$$

Then model (4) becomes a log-logistic regression model. It follows that

$$\overline{F}_0(t) = \frac{1}{1 + \nu t^\alpha}, \quad \text{and} \quad h_0(t) = \frac{\alpha\nu t^{\alpha-1}}{1 + \nu t^\alpha},$$

where $\nu = \exp(-\alpha\mu)$. In this example,

$$h(t|\mathbf{x}) = \frac{\alpha\nu t^{\alpha-1}\exp(-\alpha\mathbf{x}^T\boldsymbol{\beta})}{1+\nu t^{\alpha}\exp(-\alpha\mathbf{x}^T\boldsymbol{\beta})}.$$

A closed form for the log-likelihood function can be derived by using the explicit expression of $\overline{F}_0(t)$ and $h_0(t)$. In this example, $\boldsymbol{\theta} = (\lambda, \alpha)^T$.

3 Variable selection for Cox's models

The Cox proportional hazard model assumes

$$h(t|\mathbf{x}) = h_0(t)\exp(\mathbf{x}^T\boldsymbol{\beta}), \tag{7}$$

where the baseline hazard function $h_0(t)$ is an unspecified function. To present explicitly the likelihood function of the observed data $\{(\mathbf{x}_i, Z_i, \delta_i) : i = 1, \cdots, n\}$ from Cox's proportional hazards model, more notation is needed. Let $t_1^0 < \cdots < t_N^0$ denote the ordered observed failure times. Let (j) provide the label for the item falling at t_j^0 so that the covariates associated with the N failures are $\mathbf{x}_{(1)}, \cdots, \mathbf{x}_{(N)}$. Let R_j denote the risk set right before the time t_j^0 : $R_j = \{i : Z_i \geq t_j^0\}$. The likelihood in (3) becomes

$$L = \prod_{i=1}^{N} h_0(Z_{(i)})\exp(\mathbf{x}_{(i)}^T\boldsymbol{\beta})\prod_{i=1}^{n}\exp\{-H_0(Z_i)\exp(\mathbf{x}_i^T\boldsymbol{\beta})\},$$

where $H_0(\cdot)$ is the cumulative baseline hazard function. The corresponding penalized log-likelihood function is

$$\sum_{i=1}^{N}[\log\{h_0(Z_{(i)})\} + \mathbf{x}_{(i)}^T\boldsymbol{\beta}] - \sum_{i=1}^{n}\{H_0(Z_i)\exp(\mathbf{x}_i^T\boldsymbol{\beta})\} - n\sum_{j=1}^{d}p_{\lambda_n}(|\beta_j|). \tag{8}$$

Since the baseline hazard and cumulative hazard functions are unknown and have not been parameterized, the penalized log-likelihood function (8) is not ready for optimization yet. Following Breslow's idea, consider the "least informative" nonparametric modeling for $H_0(\cdot)$, in which $H_0(t)$ has a possible jump h_j at the observed failure time t_j^0. More precisely, let $H_0(t) = \sum_{j=1}^{N} h_j I(t_j^0 \leq t)$. Then

$$H_0(Z_i) = \sum_{j=1}^{N} h_j I(i \in R_j). \tag{9}$$

Using (9), the logarithm of penalized likelihood function of (8) becomes

$$\sum_{j=1}^{N}\{\log(h_j) + \mathbf{x}_{(j)}^T\boldsymbol{\beta}\} - \sum_{i=1}^{n}\{\sum_{j=1}^{N} h_j I(i \in R_j)\exp(\mathbf{x}_i^T\boldsymbol{\beta})\} - n\sum_{j=1}^{d}p_{\lambda_n}(|\beta_j|). \tag{10}$$

Taking the derivative with respect to h_j and setting it to be zero, we obtain that

$$\hat{h}_j = \{\sum_{i \in R_j} \exp(\mathbf{x}_i^T \boldsymbol{\beta})\}^{-1}. \tag{11}$$

Substituting $\hat{h}_j$ into (10), we get the penalized partial likelihood

$$\sum_{j=1}^{N} [\mathbf{x}_{(j)}^T \boldsymbol{\beta} - \log\{\sum_{i \in R_j} \exp(\mathbf{x}_i^T \boldsymbol{\beta})\}] - n \sum_{j=1}^{d} p_{\lambda_n}(|\beta_j|) \stackrel{\text{def}}{=} \ell_c(\boldsymbol{\beta}) - n \sum_{j=1}^{d} p_{\lambda_n}(|\beta_j|), \tag{12}$$

after dropping a constant term "$-N$". When $p_\lambda(\cdot) \equiv 0$, (12) is the partial likelihood function (Cox (1975)). Thus, the penalized likelihood indeed is the penalized partial likelihood. The penalized likelihood estimate of $\boldsymbol{\beta}$ is obtained via maximizing (12) with respect to $\boldsymbol{\beta}$.

Numerical comparison in Fan & Li (2002) shows that the SCAD performs as well as the oracle estimate, and outperforms the penalized likelihood with the L_1 penalty and the best subset selection with the BIC. This oracle property is further demonstrated by the following asymptotic formulation.

Let $\boldsymbol{\beta}_0$ be the true value of $\boldsymbol{\beta}_0$ and $\boldsymbol{\beta}_0 = (\beta_{10}, \cdots, \beta_{d0})^T = (\boldsymbol{\beta}_{10}^T, \boldsymbol{\beta}_{20}^T)^T$. Without loss of generality, assume that $\boldsymbol{\beta}_{20} = \mathbf{0}$. Denote by s the number of the component of $\boldsymbol{\beta}_1$. Fan & Li (2002) first showed that under certain regularity conditions, if $\lambda_n \to 0$, then there exists a local maximizer $\hat{\boldsymbol{\beta}}$ of the SCAD penalized partial likelihood function in (12) such that $\|\hat{\boldsymbol{\beta}} - \boldsymbol{\beta}_0\| = O_P(n^{-1/2})$. They further proved that if $\lambda_n \to 0$ and $\sqrt{n}\lambda_n \to \infty$, then under certain regularity conditions, with probability tending to 1, the root n consistent local maximizer $\hat{\boldsymbol{\beta}} = (\hat{\boldsymbol{\beta}}_1^T, \hat{\boldsymbol{\beta}}_2^T)^T$ of the penalized partial likelihood in (12) with the SCAD penalty must satisfy

(i) (**Sparsity**) $\hat{\boldsymbol{\beta}}_2 = \mathbf{0}$;
(ii) (**Asymptotic normality**)

$$\sqrt{n}(\hat{\boldsymbol{\beta}}_1 - \boldsymbol{\beta}_{10}) \to N(\mathbf{0}, I_1^{-1}(\boldsymbol{\beta}_{10})),$$

where $I_1(\boldsymbol{\beta}_{10})$ is the first $s \times s$ submatrix of $I(\boldsymbol{\beta}_0)$, the Fisher information matrix of the partial likelihood.

Property (i) and (ii) is referred to as an oracle property, which provides a foundation for variable selection. The sparsity (i) indicates that $\hat{\boldsymbol{\beta}}_2 = 0$ is the same as the oracle estimator who knows in advance $\boldsymbol{\beta}_2 = 0$. Furthermore, the estimator $\hat{\boldsymbol{\beta}}_1$ shares the same sampling property as the oracle estimator, and is more efficient than the maximum partial likelihood estimator (without penalty). In other words, the SCAD possesses this oracle property. The oracle property holds not only for the SCAD, but also for a class of infinitely many penalty functions. But it does not hold for the L_1 penalty due to the excessive biases inherent to the L_1 penalty. As demonstrated in Cai, Fan, Li & Zhou (2004), the oracle property is also valid for the setting in which the number of covariates is allowed to depend on n and the number of nonzero coefficients, say s_n, tends to infinite as $n \to \infty$. See Fan & Peng (2004) for a formulation under general settings.

4 Variable selection for multivariate survival data

It is assumed for the Cox proportional hazards model that the survival times of subjects are independent. This assumption might be violated in some situations, in which the collected data are correlated. The well-known Cox model (Cox (1972)) is not valid in this situation because independence assumption among individuals is violated. Extensions of the Cox regression model to the analysis of multivariate failure time data include frailty model and marginal model. In this section, we extend the nonconcave penalized likelihood approach for the frailty model and the marginal model.

4.1 Frailty models

One popular approach to modeling correlated survival times is to use a frailty model. A frailty corresponds to a random block effect that acts multiplicatively on the hazard rates of all subjects in a group. In this section, we only consider the Cox proportional hazard frailty model, in which it is assumed that the hazard rate for the j-th subject in the i-th subgroup is

$$h_{ij}(t|\mathbf{x}_{ij}, u_i) = h_0(t) u_i \exp(\mathbf{x}_{ij}^T \boldsymbol{\beta}), \; i = 1, \cdots, n, j = 1, \cdots, J_i, \tag{13}$$

where the u_i's are associated with frailties, and they are a random sample from some population. It is frequently assumed that given the frailty u_i, the data in the i-th group are independent. The most frequently used distribution for frailty is the gamma distribution due to its simplicity. Assume without loss of generality that the mean of frailty is 1 so that all parameters involved are estimable. For the gamma frailty model, the density of u is

$$g(u) = \frac{\alpha^\alpha u^{\alpha-1} \exp(-\alpha u)}{\Gamma(\alpha)}.$$

From (3), the full likelihood of "pseudo-data" $\{(u_i, \mathbf{x}_{ij}, Z_{ij}, \delta_{ij}) : i = 1, \cdots, n, j = 1 \cdots, J_i\}$ is

$$\prod_{i=1}^{n} \prod_{j=1}^{J_i} [\{h(z_{ij}|\mathbf{x}_{ij}, u_i)\}^{\delta_{ij}} \overline{F}(z_{ij}|\mathbf{x}_{ij}, u_i)] \prod_{i=1}^{n} g(u_i).$$

Integrating the full likelihood function with respect to $u_1, \cdots, u_n$, the likelihood of the observed data is given by

$$L(\boldsymbol{\beta}, \boldsymbol{\theta}) = \exp\{\boldsymbol{\beta}^T (\sum_{i=1}^{n} \sum_{j=1}^{J_i} \delta_{ij} \mathbf{x}_{ij})\} \prod_{i=1}^{n} \frac{\alpha^\alpha \prod_{j=1}^{J_i} \{h_0(z_{ij})\}^{\delta_{ij}}}{\Gamma(\alpha) \{\sum_{j=1}^{J_i} H_0(z_{ij}) \exp(\mathbf{x}_{ij}^T \boldsymbol{\beta}) + \alpha\}^{(A_i + \alpha)}}, \tag{14}$$

where $\boldsymbol{\theta} = (\alpha, H)$, and $A_i = \sum_{j=1}^{J_i} \delta_{ij}$. The log-likelihood of the observed data is

$$\ell_f(\boldsymbol{\beta},\boldsymbol{\theta}) = \sum_{i=1}^{n}\{\sum_{j=1}^{J_i}\delta_{ij}\log h(z_{ij}) - [(A_i+\alpha)\log\{\sum_{j=1}^{J_i}H_0(z_{ij})\exp(\mathbf{x}_{ij}^T\boldsymbol{\beta})+\alpha\}]\}$$
$$+\sum_{i=1}^{n}\{\boldsymbol{\beta}^T(\sum_{j=1}^{J_i}\delta_{ij}\mathbf{x}_{ij}) + \alpha\log\alpha - \log\Gamma(\alpha)\}$$

Therefore the logarithm of the penalized likelihood of the observed data is

$$\ell_f\{\boldsymbol{\beta}, h(\cdot)\} - n\sum_{j=1}^{d}p_\lambda(|\beta_j|). \tag{15}$$

To eliminate the nuisance parameter $h(\cdot)$, we again employ the profile likelihood method. Consider the "least informative" nonparametric modeling for $H_0(\cdot)$:

$$H_0(z) = \sum_{l=1}^{N}\lambda_l I(z_l \le z), \tag{16}$$

where $\{z_1, \cdots, z_N\}$ are pooled observed failure times.

Substituting (16) into (15), then differentiating it with respect to λ_l, $l = 1, \cdots, N$, the root of the corresponding score function should satisfy the following equations:

$$\lambda_l^{-1} = \sum_{i=1}^{n}\frac{(A_i+\alpha)\sum_{j=1}^{J_i}I(z_l\le z_{ij})\exp(\mathbf{x}_{ij}^T\boldsymbol{\beta})}{\sum_{k=1}^{N}\lambda_k\sum_{j=1}^{J_i}I(z_k\le z_{ij})\exp(\mathbf{x}_{ij}^T\boldsymbol{\beta})+\alpha} \quad \text{for } l = 1,\cdots,N. \tag{17}$$

The above solution does not admit a close form, neither does the profile likelihood function. However, the maximum profile likelihood can be implemented as follows. With initial values for $\alpha, \boldsymbol{\beta}$ and λ_l, update $\{\lambda_l\}$ from (17) and obtain the penalized profile likelihood of (15). With known $H_0(\cdot)$ defined by (16), maximize the penalized likelihood (15) with respect to $(\alpha, \boldsymbol{\beta})$, and iterate between these two steps. When the Newton-Raphson algorithm is applied to the penalized likelihood (15), it involves the first two order derivatives of the gamma function, which may not exist for certain value of α. One approach to avoid this difficulty is the use of a grid of possible values for the frailty parameter α and finding the maxima over this discrete grid, as suggested by Nielsen, Gill, Andersen & Sørensen (1992). Our simulation experience shows that the estimate of $\boldsymbol{\beta}$ is quite empirically robust to the chosen grid of possible values for α. This profile likelihood method even without the task of variable selection provides a viable alternative approach to the EM algorithm frequently used in the frailty model.

A natural initial estimator for $\boldsymbol{\beta}$ is the maximum pseudo-partial likelihood estimates of $\boldsymbol{\beta}$ ignoring possible dependency within each group. The corresponding $h_1, \cdots, h_N$ in (11) may serve as an initial estimator for $\lambda_1, \cdots, \lambda_N$. Hence given a value of α and initial values of $\boldsymbol{\beta}$ and $\lambda_1, \cdots, \lambda_N$, update the values of $\lambda_1, \cdots, \lambda_N$ and α, $\boldsymbol{\beta}$ in turn until they converge or the penalized profile likelihood fails to change substantially. The proposed algorithm avoids optimizing a high-dimensional problem. It will give us an efficient estimate for $\boldsymbol{\beta}$. The algorithm may converge slowly or even not converge. In this situation, the idea of one-step estimator (see Bickel (1975)) provides us an alternative approach.

Fan & Li (2002) assessed the finite sample performance of the resulting estimate by extensive Monte Carlo simulation. From their numerical comparisons, it can be seen that the SCAD performs almost as well as the oracle estimator in terms of model error, and it outperforms the penalized likelihood with the L_1 penalty in terms of model complexity and model error. The performance of the SCAD is similar to the best subset selection with BIC in terms of model complexity and model error, but the computational time of SCAD is dramatically less than that of the best subset selection. Under certain regularity conditions, Fan & Li (2002) showed that if $\lambda_n \to 0$ and $\sqrt{n}\lambda_n \to \infty$, then the resulting estimate of the SCAD is root n consistent, and with probability tending to one, $\hat{\boldsymbol{\beta}}_2 = 0$ and

$$\sqrt{n}(\hat{\boldsymbol{\theta}}_1 - \boldsymbol{\theta}_{10}) \to N\left\{\mathbf{0}, \tilde{I}_1^{-1}(\boldsymbol{\theta}_{10})\right\},$$

where $\tilde{I}_1(\boldsymbol{\theta}_{10})$ consists of the first $(s+1)\times(s+1)$ submatrix of $\tilde{I}_0(\boldsymbol{\theta}_{10}, \mathbf{0})$, the Fisher information matrix of the frailty model, and $\hat{\boldsymbol{\theta}}_1 = (\hat{\alpha}, \hat{\boldsymbol{\beta}}_1^T)^T$, $\boldsymbol{\theta}_{10} = (\alpha_0, \boldsymbol{\beta}_{10}^T)^T$.

4.2 Marginal Hazard Models

The interpretations of the regression coefficients in the frailty model are different from those in the Cox model. Consequently, when the correlation among the observations is not of interest, the marginal proportional hazard models have received much attention in the recent literature because they are semiparametric models and retain the virtue of the Cox model (e.g., Wei, Lin & Weisseld (1989), Lee, Wei & Amato (1992), Liang, Self & Chang (1993), Lin (1994), Cai and Prentice (1995, 1997), Cai (1999), Spiekerman & Lin (1998) and Clegg, Cai & Sen (1999) among others).

To fix notation, suppose that there are n independent clusters and each cluster has K subjects. For each subject, J types of failure may occur. For the failure time in the case of the jth type of failure on subject k in cluster i, the marginal mixed baseline hazards model is taken as

$$h_{ijk}(t|\mathbf{x}_{ijk}(t)) = h_{0j}(t)\exp\{\boldsymbol{\beta}^T\mathbf{x}_{ijk}(t)\}, \tag{18}$$

where $\boldsymbol{\beta} = (\beta_1, \cdots, \beta_d)^T$ is a vector of unknown regression coefficients, $\mathbf{x}_{ijk}(t)$ is a possibly external time-dependent covariate vector, and $h_{0j}(t)$ and $h_0(t)$ are unspecified baseline hazard functions.

The marginal model approach does not specify correlation structure for the failure times within a cluster, hence inferences are based on a pseudo-partial likelihood approach. Under a working independence assumption (Wei, Lin & Weisseld

(1989)), i.e., assuming the independence among failure times in a cluster, we obtain the logarithm of a pseudo-partial likelihood function of the observed data $\{(\mathbf{x}_{ijk}, Z_{ijk}, \delta_{ijk}) : i = 1, \cdots, n, j = 1 \cdots, J, k = 1, \cdots, K\}$ from model (18) as following:

$$\ell_p(\boldsymbol{\beta}) = \sum_{i=1}^{n}\sum_{j=1}^{J}\sum_{k=1}^{K} \delta_{ijk} \Bigg(\boldsymbol{\beta}^T \mathbf{x}_{ijk}(Z_{ijk}) - \log\left[\sum_{l=1}^{n}\sum_{g=1}^{K} Y_{ljg}(Z_{ijk}) \exp\{\boldsymbol{\beta}^T \mathbf{x}_{ljg}(Z_{ijk})\}\right]\Bigg). \tag{19}$$

where $Y(t) = I(Z \geq t)$ be the at-risk indicator. We use a penalized pseudo-partial likelihood for model (18) which is defined as

$$\mathcal{L}(\boldsymbol{\beta}) = \ell_p(\boldsymbol{\beta}) - n\sum_{j=1}^{d} p_{\lambda_n}(|\beta_j|). \tag{20}$$

Let

$$a_n = \max\{|p'_{\lambda_n}(|\beta_{j0}|)| : \beta_{j0} \neq 0\}, \quad \text{and} \quad b_n = \max\{|p''_{\lambda_n}(|\beta_{j0}|)| : \beta_{j0} \neq 0\}. \tag{21}$$

We first show that there exists a penalized pseudo-partial likelihood estimator that converges at rate $O_P(n^{-1/2} + a_n)$, and then establish the oracle property for the resulting estimator. We only state the main theoretic results here and leave the regularity conditions in the Appendix. Technical proofs are given in Cai, Fan, Li & Zhou (2004).

Theorem 1. *Under Conditions A-D in the Appendix, if both a_n and b_n tend to 0 as $n \to \infty$, then with probability tending to one, there exists a local maximizer $\hat{\boldsymbol{\beta}}$ of $\mathcal{L}(\boldsymbol{\beta})$ defined in (6) such that $\|\hat{\boldsymbol{\beta}} - \boldsymbol{\beta}_0\| = O_P(n^{-1/2} + a_n)$.*

From Theorem 1, provided that $a_n = O(n^{-1/2})$, which can be achieved by choosing proper $\lambda'_n s$, there exists a root n consistent penalized pseudo-partial likelihood estimator. Denote by

$$\Sigma = \text{diag}\{p''_{\lambda_n}(|\beta_{10}|), \cdots, p''_{\lambda_n}(|\beta_{s0}|)\},$$

and

$$\mathbf{b} = (p'_{\lambda_n}(|\beta_{10}|)\text{sgn}(\beta_{10}), \cdots, p'_{\lambda_n}(|\beta_{s0}|)\text{sgn}(\beta_{s0})).$$

Theorem 2. *Assume that the penalty function $p_{\lambda_n}(|\beta_j|)$ satisfies that*

$$\liminf_{n\to\infty} \liminf_{\beta\to 0+} p'_{\lambda_n}(\beta)/\lambda_n > 0. \tag{22}$$

If $\lambda_n \to 0$, $\sqrt{n}\lambda_n \to \infty$ and $a_n = O(n^{-1/2})$, then under the conditions of Theorem 1, with probability tending to 1, the root n consistent local maximizer $\hat{\boldsymbol{\beta}} = (\hat{\boldsymbol{\beta}}_1^T, \hat{\boldsymbol{\beta}}_2^T)^T$ in Theorem 1 must satisfy that $\hat{\boldsymbol{\beta}}_2 = 0$, and

$$\sqrt{n}\{A_{11} + \Sigma\}\{\hat{\boldsymbol{\beta}}_1 - \boldsymbol{\beta}_{10} + (A_{11} + \Sigma)^{-1}\mathbf{b}\} \to N(0, D_{11}) \tag{23}$$

in distribution, where A_{11} and D_{11} consist of the first s columns and rows of $A(\boldsymbol{\beta}_{10}, \mathbf{0})$ and $D(\boldsymbol{\beta}_{10}, \mathbf{0})$ defined in the Appendix, respectively.

5 Practical implementation

5.1 Local quadratic approximation and standard error formula

The L_q, $(0 < q \leq 1)$, and SCAD penalty functions are singular at the origin, and they do not have continuous second order derivatives. Therefore, maximizing the nonconcave penalized likelihood is challenging. Fan & Li (2001) proposed a unified algorithm for their nonconcave penalized likelihood using a local quadratic approximation for the penalty function. The unified algorithm is ready for the penalized likelihood function

$$Q(\boldsymbol{\beta}, \boldsymbol{\theta}) = \ell(\boldsymbol{\beta}, \boldsymbol{\theta}) - n \sum_{j=1}^{d} p_{\lambda_n}(|\beta_j|). \tag{24}$$

where $\ell(\boldsymbol{\beta}, \boldsymbol{\theta})$ may be the likelihood function $\ell_a(\boldsymbol{\beta}, \boldsymbol{\theta})$ in Section 2, $\ell_c(\boldsymbol{\beta})$ for the Cox model, $\ell_f(\boldsymbol{\beta}, \boldsymbol{\theta})$ for the Cox's frailty model and $\ell_p(\boldsymbol{\beta})$ for the marginal model. Although the penalty function is singular at the origin and may not have continuous 2nd order derivative. Fan & Li (2001) propose to locally approximate using a quadratic function as follows. Set the initial value to be the maximum likelihood estimate (without penalty). Under certain regularity conditions, the MLE is root n consistent, and therefore it is close to the true value. Suppose that we are given an initial value $\boldsymbol{\beta}^0$ that is close to the minimizer of (24). If β_j^0 is very close to 0, then set $\hat{\beta}_j = 0$. Otherwise they can be locally approximated by a quadratic function as

$$[p_{\lambda_n}(|\beta_j|)]' = p'_{\lambda_n}(|\beta_j|)\text{sgn}(\beta_j) \approx \{p'_{\lambda_n}(|\beta_j^0|)/|\beta_j^0|\}\beta_j,$$

when $\beta_j \neq 0$. In other words,

$$p_{\lambda_n}(|\beta_j|) \approx p_{\lambda_n}(|\beta_j^0|) + \frac{1}{2}\{p'_{\lambda_n}(|\beta_j^0|)/|\beta_j^0|\}(\beta_j^2 - \beta_j^{02}), \text{ for } \beta_j \approx \beta_j^0. \tag{25}$$

With the aid of the quadratic approximation, the maximization of $Q(\boldsymbol{\beta}, \boldsymbol{\theta})$ can be carried out by using the Newton-Raphson algorithm.When the algorithm converges, the estimator satisfies the condition

$$\frac{\partial \ell(\hat{\boldsymbol{\beta}}^0)}{\partial \beta_j} + np'_{\lambda_n}(|\hat{\beta}_j^0|)\text{sgn}(\hat{\beta}_j^0) = 0,$$

the penalized likelihood equation, for non-zero elements of $\hat{\boldsymbol{\beta}}^0$.

Following conventional techniques in the likelihood setting, we can estimate the standard error of the resulting estimate by using the sandwich formula. Specifically, the corresponding sandwich formula can be used as an estimator for the covariance of the estimates $\hat{\boldsymbol{\beta}}_1$, the non-vanishing component of $\hat{\boldsymbol{\beta}}$. That is,

$$\{\nabla^2 \ell(\hat{\boldsymbol{\beta}}_1, \hat{\boldsymbol{\theta}}) + n\Sigma_\lambda(\hat{\boldsymbol{\beta}}_1)\}^{-1} \widehat{\mathrm{cov}}\{\nabla \ell(\hat{\boldsymbol{\beta}}_1, \hat{\boldsymbol{\theta}})\}\{\nabla^2 \ell(\hat{\boldsymbol{\beta}}_1, \hat{\boldsymbol{\theta}}) - n\Sigma_\lambda(\hat{\boldsymbol{\beta}}_1)\}^{-1}, \quad (26)$$

where

$$\Sigma_\lambda(\hat{\boldsymbol{\beta}}_1) = \mathrm{diag}\{p'_\lambda(|\hat{\beta}_1|)/|\hat{\beta}_1|, \cdots, p'_\lambda(|\hat{\beta}_s|)/|\hat{\beta}_s|, 0, \cdots, 0\},$$

where the number of zeros equals the dimension of $\boldsymbol{\theta}$, and s the dimension of $\hat{\boldsymbol{\beta}}_1$.

5.2 Selection of regularization parameters

To implement the methods described in previous sections, it is desirable to have an automatic method for selecting the thresholding parameter λ involved in $p_\lambda(\cdot)$ based on data. Here we estimate λ via minimizing an approximate generalized cross-validation (GCV) statistic (Craven & Wahba (1979)). By some straightforward calculation, the effective number of parameters for $Q(\boldsymbol{\beta}, \boldsymbol{\theta})$ in (24) in the last step of the Newton-Raphson algorithm iteration is

$$e(\lambda) = \mathrm{tr}[\{\nabla^2 \ell(\hat{\boldsymbol{\beta}}, \hat{\boldsymbol{\theta}}) + \Sigma_\lambda(\hat{\boldsymbol{\beta}})\}^{-1} \nabla^2 \ell(\hat{\boldsymbol{\beta}}, \hat{\boldsymbol{\theta}})].$$

Therefore the generalized cross-validation statistic is defined by

$$\mathrm{GCV}(\lambda) = \frac{-\ell(\hat{\boldsymbol{\beta}}, \hat{\boldsymbol{\theta}})}{n\{1 - e(\lambda)/n\}^2}$$

and $\hat{\lambda} = \mathrm{argmin}_\lambda\{\mathrm{GCV}(\lambda)\}$ is selected. The minimization can be carried out by searching over a grid of points for λ.

6 An Example

We illustrate the proposed variable selection procedures by an analysis of a data set collected in the Framingham Heart Study (FHS, Dawber (1980)). In this study, multiple failure outcomes, for instance, times to coronary heart disease (CHD) and cerebrovascular accident (CVA), were observed from the same individual. In addition, as the primary sampling unit was the family, failure times recorded are likely to be dependent for the individuals within a family.

For simplicity, we consider only time to obtain first evidence of CHD and of CVA, and analyze only data for participants in the FHS study who had an examination at age 44 or 45 and were disease-free at that examination. By disease-free we mean that there exists no history of hypertension or glucose intolerance and no previous experience of a CHD or CVA. The time origin is the time of

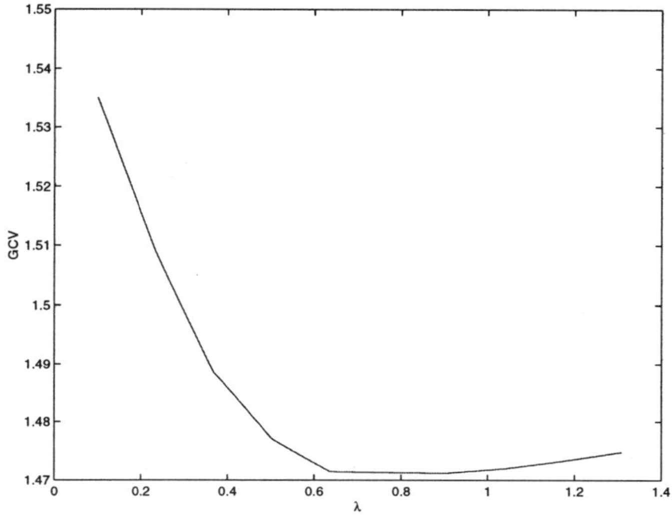

Fig. 2. *Plot of Generalized Cross-Validation for the Framingham Heart Study Analysis*

the examination at which an individual participated in the study and the follow up information is up to year 1980. The risk factors of interest are: body mass index (BMI), denoted by x_1, cholesterol level (x_2), systolic blood pressure (x_3), smoking status (x_4), coded by 1 if this individual is smoking, and 0 otherwise, gender (x_5), coded by 1 for female and 0 for male. The values of risk factors were taken from the biennial examination at which an individual was included in the sample. Because some individuals were in the study several years prior to inclusion into the data set, the waiting time, denoted by x_6, from entering the study to reaching 44 or 45 years of age was used as a covariate to account for the cohort effect. Since x_1, x_2, x_3 and x_6 are continuous covariates, they are standardized individually in our analysis.

To explore possible nonlinear effects and interaction effects of the risk factors, we include all main effects, quadratic effects and interaction effects of the risk factors and covariates in the full model. This results in a mixed baseline hazard model with 50 covariates:

$$h_{ijk}(t, \mathbf{x}_{ijk}) = h_{0j}(t) \exp\{\boldsymbol{\beta}_j^T \mathbf{x}_{ijk}\}, \tag{27}$$

where $\mathbf{x}_{ijk}$ consists of all possible linear, quadratic and interaction terms of the risk factors and covariates x_1 to x_6. Model (27) allows different baseline hazards and different regression coefficients for CHD and CVA, but an identical baseline

Table 1. Estimated Coefficients and Standard Errors for the FHS data

	CHD	CVA		CHD	CVA
Effect	$\hat{\beta}$(SE($\hat{\beta}$))	$\hat{\beta}$(SE($\hat{\beta}$))	Effect	$\hat{\beta}$(SE($\hat{\beta}$))	$\hat{\beta}$(SE($\hat{\beta}$))
x_1	0.0810 (0.1288)	0.4773 (0.2423)	$x_1 * x_5$	0 (−)	0 (−)
x_2	0.0576 (0.1200)	-0.2409 (0.2655)	$x_1 * x_6$	-0.1060 (0.0808)	0 (−)
x_3	0.4129 (0.1570)	0.2917 (0.1477)	$x_2 * x_3$	0 (−)	0 (−)
x_4	0.4754 (0.2361)	0.7077 (0.3587)	$x_2 * x_4$	0.1550 (0.1425)	0.5702 (0.3766)
x_5	-0.3666 (0.2543)	-0.1016 (0.2890)	$x_2 * x_5$	0 (−)	0 (−)
x_6	0.0249 (0.0802)	-0.1395 (0.1916)	$x_2 * x_6$	0 (−)	0 (−)
$x_1 2$	-0.0743 (0.0512)	0 (−)	$x_3 * x_4$	-0.1952 (0.1489)	0 (−)
$x_2 2$	0 (−)	-0.0768 (0.1052)	$x_3 * x_5$	-0.2054 (0.1378)	0 (−)
$x_3 2$	0 (−)	0 (−)	$x_3 * x_6$	0 (−)	0 (−)
$x_6 2$	0 (−)	0.2062 (0.1229)	$x_4 * x_5$	-0.3071 (0.3106)	0 (−)
$x_1 * x_2$	0 (−)	0 (−)	$x_4 * x_6$	0 (−)	0 (−)
$x_1 * x_3$	0 (−)	-0.2224 (0.1435)	$x_5 * x_6$	0 (−)	0.5753 (0.2545)
$x_1 * x_4$	0.1409 (0.1495)	-0.2207 (0.2628)			

hazards for siblings. A thorough analysis of this data set was also given in Cai, Fan, Li & Zhou (2004).

The maximum pseudo-partial likelihood estimate for $\boldsymbol{\beta}$ is computed. The natural logarithm of the pseudo-partial likelihood for the full model of 50 covariates is -2017.9590. Next we apply the SCAD procedure to model (7) to select significant variables. In the implementation of the SCAD procedure, since all covariates are important confounding variables, we included them in the model by not penalizing the linear main effect of x_1 to x_6. Thus, all linear effects are included in the selected model. The GCV method is used to select the regularization parameter. Figure 2 depicts the plot of GCV score versus λ. The regularization parameter λ equals 0.9053, selected by minimizing the GCV scores. The resulting estimate and standard error for $\boldsymbol{\beta}$ in the selected model is depicted in Table 1. The logarithm of the pseudo-partial likelihood for the model selected by the SCAD with the selected λ is -2022.6635. This represents an increase of the logarithm of the pseudo-partial likelihood by 10.1923 from that of the full model, which is much less than 25, the number of covariates excluded from the full model. From extension of Theorem 3 of Cai (1999), the limiting distribution of the pseudo-partial likelihood ratio statistic is a weighted sum of Chi-square distributions with 1 degree of freedom. Based on 100,000 Monte Carlo simulations, we computed the p-value, which equals 0.9926. This is in favor of the selected model. We further compared the selected model by SCAD with the one selected by the naive approach. The corresponding pseudo-partial likelihood ratio statistic is 90.4989 with p-value 0.0000 obtained by 100,000 Monte Carlo simulations. This is also in favor of the selected model by SCAD.

In another confirmation of the selected model, we compare the selected model with the linear main effect model which include only all the linear main effects of x_1 to x_6. The maximum pseudo-partial likelihood estimate for the unknown regression coefficients is computed, and the natural logarithm of the pseudo-partial likelihood for the linear main effect model is -2034.6527. The pseudo-partial likelihood ratio statistic for testing H_0: the linear main effect model versus H_1: the selected model, is 23.9783. Based on 100,000 Monte Carlo simulations, the corresponding p-value equals 0.0353. This indicates that the selected model fits the data better than the model with only the linear main effects.

Acknowledgements

Fan's research was supported by an NSF grant DMS-0354223 and an NIH grant R01 HL69720. G. Li's research was partially supported by NIH grant CA78314-03. R. Li's research was supported by an NSF grant DMS-0348869 and a National Institute on Drug Abuse (NIDA) Grant 1-P50-DA10075.

Appendix: Regularity Conditions

To facilitate the notation, let $N_{ijk}(t) = I(Z_{ijk} \le t, \delta_{ijk} = 1)$ be the counting process, and $h_{ijk}(t)$ and $M_{ijk}(t) = N_{ijk}(t) - \int_0^t Y_{ijk}(u)h_{ijk}(u)\,du$ be their corresponding marginal hazards function and marginal martingale, respectively, with respect to the filtration $\mathcal{F}_{jk}(t^-)$, where $\mathcal{F}_{jk}(t)$ is the σ-field generated by $\{N_{ijk}(u), Y_{i11}(u), ..., Y_{iJK}(u), \mathbf{x}_{i11}(u), ..., \mathbf{x}_{iJK}(u); 0 \le u \le t, i = 1, ..., n\}$. Define

$$\begin{aligned}
\mathbf{S}_{jk}^{(d)}(\boldsymbol{\beta};t) &= \frac{1}{n}\sum_{i=1}^{n} Y_{ijk}(t)\mathbf{x}_{ijk}(t)^{\otimes d}\exp\{\boldsymbol{\beta}^T\mathbf{x}_{ijk}(t)\}, \quad d=0,1,2 \\
\mathbf{S}_{j\cdot}^{(d)}(\boldsymbol{\beta};t) &= \sum_{k=1}^{K}\mathbf{S}_{jk}^{(d)}(\boldsymbol{\beta};t), \quad d=0,1,2, \\
\mathbf{E}_j(\boldsymbol{\beta};t) &= \mathbf{S}_{j\cdot}^{(1)}(\boldsymbol{\beta};t)/\mathbf{S}_{j\cdot}^{(0)}(\boldsymbol{\beta};t), \\
\mathbf{V}_j(\boldsymbol{\beta};t) &= \mathbf{S}_{j\cdot}^{(2)}(\boldsymbol{\beta},t)/\mathbf{S}_{j\cdot}^{(0)}(\boldsymbol{\beta},t) - \mathbf{E}_j(\boldsymbol{\beta};t)^{\otimes 2},
\end{aligned}$$

where $\mathbf{a}^{\otimes 0} = 1$, $\mathbf{a}^{\otimes 1} = 1$, and $\mathbf{a}^{\otimes 2} = \mathbf{a}\mathbf{a}^T$ for a vector $\mathbf{a}$.

Regularity conditions:

(A) For simplicity, assume that T_{ijk} takes values on a finite interval $[0, \tau]$, and $\int_0^\tau h_{0j}(t)\,dt < \infty$ for $j = 1, \cdots, J$.

(B) There exists a neighborhood $\mathcal{B}$ of the true value $\boldsymbol{\beta}_0$ that satisfies each of the following conditions: (1) there exists a scalar, vector, and matrix function $\mathbf{s}_{jk}^{(d)}(\boldsymbol{\beta}, t)$ $(d = 0, 1, 2)$ defined on $\mathcal{B} \times [0, \tau]$ such that $\sup_{t\in[0,\tau],\boldsymbol{\beta}\in\mathcal{B}} \|\mathbf{S}_{jk}^{(d)}(\boldsymbol{\beta}, t) - \mathbf{s}_{jk}^{(d)}(\boldsymbol{\beta}, t)\| \to 0$ in probability; (2) there exists a matrix $\mathbf{D} = \mathbf{D}(\boldsymbol{\beta})$ such that

$$\frac{1}{n}\sum_{i=1}^{n} \mathrm{var}(\mathbf{D}_i) \to \mathbf{D},$$

where

$$\mathbf{D}_i = \sum_{j=1}^{J}\sum_{k=1}^{K}\int_0^\tau \{\mathbf{x}_{ijg}(t) - \mathbf{e}_j(\boldsymbol{\beta}_0; t)\}\, dM_{ijk}(t),$$

and $\mathbf{e}_j(\boldsymbol{\beta}; t) = \left\{\sum_{k=1}^{K} \mathbf{s}_{jk}^{(2)}(\boldsymbol{\beta}; t)\right\} / \left\{\sum_{k=1}^{K} \mathbf{s}_{jk}^{(0)}(\boldsymbol{\beta}; t)\right\}$.

(C) Let $\mathbf{s}_{jk}^{(d)}$, $d = 0, 1, 2$, $\mathcal{B}$ and $\mathbf{e}_j$ be as in Condition (B) and define $\mathbf{v}_j = \left\{\sum_{k=1}^{K} \mathbf{s}_{jk}^{(2)}(\boldsymbol{\beta}, t)\right\} / \left\{\sum_{k=1}^{K} \mathbf{s}_{jk}^{(0)}(\boldsymbol{\beta}, t)\right\} - \mathbf{e}_j(\boldsymbol{\beta}; t)^{\otimes 2}$. Then for all $\boldsymbol{\beta} \in \mathcal{B}$, $t \in [0, \tau]$, $j = 1, \ldots, J$ and $k = 1, \ldots, K$: $\mathbf{s}_{jk}^{(1)}(\boldsymbol{\beta}; t) = \partial \mathbf{s}_{jk}(\boldsymbol{\beta}; t)/\partial\boldsymbol{\beta}$ and $\mathbf{s}_{jk}^{(2)}(\boldsymbol{\beta}; t) = \partial \mathbf{s}_{jk}^{(1)}(\boldsymbol{\beta}; t)/\partial\boldsymbol{\beta}$. Assume $\mathbf{s}_{jk}^{(0)}(\boldsymbol{\beta}; t)$ is bounded away from 0 on $\mathcal{B} \times [0, \tau]$, and the matrix $\sum_{j=1}^{J} \int_0^\tau \mathbf{v}_j(\boldsymbol{\beta}_0; t) \sum_{k=1}^{K} \mathbf{s}_{jk}^{(0)}(\boldsymbol{\beta}_0; t) h_{0j}(t)\,dt$ is positive definite.

(D) In probability

$$\frac{1}{n}\sum_{i=1}^{n} E\{\|\mathbf{D}_i\|^2 I(\|\mathbf{D}_i\| > \varepsilon n^{1/2})\} \to 0.$$

These conditions are adapted from Clegg, Cai & Sen (1999) and guarantee the asymptotic normality of the pseudo-partial likelihood estimator, the maximizer of $\ell(\boldsymbol{\beta})$ defined in (5). Under these conditions, there exists a sequence $\boldsymbol{\beta}_n \to \boldsymbol{\beta}_0$ as $n \to \infty$.

References

Akaike, H. (1974). A new look at the statistical model identification. *IEEE Trans. on Automatic Control* **19** 716–723.

Antoniadis, A. & Fan, J. (2001). Regularization of wavelets approximations (with discussions). *J. Amer. Statist. Assoc.* **96** 939–967.

Badgonavičius, V. & Nikulin, M. (2002). *Accelerated Life Models: Modeling & Statistical Analysis.* Chapman and Hall, New York.

Bickel, P. J. (1975). One-step Huber estimates in the linear model. *Jour. Ameri. Statist. Assoc.* **70** 428–433.

Breiman, L. (1995). Better subset regression using the nonnegative garrote. *Technometrics* **37** 373–384.

Breiman, L. (1996). Heuristics of instability and stabilization in model selection. *Ann. Statist.* **24** 2350–2383.

Cai, J. (1999). Hypothesis testing of hazard ratio parameters in marginal models for multivariate failure time data. *Lifetime Data Analysis* **5** 39–53.

Cai, J., Fan, J., Li, R. & Zhou, H. (2004). Variable Selection for Multivariate Failure Time Data. Biometrika. In press.

Cai, J. & Prentice, R. L. (1995). Estimating equations for hazard ratio parameters based on correlated failure time data. *Biometrika* **82** 151–164.

Cai, J. & Prentice, R. L. (1997). Regression estimation using multivariate time data & a common baseline hazard function model. *Lifetime Data Analysis* **3** 197–213.

Clegg, L. X., Cai, J. & Sen, P. K. (1999). A marginal mixed baseline hazards model for multivariate failure time data. *Biometrics* **55** 805–812.

Cox, D. R. (1972). Regression models and life tables (with discussion). *Jour. Roy. Statist. Soc. Ser. B* **34** 187–220.

Cox, D. R. (1975). Partial likelihood. *Biometrika* **62** 269–276.

Craven, P. & Wahba, G. (1979). Smoothing noisy data with spline functions: estimating the correct degree of smoothing by the method of generalized cross–validation. *Numer. Math.* **31** 377–403.

Dawber, T. R. (1980). *The Framingham Study, The Epidemiology of Atherosclerotic Disease*, Cambridge, MA, Harvard University Press.

Donoho, D. L. & Johnstone, I. M. (1994). Ideal spatial adaptation by wavelet shrinkage. *Biometrika* **81** 425–455.

Efron, B., Hastie, T., Johnstone, I. & Tibshirani, R. (2004). Least angle regression (with discussions). *Ann. Statist.* **32** 409–499.

Fan, J. & Li, R. (2001). Variable selection via nonconcave penalized likelihood and its oracle properties. *J. Amer. Statist. Assoc.* **96** 1348–1360.

Fan, J. & Li, R. (2002). Variable selection for Cox's proportional hazards model and frailty model. *The Annals of Statistics* **30** 74–99.

Fan, J. & Peng, H. (2004). Nonconcave penalized likelihood with a diverging number of parameters. *Annals of Statistics* **32** 928–961.

Faraggi, D. (1998). Bayesian variable selection method for censored survival data. *Biometrics* **54** 1475–1485.

Faraggi, D. & Simon, R. (1997). Large sample Bayesian inference on the parameters of the proportional hazard models. *Statistics in Medicine* **16** 2573–2585.

Frank, I. E. & Friedman, J. H. (1993). A statistical view of some chemometrics regression tools. *Technometrics* **35** 109–148.

Giudici, P., Mezzetti, M. & Muliere, P. (2003). Mixtures of products of Dirichlet processes for variable selection in survival analysis. *Journal of Statistical Planning and Inference* **111** 101–115.

Ibrahim, J. G. & Chen, M. H. (2000). Power prior distributions for regression models. *Statistical Sciences* **15** 46–60.

Ibrahim, J. G., Chen, M. H. & MacEachern, S. N. (1999). Bayesian variable selection for proportional hazards models. *Canadian Journal of Statistics* **27** 701–717.

Ibrahim, J. G., Chen, M. H. & Sinha, D. (2001). Bayesian variable selection for proportional hazards model. *Statistica Sinica* **11** 419–443.

Lee, E. W., Wei, L. J. & Amato, D. A. (1992). Cox-type regression analysis for large numbers of small groups of correlated failure time observations. In *Survival Analysis: State of the Art* (J. P. Klein & P. Goel, eds.) 237–248. Boston: Kluwer Academic Publishers.

Liang, K. Y., Self, S. G. & Chang, Y. C. (1993). Modelling marginal hazards in multivariate failure time data. *J. Royal Statist. Soc., Ser. B* **55** 441–453.

Lin, D. Y. (1994). Cox regression analysis of multivariate failure time data: The marginal approach. *Statist. in Med.* **13** 2233–2247.

Lindley, D. V. (1968). The choice of variables in multiple regression (with discussion). *Jour. Roy. Statist. Soc., B* **30** 31–66.

Mallows, C. L. (1973). Some comments on C_p. *Technometrics* **15** 661–675.

McCullagh, P. & Nelder, J. A. (1989). *Generalized Linear Models* 2nd ed. Chapman and Hall, London.

Miller, A. J. (2002). *Subset Selection in Regression* 2nd Edition Chapman and Hall, London.

Nielsen, G. G., Gill, R. D., Andersen, P. K., & Sørensen, T. I. A. A. (1992). A counting process approach to maximum likelihood estimator in frailty models. *Scandin. J. Statist.* **19** 25–43.

Schwarz, G. (1978). Estimating the dimension of a model. *Ann. Statist.* **6** 461–464.

Sinha, D., Chen, M. H. & Ghosh, S. K. (1999). Bayesian analysis and model selection for interval censored survival data. *Biometrics* **55** 585–590.

Spiekerman, C. F. & Lin, D. Y. (1998). Marginal regression models for multivariate failure time data. *Jour. Amer. Statist. Assoc.* **93** 1164–1175.

Tibshirani, R. (1996). Regression shrinkage & selection via the LASSO. *J. Royal Statist. Soc., Ser. B* **58** 267–288.

Tibshirani, R. (1997). The lasso method for variable selection in the Cox model. *Statistics in Medicine* **16** 385–395.

Volinsky, C. T. & Raftery, A. E. (2000). Bayesian information criterion for censored survival models. *Biometrics* **56** 256–262.

Wei, L. J., Lin, D. Y. & Weisseld, L. (1989). Regression analysis of multivariate incomplete failure time data by modeling marginal distributions. *J. Amer. Statist. Assoc.* **84** 1065–1073.

Empirical Likelihood in Survival Analysis

Gang Li[1], Runze Li[2], and Mai Zhou[3]

[1] Department of Biostatistics, University of California, Los Angeles, CA 90095, U.S.A. (vli@ucla.edu)
[2] Department of Statistics, The Pennsylvania State University, University Park, PA 16802-2111, U.S.A. (rli@stat.psu.edu)
[3] Department of Statistics, University of Kentucky, Lexington, KY 40506, U.S.A. (mai@ms.uky.edu)

Summary. Since the pioneer work of Thomas & Grunkemeier (1975) and Owen (1988), empirical likelihood has been developed as a powerful nonparametric inference approach and become popular in statistical literature. There are many applications of empirical likelihood in survival analysis. In this paper, we present an overview of recent developments of empirical likelihood methods for survival data. In particular, we discuss empirical likelihood results for a general mean functional of the distribution function, a functional of the hazard function, the Cox proportional hazards model, and a semiparametric accelerated failure time model.

Key words: Accelerated failure time model; censored data; Cox proportional hazards model; Wilks theorem.

2000 Mathematics Subject Classification: 62N02, 62N03, 62G20.

1 Introduction

Empirical likelihood (EL) appears to be first used by Thomas & Grunkemeier (1975) to obtain better confidence intervals involving the Kaplan-Meier estima-

Contemporary Multivariate Analysis and Experimental Design—In Honor Celebration of Professor Kai-Tai Fang's 65th birthday. Edited by Jianqing Fan and Gang Li. The World Scientific Publisher, 2005.

tor in survival analysis. Theoretical development of empirical likelihood was originated by Owen (1988) who derived nonparametric confidence regions for the mean of a random vector based on i.i.d. observations. Since the work of Owen (1988), EL method has become popular in the statistical literature and has been extended to a variety of applications including linear regression models [cf. Owen (1991) and Chen (1993, 1994)], general estimating equations (Qin & Lawless (1994)), and nonparametric regression (Chen & Qin (2000)). A comprehensive review of empirical likelihood for various complete data settings is given by Owen (2001). EL has many desirable properties. For instance, the EL based confidence interval is range preserving and transform respecting. It uses data to determine the shape and orientation of a confidence region. Another notable feature of EL is its ability to carry out a hypothesis test and construct confidence intervals without the need of estimating the variance. This feature has been appreciated particularly in survival analysis since variance estimation can be very difficult in many survival analysis problems. Because of this difficulty, many estimation procedures saw limited action in practice. EL can provide a way to circumvent the complicated variances and make many inference procedures practical.

Because the results of Owen (1988, 1991) cannot be easily extended to deal with incomplete survival data, many authors have worked to develop EL procedures for analysis of survival data. Li (1995) and Murphy (1995) gave theoretical justifications for the method of Thomas & Grunkemeier (1975). Li (1995) demonstrated that the likelihood ratio used by Thomas & Grunkemeier (1975) is a "genuine" nonparametric likelihood ratio. That is, it can be derived by considering the parameter space of all survival functions. This property is not shared by many existing EL. Li, Qin & Tiwari (1997) derived likelihood ratio-based confidence intervals for survival probabilities and for the truncation proportion under statistical setting in which the truncation distribution is either known or belong to a parametric family. Hollander, McKeague & Yang (1997) constructed simultaneous confidence bands for survival probabilities based on right-censored data using EL. Pan & Zhou (1999) illustrated the use of a particular kind of one-parameter sub-family of distribution in the analysis of EL. Einmahl & McKeague (1999) constructed simultaneous confidence tubes for multiple quantile plots based on multiple independent samples using the EL approach. Wang & Jing (2001) investigated how to apply the EL method to a class of functional of survival function in the presence of censoring by using an adjusted EL. Pan & Zhou (2002) studied the EL ratios for right censored data and with parameters that are linear functionals of the cumulative hazard function. Li & van Keilegom (2002) constructed confidence intervals and bands for the conditional survival and quantile functions using an EL ratio approach. McKeague & Zhao (2002) derived a simultaneous confidence band for the ratio of two survival functions based on independent right-censored data. Chen & Zhou (2003) extended the self consistent algorithm (Turnbull (1976)) to include a constraint on the nonparametric maximum likelihood estimator of the distribution function with doubly censored data. They further show how to construct confidence intervals and test hypothesis based on the nonparametric maximum likelihood estimator via the EL ratio.

The EL ratio has also been used to construct confidence intervals for other parameters or functionals of a population in addition to survival probabilities. For instance, Ren (2001) used weighted EL ratio to derive confidence intervals for the mean with censored data. Adimari (1997) suggested a simple way to obtain EL type confidence intervals for the mean under random censorship. Li, Hollander, McKeague & Yang (1996) derived confidence bands for quantile functions using the EL ratio approach. The EL method has also been applied for linear regression with censored data (Qin & Jing (2001a), Li & Wang (2003), Qin & Tsao (2003)). Furthermore, the EL method has been adapted for semiparametric regression models, including partial linear models (Leblanc & Crowley (1995), Shen, Shi & Wong (1999), Qin & Jing (2001b), Lu, Chen & Gan (2002), Wang & Li (2002)). Naiknimbalkar & Rajarshi (1997) proposed the EL ratio test for equality of k-medians in censored data. Chen, Leung & Qin (2003) extended the EL method for censored data with surrogate endpoints. Li (2003) developed EL methods for testing goodness-of-fit with right censored data.

This article will focus on some recent EL results for censored data where the limiting distribution in the Wilks type theorem is a pivotal quantity, like a chi square. In Section 2, we introduce in detail EL results for a mean functional with right censored data, while in Section 3 we discuss EL results for a functional of the hazard function. Section 4 discusses some computation issues for the censored EL. EL for the Cox proportional hazards regression model is studied in Section 5. Section 6 presents the EL method for a semiparametric accelerated failure time model. Finally Section 7 gives a brief discussion on EL results with other types of censored data.

2 Empirical Likelihood for a mean functional

We define a *mean functional* as a parameter θ that is determined by

$$\int g(t,\theta)dF(t) = 0 \ , \tag{1}$$

where F is the unknown cumulative distribution function (CDF) and g is a known function. For example, θ is the mean if $g(t,\theta) = t-\theta$, an s-year survival probability if $g(t,\theta) = I(t \le s) - \theta$, and the p-th quantitle if $g(t,\theta) = I(t \le \theta) - p$.

Suppose that $X_1, X_2, \cdots, X_n$ are independent and identically distributed lifetimes with CDF $F(t) = P(X_i \le t)$. Let $C_1, C_2, \cdots, C_n$ be censoring times with CDF $G(t) = P(C_i \le t)$. Assume further that the life times and the censoring times are independent. Under the random censorship model, we observe only

$$T_i = \min(X_i, C_i), \quad \delta_i = I_{[X_i \le C_i]}, \quad i = 1, \ldots, n. \tag{2}$$

The EL of the censored data pertaining to F is

$$EL(F) = \prod_{i=1}^{n} [\Delta F(T_i)]^{\delta_i} [1 - F(T_i)]^{1-\delta_i} , \tag{3}$$

where $\Delta F(s) = F(s) - F(s-)$.

It is well known that among all the cumulative distribution functions, the Kaplan & Meier (1958) estimator maximizes (3). Let us denote the Kaplan-Meier estimator by $\hat{F}_n(t)$. Although the maximum of the censored EL under constraint (1) does not always have an explicit expression, we have the following Wilks type result.

Theorem 1. *Suppose the true distribution of lifetimes satisfies constraint (1). Assume further that the asymptotic variance of* $\sqrt{n} \int g(t, \theta) d\hat{F}_n(t)$ *is positive and finite. Then, as* $n \to \infty$

$$-2 \log \frac{\sup_F EL(F)}{EL(\hat{F}_n)} \xrightarrow{\mathcal{D}} \chi^2_{(1)} , \tag{4}$$

where the sup *is taken over all the CDFs that satisfy (1) and* $F \ll \hat{F}_n$.

The proof of Theorem 1 can be found in Murphy & van der Vaart (1997) or Pan & Zhou (1999).

Counting process martingale techniques has now become a standard tool in the literature of survival analysis. Given the censored data (2), it is well known that we can define a filtration $\mathcal{F}_t$ such that

$$M_n(t) = \frac{\hat{F}_n(t) - F(t)}{1 - F(t)}$$

is a (local) martingale with respect to the filtration $\mathcal{F}_t$, see Fleming & Harrington (1991) for details. It is also known that under mild regularity conditions, $\sqrt{n} M_n(t)$ converges weakly to a time changed Brownian motion.

To develop EL methods for regression models such as the Cox model and the accelerated failure time model, we need to extend Theorem 1 to a more general setting in which $g(t)$ is replaced by a random function $g_n(t)$ satisfying the following conditions:

1. (i) $g_n(t)$ are predictable with respect to $\mathcal{F}_t$ and $g_n(t) \xrightarrow{P} g(t)$ as $n \to \infty$.
2. (ii) $\sqrt{n} \int_{-\infty}^{\infty} \{g_n(t)[1 - F_X(t)] + \int_{-\infty}^{t} g_n(s) dF_X(s)\} dM_n(t)$ converges in distribution to a zero mean normal random variable with a finite and non-zero variance.

It is worth noting that the integrand inside the curly brackets in (ii) is predictable. If we put a variable upper limit in the outside integration in (ii), then it is also a martingale. It is not difficult to give a set of sufficient conditions that will imply asymptotic normality. Usually a Lindeberg type condition is needed.

Theorem 2. *Suppose that $g_n(t)$ is a random function satisfying the above two conditions and that for each n,*

$$\int_{-\infty}^{\infty} g_n(t) dF_X(t) = 0 \ , \tag{5}$$

then

$$-2 \log \frac{\sup_F EL(F)}{EL(\hat{F}_n)} \xrightarrow{\mathcal{D}} \chi^2_{(1)}$$

where the sup *in the numerator is taken over those F that $F \ll \hat{F}_n$ and satisfy the constraint*

$$\int_{-\infty}^{\infty} g_n(t) dF(t) = 0 \ . \tag{6}$$

The proof of Theorem 2 is given in Zhou & Li (2004). In Theorem 2, it is assumed that the true distribution of X_i satisfies (5). However, the Kaplan-Meier estimator may not satisfy this condition. Generalizations of the above two Theorems for multiple constraints type similar to (1) or (6) are seen to hold, but a formal proof is tedious and not available in the published literature. The limiting distribution will be $\chi^2_{(q)}$ where q is the number of constraints.

3 Empirical Likelihood for functionals of the hazard

Hazard function is a quantity often of interest in survival analysis. For a random variable X with cumulative distribution function $F(t)$, the cumulative hazard function is defined by

$$H(t) = \int_{(-\infty,t]} \frac{dF(s)}{1 - F(s-)} .$$

Given the randomly censored data (2), a natural way to define the EL in terms of the hazard is:

$$EL(H) = \prod_{i=1}^{n} [\Delta H(T_i)]^{\delta_i} \exp(-H(T_i)). \tag{7}$$

It can be easily verified that $EL(H)$ is maximized when the hazard is the Nelson-Aalen estimator, denoted by $\hat{H}_n(t)$. The parameter θ of interest is defined by

$$\int g_n(t, \theta) dH(t) = 0, \tag{8}$$

where the meaning of the parameter θ is similar to that in (1) or (5) and the function g_n is stochastic.

Theorem 3. *Suppose that $g_n(t)$ is a sequence of predictable functions with respect to the filtration $\mathcal{F}_t$, and $g_n \xrightarrow{P} g(t)$ with*

$$0 < \int \frac{|g(x)|^m dH(x)}{(1 - F_X(x))(1 - G(x))} < \infty, \qquad m = 1, 2.$$

If the true underlying cumulative hazard function satisfies the condition (8), then

$$-2 \log \frac{\sup_H EL(H)}{EL(\hat{H}_n)} \xrightarrow{\mathcal{D}} \chi^2_{(1)} \qquad as\ n \to \infty\ ,$$

where the sup *is taken over those H that satisfy (8) and $H \ll \hat{H}_n$.*

One nice feature here is that we can use stochastic functions to define the statistics, i.e. $g(t) = g_n(t)$, as long as $g_n(t)$ is predictable with respect to $\mathcal{F}_t$. See Pan & Zhou (2002). For example $g_n(t)$ = size of risk set at time t, will produce a statistic corresponding to the one sample log-rank test, etc.

Multivariate version of the Theorem 3 can similarly be obtained with a limiting distribution of $\chi^2_{(q)}$. The q parameters are defined through q equations similar to (8) with different g and θ.

4 Computation of the censored Empirical Likelihood

Computation of the EL ratio can sometimes be reduced by the Lagrange multiplier method to the dual problem. When it does, the computation of EL is relatively easy, and is equivalent to the problem of finding the root of q nonlinear monotone equations with q unknowns. But more often than not the censored EL problem cannot be simplified by the Lagrange multiplier method. A case in point is the right censored data with a mean constraint. No reduction to the dual problem is available.

Below we discuss two computational methods when the maximization problem of the censored empirical likelihood cannot be simplified by the Lagrange multiplier method.

Sequential quadratic programming (SQP) is a general optimization procedure and a lot of related literature and software are available from the optimization field. It repeatedly approximates the target function locally by a quadratic function. SQP can be used to find the maximum of the censored EL under a (linear) constraint and the constrained NPMLE. This in turn enables us to obtain the censored empirical likelihood ratio. A drawback of this approach is that without the Lagrange multiplier reduction, the memory and computation requirement increases dramatically as the sample size increases. It needs to invert matrices of size $n \times n$. With today's computing hardware, SQP works well for small to medium sample sizes. In our own experience, it is quite fast for samples of size under 1000, but for larger sample sizes difficulty may rise.

EM algorithm has long been used to compute the NPMLE for censored data. Turnbull (1976) showed how to find NPMLE with arbitrary censored, grouped or truncated data. Zhou (2002) generalized Turnbull's EM algorithm to obtain the maximum of the censored empirical likelihood under mean constraints, and thus the censored empirical likelihood ratio can be computed. Compared to the SQP, the generalized EM algorithm can handle much larger sample sizes, up to 10,000 and beyond. The memory requirement of the generalized EM algorithm increases linearly with the sample size. The computation time is comparable to the sum of two computation problems: the same EL problem but with uncensored data (which has a Lagrange multiplier dual reduction), and Turnbull's EM for censored data NPMLE.

Many EL papers include examples and simulation results and thus various software are developed. There are two sources of publicly available software for EL: there are Splus codes and Matlab codes on the EL web site maintained by Owen but it cannot handle censored data. There is a package *emplik* for the statistical software R (Gentleman & Ihaka (1996)) written by Mai Zhou, available from CRAN. This package includes several functions that can handle EL computations for right censored data, left censored data, doubly censored data, and right censored and left truncated data.

The package *emplik* also includes functions for computing EL in the regression models discussed in Section 6. No special code is needed to compute EL in the Cox proportional hazards model, since the EL coincides with the partial likelihood, which can be computed using existing softwares.

5 Cox proportional hazards regression model

For survival data, the most popular model is the Cox model. It is known that the partial likelihood ratio of Cox (1972, 1975) can be interpreted as the (profile) empirical likelihood ratio; see, e.g., Pan (1997) and Murphy & van der Vaart (2000).

Let X_i, $i = 1, \cdots, n$ be independent lifetimes and z_i, $i = 1, \cdots, n$ be the associated covariates. The Cox model assumes that

$$h(t|z_i) = h_0(t) \exp(\beta z_i),$$

where $h_0(t)$ is the unspecified baseline hazard function and β is a parameter.

The contribution to the empirical likelihood function from the ith observation (T_i, δ_i) is

$$(\Delta H_i(T_i))^{\delta_i} \exp\{-H_i(T_i)\},$$

where $H_i(t) = H_0(t)\exp(z_i\beta)$. The empirical likelihood function is then the product of the above over i:

$$EL^c(H_0, \beta) = \prod_{i=1}^{n} [\Delta H_0(T_i)\exp(z_i\beta)]^{\delta_i} \exp\{-H_0(T_i)\exp(z_i\beta)\} .$$

It can be verified that for any given β the EL^c is maximized at the so called Breslow estimator, $H_0 = \hat{H}_n^{\beta}$. Also, by definition, the maximum of $EL^c(\hat{H}_n^{\beta}, \beta)$ with respect to β is obtained at the Cox partial likelihood estimator of the regression parameter. Denote the Cox partial likelihood estimator of β by $\hat{\beta}_c$.

Theorem 4. *Under the conditions that will guarantee the asymptotic normality of the Cox maximum partial likelihood estimator as in Andersen & Gill (1982), we have the following empirical likelihood ratio result:*

$$-2\log \frac{EL^c(\hat{H}_n^{\beta_0}, \beta_0)}{\sup_{\{\beta,\ H_0\}} EL^c(H_0, \beta)} = I(\xi)(\beta_0 - \hat{\beta}_c)^2 , \tag{9}$$

where $I(\cdot)$ is the information matrix as define in Andersen & Gill (1982), ξ is between β_0 and $\hat{\beta}_c$, and the sup *in the denominator is over all β and hazard $H_0 \ll \hat{H}_n$. It then follows easily that the right hand side of (9) converges in distribution to $\chi^2_{(1)}$ as $n \to \infty$.*

The proof of Theorem 4 was given in Pan (1997). Zhou (2003) further studied EL inference for the Cox model along the lines of the above discussion. He obtained the Wilks theorem of the EL for estimating/testing β when some partial information for the baseline hazard is available. The (maximum EL) estimator of β is more efficient than $\hat{\beta}_c$ due to the extra information on the baseline hazard.

6 Accelerated Failure Time model

The semiparametric accelerated failure time (AFT) model is a linear regression model where the responses are the logarithm of the survival times and the error term distribution is unspecified. It provides a useful alternative to the popular Cox proportional hazards model for analyzing censored survival data [cf. Wei (1992)]. AFT models are sometimes more natural than the Cox model [cf. see Reid (1994)].

For simplicity, we denote X_i to be the logarithm of the lifetime for subject i. Suppose

$$X_i = \beta^t z_i + \epsilon_i \qquad i = 1, \ldots, n;$$

where ϵ_i's are i.i.d. random errors, β is the regression parameter to be estimated and z_i is the vector of covariates for subject i. For subject i, let C_i be the censoring time and assume that C_i and X_i are independent. Due to censoring, we observe only

$$T_i = \min(X_i, C_i), \quad \delta_i = I_{[X_i \le C_i]}, \quad z_i, \quad i = 1, \ldots, n. \tag{10}$$

For any candidate estimate b of β, we define

$$e_i(b) = T_i - b^t z_i \ .$$

When $b = \beta$, the $e_i(\beta)$'s are the censored sample for the ϵ_i's.

Two different approaches of EL analysis of the AFT model are available in the literature. The first approach is characterized by its definition of the EL as

$$EL(AFT) = \prod_{i=1}^{n} p_i \ . \tag{11}$$

However, this is a bona fide EL only for iid uncensored data. Similar to Owen (1991), this $EL(AFT)$ is coupled with the least squares type estimating equations

$$\sum_{i=1}^{n} z_i(T_i^* - \beta^t z_i) = 0$$

where T_i^* is defined by either the synthetic data approach:

$$T_i^* = \frac{\delta_i T_i}{1 - G(T_i)} \ ;$$

or the Buckley-James approach:

$$T_i^* = \delta_i T_i + (1 - \delta_i) E(X_i | T_i, \beta) \ .$$

Both definition of T_i^* are based on the observation that $E(T_i^*) = E(X_i)$. Unfortunately, the censoring distribution function G in the synthetic data approach is unknown and is typically replaced by a Kaplan-Meier type estimator. In the Buckley-James approach the conditional expectation depends on the unknown error distribution and also need to be estimated. These substitution, however, makes the T_i^* dependent on each other and careful analysis show that the log EL(AFT) ratio has a limiting distribution characterized by linear combinations of chi squares, with the coefficients need to be estimated. See Qin & Jing (2001a), Li & Wang (2003) and Fang, Li & Qin (2004) for details.

The second approach of EL for the censored AFT model defines the EL as

$$EL(error) = \prod_{i=1}^{n} p_i^{\delta_i} [1 - \sum_{e_j \le e_i} p_j]^{1-\delta_i} \ . \tag{12}$$

This EL may be viewed as the censored EL for the iid errors in the AFT model. It is more natural in our opinion since it reflects the censoring.

Zhou & Li (2004) first noted that the least squares estimation equation with the Buckley-James approach can be written as

$$0 = \sum_i \delta_i e_i(b) \left[z_i + \sum_{k<i,\ \delta_k=0} z_k \frac{\Delta \hat{F}_n(e_i)}{1-\hat{F}_n(e_k)} \right] , \tag{13}$$

where $\hat{F}_n$ is the Kaplan-Meier estimator computed from $(e_i,\ \delta_i)$, $i = 1, \ldots, n$.

They then proposed to use the following estimation equations

$$0 = \sum_i e_i(b) \frac{z_i + \sum_{k<i,\ \delta_k=0} z_k \frac{\Delta \hat{F}_n(e_i)}{1-\hat{F}_n(e_k)}}{n\Delta \hat{F}_n(e_i)} \delta_i p_i , \tag{14}$$

with the $EL(error)$ defined earlier.

With the aid of Theorem 2, Zhou and Li proved the following theorem.

Theorem 5. *Suppose that in the censored AFT model ϵ_i are iid with a finite second moment. Under mild regularity conditions on the censoring, we have, as $n \to \infty$*

$$-2 \log \frac{\sup EL(error)}{\sup EL(error)} \xrightarrow{\mathcal{D}} \chi^2_{(1)} ,$$

where the sup *in the numerator is taken over $b = \beta$ and all probabilities p_i that satisfy the estimating equations (14); the* sup *in the denominator is taken over $b =$ Buckley-James estimator and all probabilities p_i.*

A multivariate version of the above theorem can also be established. M-estimation of β for the censored AFT model is also discussed by Zhou & Li (2004).

7 Other Applications

Empirical likelihood method is applicable to many other types of censored data. However, fewer results are available due to technical difficulties. Li (1996) and Li, Qin & Tiwari (1997) studied EL for left truncated data. Similar results for the left truncated and right censored data should also hold. Murphy & van der Vaart (1997) described a general framework for studying EL. In particular, they showed that for doubly censored data, where the lifetimes are subject to censoring from above and below, EL results similar to Theorem 1 hold. Huang (1996) gave EL results for a proportional hazards model with current status data.

Acknowledgements

G. Li's research was partially supported by NIH grant CA78314-03. R. Li's work was supported by NSF grants DMS 0102505 and DCCF 0426628.

References

Adimari, G. (1997). Empirical likelihood type confidence intervals under random censorship. *Annals of the Institute of Statistical Mathematics* **49** 447–466.

Andersen, P. K. & Gill, R. D. (1982). Cox's Regression Model for Counting Processes: a Large Sample Study. *Annals of Statistics* **10** 1100–1200.

Chen, S. X. (1993). On the accuracy of empirical likelihood confidence regions for linear regression model. *Ann. Inst. Statist. Math.* **45** 621–637.

Chen, S. X. (1994). Empirical likelihood confidence intervals for linear regression coefficients. *J. Multivariate Anal.* **49** 24–40.

Chen, S. X. & Qin, Y. S. (2000). Empirical likelihood confidence intervals for local smoothers. *Biometrika* **87** 946–953.

Chen, K. & Zhou, M. (2003). Non–parametric hypothesis testing and confidence intervals with doubly censored data. *Lifetime Data Analysis* **9** 71–91.

Chen, S. X., Leung, D. H. Y. & Qin, J. (2003). Information recovery in a study with surrogate endpoints. *Journal of American Statistical Association* **98** 1052–1062.

Cox, D. R. (1972). Regression Models and Life Tables (with discussion). *J. Roy. Statist. Soc B.* **34** 187–220.

Cox, D. R. (1975). Partial Likelihood. *Biometrika* **62** 269–276.

Einmahl, J. H. J. & McKeague, I. W. (1999). Confidence tubes for multiple quantile plots via empirical likelihood. *Annals of Statistics* **27** 1348–1367.

Fang, K. T., Li, G. & Qin, H. (2004). Empirical Likelihood Inference in Linear Regression for Right Censored Survival Data. *Manuscript.*

Fleming, T. & Harrington, D. (1991). *Counting Processes and Survival Analysis.* Wiley, New York.

Gentleman, R. & Ihaka, R. (1996). R: A Language for data analysis and graphics. *Journal of Computational and Graphical Statistics* **5** 299–314.

Hollander, M., McKeague, I. W. & Yang, J. (1997). Likelihood ratio–based confidence bands for survival functions. *Journal of American Statistical Association* **92** 215–226.

Huang, J. (1996). Efficient estimation for the proportional hazards model with interval censoring. *Annals of Statistics* **24** 540–568.

Kaplan, E. & Meier, P. (1958). Non–parametric Estimator From Incomplete Observations. *Journal of American Statistical Association* **53** 457–481.

Leblanc, M. & Crowley, J. (1995). Semiparametric regression functions. *Journal of American Statistical Association* **90** 95–105.

Li, G. (1995). On nonparametric likelihood ratio estimation of survival probabilities for censored data. *Statist. & Prob. Letters* **25** 95–104.

Li, G. (1996). Nonparametric likelihood ratio estimation of probabilities for truncated data. *Journal of American Statistical Association* **90** 997–1003.

Li, G. (2003). A nonparametric likelihood ratio approach to testing goodness of fit for survival data. *Journal of Multivariate Analysis* **86** 166–182.

Li, G., Hollander, M., McKeague, I. W. & Yang, J. (1996). Nonparametric likelihood ratio confidence bands for quantile functions from incomplete survival data. *Annals of Statistics* **24** 628–640.

Li, G., Qin, J. & Tiwari, R. C. (1997). Semiparametric likelihood ratio–based inferences for truncated data. *Journal of American Statistical Association* **92** 236–245.

Li, G. & van Keilegom, I. (2002). Likelihood ratio confidence bands in non–parametric regression with censored data. *Scandinavian Journal of Statistics* **29** 547–562.

Li, G. & Wang, Q. H. (2003). Empirical likelihood regression analysis for right censored data. *Statistica Sinica* **13** 51–68.

Lu, J. C., Chen, D. & Gan, N. C. (2002). Semi–parametric modelling and likelihood estimation with estimating equations. *Australian & New Zealand Journal of Statistics* **44** 193–212.

McKeague, I. W. & Zhao, Y. C. (2002). Simultaneous confidence bands for ratios of survival functions via empirical likelihood. *Statistics & Probability Letters* **60** 405–415.

Murphy, S. A. (1995). Likelihood ratio–based confidence intervals in survival analysis. *Journal of the American Statistical Association* **90** 1399–1405.

Murphy, S. A. & van der Vaart, A. (1997). Semiparametric likelihood ratio inference. *Annals of Statistics* **25** 1471–509.

Murphy, S. A. & van der Vaart, A. (2000). On profile likelihood. *Journal of American Statistical Association*, **90**, 1399–1405.

Naiknimbalkar, U. V. & Rajarshi, M. B. (1997). Empirical likelihood ratio test for equality of k medians in censored data. *Statistics & Probability Letters* **34** 267–273.

Owen, A. B. (1988). Empirical likelihood ratio confidence intervals for a single functional. *Biometrika* **75** 237–49.

Owen, A. B. (1991). Empirical Likelihood for Linear Models. *Annals of Statistics* **19** 1725–1747.

Owen, A. B. (2001). *Empirical likelihood.* Chapman & Hall, London.

Pan, X. R. (1997). Empirical Likelihood Ratio Method for Censored Data. *Ph.D. Dissertation*, University of Kentucky, Department of Statistics.

Pan, X. R. & Zhou, M. (1999). Using one parameter sub-family of distributions in empirical likelihood with censored data. *Journal Statistical Planning & Inference* **75** 379–392.

Pan, X. R. & Zhou, M. (2002). Empirical likelihood in terms of cumulative hazard function for censored data. *Journal of Multivariate Analysis* **80** 166–188.

Qin, G. S. & Jing, B. Y. (2001a). Empirical likelihood for censored linear regression. *Scand J of Statist* **28** 661–673.

Qin, G. S. & Jing, B. Y. (2001b). Censored partial linear models and empirical likelihood. *Journal of Multivariate Analysis* **78** 37–61.

Qin, J. & Lawless, J. F. (1994). Empirical likelihood & general estimating equations. *Annals of Statistics* **22** 300–325.

Qin, G. S. & Tsao, M. (2003). Empirical likelihood inference for median regression models for censored survival data. *Journal of Multivariate Analysis* **85** 416–430.

Reid, N. (1994). A conversation with Sir David Cox. *Statist. Sci.* **9** 439–455.

Ren, J.J. (2001). Weighted empirical likelihood ratio confidence intervals for the mean with censored data. *Annals of the Institute of Statistical Mathematics* **53** 498–516.

Shen, X. T., Shi, J. & Wong, W. H. (1999). Random sieve likelihood & general regression models. *Journal of American Statistical Association* **94** 835–846.

Thomas, D. R. & Grunkemeier, G. L. (1975). Confidence Interval Estimation of Survival Probabilities for Censored Data. *Journal of the American Statistical Association* **70** 865–871.

Turnbull, B. (1976). The empirical distribution function with arbitrarily grouped, censored and truncated data. *Journal of Royal Statistical Society, Series, B* **38** 290–295.

Wang, Q. H. & Jing, B. Y. (2001). Empirical likelihood for a class of functionals of survival distribution with censored data. *Annals of the Institute of Statistical Mathematics* **53** 517–527.

Wang, Q. H. & Li, G. (2002). Empirical likelihood semiparametric regression analysis under random censorship. *Journal of Multivariate Analysis* **83** 469–486.

Wei, L. J. (1992). The accelerated failure time model: a useful alternative to the Cox regression model in survival analysis. *Statist. Med.* **11** 1871–79.

Zhou, M. (2002). Empirical likelihood ratio with arbitrarily censored/truncated data by a modified EM algorithm. Tech Report. University of Kentucky. To appear in *Journal of Computational and Graphical Statistics*

Zhou, M. (2003). Cox proportional hazards model with additional information on the baseline. Tech Report. University of Kentucky.

Zhou, M. & Li, G. (2004). Empirical likelihood analysis of the Buckley–James estimator. Tech Report. UCLA.

Statistical Analysis for Tumor Xenograft Experiments in Drug Development

Ming Tan, Hong-Bin Fang and Guo-Liang Tian

Division of Biostatistics, University of Maryland Greenebaum Cancer Center, 22 South Greene Street, Baltimore, MD 21201, U.S.A. (`mtan@umm.edu`)

Summary. In cancer drug development, demonstrated efficacy in tumor xenograft experiments on severe combined immunodeficient mice who are grafted with human tumor tissues or cells is an important step to bring a promising compound to human. These experiments also demonstrated a good correlation in efficacy with clinical outcomes. A key outcome variable is tumor volumes measured over a period of time, while mice are treated with certain treatment regimens. To analyze such data from xenograft experiments and evaluate the efficacy of a new drug, some statistical methods have been developed in literature. However, a mouse may die during the experiment or may be sacrificed when its tumor volume reaches a threshold. A tumor may be suppressed its tumor burden (volume) may become undetectable for some time but regrow and its tumor burden (volume) may become (e.g., $< 0.01cm^3$) undetectable at times. Thus, incomplete repeated measurements arise. Because of the small sample sizes in these experiments, asymptotic inferences are usually questionably. In addition, were the tumor-bearing mice not treated, the tumors would keep growing until the mice die or are sacrificed. This intrinsic growth of tumor in the absence of treatment constrains the parameters in the statistical model and causes further difficulties in statistical analysis. In this paper, we review the recent advance in statistical inference accounting for these statistical challenges. Furthermore, we develop a multivariate random effects model with constrained parameters for multiple tumors in xenograft experiments. A real xenograft study on the antitumor agent exemestane, an aromatase inhibitor, combined with tamoxifen against the postmenopausal breast cancer is analyzed using the proposed methods.

Contemporary Multivariate Analysis and Experimental Design—In Honor Celebration of Professor Kai-Tai Fang's 65th birthday. Edited by Jianqing Fan and Gang Li. The World Scientific Publisher, 2005.

Key words: Bayesian analysis; ECM algorithm; longitudinal data; missingness; MLE; restricted parameter estimation; truncation; tumor xenograft models.

2000 Mathematics Subject Classification: 62F15, 62F30, 62H12, 62H15, 62J05, 62P10

1 Introduction

Most human tumors have gone undetected clinically for the greater part of their growth. It is thus necessary to study human tumors in an experimental situation which requires the approach of xenografting, where human tumor tissues (e.g., sliced tissue blocks, or tumor cells) are grown in experimental animals. The most widely used animal model is the mice (especially, the severe combined immunodeficient (*scid*) mice) model. Because anti-tumor activity in the xenograft correlates well with patient response, it is an important step to demonstrate such activity in xenografts to bring a promising compound to human.

In a typical xenograft experiment, treatment is initiated when the diameter of tumor reaches certain level (e.g., $0.5cm$). Several treatment regimens are administered and the outcome variables such as tumor volumes are measured (using, e.g., the Maxcal digital caliper) at the start of the treatment and regularly in a given period of follow-up time. Measurements are transmitted directly to a computer. The renewed interest stems from humane and cost considerations as well as scientific considerations for analyzing the data efficiently and make inference properly. A general statistical guideline for the design and analysis of experiments involving animals is proposed by Festing & Altman (2002) and Rygaard & Spang–Thomsen (1997). An optimal design and a sample size formula for tumor xenograft models are developed in Tan, Fang, Tian & Houghton (2003). Without missing data, the statistical analysis for tumor growth experiments has been studied by Heitjan (1991) and Heitjan, Manni & Santen (1993).

However, the statistical analysis of such longitudinal data presents several challenges for a number of reasons. First, sample sizes in xenograft experiments are usually small because of cost, graft failures due to body rejection or some xeno-antigens. Therefore, small sample inference procedures are needed. Second, in xenograft experiments, the tumor growth depends on initial volumes. If no treatment were given, tumors in mice would keep growing until the tumor-bearing mice die or are sacrificed. Thus, estimating antitumor activity should adjust for the intrinsic tumor growth in the absence of treatment, which thus constrains the regression coefficients. Finally, missing data is hard to avoid in these experiments because a mouse may die of toxicity or may be sacrificed when its tumor volume reaches a set criterion (*i.e.* quadruples) or the tumor volume becomes unmeasurable (for example, when it is less than $0.01cm^3$).

Although the analysis of incomplete longitudinal data has attracted a great deal of attention in the literature (see *e.g.* Diggle & Kenward (1994); Little (1995); Hogan & Laird (1997); and Wu & Follmann (1999)), existing methods for informative missingness do not account for the constraints in the model parameters. To account for special features of data arisen from xenograft experiments such as moderate samples and informative censoring, Tan, Fang, Tian & Houghton (2002) developed a t-test via the EM algorithm and a Bayesian approach for testing difference between two treatment regimens in xenograft experiments. Tan, Fang, Tian & Houghton (2004) proposed a class of regression models with constrained parameters and Fang, Tian & Tan (2004) considered a Bayesian hierarchical model accounting for the parameter constraints. In this paper, we synthesize recent advances in statistical inference accounting for these challenges and further develop a random-effects model with constrained parameters for multiple tumors.

In the rest of this article, we first summarize and formulate the statistical models for xenograft experiments in Section 2. The parameters are estimated using two inter-related methods, a maximum likelihood approach based on the ECM algorithm and a Bayesian approach that takes advantage of the likelihood results in Section 3. The comparison of two treatments is discussed in Section 4. Section 5 develops a multivariate random-effects model with constrained parameters for multiple tumors in xenograft experiments. We illustrate the method with the xenograft experiments in breast cancer therapy when multiple tumors are grown simultaneously. We conclude with a discussion in Section 6.

2 Statistical Models

To model the antitumor activity, let q be the number of agents and $t_0 < t_1 < \cdots < t_m$ the prespecified follow-up times. For subject i $(i = 1, \ldots, n)$, let y_{i0} and $(y_{i1}, \ldots, y_{im})^T$ denote the initial tumor volume at t_0 and the tumor volumes at $(t_1, \ldots, t_m)$ in logarithm scale, respectively. Let $z_{ij}^{(k)}$ denote the cumulative dose of agent $k\,(k = 1, \ldots, q)$ administered to subject i up to time t_j for $j = 1, \ldots, m$. Further consider $(p-q)$ interaction terms of the q agents. The corresponding doses are denoted by $z_{ij}^{(k)}$, where $k = q+1, \ldots, p$. Therefore, $\mathbf{z}_i^{(k)} = (z_{i1}^{(k)}, \ldots, z_{im}^{(k)})^T$ denote the doses of agent $k\,(k = 1, \ldots, p)$ given to subject i. Treating the initial value y_{i0} as a covariate, we obtain a known $m \times (p+1)$ covariate matrix $\mathbf{Z}_i = (y_{i0}\mathbf{1}_m, \mathbf{z}_i^{(1)}, \ldots, \mathbf{z}_i^{(p)})$, where $\mathbf{1}_m$ denotes the m-vector of component 1.

Note that all initial values $\{y_{i0}\}$ are completely observed but $(y_{i1}, \ldots, y_{im})^T$ may be not. Due to early withdrawal, we only observe $\mathbf{y}_i = (y_{i1}, \ldots, y_{in_i})^T$. If the tumor volume is below some threshold a (e.g., $0.01cm^3$), then it is unmeasurable and the informative censoring occurs. Therefore, the observation $\mathbf{y}_i$ consists of the observed part $\mathbf{y}_{i,\text{obs}}$ with length p_i and the informative censoring part $\mathbf{y}_{i,\text{inf}}$ with length $n_i - p_i$, of which all components are less than the threshold a. Fang, Tian

& Tan (2004) proposed the following linear mixed-effects model for the dose-effect relationship

$$\mathbf{y}_i = \boldsymbol{\gamma}_i + \mathbf{X}_i\boldsymbol{\beta} + \mathbf{W}_i\mathbf{b}_i + \mathbf{e}_i, \qquad i = 1, \ldots, n, \tag{2.1}$$

where $\boldsymbol{\gamma}_i$ is unknown intercept, $\boldsymbol{\beta} = (\beta_0, \beta_1, \ldots, \beta_p)^T$ is the fixed-effects, $\mathbf{X}_i$ is the $n_i \times (p+1)$ submatrix of $\mathbf{Z}_i$ by removing the last $(m - n_i)$ rows, $\mathbf{W}_i$ is the $n_i \times (q+1)$ submatrix of $\mathbf{X}_i$ by removing the last $(p-q)$ columns, $\mathbf{b}_i$ is the $(q+1) \times 1$ random-effects, and $\mathbf{e}_i$ is the error vector. We further assume that $\mathbf{b}_i \sim N_{q+1}(\mathbf{0}, \mathbf{D})$, $\mathbf{e}_i \sim N_{n_i}(\mathbf{0}, \sigma^2\mathbf{I}_{n_i})$, and $\mathbf{b}_i$ is independent of $\mathbf{e}_i$, where $\mathbf{D}$ and σ^2 are unknown $(q+1) \times (q+1)$ matrix and variance, respectively, and $\mathbf{I}_{n_i}$ is the identity matrix.

In model (2.1), the intercept $\boldsymbol{\gamma}_i = (\gamma_1, \ldots, \gamma_{n_i})^T$ reflects the intrinsic growth of tumor when no treatment was given. To characterize the unperturbed tumor growth, some parametric models were used, for example, Norton & Simon (1977) used a Gompertzian curve and Heitjan (1991) proposed a family of parametric models that contains the Gompertz and several other popular growth models. However, those parametric models do not always fit well across the entire curves for certain tumors, especially for different tumor growth curves even in the same xenograft experiments (*e.g.* untreated growth curves for breast cancer in Section 5 below). Then, we propose a regression model with nondecreasing intercept parameters for unperturbed tumor growth here. Since the tumor volumes in the control group are non-decreasing, a reasonable restriction on $\boldsymbol{\gamma}_i$ is

$$\gamma_1 \le \gamma_2 \le \cdots \le \gamma_{n_i}. \tag{2.2}$$

$\boldsymbol{\gamma}_i$ is an n_i-dimensional vector with varying length. To effectively deal with the restricted parameter problem (2.2), we make a transformation $\boldsymbol{\gamma}_i = \mathbf{V}_i\boldsymbol{\alpha}$, where $\mathbf{V}_i$ is the $n_i \times m$ submatrix of $\mathbf{V}$ by removing its last $m - n_i$ rows, and $\mathbf{V} = (v_{\ell\ell'})$ is the $m \times m$ lower-triangle matrix with $v_{\ell\ell'} = 1$ for $\ell \ge \ell'$ and 0 otherwise, and $\boldsymbol{\alpha} = (\alpha_1, \ldots, \alpha_m)^T \in I\!R \times I\!R_+^{m-1}$, where

$$I\!R \times I\!R_+^{m-1} = \{\boldsymbol{\alpha} : -\infty < \alpha_1 < +\infty,\ \alpha_j \ge 0,\ j = 2, \ldots, m\}. \tag{2.3}$$

Therefore, the model (2.1) can be rewritten in a hierarchical form

$$\begin{aligned} \mathbf{y}_i|\mathbf{b}_i &\overset{\text{ind}}{\sim} N_{n_i}(\mathbf{V}_i\boldsymbol{\alpha} + \mathbf{X}_i\boldsymbol{\beta} + \mathbf{W}_i\mathbf{b}_i,\ \sigma^2\mathbf{I}_{n_i}), \\ \mathbf{b}_i &\overset{\text{iid}}{\sim} N_{q+1}(\mathbf{0},\ \mathbf{D}), \quad i = 1, \ldots, n, \end{aligned} \tag{2.4}$$

In tumor xenograft experiments, mice often from the same strain are used and they are virtually genetically identical. Then the random-effects among mice may be omitted and Tan, Fang, Tian & Houghton (2004) proposed the following linear regression model with constrained parameters

$$\tilde{\mathbf{y}}_i = \boldsymbol{\gamma} + \mathbf{Z}_i\boldsymbol{\beta} + \boldsymbol{\varepsilon}_i, \qquad i = 1, \ldots, n, \tag{2.5}$$

where $\tilde{\mathbf{y}}_i = (y_{i1}, \ldots, y_{im})^T$, $\boldsymbol{\gamma} = (\gamma_1, \ldots, \gamma_m)^T$ and

$$\gamma_1 \leq \gamma_2 \leq \cdots \leq \gamma_m. \tag{2.6}$$

The error term ε_i is assumed to have the m-dimensional normal distribution with mean $\mathbf{0}$ and covariance matrix $\boldsymbol{\Sigma}_m = \sigma^2 \mathbf{R}_m$ and $\mathbf{R}_m = (r_{jl})$ has the Toeplitz correlation structure,

$$r_{jl} = \text{Corr}(y_{ij}, y_{il}) = \rho^{|j-l|}, \qquad j,\ l = 1, \ldots, m, \qquad \rho \in (-1, 1) \tag{2.7}$$

ρ is the correlation between successive measurements on the same subject and the correlation structure suggests that tumor volumes at two closer time points have higher correlation than those at two time points.

Although the normality assumption on tumor volume in *log*-scale is often appropriate and convenient for statistical inference, one natural question is how to actually test the multinormality of incomplete longitudinal data with small sample size. For the complete data, Liang, Li, Fang & Fang (2000) proposed the projection tests for multivariate normality based on the properties of left-spherical matrix distributions and affine invariant statistics. Especially, in the case of the small sample size n and the large dimension m ($m \leq n$), the projection tests are still effective in testing multinormality if one chooses the projection dimension $q < \min(m, n-1)$. Using multiple imputations and the projection test, Tan, Fang, Tian & Wei (2005) developed a test procedure for the multinormality of incomplete longitudinal data with small sample size. In addition, historic data of xenograft models (for a series of drugs in a host of cell lines) are usually available within a developmental therapeutics program and should be utilized to assist model checking and the selection of a transformation to approximate normality.

3 Estimation of Parameters

In this section, we focus on statistical inference for model (2.1) with restriction (2.2). The methods proposed below are adoptable for model (2.5) with parameter constraint (2.6). We denote the unknown population parameters by $\boldsymbol{\theta} = (\boldsymbol{\sigma}, \boldsymbol{\alpha})$, where $\boldsymbol{\sigma} = (\mathbf{D}, \boldsymbol{\beta}, \sigma^2)$. Further denote the observed *part* by $\mathbf{Y}_{\text{obs}} = \{\mathbf{y}_{i,\text{obs}} : i = 1, \ldots, n\}$, the informative censoring part by $\mathbf{Y}_{\text{inf}} = \{\mathbf{y}_{i,\text{inf}} : i = 1, \ldots, n\}$, and the observed *data* by $\mathbf{Y}^*_{\text{obs}} = \{\mathbf{Y}_{\text{obs}}, \boldsymbol{\Delta}\}$, where $\boldsymbol{\Delta} = \{\boldsymbol{\delta}_i : i = 1, \ldots, n\}$, and $\boldsymbol{\delta}_i$ is a vector of indicator whose components equal to 1 if the corresponding component of $\mathbf{y}_i$ is less than a, otherwise 0. We treat both $\mathbf{y}_{i,\text{inf}}$ and $\mathbf{b}_i$ as missing data, and denote them by $\mathbf{Y}_{\text{mis}} = \{(\mathbf{y}_{i,\text{inf}}, \mathbf{b}_i) : i = 1, \ldots, n\}$.

From (2.4), the likelihood function of $\boldsymbol{\theta}$ for the complete-data $\mathbf{Y}_{\text{com}} = \{(\mathbf{y}_i, \mathbf{b}_i) : i = 1, \ldots, n\}$ is proportional to

$$(\sigma^2)^{-\frac{N}{2}} \exp\left\{-\frac{\sum_{i=1}^n \text{tr}\, \mathbf{M}_i}{2\sigma^2}\right\} \cdot |\mathbf{D}|^{-\frac{n}{2}} \exp\left\{-\frac{\text{tr}\,(\mathbf{D}^{-1} \sum_{i=1}^n \mathbf{b}_i \mathbf{b}_i^T)}{2}\right\}, \tag{3.1}$$

where $N = \sum_{i=1}^n n_i$ and $\mathbf{M}_i$ is an $n_i \times n_i$ matrix defined by

$$\mathbf{M}_i = (\mathbf{y}_i - \mathbf{V}_i\boldsymbol{\alpha} - \mathbf{X}_i\boldsymbol{\beta} - \mathbf{W}_i\mathbf{b}_i)(\mathbf{y}_i - \mathbf{V}_i\boldsymbol{\alpha} - \mathbf{X}_i\boldsymbol{\beta} - \mathbf{W}_i\mathbf{b}_i)^T. \tag{3.2}$$

We develop two inter-related methods, a maximum likelihood approach based on the ECM algorithm and a Bayesian approach to estimate the dose-response relationship while accounting for the informative censoring and the constrained parameters. Posteriors in the Bayesian analysis are computed straightforwardly by taking advantage of the MLEs using the *inverse Bayes formulae* (IBF) sampling procedure (Tan, Tian & Ng (2003)).

3.1. Maximum Likelihood Estimation via the ECM Algorithm

From (3.1), given the complete-data $\mathbf{Y}_{\text{com}}$, the MLE of $\mathbf{D}$ depends only on $\{\mathbf{b}_i\}$ and does not involve the other parameters $(\boldsymbol{\beta}, \sigma^2, \boldsymbol{\alpha})$,

$$\hat{\mathbf{D}} = \frac{1}{n}\sum_{i=1}^{n} \mathbf{b}_i\mathbf{b}_i^T. \tag{3.3}$$

Given $\mathbf{Y}_{\text{com}}$, the MLEs of $(\boldsymbol{\beta}, \sigma^2, \boldsymbol{\alpha})$ are determined by the following equations:

$$\boldsymbol{\beta} = \left(\sum_{i=1}^{n} \mathbf{X}_i^T\mathbf{X}_i\right)^{-1} \sum_{i=1}^{n} \mathbf{X}_i^T(\mathbf{y}_i - \mathbf{V}_i\boldsymbol{\alpha} - \mathbf{W}_i\mathbf{b}_i), \tag{3.4}$$

and

$$\sigma^2 = \frac{1}{N}\sum_{i=1}^{n} \operatorname{tr}\mathbf{M}_i, \qquad \alpha_1 = (v_1 - \boldsymbol{\Omega}_{12}\boldsymbol{\alpha}_{-1})/\omega_{11}, \tag{3.5}$$

$$\boldsymbol{\alpha}_{-1} = \arg\min_{\boldsymbol{\alpha}_{-1}\geq 0} \left\{\boldsymbol{\alpha}_{-1}^T\boldsymbol{\Omega}_{22}\boldsymbol{\alpha}_{-1} - 2\boldsymbol{\alpha}_{-1}^T(\mathbf{v}_{-1} - \alpha_1\boldsymbol{\Omega}_{21})\right\}, \tag{3.6}$$

where $\operatorname{argmin}\{f(x)\}$ means the value of x at which $f(x)$ arrives its minimum, $\boldsymbol{\alpha}_{-1} = (\alpha_2, \ldots, \alpha_m)^T$, $\mathbf{M}_i$ is given by (3.2), and

$$\boldsymbol{\Omega} = \begin{pmatrix} \omega_{11} & \boldsymbol{\Omega}_{12} \\ \boldsymbol{\Omega}_{21} & \boldsymbol{\Omega}_{22} \end{pmatrix} \equiv \textstyle\sum_{i=1}^{n} \mathbf{V}_i^T\mathbf{V}_i,$$

$$\tag{3.7}$$

$$\mathbf{v} = \begin{pmatrix} v_1 \\ \mathbf{v}_{-1} \end{pmatrix} \equiv \textstyle\sum_{i=1}^{n} \mathbf{V}_i^T(\mathbf{y}_i - \mathbf{X}_i\boldsymbol{\beta} - \mathbf{W}_i\mathbf{b}_i).$$

To obtain the solution of $\boldsymbol{\alpha}_{-1}$ in (3.6), we use a novel EM algorithm via data augmentation proposed by Tian, Fang & Tan (2004). Equation (3.6) is equivalent to

$$\boldsymbol{\alpha}_{-1} = \arg\min_{\boldsymbol{\alpha}_{-1}\geq 0} (\boldsymbol{\alpha}_{-1} - \tilde{\mathbf{v}})^T\boldsymbol{\Omega}_{22}(\boldsymbol{\alpha}_{-1} - \tilde{\mathbf{v}}) = \arg\min_{\boldsymbol{\alpha}_{-1}\geq 0} \|\boldsymbol{\mu} - \mathbf{A}\boldsymbol{\alpha}_{-1}\|^2,$$

where $\tilde{\mathbf{v}} = \mathbf{v}_{-1} - \alpha_1\boldsymbol{\Omega}_{21}$, $\mathbf{A} = (a_{ik})$ is the upper triangular matrix with positive diagonal elements from the Cholesky decomposition of $\boldsymbol{\Omega}_{22}$ such that

$\boldsymbol{\Omega}_{22} = \mathbf{A}^T\mathbf{A}$ and $\boldsymbol{\mu} = (\mu_2, \ldots, \mu_m)^T = \mathbf{A}\tilde{\mathbf{v}}$. Given the current estimate $\boldsymbol{\alpha}_{-1}^{(t)} = (\alpha_2^{(t)}, \ldots, \alpha_m^{(t)})^T$, the E-step calculates

$$S_k^{(t)} = \sum_{i=2}^m a_{ik}\left[a_{ik}\alpha_k^{(t)} + \frac{\mu_i - \sum_{\ell=2}^m a_{i\ell}\alpha_\ell^{(t)}}{m-1}\right] \Big/ \sum_{i=2}^m a_{ik}^2, \quad k = 2, \ldots, m,$$

and the M-step updates

$$\alpha_k^{(t+1)} = \max\{0, S_k^{(t)}\}, \quad k = 2, \ldots, m.$$

The parameters $\boldsymbol{\theta}$ can be estimated via (3.3)-(3.6) using the EM algorithm. Because of missing data, the MLE involves an iterative algorithm and the conventional EM algorithm (Dempster, Laird & Rubin (1977)) does not apply. Thus, we use the ECM algorithm (Meng & Rubin (1993)) to obtain MLEs of $\boldsymbol{\theta}$. Let $\boldsymbol{\theta}^{(t)} = (\mathbf{D}^{(t)}, \boldsymbol{\beta}^{(t)}, \sigma^{2(t)}, \boldsymbol{\alpha}^{(t)})$ be the estimates from the t-th iteration of the CM steps. The $(t+1)$-th E-step computes the conditional expected values of the complete-data sufficient statistics:

$$E(\mathbf{Y}_{\text{mis}}|\mathbf{Y}_{\text{obs}}^*, \boldsymbol{\theta}^{(t)}), \qquad E(\mathbf{Y}_{\text{mis}}\mathbf{Y}_{\text{mis}}^T|\mathbf{Y}_{\text{obs}}^*, \boldsymbol{\theta}^{(t)}), \tag{3.8}$$

which are calculated by the method proposed in Tian, Fang & Tan (2004). Then we perform the first CM-step, which calculates $(\mathbf{D}^{(t)}, \boldsymbol{\beta}^{(t+1)})$ using (3.3) and (3.4) with $\mathbf{Y}_{\text{mis}}$ and $\mathbf{Y}_{\text{mis}}\mathbf{Y}_{\text{mis}}^T$ being replaced by the corresponding conditional expectations in (3.8), respectively. Having obtained $(\mathbf{D}^{(t)}, \boldsymbol{\beta}^{(t+1)})$, we then perform the second CM-step, which calculates $(\sigma^{2(t+1)}, \boldsymbol{\alpha}^{(t+1)})$ based on (3.5) and (3.6) where $\mathbf{Y}_{\text{mis}}$ and $\mathbf{Y}_{\text{mis}}\mathbf{Y}_{\text{mis}}^T$ are replaced with the corresponding conditional expectations in (3.8), respectively. The algorithm is iterated until $\|\boldsymbol{\theta}^{(t+1)} - \boldsymbol{\theta}^{(t)}\|$ is sufficiently small. Assume that the ECM algorithm converged at the $(t+1)$-th iteration, then the MLE $\hat{\boldsymbol{\theta}} = \boldsymbol{\theta}^{(t+1)}$. The observed information matrix is given by (Louis (1982))

$$-E\left\{\frac{\partial^2 \ell(\boldsymbol{\theta}|\mathbf{Y}_{\text{com}})}{\partial\boldsymbol{\theta}\partial\boldsymbol{\theta}^T}\right\}\bigg|_{\boldsymbol{\theta}=\hat{\boldsymbol{\theta}}} - \text{Var}\left\{\frac{\partial \ell(\boldsymbol{\theta}|\mathbf{Y}_{\text{com}})}{\partial\boldsymbol{\theta}}\right\}\bigg|_{\boldsymbol{\theta}=\hat{\boldsymbol{\theta}}} \tag{3.9}$$

where the expectation and variance are with respect to $f(\mathbf{Y}_{\text{mis}}|\mathbf{Y}_{\text{obs}}^*, \hat{\boldsymbol{\theta}})$. Standard errors are equal to the square root of the diagonal elements of the inverse of the estimated information matrix.

3.2. Bayesian Analysis via the IBF Sampling

The likelihood-based statistical inferences depend on the large sample theories and the ability to compute standard errors of the estimates (Louis (1982); Oakes (1999)). Thus, for small to moderate sample sizes, the Bayesian approach is an appealing alternative. The key to Bayesian analysis is to compute posteriors. Although the *Markov chain Monte Carlo* (MCMC) method is applicable to the present model, the burden of proof is shifted to the assessment of the convergence of the Markov chain to its stationary distribution. Furthermore, convergence is slowed substantially since each cycle of the MCMC requires another

MCMC to generate dependent samples from *truncated multivariate normal distributions* (TMVND) induced by the restricted parameters (Breslaw (1994); Robert (1995)). In addition, the slowness is acute if the components of the truncated multivariate normal vector are highly correlated. Therefore, we adopt a non-iterative sampling procedure, the IBF sampler (Tan, Tian & Ng (2003)), to obtain *independently and identically distributed* (iid) samples approximately from the observed posterior distribution, thus eliminating the convergence problem associated with MCMC.

We adopt the common prior of inverse Wishart distribution on $\mathbf{D}$: $\mathbf{D} \sim \mathrm{IW}_{q+1}(\nu_0,\, \boldsymbol{\Lambda}_0^{-1})$ with density

$$\mathrm{IW}_{q+1}(\mathbf{D}|\nu_0,\, \boldsymbol{\Lambda}_0^{-1}) \propto |\mathbf{D}|^{-(\nu_0+q+2)/2} \exp\left\{-\frac{1}{2}\,\mathrm{tr}\,(\boldsymbol{\Lambda}_0\mathbf{D}^{-1})\right\},$$

a diffuse prior on $\boldsymbol{\beta}$: $\boldsymbol{\beta} \sim N_{p+1}(\mathbf{0}, \mathbf{B}_0^{-1})$ with $\mathbf{B}_0 \to 0$, an inverse gamma prior on σ^2: $\sigma^2 \sim \mathrm{IG}(\frac{q_0}{2}, \frac{\lambda_0}{2})$ with density

$$\mathrm{IG}\left(u \,\middle|\, \frac{q_0}{2}, \frac{\lambda_0}{2}\right) = \frac{(\lambda_0/2)^{q_0/2}}{\Gamma(q_0/2)} u^{-1-q_0/2} \exp\left\{-\frac{\lambda_0}{2u}\right\},$$

where q_0 and λ_0 are known constants, and a diffuse prior of truncated m-dimensional normal distribution on $\boldsymbol{\alpha}$: $\boldsymbol{\alpha} \sim TN_m(\mathbf{0}, \mathbf{A}_0^{-1};\, I\!R \times I\!R_+^{m-1})$ with $\mathbf{A}_0 \to 0$, where the TMVND is formally defined in Tian, Fang & Tan (2004). We further assume that these priors are independent. Then the complete-data posterior distribution is given by

$$f(\boldsymbol{\theta}|\mathbf{Y}_{\mathrm{com}}) = f(\boldsymbol{\sigma}|\mathbf{Y}_{\mathrm{com}}, \boldsymbol{\alpha}) \times f(\boldsymbol{\alpha}|\mathbf{Y}_{\mathrm{com}}) \tag{3.10}$$

and

$$\begin{aligned} f(\boldsymbol{\sigma}|\mathbf{Y}_{\mathrm{com}}, \boldsymbol{\alpha}) &\\ &= f(\mathbf{D}|\mathbf{Y}_{\mathrm{com}}, \boldsymbol{\alpha}) \times f(\boldsymbol{\beta}|\mathbf{Y}_{\mathrm{com}}, \sigma^2, \boldsymbol{\alpha}) \times f(\sigma^2|\mathbf{Y}_{\mathrm{com}}, \boldsymbol{\alpha}) \\ &= \mathrm{IW}_{q+1}(\nu_0+n, \boldsymbol{\Lambda}^{-1}) \times N_{p+1}(\boldsymbol{\beta}|\boldsymbol{\beta}_0(\boldsymbol{\alpha}), \sigma^2\mathbf{X}) \times \mathrm{IG}\left(\sigma^2 \,\middle|\, \tfrac{q^*}{2}, \tfrac{\lambda^*}{2}\right), \end{aligned} \tag{3.11}$$

where $q^* = q_0 + N - p - 1$, $\lambda^* = \lambda_0 + s(\boldsymbol{\alpha})$,

$$\boldsymbol{\Lambda} = \boldsymbol{\Lambda}_0 + \textstyle\sum_{i=1}^n \mathbf{b}_i\mathbf{b}_i^T,$$

$$\boldsymbol{\beta}_0(\boldsymbol{\alpha}) = \mathbf{X}\textstyle\sum_{i=1}^n \mathbf{X}_i^T(\mathbf{y}_i - \mathbf{V}_i\boldsymbol{\alpha} - \mathbf{W}_i\mathbf{b}_i),$$

$$s(\boldsymbol{\alpha}) = \textstyle\sum_{i=1}^n \|\mathbf{y}_i - \mathbf{V}_i\boldsymbol{\alpha} - \mathbf{W}_i\mathbf{b}_i\|^2 - \boldsymbol{\beta}_0^T(\boldsymbol{\alpha})\mathbf{X}^{-1}\boldsymbol{\beta}_0(\boldsymbol{\alpha}),$$

and $\mathbf{X} = (\sum_{i=1}^n \mathbf{X}_i^T\mathbf{X}_i)^{-1}$.

To obtain iid samples approximately from the observed posterior distribution $f(\boldsymbol{\theta}|\mathbf{Y}^*_{\text{obs}})$, we use the IBF sampling method (Tan, Tian & Ng (2003)) which utilizes the following formula

$$f(\boldsymbol{\theta}|\mathbf{Y}^*_{\text{obs}}) \propto \frac{f(\boldsymbol{\theta}|\mathbf{Y}^*_{\text{obs}}, Z_0)}{f(Z_0|\mathbf{Y}^*_{\text{obs}}, \boldsymbol{\theta})}, \tag{3.12}$$

where $Z_0 = E(\mathbf{Y}_{\text{mis}}|\mathbf{Y}^*_{\text{obs}}, \hat{\boldsymbol{\theta}})$ and $\hat{\boldsymbol{\theta}}$ denotes the MLE of $\boldsymbol{\theta}$. The IBF sampling via sampling/importance resampling method (Rubin (1988)) is as follows:

(i) Draw J independent samples of $\boldsymbol{\theta}$ from $f(\boldsymbol{\theta}|\mathbf{Y}^*_{\text{obs}}, Z_0)$, denoted by $\{\boldsymbol{\theta}^{(j)}\}_1^J$;
(ii) Calculate the weights $\omega_j = f^{-1}(Z_0|\mathbf{Y}^*_{\text{obs}}, \boldsymbol{\theta}^{(j)})/\sum_{\ell=1}^J f^{-1}(Z_0|\mathbf{Y}^*_{\text{obs}}, \boldsymbol{\theta}^{(\ell)})$ for $j = 1, \ldots, J$;
(iii) Choose a subset from $\{\boldsymbol{\theta}^{(j)}\}_1^J$ via resample *without* replacement from the discrete distribution on $\{\boldsymbol{\theta}^{(j)}\}_1^J$ with probabilities $\{\omega_j\}_1^J$ to obtain an iid sample of size $M\ (< J)$ approximately from the observed posterior $f(\boldsymbol{\theta}|\mathbf{Y}^*_{\text{obs}})$ with the approximation "improving" as J increases (Smith & Gelfand (1992)).

Sampling from $f(\boldsymbol{\theta}|\mathbf{Y}^*_{\text{obs}}, Z_0)$ can be obtained via (3.10), in which sampling from $f(\boldsymbol{\sigma}|\mathbf{Y}_{\text{com}}, \boldsymbol{\alpha})$ in (3.11) is straightforward. To obtain iid samples approximately from $f(\boldsymbol{\alpha}|\mathbf{Y}_{\text{com}})$, similar to (3.12), we have

$$f(\boldsymbol{\alpha}|\mathbf{Y}_{\text{com}}) \propto \frac{f(\boldsymbol{\alpha}|\mathbf{Y}_{\text{com}}, \hat{\boldsymbol{\sigma}})}{f(\hat{\boldsymbol{\sigma}}|\mathbf{Y}_{\text{com}}, \boldsymbol{\alpha})}, \tag{3.13}$$

where $\hat{\boldsymbol{\sigma}}$ is the MLE of $\boldsymbol{\sigma} = (\mathbf{D}, \boldsymbol{\beta}, \sigma^2)$. Note that $f(\hat{\boldsymbol{\sigma}}|\mathbf{Y}_{\text{com}}, \boldsymbol{\alpha})$ denotes the value of the conditional density $f(\boldsymbol{\sigma}|\mathbf{Y}_{\text{com}}, \boldsymbol{\alpha})$ evaluated at $\boldsymbol{\sigma} = \hat{\boldsymbol{\sigma}}$, from (3.11). However, the random vector $\boldsymbol{\alpha}$ given $(\mathbf{Y}_{\text{com}}, \boldsymbol{\sigma})$ has a truncated multivariate normal distribution,

$$f(\boldsymbol{\alpha}|\mathbf{Y}_{\text{com}}, \boldsymbol{\sigma}) = TN_m(\boldsymbol{\alpha}|\boldsymbol{\Omega}^{-1}\mathbf{v},\ \sigma^2\boldsymbol{\Omega}^{-1};\ \mathbb{R} \times \mathbb{R}_+^{m-1}), \tag{3.14}$$

where the rectangle $\mathbb{R} \times \mathbb{R}_+^{m-1}$ is defined in (2.3), $\mathbf{v}$ and $\boldsymbol{\Omega}$ are given by (3.7). Then the sampling approach for truncated multivariate normal distributions in Tian, Fang & Tan (2004) can be used.

To calculate the denominator $f(Z_0|\mathbf{Y}^*_{\text{obs}}, \boldsymbol{\theta})$ in (3.12), we now consider the conditional predictive distribution of the missing data $\mathbf{Y}_{\text{mis}}$ for given $(\mathbf{Y}^*_{\text{obs}}, \boldsymbol{\theta})$ since $f(Z_0|\mathbf{Y}^*_{\text{obs}}, \boldsymbol{\theta}) = f(\mathbf{Y}_{\text{mis}}|\mathbf{Y}^*_{\text{obs}}, \boldsymbol{\theta})|_{\mathbf{Y}_{\text{mis}}=Z_0}$. From (2.4), we have $(\mathbf{y}_i^T, \mathbf{b}_i^T)^T \overset{\text{ind}}{\sim} N_{n_i+q+1}(\boldsymbol{\mu}_i,\ \boldsymbol{\Sigma}_i)$, where

$$\boldsymbol{\mu}_i = \begin{pmatrix} \mathbf{V}_i\boldsymbol{\alpha} + \mathbf{X}_i\boldsymbol{\beta} \\ \mathbf{0} \end{pmatrix}, \quad \boldsymbol{\Sigma}_i = \begin{pmatrix} \mathbf{W}_i\mathbf{D}\mathbf{W}_i^T + \sigma^2\mathbf{I}_{n_i} & \mathbf{W}_i\mathbf{D} \\ \mathbf{D}\mathbf{W}_i^T & \mathbf{D} \end{pmatrix}.$$

Denote the conditional density of $\mathbf{b}_i$ given $(\mathbf{Y}^*_{\text{obs}}, \boldsymbol{\theta})$ by $f_1(\mathbf{b}_i|\mathbf{Y}^*_{\text{obs}}, \boldsymbol{\theta})$ which is a $(q+1)$-dimensional normal density, and the conditional density of $\mathbf{y}_{i,\text{inf}}$ given $(\mathbf{Y}^*_{\text{obs}}, \mathbf{b}_i, \boldsymbol{\theta})$ by $f_2(\mathbf{y}_{i,\text{inf}}|\mathbf{Y}^*_{\text{obs}}, \mathbf{b}_i, \boldsymbol{\theta})$ which is a truncated $(n_i - p_i)$-dimensional normal density,

$$f_2(\mathbf{y}_{i,\inf}|\mathbf{Y}^*_{\mathrm{obs}}, \mathbf{b}_i, \boldsymbol{\theta}) = TN_{n_i-p_i}\Big(\mathbf{y}_{i,\inf}\Big|\tilde{\boldsymbol{\mu}}_i,\ \sigma^2\mathbf{I}_{n_i-p_i};\ \mathcal{A}_{n_i-p_i}(a)\Big), \quad (3.15)$$

where $\tilde{\boldsymbol{\mu}}_i = (\tilde{\mu}_{i,p_i+1}, \ldots, \tilde{\mu}_{in_i})^T = E(\mathbf{y}_{i,\inf}|\mathbf{Y}^*_{\mathrm{obs}}, \mathbf{b}_i, \boldsymbol{\theta})$ and $\mathcal{A}_{n_i-p_i}(a) = \{\mathbf{y}_{i,\inf} : \mathbf{y}_{i,\inf} < a\}$. Noting that $\sigma^2\mathbf{I}_{n_i-p_i}$ is a diagonal matrix, then the components of $\mathbf{y}_{i,\inf} = (y_{i,p_i+1}, \ldots, y_{in_i})^T$ given $(\mathbf{Y}^*_{\mathrm{obs}}, \mathbf{b}_i, \boldsymbol{\theta})$ are independent and distributed as truncated univariate normal, i.e., $f_2(\mathbf{y}_{i,\inf}|\mathbf{Y}^*_{\mathrm{obs}}, \mathbf{b}_i, \boldsymbol{\theta}) = \prod_{k=p_i+1}^{n_i} TN_1(y_{ik}|\tilde{\mu}_{ik},\ \sigma^2;\ (-\infty, a))$. Hence, the conditional predictive density of $\mathbf{Y}_{\mathrm{mis}}$ is given by

$$f(\mathbf{Y}_{\mathrm{mis}}|\mathbf{Y}^*_{\mathrm{obs}}, \boldsymbol{\theta}) = \prod_{i=1}^{n}\Big\{f_1(\mathbf{b}_i|\mathbf{Y}^*_{\mathrm{obs}}, \boldsymbol{\theta}) \times f_2(\mathbf{y}_{i,\inf}|\mathbf{Y}^*_{\mathrm{obs}}, \mathbf{b}_i, \boldsymbol{\theta})\Big\}. \quad (3.16)$$

4 Comparison of Treatment Effects

To compare two treatments in xenograft experiments, a common method is to test the treatment diffierence at a fixed points using a t-test or a Mann-Whitney test, an ANOVA F-test or a Kruskal-Wallis test, indicating all the times at which differences were significant (*e.g.* Kasprzyk, Song, Di Fiore & King (1992)). For example, a test may be performed at the final measurement time or the final time when a substantial fraction of the animals were alive (Sakaguchi, Maehara, Baba, Kusumoto, Sugimachi & Newman (1992)). The method to analyse animal survival times in addition to tumor volumes was also used in literature (Honghton *et al.* 2000). Based on a comparison of those methods, Heitjan, Manni & Santen (1993) indicated that the methods commonly used are deficient in that they have either low power or misleading type I error rates and proposed a multivariate method to improving the efficiency of testing. However, this multivariate test method does not account for the missing data and large sample sizes are necessary.

A major goal in the xenograft model is to assess the effectiveness of treatment regimens, e.g., the mean tumor sizes at different time points as opposed to a growth curve analysis where the nonlinearity of tumor responses over the follow-up period and the limited amount of data in xenograft models preclude a growth curve characterization. When the goal of the study is to compare two treatment groups, a test can be derived as a special case in the random-effects model. However, a simpler approach is to use a modified t-test and the Bayesian hypothesis testing.

Consider a longitudinal study with m prespecified follow-up times $t_1 < t_2 < \cdots < t_m$ for n subjects. Let $\mathbf{Y}_i^{(k)} = (Y_{i1}^{(k)}, \ldots, Y_{im}^{(k)})^T$ be an $m \times 1$ vector of outcomes which are the tumor volumes (in log-scale) from the ith subject in the kth group, $i = 1, 2, \ldots, n_k$, $k = 1, 2$, and $n = n_1 + n_2$. Based on the assumptions in Section 2, we focus on the model where $\mathbf{Y}_i^{(k)}$ has a multivariate normal distribution with mean vector and covariance matrix of the toeplitz form

$$E(\mathbf{Y}_i^{(k)}) = \boldsymbol{\mu}^{(k)} \,\hat{=}\, (\mu_1^{(k)}, \ldots, \mu_m^{(k)})^T,$$
$$\mathrm{Cov}(\mathbf{Y}_i^{(k)}) = \boldsymbol{\Sigma}_m = \sigma^2 \mathbf{R}_m, \tag{4.1}$$

respectively, $k = 1, 2$, where $\mathbf{R}_m$ is defined in (2.7).

Our goal is to test hypotheses

$$H_0: \quad \boldsymbol{d}^T \boldsymbol{\mu}^{(1)} = \boldsymbol{d}^T \boldsymbol{\mu}^{(2)} \qquad \text{versus} \qquad H_1: \quad \boldsymbol{d}^T \boldsymbol{\mu}^{(1)} < \boldsymbol{d}^T \boldsymbol{\mu}^{(2)},$$

where $\boldsymbol{d} = (d_1, \ldots, d_m)^T$ is a known contrast vector which is chosen based on the scientific goal of the study. For example, if we want to find out if a different dosing schedule can decrease the total tumor volumes (or area under the tumor growth curve) further than does another schedule in the xenograft models, the contrast vector $\boldsymbol{d}$ would be the m-dimensional vector with unit component. If our goal is to compare tumor response after the first course of treatment (consisting of multiple doses of a drug), the components of the contrast vector $\boldsymbol{d}$ should be weighted appropriately. To test the hypotheses H_0, Tan, Fang, Tian & Houghton (2002) proposed two approaches, one is a heuristic t-test and a Bayesian hypothesis test.

4.1 Exact Test Based on the EM Algorithm

If we had the complete-data $\mathbf{Y}_{\text{com}} = \{\mathbf{Y}_i^{(k)} : i = 1, \ldots, n_k,\ k = 1, 2\}$, where $\mathbf{Y}_i^{(k)} \sim N_m(\boldsymbol{\mu}^{(k)}, \sigma^2 \mathbf{R}_m)$, we could simply use the t-statistic to test H_0,

$$t = \frac{\boldsymbol{d}^T \bar{\mathbf{Y}}^{(1)} - \boldsymbol{d}^T \bar{\mathbf{Y}}^{(2)}}{\sqrt{\boldsymbol{d}^T (S^{(1)} + S^{(2)}) \boldsymbol{d}}} \sqrt{\frac{n_1 n_2 (n_1 + n_2 - 2)}{n_1 + n_2}}, \tag{4.2}$$

where $\bar{\mathbf{Y}}^{(k)} = \frac{1}{n_k} \sum_{i=1}^{n_k} \mathbf{Y}_i^{(k)}$, $S^{(k)} = \sum_{i=1}^{n_k} \mathbf{Y}_i^{(k)} \mathbf{Y}_i^{(k)T} - n_k \bar{\mathbf{Y}}^{(k)} \bar{\mathbf{Y}}^{(k)T} = (1 - \frac{1}{n_k}) \sum_{i=1}^{n_k} \mathbf{Y}_i^{(k)} \mathbf{Y}_i^{(k)T} - \frac{1}{n_k} \sum_{i \neq j} \mathbf{Y}_i^{(k)} \mathbf{Y}_j^{(k)T}$, for $k = 1, 2$. If H_0 holds, then the t-statistic in (4.2) has t-distribution with degree of freedom $n_1 + n_2 - 2$.

When there are missing data, we use the EM algorithm to obtain maximum likelihood estimates of the parameters of interest $\boldsymbol{\phi} = (\boldsymbol{\mu}^{(1)}, \boldsymbol{\mu}^{(2)}, \sigma^2, \rho)^T$ (cf. Section 3.1 above). Denote the MLE of $\boldsymbol{\phi}$ by $\hat{\boldsymbol{\phi}}$. Then, the t-statistic in (4.1) is calculated with $\mathbf{Y}_i^{(k)}$ and $\mathbf{Y}_i^{(k)} \mathbf{Y}_i^{(k)T}$ being replaced by the corresponding conditional expectations

$$E(\mathbf{Y}_i^{(k)} | \mathbf{Y}_{\text{obs}}^*, \hat{\boldsymbol{\phi}}) \quad \text{and} \quad E(\mathbf{Y}_i^{(k)} \mathbf{Y}_i^{(k)T} | \mathbf{Y}_{\text{obs}}^*, \hat{\boldsymbol{\phi}}), \tag{4.3}$$

respectively, where $\mathbf{Y}_{\text{obs}}^*$ is defined as in Section 3.

4.2 Bayesian Test

The Bayesian hypothesis testing requires calculating the observed posterior probability of the one-sided alternative hypothesis $H_1 : \mathbf{d}^T \boldsymbol{\mu}^{(1)} < \mathbf{d}^T \boldsymbol{\mu}^{(2)}$, namely,

$$\Pr\{H_1|\mathbf{Y}^*_{\text{obs}}\} = \Pr\{\mathbf{d}^T(\boldsymbol{\mu}^{(1)} - \boldsymbol{\mu}^{(2)}) < 0|\mathbf{Y}^*_{\text{obs}}\}$$
$$\doteq \tfrac{1}{L}\textstyle\sum_{\ell=1}^{L} I(\mathbf{d}^T(\boldsymbol{\mu}^{(1,\ell)} - \boldsymbol{\mu}^{(2,\ell)}) < 0), \tag{4.4}$$

where $\{(\boldsymbol{\mu}^{(1,\ell)}, \boldsymbol{\mu}^{(2,\ell)}) : \ell = 1, \ldots, L\}$ is an iid sample from the observed posterior density $f(\boldsymbol{\mu}^{(1)}, \boldsymbol{\mu}^{(2)}|\mathbf{Y}^*_{\text{obs}})$ and $I(\cdot)$ denotes the indicator function. If the posterior probability of H_1 is greater than or equal to certain level (e.g., 95%), we reject the null hypothesis. We chose posterior probability (over the Bayes factor) for its ease of interpretation to biologists. For incomplete data, the samples from the observed posterior density $f(\boldsymbol{\mu}^{(1)}, \boldsymbol{\mu}^{(2)}|\mathbf{Y}^*_{\text{obs}})$ can be obtained using the IBF sampler (cf. Section 3.2).

With an analysis of a real study on xenograft models for two new anti-cancer agents temozolomide and irinotecan, Tan, Fang, Tian & Houghton (2002) shows that these two test approaches are in concordance. The main advantage of the presented methods is that they are valid for small samples which occur animal studies for cancer drug development.

5 Models for Multiple Tumors

In some tumor systems, multiple tumors are grown in each mouse. For example, preclinical studies evaluating the anti-tumor effects of exemestane and tamoxifen for postmenopausal breast cancer, aromatase-transfected human breast cancer cells (MCF-7Ca) were inoculated into the ovariectomized athymic mice. Each mouse received subcutaneous injections at two sites on each flank with $0.1ml$ of cell suspension (2.5×10^7 cells/ml) and four tumors were grown in each mouse (see Jelovac, Macedo, Handratta, Long, Goloubeva & Brodie (2004).

In this section, we extend the proposed methods to xenograft models for multiple tumors where a mouse is simultaneously grafted several human tumor cells. To model the antitumor activity, let q be the number of agents and $t_0 < t_1 < \cdots < t_m$ the prespecified follow-up times. Denote K the number of tumors which were grafted in each mouse. For mouse $i (i = 1, \ldots, n)$, let y_{i0k} and y_{ijk} be the initial tumor volume at t_0 and the tumor volume of k-th tumor at t_j of mouse i (in log scale), respectively. Due to early withdrawal, we only observe $\mathbf{y}_{ik} = (y_{i1k}, \ldots, y_{in_ik})^T$. Let $z_{ij}^{(h)}$ denote the cumulative dose of agent $h\,(h = 1, \ldots, q)$ administered to subject i up to time t_j for $j = 1, \ldots, m$. Further consider $(p - q)$ interaction terms induced by the q agents. The corresponding doses are denoted by $z_{ij}^{(h)}$, where $h = q + 1, \ldots, p$. Therefore, $\mathbf{z}_i^{(h)} = (z_{i1}^{(h)}, \ldots, z_{im}^{(h)})^T$ denotes the doses of agent $h\,(h = 1, \ldots, p)$ given to subject i. Treating the initial value y_{i0k} as a covariate, we obtain a known $m \times p$ covariate matrix $\mathbf{Z}_{ik} = (y_{i0k}\mathbf{1}_m, \mathbf{z}_i^{(1)}, \ldots, \mathbf{z}_i^{(p)})$. Then, the dose-response relationship can be modelled by the following hierarchical models

$$\mathbf{y}_{ik} = \boldsymbol{\gamma}_{ik} + \mathbf{X}_{ik}\boldsymbol{\beta}_{ik} + \boldsymbol{\varepsilon}_{ik},$$
$$\boldsymbol{\beta}_{ik} = \boldsymbol{\beta} + \mathbf{b}_{ik}, \quad \text{for } k = 1, \ldots, K; i = 1, \ldots, n, \tag{5.1}$$

where $\mathbf{X}_{ik}$ is the $n_i \times (p+1)$ matrix consisting of the first n_i rows of $\mathbf{Z}_{ik}$, $\boldsymbol{\beta}$ is a $(p+1)$-dimensional unknown parmeter vector. Assume that the errors $\mathbf{b}_{ik}$ and $\boldsymbol{\varepsilon}_{ik}$ are independent and

$$\mathbf{B}_i = (\mathbf{b}_{i1}, \ldots, \mathbf{b}_{iK})^T \overset{\text{iid}}{\sim} N_{K\times(p+1)}(\mathbf{0}, \mathbf{R} \times \mathbf{D}),$$
$$\mathbf{E}_i = (\boldsymbol{\varepsilon}_{i1}, \ldots, \boldsymbol{\varepsilon}_{iK})^T \overset{\text{iid}}{\sim} N_{K\times n_i}(\mathbf{0}, \sigma^2(\mathbf{I}_K \otimes \mathbf{I}_{n_i})), \tag{5.2}$$

for $i = 1, \ldots, n$, where $\mathbf{R}$ and $\mathbf{D}$ are K and $(p+1)$ order positive definite matrices, respectively.

In model (5.1), the intercept $\boldsymbol{\gamma}_{ik} = (\gamma_{1k}, \ldots, \gamma_{n_ik})^T)$ reflects the intrinsic growth of tumor when no drug is given. This results in an order constraint on some regression coefficients, i.e. for each k,

$$\gamma_{1k} \leq \cdots \leq \gamma_{n_ik}. \tag{5.3}$$

According to the description in Section 2, we make a transformation $\boldsymbol{\gamma}_{ik} = \mathbf{V}_i\boldsymbol{\alpha}_k$ for the restricted parameter estimation, where $\boldsymbol{\alpha}_k = (\alpha_{1k}, \cdots, \alpha_{mk})^T \in I\!R \times I\!R_+^{m-1}$ and $\mathbf{V}_i$ is the $n_i \times m$ matrix defined in Section 2. The Bayesian analysis via IBF sampling proposed in Section 3 is employed to estimate the parameters.

We now analyze the anti-tumor efficacy of steroidal aromatase inhibitor exemestane alone or in combination with the antiestrongen tamoxifen in the xenograft models of postmenopausal breast cancer. Thirty-three female BALB/c athymic ovariectomized mice have been successfully cultured with subcutaneous transplant of tumors. Treatment started when one of the 4 tumors reached a measurable sizes ($\sim 0.03cm^3$) about 4 weeks after inoculation. At the start of treatment, the upper left tumor of 8 mice, the lower left tumor of one mouse, the upper right tumor of 5 mice and the lower tumor of 5 mice were unmeasurable ($< 0.03cm^3$). Mice were assigned to different treatment groups (5 mice per group) with different dosage of single drugs or their combinations, *e.g.* the mouse received 100 μg/day tamoxifen in gourp I, 100 μg/day exemestane in group II, 250 μg/day exemestane in group III, 100 μg/day exemestane combined with 100 μg/day tamoxifen in group IV, 250 μg/day exemestane combined with 100 μg/day tamoxifen in group V, and no treated in the control group. Tumors were measured weekly over a given period of time (e.g. 9 weeks). Of 33 mice, the tumor volumes of 19 mice were complete observed and 14 mice were incomplete observed. About 11% of these tumor volumes are below $0.03cm^3$ and were considered not measurable. Figure 1 shows the plots for tumor volumes of mice in group II.

Let y_{i0k} and $\mathbf{y}_{ik} = (y_{i1k}, \ldots, y_{i9k})^T$ be the logarithmic transformations of the initial tumor volume and a 9-dimensional vector of tumor volumes at each week from the k-tumor of the i-th mouse, respectively. Then, y_{ijk} is truncated if $y_{ijk} < \log 0.03$. Let $x_{ij}^{(1)}$ and $x_{ij}^{(2)}$ be the cumulative weekly total doses of tamoxifen and exemestane respectively, which the i-th mouse received at j-th week, $(j = 1, 2, \ldots, 9)$. To consider the synergism of the two drugs, we denote $x_{ij}^{(3)} = \sqrt{x_{ij}^{(1)} x_{ij}^{(2)}}$ as the interaction term (cf. Finney (1971)). Let $\mathbf{x}_i^{(u)} = (x_{i1}^{(u)}, \ldots, x_{i9}^{(u)})^T$, $u = 1, 2, 3$. Then, the model (5.1) is proposed to study the activity of exemestane combined with tamoxifen against the postmenopausal breast cancer in xenograft

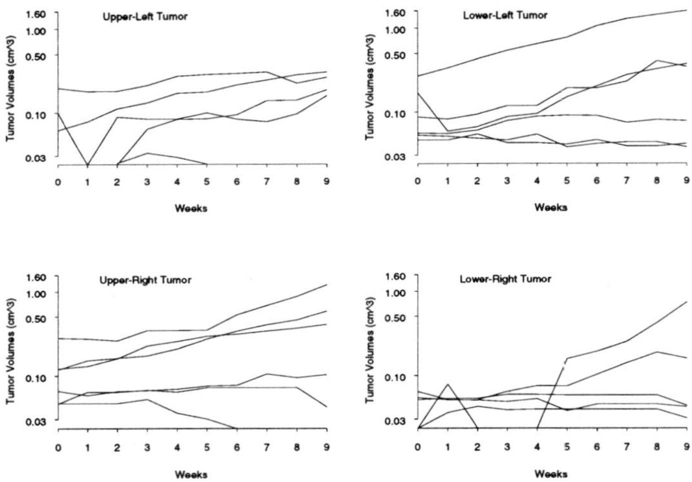

Figure 1.

Fig. 1. Aromatase-transfected breast cancer tumor volumes(cm^3) measured weekly for 9 weeks in treatment with 100 μg/day exemestane.

experiments,

$$\begin{aligned}\mathbf{y}_{ik} &= \boldsymbol{\gamma}_{ik} + \mathbf{X}_{ik}\boldsymbol{\beta} + \mathbf{X}_{ik}\mathbf{b}_{ik} + \boldsymbol{\varepsilon}_{ik}, \\ &= \mathbf{V}_i\boldsymbol{\alpha}_k + y_{i0k}\mathbf{1}_9\beta_0 + \mathbf{x}_i^{(1)}\beta_1 + \mathbf{x}_i^{(2)}\beta_2 + \mathbf{x}_i^{(3)}\beta_3 + \mathbf{X}_{ik}\mathbf{b}_{ik} + \boldsymbol{\varepsilon}_{ik},\end{aligned} \tag{5.4}$$

for $k = 1, \ldots, 4$ and $i = 1, \ldots, 33$, where $\mathbf{X}_{ik} = (y_{i0k}\mathbf{1}_9, \mathbf{x}_i^{(1)}, \mathbf{x}_i^{(2)}, \mathbf{x}_i^{(3)})$, $\boldsymbol{\beta} = (\beta_0, \beta_1, \beta_2, \beta_3)^T$ are the fixed effects, $\mathbf{b}_{ik} = (b_{i1k}, b_{i2k}, b_{i3k}, b_{i4k})^T$ are the random effects, and $\boldsymbol{\varepsilon}_{ik}$ is the error vector.

As alluded to earlier to use the IBF sampler, we use the ECM algorithm to obtain the MLE $\hat{\boldsymbol{\theta}} = (\hat{\mathbf{D}}, \hat{\mathbf{R}}, \hat{\sigma}^2, \hat{\boldsymbol{\beta}}, \hat{\boldsymbol{\alpha}}_k; k = 1,2,3,4)$ at 120 iterations and calculate $Z_0 = E(\mathbf{Y}_{\text{mis}}|\mathbf{Y}^*_{\text{obs}}, \hat{\boldsymbol{\theta}})$. Note some initial tumor volumes were unmeasurable. For the unmeasurable y_{i0k}, we draw a sample from the truncated normal distribution $TN(\mu_k, \sigma_k^2; (-\infty, \log 0.03))$ at each iteration, where μ_k and σ_k^2 are the estimated mean and variance from the observations of the k-tumor initial volumes ($k = 1,2,3,4$). We take the diffuse (or flat) priors for $\boldsymbol{\theta}$ as same as in Section 3.2. With $J = 8000$ and $M = 5000$, the IBF sampling based on (3.11) gives an iid sample of size 5000 approximately from the observed posterior distribution $f(\boldsymbol{\theta}|\mathbf{Y}^*_{\text{obs}})$. The posterior means of $\boldsymbol{\alpha}$ are: $\boldsymbol{\alpha}_1 = (-1.3876, 0.1469, 0.0656, 0.0467, 0.0419, 0.0404, 0.0398, 0.0456, 0.0678)^T$, $\boldsymbol{\alpha}_2 = (-1.2772, 0.0628, 0.0408, 0.0329, 0.0302, 0.0307, 0.0330, 0.0403, 0.0594)^T$, $\boldsymbol{\alpha}_3 = (-1.4264, 0.0976, 0.0685, 0.0629, 0.0609, 0.0613, 0.0624, 0.0703, 0.0891)^T$, $\boldsymbol{\alpha}_4 = (-1.4818, 0.0741, 0.0513, 0.0428, 0.0443, 0.0481, 0.0514, 0.0526, 0.0783)^T$. Figure 2 shows the intrinsic growth of untreated tumors which depends on the initial tumor volume. When no treatment was given, the tumors have different growth curves and the upper tumors have a faster growth than the lower tumors.

The posterior estimates of the parameters of interest are given in Table 1. The analysis results show that even with a more powerful test, both drugs exemestane and tamoxifen have low activity against the aromatase-transfected breast cancer tumor. Tamoxifen is more effective than exemestane in suppressing tumor growth. However, there is a significant synergism between these two drugs. The combination of two drugs is significantly effective than the individule drugs against the breast cancer.

Table 1. Bayesian Estimates of the Parameters

Parameter	Posterior Mean	Posterior SD	95% CI	p-value
β_0	0.46223	0.010254	$[0.44256, 0.48248]$	< 0.0001
β_1	-2.75×10^{-6}	0.000087	$[-0.00017, 0.00017]$	0.4873
β_2	4.16×10^{-6}	0.000038	$[-0.00007, 0.00008]$	0.4362
β_3	-2.43×10^{-4}	0.000090	$[-0.00042, -0.00007]$	0.0035
σ^2	0.39491	0.016403	$[0.36412, 0.42810]$	< 0.0001

6 Discussions

We have developed a class of multivariate random effects models to characterize the dose-response relationship in xenograft experiments with incomplete (missing at random and informative censoring) longitudinal data and the monotonicity of

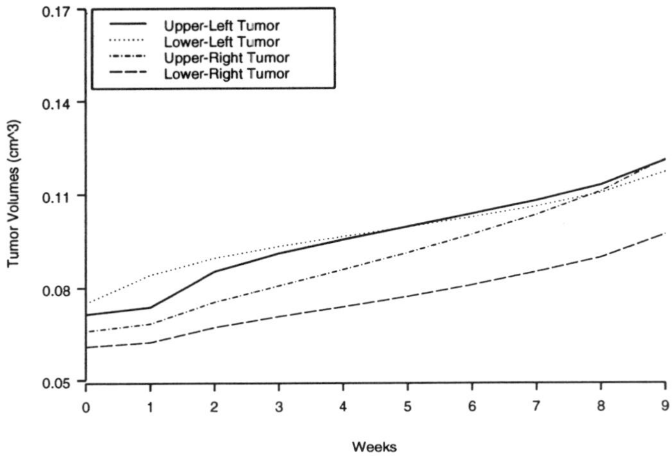

Figure 2.

Fig. 2. The predicted growth curves of untreated aromatase-transfected breast cancer tumor in xenograft experiments.

model parameters. Two approaches, MLE and Bayesian analysis, are proposed for model estimation. The main advantage of the presented methods is that they are valid for small samples which are common in the translational stage of the drug development process. The flat priors used in the Bayesian analysis imply the conclusions are similar to those from MLE. Although we have focused on the Toeplitz correlation structure, the simpler compound symmetry structure may be justified for within-cluster correlation in some cases. On the other hand, when the sample sizes are moderate, more complicated covariance structures may be incorporated in models (2.5) and (4.1).

Although we have focused on the type of xenograft experiments in our developmental therapeutics programs, we believe that the model formulation and estimation methods can be adopted to other animal experiments in research. Because the mechanism of the compound in development is better understood and some drugs are designed based on molecular targets resulted from the vast progresses in molecular and cellular biology and genetics in the last decade, coupled with the need for protecting human research subjects, tumor xenograft models play an important role in the translational research of bringing laboratory advances to clinic. The statistical models and methods proposed in this article serve as a basis for further development of methods to fully utilize the costly data and provide relevant information for the design of subsequent clinical trials.

Acknowledgments

The authors thank Dr. Angela Brodie at University of Maryland for permission to use her data as an example. The first author also wants to thank Prof. K.T. Fang for his encouragement that led to the introduction of a Biostatistics session to the 1997 International Symposium on Contemporary Multivariate Analysis and it Applications and for many thought-provoking conversations on experimental design, applied statistics and traditional Chinese medicine with him since then. This research was supported in part by US National Institutes of Health grant CA106767.

References

Breslaw, J. A. (1994). Random sampling from a truncated multivariate normal distribution. *Applied Mathematics Letters* **7** 1–6.

Diggle, P. J. & Kenward, M. G. (1994). Informative dropout in longitudinal data analysis (with discussion). *Applied Statistics* **43** 49–93.

Dempster, A. P., Laird, N. M. & Rubin, D. B. (1977). Maximum likelihood from incomplete data via the EM algorithm (with discussion). *Journal of the Royal Statistical Society, Series B* **39** 1–38.

Fang, H. B., Tian, G. L. & Tan, M. (2004). Hierarchical models for tumor xenograft experiments in drug development. *Journal of Biopharmaceutical Statiatics* **14** No.4 1–15.

Festing, M. F. W. & Altman, D. G. (2002). Guidelines for the design and statistical analysis of experiments using laboratory animals. *Institutre for Laboratory Animal Research Journal* **43** No.4 244–258.

Finney, D. J. *Probit Analysis*, 3rd edn. Cambridge University Press: London, 1971.

Heitjan, D. F. (1991). Generalized Norton-Simon models of tumour growth. *Statistics in Medicine* **10** 1075–1088.

Heitjan, D. F., Manni, A. & Santen, R. (1993). Statistical analysis of in vivo tumor growth experiments. *Cancer Research* **53** 6042–6050.

Hogan, J. W. & Laird, N. M. (1997). Mixture models for the joint distribution of repeated measures and event times. *Statistics in Medicine* **16** 239–258.

Jelovac, D., Macedo, L., Handratta, V., Long, B. L., Goloubeva, O. G., & Brodie, A. M. H. (2004). Preclinical Studies Evaluating the Anti–tumor Effects of Exemestane Alone or Combined with Tamoxifen in a Postmenopausal Breast Cancer Model. to be submitted.

Kasprzyk, P. G., Song, S. U., Di Fiore, P. P. & King, C. R. (1992). Therapy of an animal model of human gastric cancer using a combination of anti–*erb*B–2 monoclonal antibodies. *Cancer Research* **52** 2771–2776.

Liang, J., Li, R., Fang, H. B., & Fang, K. T. (2000). Testing multinormality based on low–dimensional projection. *Journal of Statistical Planning and Inference* **86** 129–141.

Little, R. J. A. (1995). Modeling the drop–out mechanism in repeated–measures studies. *Journal of American Statistical Association* **90** 1112–1121.

Louis, T. A. (1982). Finding observed information using the EM algorithm. *Journal of the Royal Statististical Society B* **44** 98–130.

Meng, X. L. & Rubin, D. B. (1993). Maximum likelihood estimation via the ECM algorithm: a general framework. *Biometrika* **80** 267–278.

Norton, L. & Simon, R. (1977). Growth curve of an experimental solid tumor following radiotherapy. *Journal of the National Cancer Institute* **58** 1735–1741.

Oakes, D. (1999). Direct calculation of the information matrix via the EM algorithm. *Journal of the Royal Statistical Society, Series B* **61** 479–482.

Robert, C. P. (1995). Simulation of truncated normal variable. *Statistics and Computing* **5** 121–125.

Rubin, D. B. (1988). Using the SIR algorithm to simulate posterior distributions (with discussion). In *Bayesian Statistics, 3* (J. M. Bernardo *et al.*, eds.) 395–402. Oxford University Press, Oxford.

Rygaard, K. & Spang–Thomsen, M. (1997). Quantitation and Gompertzian analysis of tumor growth. *Breast Cancer Research and Treatment* **46** 303–312.

Sakaguchi, Y., Maehara, Y., Baba, H., Kusumoto, T., Sugimachi, K. & Newman, R. (1992). Flavone acetic acid increases the antitumor effect of hyperthermia in mice. *Cancer Research* **52** 3306–3309.

Smith, A. F. M. & Gelfand, A. E. (1992). Bayesian statistics without tears: a sampling–resampling perspective. *The American Statistician* **46** 84–88.

Tan, M., Fang, H. B., Tian, G. L., & Houghton, P. J. (2002). Small sample inference for incomplete longitudinal data with trunction and censoring in tumor xenograft models. *Biometrics* **58** 612–620.

Tan, M., Fang, H. B., Tian, G. L. & Houghton, P. J. (2003). Experimental design and sample size determination for testing synergism in drug combination studies based on uniform measures. *Statistics in Medicine* **22** No.13 2091–2100.

Tan, M., Fang, H. B., Tian, G. L. & Houghton, P. J. (2004). Repeated–measures models with constrained parameters for incomplete data in tumor xenograft experiments. *Statistics in Medicine* **23** in press.

Tan, M., Fang, H. B., Tian, G. L. & Wei, G. (2005). Testing multivariate normality in incomplete data of small sample size. *Journal of Multivariate Analysis* **93** 164–179.

Tan, M., Tian, G. L. & Ng, K. W. (2003). A noniterative sampling method for computing posteriors in the structure of EM–type algorithms. *Statistica Sinica* **13** No.3 625–639.

Tian, G. L., Fang, H. B. & Tan, M. (2004). Efficient algorithms for generating truncated multivariate normal distributions. *Journal of Statistical Planning and Inference.* in revision.

Wu, M. C. & Follmann, D. A. (1999). Use of summary measures to adjust for informative missingness in repeated measures data with random effects. *Biometrics* **55** 75–84.

Estimating Secondary Parameters after Termination of a Multivariate Group Sequential Test

Chengqing Wu[1], Aiyi Liu[2], and Kai F. Yu[3]

[1] National Institute of Child Health and Human Development,DHHS, U.S.A. University of Science and Technology of China, PRC. (wuch@mail.nih.gov)
[2] National Institute of Child Health and Human Development, DHHS, U.S.A. (liua@mail.nih.gov)
[3] National Institute of Child Health and Human Development,DHHS, U.S.A. (yukf@mail.nih.gov)

Summary. We consider estimation of secondary parameters following a group sequential test, with stopping regions determined by testing hypotheses concerning a set of primary parameters. We derive statistics that are jointly sufficient for the primary and secondary parameters and show that the maximum likelihood estimators remain unchanged but no longer possess unbiasedness and minimum variance. We construct bias-reduced and unbiased estimators for the vector of secondary parameters and show them to substantially reduce the bias and improve the precision of estimation.

Key words: Bias and bias-reduction; Correlated endpoints; Medical trials; Minimum variance; Primary and secondary endpoints; Restricted completeness; Truncation - adaptation.

2000 Mathematics Subject Classification: 62L12, 62F10

Contemporary Multivariate Analysis and Experimental Design—In Honor Celebration of Professor Kai-Tai Fang's 65th birthday. Edited by Jianqing Fan and Gang Li. The World Scientific Publisher, 2005.

1 Introduction

Justified by administrative, ethical and cost concerns, group sequential testing procedures are frequently used in clinical trials. Most often the hypotheses involve only a real-valued parameter (as a measure of treatment difference in a single primary endpoint), and considerable amount of work has been done in the design and analysis of sequentially testing such hypotheses: an excellent review of the past development can be found (Jennison & Turnbull 2000) and (Whitehead 1997).

In some studies the effectiveness of the treatments is comprehensive and has to be determined by a number of endpoints, often equally important. Early stopping of the trial is often guided by monitoring these endpoints simultaneously, rather than by one single endpoint; See (Jennison & Turnbull 1993) and (Todd 1999) for practical examples. Unlike sequential tests with a single endpoint, sequential tests with multiple endpoints have received relatively less attention, perhaps due to their relatively less frequent use, and technical complexity in both design and analysis. To date, only a number of publications have been found in the literature that address the design issues of such tests. See (Siegmund 1980) (Geller & Pocock 1988)(Tang, Gnecco & Geller 1989)(Tang, Gnecco & Geller 1989b)(Lin 1991) (Jennison & Turnbull 1991)(Geller, Gnecco & Tang 1992)(Su & Lachin 1992) (Tang, Geller & Pocock 1993) and (Tang & Geller 1999). To our best knowledge, analysis following these tests such as estimation, confidence regions, p-values, and inference on secondary endpoints, has not been adequately addressed. Only recently (Liu, Wu, Yu & Yuan 2004) investigated estimation of the primary parameters after a group sequential test with multiple endpoints, assuming no secondary endpoints to be analyzed.

The need to analyze secondary endpoints collected during the course of the trial further complicates the situation, since in general these secondary endpoints are correlated with the primary ones and hence analysis of these secondary endpoints has to take into account the correlation along with the sequential sampling based on the primary endpoints. Failing to do so introduces bias to the conventional inference procedures, as well demonstrated in the case of univariate sequential testing ((Whitehead 1986a)(Whitehead 1986b)(Liu & Hall 2001)(Whitehead, Todd & Hall 2000)(Hall & Yakir 2003)).

The present paper investigates estimation of secondary parameters following a multivariate group sequential test, and whether it has any effect on the estimation of the primary parameters. The context is that a p-variate random vector X and a q-variate random vector Y, jointly normally distributed, with $X \sim N_p(\boldsymbol{\mu}, I_p)$, $Y \sim N_q(\boldsymbol{\delta}, I_q)$, and $p \times q$ correlation matrix Ω (assumed known), are of interest. The random stopping of the sampling process is based solely on observations of X according to a group sequential testing procedure concerning the (primary) mean vector $\boldsymbol{\mu}$. Inference such as estimation of the (secondary) mean parameter $\boldsymbol{\delta}$ is conducted only after the sampling process stops. Such paradigm provides large-sample approximation to most random sampling problems with the exception of survival type data; See (Hall & Yakir 2003). In Section 2 we introduce multivariate group sequential tests with random sampling models incorporating secondary endpoints, and derive statistics with their joint density function that are minimally jointly sufficient for $\boldsymbol{\mu}$, and $\boldsymbol{\delta}$. In Section 3 we consider joint estimation of parameters, focusing on $\boldsymbol{\delta}$ in particular. We show that the usual sample means continue to be the maximum likelihood estimators but are biased, and sometimes badly so, with inflated variance. We propose two estimators to reduce the bias, one derived by (Whitehead 1986a)(Whitehead 1986b) bias-adjusted approach and the other by Rao-Blackwell method of conditioning, and demonstrate their outperformance over the maximum likelihood estimator in bias and precision. In Section 4 we address the issue of completeness and minimum variance. We show that the family of distributions of the sufficient statistics is not complete, and that there exist infinitely many unbiased estimators of the secondary vector of parameters, $\boldsymbol{\delta}$, depending on the sufficient statistic and, moreover, none has uniformly minimum variance. In Section 5 we go on to show that the family of distributions of the sufficient statistics satisfies restricted completeness when confining to statistics that are independent of future stopping criteria. Moreover, the Rao-Blackwell estimator has uniformly minimum variance among unbiased estimators of such kind. Finally we present some discussions in Section 6, with technical details and proofs given in the appendix.

2 Multivariate group sequential tests incorporating secondary endpoints: The random sampling model

We extend the random sampling model of(Whitehead 1986b), see also (Whitehead, Todd & Hall 2000), to multivariate group sequential test setting. Throughout, vectors are written as column vectors, and A' stands for the transpose of a vector or matrix A.

Denote respectively the primary and secondary responses of the ith object by $\mathbf{X}_i = (X_{i1}, \cdots, X_{ip})'$ and $\mathbf{Y}_i = (Y_{i1}, \cdots, Y_{iq})'$. Suppose observations are taken in groups with g_k observations in the kth group. Define $n_k = g_1 + g_2 + \ldots + g_k$, the cumulative sample size at stage k, and let $(\mathbf{X}_{n_{k-1}+1}, \mathbf{Y}_{n_{k-1}+1}), \ldots, (\mathbf{X}_{n_k}, \mathbf{Y}_{n_k})$ be the g_k observations in the kth group. Further, we define

$$\mathbf{S}_k = \sum_{i=1}^{n_k} \mathbf{X}_i, \quad \mathbf{Z}_k = \sum_{i=1}^{n_k} \mathbf{Y}_i \ ,$$

the cumulative sample sums of $\mathbf{X}$s and $\mathbf{Y}$s at stage k, respectively.

Now suppose that the null hypothesis of primary interest is $H_o : \boldsymbol{\mu} = \mathbf{0}$. With the above sampling scheme, a K-stage group sequential test concerning H_o in general can be specified by a partition of the outcome space of $\mathbf{S}_k$ into a continuation region $\mathcal{C}_k$ and a stopping region $\mathcal{B}_k$, for each stage k, $k = 1, \cdots, K$. The test stops the sampling process at the kth analysis if

$$\mathbf{S}_j \in \mathcal{C}_j, j \leq k-1, \mathbf{S}_k \in \mathcal{B}_k. \tag{1}$$

The last stopping region, $\mathcal{B}_K$, is set to be R^p, the p-dimensional Euclidean space, to force the sampling process to stop at the Kth analysis, if it does not so at an earlier stage. Note that the continuation and stopping regions are solely determined by the primary endpoints X. Common methods to compute these regions include sequential χ^2 tests and sequential Bonferroni procedures. See(Siegmund 1980) (Jennison & Turnbull 1991)(Jennison & Turnbull 2000), among others.

Upon termination of the sequential test, a triplet $(M, \mathbf{S}_M, \mathbf{Z}_M)$ is observed, where M denotes the (random) number of analyses performed. Without ambiguity, we will omit the subscription and simply write the two sample sums as $\mathbf{S}$ (for $\mathbf{S}_M$) and $\mathbf{Z}$ (for $\mathbf{Z}_M$). Note that for each fixed k, $(\mathbf{S}_k, \mathbf{Z}_k)$ is jointly sufficient for $(\boldsymbol{\mu}, \boldsymbol{\delta})$. From (Blackwell 1947), the statistic $\mathbf{U} = (M, \mathbf{S}, \mathbf{Z})$ is then jointly sufficient for $(\boldsymbol{\mu}, \boldsymbol{\delta})$; See also (Lehmann & Stein 1950). Hence following the sequential test it suffices to base solely on $\mathbf{U}$ to make secondary inference on $(\boldsymbol{\mu}, \boldsymbol{\delta})$.

The density of $\mathbf{U}$ is given below.

Theorem 1. *Let $p_0(k, \mathbf{s})$ be the density of $(M, \mathbf{S})$ at $\boldsymbol{\mu} = 0$, and $\phi_A(\cdot)$ be the density of $N(\mathbf{0}, A)$. Then the density of $\mathbf{U} = (M, \mathbf{S}, \mathbf{Z})$ is given by*

$$p_{\boldsymbol{\mu},\boldsymbol{\delta}}(k, \mathbf{s}, \mathbf{z}) = p_{0,0}(k, \mathbf{s}, \mathbf{z}) \exp\left\{ -\frac{1}{2} Q_{\boldsymbol{\mu},\boldsymbol{\delta}}(k, \mathbf{s}, \mathbf{z}) \right\}, \tag{2}$$

where $p_{0,0}(k, \mathbf{s}, \mathbf{z}) = p_0(k, \mathbf{s}) \phi_{n_k \Delta}(\mathbf{z} - \Omega' \mathbf{s})$, and

$$Q_{\boldsymbol{\mu},\boldsymbol{\delta}}(k, \mathbf{s}, \mathbf{z}) = n_k \boldsymbol{\mu}' \boldsymbol{\mu} - 2\mathbf{s}' \boldsymbol{\mu} + n_k \boldsymbol{\theta}' \Delta^{-1} \boldsymbol{\theta} - 2(\mathbf{z} - \Omega' \mathbf{s})' \Delta^{-1} \boldsymbol{\theta} \tag{3}$$

with $\Delta = I_q - \Omega' \Omega$ and $\boldsymbol{\theta} = \boldsymbol{\delta} - \Omega' \boldsymbol{\mu}$.

3 Estimating secondary parameters

3.1 *Precision of a class of estimators*

We will consider in the rest of the section three estimators of $\boldsymbol{\delta}$, the maximum likelihood estimator, a bias-reduced estimator, and an unbiased estimator; their parallels have been extensively studied in univariate group sequential test literature. Each of these three estimators is a member of a class $\mathcal{U}$ of estimators of $\boldsymbol{\delta}$ being defined as

$$\mathcal{U} = \left\{ \hat{\boldsymbol{\delta}} : \hat{\boldsymbol{\delta}} = \frac{\mathbf{Z}}{n_M} - \Omega' \{ \frac{\mathbf{S}}{n_M} - \hat{\boldsymbol{\mu}}(M, \mathbf{S}) \} \right\}. \tag{4}$$

Note that we require $\hat{\boldsymbol{\mu}}$ depend only on observations from the primary endpoints.

A key feature of an estimator $\hat{\boldsymbol{\delta}}$ in $\mathcal{U}$ is that its bias and variance can be evaluated through that of $\hat{\boldsymbol{\mu}}$. Indeed, we have

Lemma 1. *Consider an estimator $\hat{\boldsymbol{\delta}}$ in* (4) *and let Δ be defined as in Theorem 1. Then*

$$\text{bias}(\hat{\boldsymbol{\delta}}) = \Omega'\text{bias}(\hat{\boldsymbol{\mu}}), \tag{5}$$

$$\text{Cov}(\hat{\boldsymbol{\delta}}) = \Omega'\text{Cov}(\hat{\boldsymbol{\mu}})\Omega + \Delta \text{E}\left(\frac{1}{\text{n}_\text{M}}\right). \tag{6}$$

Lemma 1 is the multivariate version of similar observations in (Whitehead 1986b) and (Liu & Hall 2001) for univariate sequential tests. Write $\text{MSEM}(\hat{\boldsymbol{\delta}}) = E\{(\hat{\boldsymbol{\delta}} - \boldsymbol{\delta})(\hat{\boldsymbol{\delta}} - \hat{\boldsymbol{\delta}})'\}$, the mean square error matrix of $\hat{\boldsymbol{\delta}}$. Then from (5) and (6) we have

$$\text{MSEM}(\hat{\boldsymbol{\delta}}) = \Omega'\text{MSEM}(\hat{\boldsymbol{\mu}})\Omega + \Delta E\left(\frac{1}{n_M}\right). \tag{7}$$

The mean squared error $\text{MSE}(\hat{\boldsymbol{\delta}}) = E\{(\hat{\boldsymbol{\delta}} - \boldsymbol{\delta})'(\hat{\boldsymbol{\delta}} - \hat{\boldsymbol{\delta}})\}$ of $\hat{\boldsymbol{\delta}}$ is then obtained by taking the trace of the matrices in (7).

3.2 *Maximum likelihood estimation*

The maximum likelihood estimators of $\boldsymbol{\mu}$ and $\boldsymbol{\delta}$ can be obtained by minimizing (3) with respect to $(\boldsymbol{\mu}, \boldsymbol{\theta})$ and $\boldsymbol{\delta} = \boldsymbol{\theta} + \Omega'\boldsymbol{\mu}$. It follows that

$$\hat{\boldsymbol{\mu}}_{ML} = \mathbf{S}/n_M, \, \hat{\boldsymbol{\delta}}_{ML} = \mathbf{Z}/n_M.$$

Note that $\hat{\boldsymbol{\delta}}_{ML}$ has form (4) with $\hat{\boldsymbol{\mu}} = \hat{\boldsymbol{\mu}}_{ML}$. Hence the maximum likelihood estimators remain unchanged, being simply the same as that in a fixed sample test. From Lemma 1, however, their statistical properties, from a frequentist's point of view, change dramatically due to the sequential nature of the sampling process—they are biased estimators, often with inflated variance.

3.3 *Bias-Reduced estimation*

Focusing on a sequential probability ratio or triangular test, (Whitehead 1986a)(Whitehead 1986b) proposed a general approach to reduce the bias of the maximum likelihood estimators. His method was later extended, for estimation of primary endpoints, to univariate group sequential test setting by (Emerson & Fleming 1990) and to multivariate group sequential test setting by (Liu, Wu, Yu & Yuan 2004). Here we utilize the method to obtain bias-reduced estimation, jointly for the primary and secondary parameters. Write $\boldsymbol{\eta} = (\boldsymbol{\mu}', \boldsymbol{\delta}')'$, and denote by $\mathbf{b}_{\hat{\eta}}(\boldsymbol{\eta})$ the bias of an estimator $\hat{\boldsymbol{\eta}}$ of $\boldsymbol{\eta}$, (and similarly for other estimators).

Following (Whitehead 1986a), the bias-adjusted estimator $\hat{\boldsymbol{\eta}}_w$ of $\boldsymbol{\eta}$ satisfies the equation $\mathbf{b}_{\hat{\eta}_{ML}}(\hat{\boldsymbol{\eta}}_w) + \hat{\boldsymbol{\eta}}_w = \hat{\boldsymbol{\eta}}_{ML}$. Since $\mathbf{b}_{\hat{\eta}_{ML}} = (I, \Omega)'\mathbf{b}_{\hat{\mu}_{ML}}$, it follows that

$$\mathbf{b}_{\hat{\mu}_{ML}}(\hat{\boldsymbol{\mu}}_w) + \hat{\boldsymbol{\mu}}_w = \hat{\boldsymbol{\mu}}_{ML}, \quad \hat{\boldsymbol{\delta}}_w = \hat{\boldsymbol{\delta}}_{ML} - \Omega'(\hat{\boldsymbol{\mu}}_{ML} - \hat{\boldsymbol{\mu}}_w). \tag{8}$$

These are the multivariate versions of Whitehead's (1986a, b) bias-adjusted estimators. Again from Lemma 1, the bias and variance of $\hat{\boldsymbol{\delta}}_w$ is determined by that of the bias-reduced estimator $\hat{\boldsymbol{\mu}}_w$; the latter has been numerically studied in (Liu, Wu, Yu & Yuan 2004).

Computation of $\hat{\boldsymbol{\mu}}_w$ and thus $\hat{\boldsymbol{\delta}}_w$ can be done using Newton-Raphson iterative method; See (Liu, Wu, Yu & Yuan 2004) for details.

3.4 *Unbiased estimation*

Unbiased estimators eliminate bias and hence have received substantial attention in the case of univariate group sequential tests; See (Emerson & Fleming 1990) (Emerson & Kittelson 1997) (Liu & Hall 1999) (Liu & Hall 2001), among others. Following (Emerson & Fleming 1990) approach, we define a Rao-Blackwell type estimator of $\boldsymbol{\delta}$ as

$$\hat{\boldsymbol{\delta}}_{RB} = E\left(\frac{\mathbf{Z}_1}{n_1}\middle| M, \mathbf{S}, \mathbf{Z}\right). \tag{9}$$

That $\hat{\boldsymbol{\delta}}_{RB}$ is unbiased for $\boldsymbol{\delta}$ is obvious since $\mathbf{Z}/n_1$ is unbiased for $\boldsymbol{\delta}$.

Let $\hat{\boldsymbol{\mu}}_{RB} = E(\mathbf{S}_1/n_1|M, \mathbf{S}, \mathbf{Z})$ be the Rao-Blackwell estimator of $\boldsymbol{\mu}$. Note that for each k, $\mathbf{W}_k = \mathbf{Z}_k - \Omega'\mathbf{S}_k$ follows a q-variate normal distribution with mean vector $n_k(\boldsymbol{\delta} - \Omega'\boldsymbol{\mu})$ and covariance matrix $n_k\Delta$, and is independent of $\mathbf{S}_i$, $i \leq K$. It then follows that $\hat{\boldsymbol{\mu}}_{RB} = E(\mathbf{S}_1/n_1|M, \mathbf{S}, \mathbf{W}) = E(\mathbf{S}_1/n_1|M, \mathbf{S})$, the Rao-Blackwell estimator depending on the primary endpoints ((Liu, Wu, Yu & Yuan 2004)).

The Rao-Blackwell estimator $\hat{\boldsymbol{\delta}}_{RB}$ is also a member of (4). Indeed, we have

$$\hat{\boldsymbol{\delta}}_{RB} = \hat{\boldsymbol{\delta}}_{ML} - \Omega'(\hat{\boldsymbol{\mu}}_{ML} - \hat{\boldsymbol{\mu}}_{RB}). \tag{10}$$

A proof of (10) is presented in the appendix.

Therefore the performance of $\hat{\boldsymbol{\delta}}_{RB}$ depends on that of $\hat{\boldsymbol{\mu}}_{RB}$. (Liu, Wu, Yu & Yuan 2004) derived recursive formula for numerical computation of $\hat{\boldsymbol{\mu}}_{RB}$ which then lead to computation of $\hat{\boldsymbol{\delta}}_{RB}$.

3.5 *Comparison of estimators*

Consider estimators in (4). From (5) and (6) it follows that an improved estimator of $\boldsymbol{\mu}$ results in an improved estimator of $\boldsymbol{\delta}$. While it is difficult to draw any analytical conclusion regarding comparison of estimators, some limited numerical results in (Liu, Wu, Yu & Yuan 2004) showed that both $\hat{\boldsymbol{\mu}}_w$ and $\hat{\boldsymbol{\mu}}_{RB}$ have smaller bias and variance than $\hat{\boldsymbol{\mu}}_{ML}$. Subsequently, both $\hat{\boldsymbol{\delta}}_w$ and $\hat{\boldsymbol{\delta}}_{RB}$ are better estimators than $\hat{\boldsymbol{\delta}}_{ML}$ in terms of quadratic loss.

The bias-adjusted estimator $\hat{\boldsymbol{\mu}}_w$, though possessing some bias, appears to have smaller mean squared error than the Rao-Blackwell estimator $\hat{\boldsymbol{\mu}}_{RB}$. However, the former depends on the knowledge of stopping criteria beyond the stopped stage M while the latter does not. The same observations apply to estimation of $\boldsymbol{\delta}$, and are, not surprisingly, in agreement with that in (Emerson & Fleming 1990)and (Liu & Hall 2001).

4 The issue of completeness and minimum variance

It is well known that in a fixed-size sampling the sample sums $(\mathbf{S}, \mathbf{Z})$ are sufficient and complete statistics for inference on $(\boldsymbol{\mu}, \boldsymbol{\delta})$, and the sample means are the unique unbiased estimators depending on the sample sums and hence have uniformly minimum variance among all unbiased estimators. (Liu & Hall 2001) showed that completeness fails in a univariate group sequential test. We demonstrate here that the same conclusion holds for multivariate group sequential tests. To do so, we construct all zero-mean, identically in $(\boldsymbol{\mu}, \boldsymbol{\delta})$, statistics based on the sufficient statistics $(M, \mathbf{S}, \mathbf{Z})$, and subsequently claim that the family of distributions of $(M, \mathbf{S}, \mathbf{Z})$ is not complete. Write a statistic $\mathbf{h} = \mathbf{h}(M, \mathbf{S}, \mathbf{Z})$ as $\mathbf{h}(k, \mathbf{s}, \mathbf{z})$ at $M = k, \mathbf{S} = \mathbf{s}, \mathbf{Z} = \mathbf{z}$. We have

Lemma 2. *A statistic* $\mathbf{h} = \mathbf{h}(M, \mathbf{S}, \mathbf{Z})$ *has mean zero if and only if it satisfies*

$$\mathbf{h}(K, \mathbf{s}, \mathbf{z}) = -\frac{1}{p_0(K, \mathbf{s})\phi_{n_K\Delta}(\mathbf{z} - \Omega'\mathbf{s})}\left[\sum_{k=1}^{K-1} \int_{\mathcal{B}_k} \int_{R^q} \mathbf{h}(k, \mathbf{s}_1, \mathbf{z}_1) p_0(k, \mathbf{s}_1) \right.$$
$$\left. \times \phi_{n_k\Delta}(\mathbf{z}_1 - \Omega'\mathbf{s}_1)\phi_{m_k I_p}(\mathbf{s} - \mathbf{s}_1)\phi_{m_k\Delta}\{\mathbf{z} - \Omega'\mathbf{s} - (\mathbf{z}_1 - \Omega'\mathbf{s}_1)\} d\mathbf{z}_1 d\mathbf{s}_1 \right]. \quad (11)$$

This is an extension to the zero mean statistics based on $(M, \mathbf{S})$ in (Liu, Wu, Yu & Yuan 2004), because

$$\phi_{n_K\Delta}(\mathbf{t}) = \int_{R^q} \phi_{n_k\Delta}(\mathbf{t}_1)\phi_{m_k\Delta}(\mathbf{t} - \mathbf{t}_1) d\mathbf{t}_1.$$

Lemma 2 leads directly to

Theorem 2. *For any parameter space in* R^{p+q} *that has an open set, the family of distributions of* $\mathbf{U} = (M, \mathbf{S}, \mathbf{Z})$ *in a group sequential test is not complete.*

Adding zero-mean statistics to the Rao-Blackwell estimator $\hat{\boldsymbol{\eta}}_{RB} = (\hat{\boldsymbol{\mu}}'_{RB}, \hat{\boldsymbol{\mu}}'_{RB})'$ yields the whole class of unbiased estimators of $\boldsymbol{\eta} = (\boldsymbol{\mu}', \boldsymbol{\delta}')'$, indeed, infinitely many of them. The following theorem shows that no unbiased estimators have uniformly minimum variance. For a proof, see the appendix.

Theorem 3. *Consider estimation of* $\boldsymbol{\beta} = \mathbf{C}'\boldsymbol{\eta}$ *following a multivariate group sequential test, where* $\mathbf{C} \neq \mathbf{0}$ *is an* $(p+q)\times l$ *(non-random) matrix. Then among all unbiased estimators of* $\boldsymbol{\beta}$*, none has uniformly minimum variance. In particular, uniformly minimum variance unbiased estimators do not exist for* $\boldsymbol{\mu}$ *or* $\boldsymbol{\delta}$.

5 Completeness and unbiasedness restricted to certain statistics

In the previous section, we have shown that the family of distributions of $\mathbf{U}$ is not complete and that no unbiased estimators, the Rao-Blackwell estimators in particular, possess minimum variance. It is then natural to ask whether $\mathbf{U}$ is complete among a meaningful subset of statistics—the so called restricted completeness—and whether the Rao-Blackwell estimators, or any other estimators, have uniformly minimum variance among a reasonable class of unbiased estimators.

We confine attention to truncation-adaptable statistics, and subsequently, truncation-adaptable unbiased estimators. The concept, which requires that an inference after a stopped sequential test be independent of the future stopping boundaries, was initially developed in (Liu & Hall 1999), and subsequently investigated in (Liu & Hall 2001), and(Liu, Wu, Yu & Yuan 2004).

First, we give some notations. Let $\mathcal{D}$ represent a group sequential test design with stopping regions $\mathcal{B}_k$, $1 \leq k \leq K$ $(\mathcal{B}_K = R^p)$, for testing a hypothesis concerning the multivariate mean vector $\boldsymbol{\mu}$. For each $k < K$, a "shorter" group sequential test design, denoted by $\mathcal{D}^k$ and called the truncated design of $\mathcal{D}$ at the kth stage, is obtained by retaining the first $k-1$ stopping regions but closing the kth stopping region of $\mathcal{D}$. Expectation under the truncated design $\mathcal{D}_k$ will be written as E_k, while the notation E is reserved for taking expectation under $\mathcal{D}$.

Write as $\mathbf{h}(k, X_1, \ldots, X_{n_k}, Y_1, \ldots, Y_{n_k})$ the statisic $\mathbf{h}$ (scale, vector or matrix) depending on the whole data set when $M = k$. Note that for each k, $\mathbf{h}$ is defined only for Xs subject to (1). Corresponding to each truncated design $\mathcal{D}_k$, $\mathbf{h}$ can also be truncated to yield a statistic $\mathbf{h}_k$ with respect to $\mathcal{D}_k$ with $\mathbf{h}_k(k, \ldots)$ extended to $\mathbf{S}_k \in \mathcal{C}_k$ and $\mathbf{h}_k(i, X_1, \ldots, X_{n_i}, Y_1, \ldots, Y_{n_i}) = \mathbf{h}(i, X_1, \ldots, X_{n_i}, Y_1, \ldots, Y_{n_i})$ for each $1 \leq i \leq k$ on the region subject to $\{\mathbf{S}_j \in \mathcal{C}_j, j \leq i-1, \mathbf{S}_i \in \mathcal{B}_i\}$. We will simply call such $\mathbf{h}_k$ an extension of $\mathbf{h}$ to the domain $\mathcal{C}_k$.

Definition 1. *We call a statistic* $\mathbf{h}$ *with respect to* $\mathcal{D}$ *truncation-adaptable, if for each* k*, there exists an extension* $\mathbf{h}_k$ *of* $\mathbf{h}$ *to the domain* $\mathcal{C}_k$ *such that* $E_k(\mathbf{h}_k) = E(\mathbf{h})$. *In particular, if* $E(\mathbf{h}) = \boldsymbol{\delta}$*, then* $\mathbf{h}$ *is said to be a truncation-adaptable unbiased estimator of* $\boldsymbol{\delta}$.

It is worth pointing out that truncation-adaptable statistics are closed under linear transformations, namely a linear combination of a number of truncation-adaptable statistics is also truncation-adaptable. Further, we have

Theorem 4. *Let* $\mathbf{h}$ *be a truncation-adaptable statistic, possibly depending on the whole data set. Then*

(i) *The statistic* $\tilde{\mathbf{h}} = E(\mathbf{h}|M, \mathbf{S}, \mathbf{Z})$ *is also truncation-adaptable.*

(ii) *If* $\mathbf{h}$ *depends solely on* $(M, \mathbf{S}, \mathbf{Z})$ *and* $E(\mathbf{h}) = \mathbf{0}$ *uniformly in* $\boldsymbol{\mu}$ *and* $\boldsymbol{\delta}$, *then* $\mathbf{h} = \mathbf{0}$, *except for a zero-measure set with respect to the density* (2).

Therefore, the family of distributions of the sufficient statistic $\mathbf{U} = (M, \mathbf{S}, \mathbf{Z})$ is *truncation-adaptably complete*; Among truncation adaptable statistics depending on $\mathbf{U}$, only the null function has mean zero!

Subsequently the Rao-Blackwell estimators $\hat{\boldsymbol{\mu}}_{RB}$ and $\hat{\boldsymbol{\delta}}_{RB}$ are respectively the uniformly minimum variance truncation-adaptable unbiased estimator of $\boldsymbol{\mu}$ and $\boldsymbol{\delta}$ depending on $\mathbf{U}$. This follows from (i) $\mathbf{S}_1/n_1$ and $\mathbf{Z}_1/n_1$ are respectively truncation-adaptable unbiased estimator of $\boldsymbol{\mu}$ and $\boldsymbol{\delta}$, and hence so are $\hat{\boldsymbol{\mu}}_{RB}$ and $\hat{\boldsymbol{\delta}}_{RB}$ by (i) of Theorem 4; (ii) truncation-adaptable unbiased estimators depending on $\mathbf{U}$ are identical, up to a zero-measure set, by (ii) of Theorem 4; And (iii) By the Rao-Blackwell theorem and (i) of Theorem 4, $\hat{\boldsymbol{\mu}}_{RB}$ and $\hat{\boldsymbol{\delta}}_{RB}$ have smaller variance than any other truncation-adaptable unbiased estimators. We summarize more generally in

Theorem 5. *Consider a parameter* $\boldsymbol{\beta} = \mathbf{C}'\boldsymbol{\eta}$ *as in Theorem 3, and define* $\hat{\boldsymbol{\beta}}_{RB} = \mathbf{C}'\hat{\boldsymbol{\eta}}_{RB}$. *Then* $\hat{\boldsymbol{\beta}}_{RB}$ *has uniformly minimum variance among all truncation-adaptable unbiased estimators of* $\boldsymbol{\beta}$.

6 Discussion

Judging by a frequentist's criteria, conventional statistical inference procedures, such as those based on likelihood functions, that are shown efficient under fixed-size sampling may be inappropriate when used in a sequential sampling setting, and the post-stopping inference of multivariate sequential tests shows no exception, as being demonstrated in the present paper and in (Liu, Wu, Yu & Yuan 2004). Unbiasedness, minimum variance, and completeness all fail in the presence of sequential sampling. Furthermore, the sampling no longer possesses uniformly minimum variance unbiased estimators. It remains to be seen whether other fixed-size properties, such as admissibility of estimators, are also lost in sequential sampling.

We investigated three estimation procedures following termination of the sequential tests, a bias-adjusted approach, the Rao-Blackwell method, and the truncation-adaptation criterion, all result in improved estimation as compared to the maximum likelihood procedure for the secondary parameters. We further note that under these procedures the presence of secondary endpoints does not provide improvement for estimation of the primary parameters; Estimation for the primary parameters can be based solely on the stopping time and the primary endpoints. We suspect that any reasonable estimation procedures should inherit the same characteristic.

All along the correlation matrix Ω is assumed to be known. When Ω is unknown, we can still obtain efficient estimation for $\boldsymbol{\delta}$ by simply replacing it by its usual estimator $\widehat{\Omega} = \sum_{i=1}^{N_M}(\mathbf{X}_i - \hat{\boldsymbol{\mu}}_{ML})'(\mathbf{Y}_i - \hat{\boldsymbol{\delta}}_{ML})\}/(N_M - 1)$ in (4). Following conditional argument it can be shown that such two-stage estimation procedure does not change the bias, but increase the variance of the estimators. The same bias results from the fact that $\widehat{\Omega}$ is independent of then sample sums.

Appendix Technical Details

A.1 *Proof of Theorem 1*

Given $M = k, \mathbf{S}_k = \mathbf{s}$, the conditional distribution of $\mathbf{Z}$ is $N_q(n_k\boldsymbol{\delta} + \Omega'(\mathbf{s} - n_k\boldsymbol{\mu}), n_k\Delta)$, see (Rao 1965). It follows that

$$p_{\boldsymbol{\mu},\boldsymbol{\delta}}(k, \mathbf{s}, \mathbf{z}) = p_{\boldsymbol{\mu}}(k, \mathbf{s})\phi_{n_k\Delta}(\mathbf{z} - n_k\boldsymbol{\delta} - \Omega'(\mathbf{s} - n_k\boldsymbol{\mu})) \quad .$$

We hence obtain (2) by noting that

$$\phi_{n_k\Delta}(\mathbf{z}-n_k\boldsymbol{\delta}-\Omega'(\mathbf{s}-n_k\boldsymbol{\mu})) = \phi_{n_k\Delta}(\mathbf{z}-\Omega'\mathbf{s})\exp\{-\frac{1}{2}n_k\boldsymbol{\theta}'\Delta^{-1}\boldsymbol{\theta}+(\mathbf{z}-\Omega'\mathbf{s})'\Delta^{-1}\boldsymbol{\theta}\},$$

and $p_{\boldsymbol{\mu}}(k, \mathbf{s}) = p_0(k, \mathbf{s})\exp\{-\frac{1}{2}n_k\boldsymbol{\mu}'\boldsymbol{\mu} + \mathbf{s}'\boldsymbol{\mu}\}$ according to (Liu, Wu, Yu & Yuan 2004).

A.2 *Proof of Lemma 1*

Recall (Section 3.4) that for each k, $\mathbf{W}_k = \mathbf{Z}_k - \Omega'\mathbf{S}_k \sim N_q(n_k(\boldsymbol{\delta} - \Omega'\boldsymbol{\mu}), n_k\Delta)$ and is independent of $\mathbf{S}_i$, $i \leq K$. Since the stopping stage M depends only on $\mathbf{S}_1, \mathbf{S}_2, \cdots$, $\mathbf{W}_k$ is independent of $(M, \mathbf{S})$. For a $\hat{\boldsymbol{\delta}}$ in (4) we have

$$\begin{aligned} \text{bias}(\hat{\boldsymbol{\delta}}) &= E\left\{\frac{\mathbf{Z}}{n_M} - \Omega'\left(\frac{\mathbf{S}}{n_M} - \hat{\boldsymbol{\mu}}\right)\right\} - \boldsymbol{\delta} \\ &= \Omega' E(\hat{\boldsymbol{\mu}}) + E\left\{E\left(\frac{\mathbf{W}_M}{n_M}\middle| M\right)\right\} - \boldsymbol{\delta} \\ &= \Omega' E(\hat{\boldsymbol{\mu}}) - \Omega'\boldsymbol{\mu} = \Omega'\mathbf{b}_{\hat{\boldsymbol{\mu}}} \ , \end{aligned}$$

and

$$\begin{aligned}
\mathrm{Cov}(\hat{\boldsymbol{\delta}}) &= \mathrm{Cov}\left(\frac{\mathbf{W}_M}{n_M} + \Omega'\hat{\boldsymbol{\mu}}\right) \\
&= \mathrm{Cov}(\Omega'\hat{\boldsymbol{\mu}}) + \mathrm{Cov}\left(\frac{\mathbf{W}_M}{n_M}\right) + 2\mathrm{Cov}\left(\frac{\mathbf{W}_M}{n_M}, \Omega'\hat{\boldsymbol{\mu}}\right) \\
&= \Omega'\mathrm{Cov}(\hat{\boldsymbol{\mu}})\Omega + \Delta E\left(\frac{1}{n_M}\right).
\end{aligned}$$

A.3 *Proof of* (10)

Let $\mathbf{W}$ be as previously defined. Then from (9),

$$\begin{aligned}
\hat{\boldsymbol{\delta}}_{RB} &= E\left(\frac{\mathbf{W}_1 + \Omega'\mathbf{S}_1}{n_1}\middle| M, \mathbf{S}_M, \mathbf{W}_M\right) \\
&= E\left(\frac{\mathbf{W}_1}{n_1}\middle| M, \mathbf{W}_M\right) + E\left(\frac{\Omega'\mathbf{S}_1}{n_1}\middle| M, \mathbf{S}_M\right) \\
&= \frac{\mathbf{W}_M}{n_M} + \Omega'\hat{\boldsymbol{\mu}}_{RB} = \hat{\boldsymbol{\delta}}_{ML} - \Omega'(\hat{\boldsymbol{\mu}}_{ML} - \hat{\boldsymbol{\mu}}_{RB}).
\end{aligned}$$

A.4 *Proof of Lemma 2*

For an $l \times l$ positive matrix B and an l-variate vector γ, integrating to 1 the density function of $N_l(B\gamma, B)$ to obtain

$$\exp\{\frac{1}{2}\gamma' B\gamma\} = \int_{R^l} \phi_B(s)\exp(s'\gamma)ds. \qquad \text{(A.1)}$$

Let $\mathbf{h}(k, \mathbf{s}, \mathbf{z})$ be zero mean statistic, that is, from (2),

$$\begin{aligned}
\mathbf{0} = &\sum_{k=1}^{K} \exp\left(-\frac{1}{2}n_k\boldsymbol{\mu}'\boldsymbol{\mu} - \frac{1}{2}n_k\boldsymbol{\theta}'\Delta^{-1}\boldsymbol{\theta}\right) \\
&\times \int_{\mathcal{B}_k}\int_{R^q} \mathbf{h}(k, \mathbf{s}, \mathbf{z})p_0(k, \mathbf{s})\phi_{n_k\Delta}(\mathbf{z} - \Omega'\mathbf{s})\exp\{\mathbf{s}'\boldsymbol{\mu} + (\mathbf{z} - \Omega'\mathbf{s})'\Delta^{-1}\boldsymbol{\theta}\}d\mathbf{z}d\mathbf{s}\ .
\end{aligned}$$

Then with $\mathbf{t} = \mathbf{z} - \Omega'\mathbf{s}$, $\boldsymbol{\eta} = \Delta^{-1}\boldsymbol{\theta}$, and $m_k = n_K - n_k$ this yield

$$\begin{aligned}
&\int_{R^p}\int_{R^q} \mathbf{h}(K, \mathbf{s}, \mathbf{t} + \Omega'\mathbf{s})p_0(K, \mathbf{s})\phi_{n_K\Delta}(\mathbf{t})\exp(\mathbf{s}'\boldsymbol{\mu} + \mathbf{t}'\boldsymbol{\eta})d\mathbf{t}d\mathbf{s} \\
&= -\sum_{k=1}^{K-1}\exp\left(\frac{1}{2}m_k\boldsymbol{\mu}'\boldsymbol{\mu} + \frac{1}{2}m_k\boldsymbol{\eta}'\Delta\boldsymbol{\eta}\right)\int_{\mathcal{B}_k}\int_{R^q} \mathbf{h}(k, \mathbf{s}, \mathbf{t} + \Omega'\mathbf{s})p_0(k, \mathbf{s}) \\
&\times\phi_{n_k\Delta}(\mathbf{t})\exp(\mathbf{s}'\boldsymbol{\mu} + \mathbf{t}'\boldsymbol{\eta})d\mathbf{t}d\mathbf{s}.
\end{aligned}$$

It follows from (A.1) that

$$\int_{R^p}\int_{R^q} \mathbf{h}(K,\mathbf{s},\mathbf{t}+\Omega'\mathbf{s})p_0(K,\mathbf{s})\phi_{n_K\Delta}(\mathbf{t})\exp(\mathbf{s}'\boldsymbol{\mu}+\mathbf{t}'\boldsymbol{\eta})d\mathbf{t}d\mathbf{s}$$
$$= \int_{R^p}\int_{R^q}\exp(\mathbf{s}'\boldsymbol{\mu}+\mathbf{t}'\boldsymbol{\eta})d\mathbf{t}d\mathbf{s}\Bigg\{-\sum_{k=1}^{K-1}\int_{\mathcal{B}_k}\int_{R^q}\mathbf{h}(k,\mathbf{s}_1,\mathbf{t}_1+\Omega'\mathbf{s}_1)p_0(k,\mathbf{s}_1)\phi_{n_k\Delta}(\mathbf{t}_1)$$
$$\times\phi_{m_kI_p}(\mathbf{s}-\mathbf{s}_1)\phi_{m_k\Delta}(\mathbf{t}-\mathbf{t}_1)d\mathbf{t}_1d\mathbf{s}_1\Bigg\}\ .$$

By the uniqueness theorem of Laplace transforms, we have

$$\mathbf{h}(K,\mathbf{s},\mathbf{t}+\Omega'\mathbf{s})p_0(K,\mathbf{s})\phi_{n_K\Delta}(\mathbf{t}) = -\sum_{k=1}^{K-1}\int_{\mathcal{B}_k}\int_{R^q}\mathbf{h}(k,\mathbf{s}_1,\mathbf{t}_1+\Omega'\mathbf{s}_1)p_0(k,\mathbf{s}_1)$$
$$\times\phi_{n_k\Delta}(\mathbf{t}_1)\phi_{m_kI_p}(\mathbf{s}-\mathbf{s}_1)\phi_{m_k\Delta}(\mathbf{t}-\mathbf{t}_1)d\mathbf{t}_1d\mathbf{s}_1,$$

yielding (11) .

On the other hand, if $\mathbf{h}(\cdot,\cdot,\cdot)$ is any function for which the integrals in (11) are finite for each $k < K$ and (11) holds, then $E_{\boldsymbol{\mu},\boldsymbol{\delta}}\mathbf{h}(M,\mathbf{S},\mathbf{Z}) = \mathbf{0}$.

A.5 *Proof of Theorem 3*

Suppose $\hat{\boldsymbol{\beta}}(M,\mathbf{S},\mathbf{Z})$ has uniformly minimum variance among unbiased estimators. Let $\mathbf{h}(M,\mathbf{S},\mathbf{Z})$ be a zero-mean statistic. Then from (Lehmann & Casella 1998, p. 85), $E_{\boldsymbol{\eta}}\{\hat{\boldsymbol{\beta}}(M,\mathbf{S},\mathbf{Z})\mathbf{h}'(M,\mathbf{S},\mathbf{Z})\} = \mathbf{0}$, indenticaly in $\boldsymbol{\eta}$. Applying Lemma 2 respectively to $\mathbf{h}$ and $\hat{\boldsymbol{\beta}}\mathbf{h}'$ we have

$$\hat{\boldsymbol{\beta}}(K,\mathbf{s},\mathbf{z})\mathbf{h}'(K,\mathbf{s},\mathbf{z})$$
$$= -\frac{1}{p_0(K,\mathbf{s})\phi_{n_K\Delta}(\mathbf{z}-\Omega'\mathbf{s})}\Bigg[\sum_{k=1}^{K-1}\int_{\mathcal{B}_k}\int_{R^q}\hat{\boldsymbol{\beta}}(k,\mathbf{s}_1,\mathbf{z}_1)\mathbf{h}'(K,\mathbf{s}_1,\mathbf{z}_1)p_0(k,\mathbf{s}_1)$$
$$\times\phi_{n_k\Delta}(\mathbf{z}_1-\Omega'\mathbf{s}_1)\phi_{m_kI_p}(\mathbf{s}-\mathbf{s}_1)\phi_{m_k\Delta}\{\mathbf{z}-\Omega'\mathbf{s}-(\mathbf{z}_1-\Omega'\mathbf{s}_1)\}d\mathbf{t}_1d\mathbf{s}_1\Bigg]$$
$$= -\frac{1}{p_0(K,\mathbf{s})\phi_{n_K\Delta}(\mathbf{z}-\Omega'\mathbf{s})}\Bigg[\sum_{k=1}^{K-1}\int_{\mathcal{B}_k}\int_{R^q}\hat{\boldsymbol{\beta}}(K,\mathbf{s},\mathbf{z})\mathbf{h}'(K,\mathbf{s}_1,\mathbf{z}_1)p_0(k,\mathbf{s}_1)$$
$$\times\phi_{n_k\Delta}(\mathbf{z}_1-\Omega'\mathbf{s}_1)\phi_{m_kI_p}(\mathbf{s}-\mathbf{s}_1)\phi_{m_k\Delta}\{\mathbf{z}-\Omega'\mathbf{s}-(\mathbf{z}_1-\Omega'\mathbf{s}_1)\}d\mathbf{t}_1d\mathbf{s}_1\Bigg],$$

yielding

$$0 = \int_{\mathcal{B}_k}\int_{R^q}\left(\hat{\boldsymbol{\beta}}(K,\mathbf{s},\mathbf{z})-\hat{\boldsymbol{\beta}}(k,\mathbf{s}_1,\mathbf{z}_1)\right)\mathbf{h}'(K,\mathbf{s}_1,\mathbf{z}_1)p_0(k,\mathbf{s}_1)$$
$$\times\phi_{n_k\Delta}(\mathbf{z}_1-\Omega'\mathbf{s}_1)\phi_{m_kI_p}(\mathbf{s}-\mathbf{s}_1)\phi_{m_k\Delta}\{\mathbf{z}-\Omega'\mathbf{s}-(\mathbf{z}_1-\Omega'\mathbf{s}_1)\}d\mathbf{t}_1d\mathbf{s}_1 \quad (A.2)$$

Since this must hold for any function $\mathbf{h}(k, \mathbf{s}, \mathbf{z})$, it follows that $\hat{\beta}(K, \mathbf{s}, \mathbf{z}) = \hat{\beta}(k, \mathbf{s}_1, \mathbf{z}_1)$ for all $(\mathbf{s}, \mathbf{z}) \in R^{p+q}$ and $(\mathbf{s}_1, \mathbf{z}_1) \in \mathcal{B}_k \times R^q$. Hence $\hat{\beta}(k, \mathbf{s}, \mathbf{z})$ is constant vector for all $(\mathbf{s}, \mathbf{z}) \in \mathcal{B}_k \times R^q$ for all $k \leq K$, and this contradict the unbiasedness of $\hat{\beta}(M, \mathbf{S}, \mathbf{Z})$. Therefore $\boldsymbol{\beta}$ can not be of minimum variance.

A.6 *Proof of Theorem 4*

Extend $\mathbf{h}$ to the domain $\mathcal{C}_k$ to obtain $\mathbf{h}_k$. Let $\tilde{\mathbf{h}}_k = E_k(\mathbf{h}_k|M, S_M)$ with $M \leq k$ and S_k being replaced by R^p. Then $\tilde{\mathbf{h}}_k$ is an extension of $\tilde{\mathbf{h}}$ to the domain $\mathcal{C}_k$ and $E_k(\tilde{\mathbf{h}}_k) = E(\tilde{\mathbf{h}})$. This proves truncation-adaptability of $\tilde{\mathbf{h}}$.

We now prove (ii). With zero-mean and truncation-adaptability, we have, for $k = 1$,

$$\int_{R^p} \int_{R^q} p_{\boldsymbol{\mu},\boldsymbol{\delta}}(1, \mathbf{s}, \mathbf{z})\mathbf{h}(1, \mathbf{s}, \mathbf{z}) d\mathbf{z} d\mathbf{s} = \mathbf{0}. \tag{A.3}$$

Since $\mathcal{D}^1$ is a non-sequential design with sample size n_1, by the completeness of the family of distributions of $(\mathbf{S}_1, \mathbf{Z}_1)$, (A.3) possess a unique solution for $\mathbf{h}$, that is $h = \mathbf{0}$.

Suppose $\mathbf{h}(j, \mathbf{s}, \mathbf{z}) = \mathbf{0}, j = 1, \cdots, k < K$. Then with truncation to $k+1$, $\mathbf{h}(k+1, \mathbf{s}, \mathbf{z})$ must satisfy the equation

$$\int_{R^p} \int_{R^q} \mathbf{h}(k+1, \mathbf{s}, \mathbf{z}) p_{\boldsymbol{\mu},\boldsymbol{\delta}}(k+1, \mathbf{s}, \mathbf{z}) d\mathbf{z} d\mathbf{s} = \mathbf{0},$$

that is,

$$\int_{R^p} \int_{R^q} \mathbf{h}(k+1, \mathbf{s}, \mathbf{t} + \Omega' \mathbf{s}) p_0(k+1, \mathbf{s}) \phi_{n_{k+1}\Delta}(\mathbf{t}) \exp(\boldsymbol{\mu}' \mathbf{s} + \mathbf{t}' \boldsymbol{\eta}) d\mathbf{t} d\mathbf{s} = \mathbf{0}.$$

By the uniqueness of Laplace transforms, we have $\mathbf{h}(k+1, \mathbf{s}, \mathbf{z}) = \mathbf{0}$. The proof hence follows by induction.

References

Blackwell, D. (1947). Conditional Expectation and Unbiased Sequential Estimation. *Annals of Mathematical Statistics* **18** 105–110.

Emerson, S. S. & Fleming, T. R. (1990). Parameter estimation following sequential hypothesis testing. *Biometrika* **77** 875–892.

Emerson, S. S. & Kittelson, J. M. (1997). A computationally simpler algorithm for the UMVUE of a normal mean following a group sequential trial. *Biometrics* **53** 365–369.

Geller, N. L., Gnecco, C. & Tang, D. I. (1992). Group sequential designs for trials with multiple endpoints. *Biopharmaceutical Sequential Statistical Applications* 29–33.

Geller, N. L. & Pocock, S. J. (1988). Design and analysis of clinical trials with group sequential stopping rules. *Biopharmaceutical Statistics for Drug Development* 489–508.

Hall, W. J. & Yakir, B. (2003). Inference about a secondary process after a sequential trial. *Biometrika* **90** 597–611.

Jennison, C. & Turnbull, B. W. (1991). Exact calculations for sequential t, χ^2 and F tests. *Biometrika* **78** 133–141.

Jennison, C. & Turnbull, B. W. (1993). Group sequential tests for bivariate response: Interim analyses of clinical trials with both efficacy and safety endpoints. *Biometrics* **49** 741–752.

Jennison, C. & Turnbull, B. W. (2000). *Group Sequential Methods with Applications to Clinical Trials.* Boca Raton, FL: Chapman & Hall/CRC.

Lehmann, E. L. (1950)*Notes on the Theory of Estimation.* Recorded by C. Blyth. Associated Students Store, University of California, Berkeley.

Lehmann, E. L. & Casella, G. (1998). *Theory of point estimation.* Springer-Verlag Inc (Berlin; New York).

Lehmann, E. L. & Stein, C. (1950). Completeness In the Sequential Case. *Annals of Mathematical Statistics* **21** 376–385.

Lin, D. Y. (1991). Nonparametric sequential testing in clinical trials with incomplete multivariate observations. *Biometrika* **78** 123–131.

Liu, A. & Hall, W. J. (1999). Unbiased estimation following a group sequential test. *Biometrika* **86** 71–78.

Liu, A. & Hall, W. J. (2001). Unbiased estimation of secondary parameters following a sequential test. *Biometika* **88** 895–900.

Liu, A., Wu, C., Yu, K. & Yuan, V. (2004). Estimation following a multivariate group sequential test. submitted to *Journal of the American Statistical Association.*

Rao, C. R. (1965). *Linear statistical inference and its applications.* John Wiley & Sons, Inc.

Siegmund, D. (1980). Sequential χ^2 and F tests and the related confidence intervals. *Biometrika* **67** 389–402.

Su, J. Q. & Lachin, J. M. (1992). Group sequential distribution-free methods for the analysis of multivariate observations. *Biometrics* **48** 1033–1042.

Tang, D. I. & Geller, N. L. (1999). Closed testing procedures for group sequential clinical trials with multiple endpoints. *Biometrics* **55** 1188–1192.

Tang, D. I., Gnecco, C. & Geller, N. L. (1989a). An approximate likelihood ratio test for a normal mean vector with nonnegative components with application to clinical trials. *Biometrika* **76** 577–583.

Tang, D. I., Gnecco, C. & Geller, N. L. (1989b). Design of group sequential clinical trials with multiple endpoints. *Journal of the American Statistical Association* **84** 776-779.

Tang, D. I., Geller, N. L. & Pocock, S. J. (1993). On the design and analysis of randomized clinical trials with multiple endpoints. *Biometrics* **49** 23–30

Todd, S. (1999). Sequential designs for monitoring two endpoints in a clinical trial. *Drug Information Journal* **33** 417–426.

Whitehead, J. (1986a). On the bias of maximum likelihood estimation following a sequential test. *Biometrika* **73** 573–581.

Whitehead, J. (1986b). Supplementary analysis at the conclusion of a sequential clinical trial. *Biometrics* **42** 461–471.

Whitehead, J. (1997). *The Design and Analysis of Sequential Clinical Trials*, revised 2nd ed. New York: Wiley.

Whitehead, J., Todd, S. & Hall, W. J. (2000). Confidence intervals for secondary parameters following a sequential test. *Journal of the Royal Statistical Society, Series B* **62** 731–745.

Part IV

Advance in Statistics

Penalized M-Estimation-Based Model Selection for Regression by Cross-Validation

Yuehua Wu

Department of Mathematics and Statistics, York University, 4700 Keele Street, Toronto, Ontario, Canada(wuyh@mathstat.yorku.ca)

Summary. We consider the problem of model (or variable) selection in the linear regression model based on M-estimation and cross-validation with an added penalty term for penalizing overfitting. Under some conditions, the new criterion is shown to be strongly consistent in the sense that with probability one, for all large n, the criterion chooses the smallest true model. The penalty function denoted by C_n depends on the sample size n and is chosen to ensure the consistency in the selection of the smallest true model. There are various choices of C_n suggested in the literature on model selection. In this paper we show by simulation that for small and medium size samples, a particular choice of C_n based on observed data, which makes it random, provides satisfactory performance compared with fixed choices of C_n.

Key words: Model selection, M-estimation, cross-validation, strong consistency, data-oriented penalty.

2000 Mathematics Subject Classification: 62J05, 62F35, 62F12

1 Introduction

Consider the linear regression model

$$y_i = \boldsymbol{x}_i'\boldsymbol{\beta} + e_i, \quad i = 1, \ldots, n, \tag{1}$$

Contemporary Multivariate Analysis and Experimental Design—In Honor Celebration of Professor Kai-Tai Fang's 65th birthday. Edited by Jianqing Fan and Gang Li. The World Scientific Publisher, 2005.

where $\boldsymbol{x}_i'$ denotes the ith row of an $n \times p$ design matrix $X_n = (\boldsymbol{x}_1, \ldots, \boldsymbol{x}_n)'$, $\boldsymbol{\beta}$ is a p-vector of unknown regression parameters, and $e_1, e_2, \ldots$, are independently and identically distributed random variables. Each component of $\boldsymbol{\beta}$ may be zero or nonzero. Each subset $\mathcal{M}$ of $\{1, 2, \cdots, p\}$ is called a sub-model. It is obvious that there are 2^p possible sub-models for the multiple regression problem. A sub-model is called a true model if $\beta_i = 0$ for all $i \notin \mathcal{M}$. The problem is to find the smallest true model which is defined to be the one whose all proper sub-models are not true models. There is considerable literature on how to solve this problem; see the book on model selection by McQuarrie & Tsai (1998) or the review paper by Rao & Wu (2001) among others.

Cross-validation is a method for model selection in terms of the predictive ability of the models. Suppose that n data points are available. A model is to be selected from a class of models. First, hold one data point and use the rest of $n-1$ data points to fit a model. Then check the predictive ability of the model in terms of the withheld data point. Perform this procedure for all data points. Select the model with the best average predictive ability. This procedure is described as the LOO (leave one out) method. There is considerable literature on this method; see Stone (1974), Shao (1993), and the book by McQuarrie & Tsai (1998) among others. When the number of predictors in any regression model under consideration is fixed, this type of cross-validation is not consistent and it can be shown that it is equivalent to Akaike information criterion (see, e.g. Li (1987)). Some consistent generalized cross-validation methods have been proposed in Shao (1993), where $k(n)$ data points are used to fit the model and the rest data points are used for assessing the predictive ability. As another approach, Rao & Wu (2004) proposed the penalized LOO, which has been shown to be consistent.

Note that least squares cross-validation is very sensitive to outliers and departures from normality assumption on error distribution. Hence, Ronchetti, Field & Blanchard (1997) proposed a robust model selection technique for regression based on cross-validation to overcome this weakness, which, jointly with Rao & Wu (2004), inspires us to propose the penalized M-estimation based model selection by cross-validation. It can be shown that under certain conditions, the proposed method is consistent.

The paper is arranged as follows: In Section 2, we introduce the penalized M-estimation based model selection by cross-validation. The simulation study is provided in Section 3. Consistency of the proposed criterion is presented in the appendix.

Let A be a $p \times p$ matrix and $\boldsymbol{a}$ be a p-vector. In the rest of this paper, $A(-i)$ denotes a matrix consisting of all the columns of A excluding its ith column, and $\boldsymbol{a}(-i)$ denotes a vector consisting of all the elements of $\boldsymbol{a}$ excluding its ith element.

2 Penalized M-estimation based model selection by cross-validation

Since each component of $\boldsymbol{\beta}$ in (1) may be zero or nonzero, let us consider the leave-one-out approach (see Rao & Wu (1989) among others) for selecting the smallest true model, which is computationally economic.

Consider the model

$$\tilde{\boldsymbol{y}}_n = X_n(-k)\boldsymbol{\beta}(-k) + \boldsymbol{e}_n. \tag{2}$$

Let $\hat{\boldsymbol{\beta}}_{-i}$ be the M-estimate of $\boldsymbol{\beta}$ in the model (1) based on the data $\{(y_1, \boldsymbol{x}_1), \ldots, (y_{i-1}, \boldsymbol{x}_{i-1}), (y_{i+1}, \boldsymbol{x}_{i+1}), \ldots, (y_n, \boldsymbol{x}_n)\}$ and let $\hat{\boldsymbol{\beta}}_{-i}(-k)$ be the M-estimate of $\boldsymbol{\beta}$ in the model (2) based on the data $\{(y_1, \boldsymbol{x}_1(-k)), \ldots, (y_{i-1}, \boldsymbol{x}_{i-1}(-k)), (y_{i+1}, \boldsymbol{x}_{i+1}(-k)), \ldots, (y_n, \boldsymbol{x}_n(-k))\}$ so that

$$\hat{\boldsymbol{\beta}}_{-i} = \arg\min_{\boldsymbol{\beta}} \sum_{j\neq i} \rho(y_j - \boldsymbol{x}_j'\boldsymbol{\beta}), \tag{3}$$

and

$$\hat{\boldsymbol{\beta}}_{-i}(-k) = \arg\min_{\boldsymbol{\beta}(-k)} \sum_{j\neq i} \rho(y_j - \boldsymbol{x}_j(-k)'\boldsymbol{\beta}(-k)), \tag{4}$$

respectively, where ρ is a convex discrepancy function. Define, for $1 \leq k \leq p$,

$$RC_n^{CV}(-k) = \sum_{i=1}^{n} \rho(y_i - \boldsymbol{x}_i(-k)'\hat{\boldsymbol{\beta}}_{-i}(-k)) - \sum_{i=1}^{n} \rho(y_i - \boldsymbol{x}_i'\hat{\boldsymbol{\beta}}_{-i}).$$

Then, choose the model as

$$\begin{cases} \beta_k = 0, & \text{if } RC_n^{CV}(-k) < C_n, \\ & \qquad\qquad \text{for } k = 1, \ldots, p, \\ \beta_k \neq 0, & \text{if } RC_n^{CV}(-k) \geq C_n, \end{cases} \tag{5}$$

where $\{C_n\}$ is a sequence of constants. This criterion will be called RC in the rest of this paper. It will be shown in the appendix that the criterion RC is consistent under certain conditions.

In the above criterion, it can be seen that C_n decides how large the penalty is. It is clear that how to choose C_n is a very important task. Note that the model will tend to overfitting if the penalty is small and in contrast the model will be prone to underefitting if the penalty is large. Since the criterion RC with a fixed choice of C_n may not perform well in various situations, there is a need to find a data-oriented penalty so that a procedure with its use will have satisfactory performance. Such efforts can be found in Rao & Wu (1989) and Shao (1998) among others.

Recently, Bai, Rao & Wu (1999) proposed a method of constructing a data-oriented penalty. Based on the same idea, Wu (2001) proposed a method of constructing a data-oriented penalty for an M-estimation-based model selection criterion. The main idea is as follows: First, obtain a consistent estimate of the regression parameters in the model of full size and compute the residuals. Then construct p pseudo models with the use of the residuals. To guarantee all the regression coefficients in the pseudo models are recognizable, small regression coefficients are replaced by a chosen threshold value. Find the differences between the pseudo models and all its sub-models. Then construct the penalty based on these differences so that the pseudo models are selected with large possibilities. For details, see Bai, Rao & Wu (1999) and Wu (2001). Rao & Wu (2004) modified this type of procedures. Instead of using p pseudo models for constructing a data-oriented penalty, only one pseudo model is employed in Rao & Wu (2004) so that the computing time is reduced. In this paper, we will adopt their approach and hence construct a data-oriented penalty for small and medium size samples. It will be shown that this type of penalty works well by simulation in the next section.

For the regression model (1), let a sequence of experimental measurements $\{(y_1, \boldsymbol{x}_1), \ldots, (y_n, \boldsymbol{x}_n)\}$ be available. The details of the procedure for constructing a data-oriented penalty are as follows:

1. Compute the M-estimate $\hat{\boldsymbol{\beta}}_n$ of $\boldsymbol{\beta}$ based on the data $\{(y_1, \boldsymbol{x}_1), \ldots, (y_n, \boldsymbol{x}_n)$ such that
$$\hat{\boldsymbol{\beta}} = \arg\min_{\boldsymbol{\beta}} \sum_{j=1}^{n} \rho(y_j - \boldsymbol{x}_j'\boldsymbol{\beta}).$$
2. Compute $\hat{\mathbf{e}}_n = \boldsymbol{y}_n - X_n\hat{\boldsymbol{\beta}}_n$.
3. Let $\overline{\boldsymbol{\beta}}_n = (\overline{\beta}_{1n}, \ldots, \overline{\beta}_{pn})'$ be defined as follows:
$$\overline{\beta}_{in} = \begin{cases} \hat{\beta}_{in}, & \text{if } |\hat{\beta}_{in}| \geq \kappa, \\ \kappa \text{sign}(\widetilde{\beta}_{in}), & \text{if } 0 < |\hat{\beta}_{in}| < \kappa, \\ \kappa, & \text{if } |\hat{\beta}_{in}| = 0, \end{cases} \quad \text{for } i = 1, \ldots, p, \tag{6}$$
where the constant κ is a suitably chosen threshold value.
4. Construct a pseudo model: $\boldsymbol{u} = X_n\overline{\boldsymbol{\beta}}_n + \hat{\mathbf{e}}_n, \quad h = 1, \ldots, p$. Compute the M-estimate $\tilde{\boldsymbol{\beta}}_{-i}$ of $\boldsymbol{\beta}$ for the data $\{(u_1, \boldsymbol{x}_1), \ldots, (u_{i-1}, \boldsymbol{x}_{i-1}), (u_{i+1}, \boldsymbol{x}_{i+1}), \ldots, (u_n, \boldsymbol{x}_n)\}$ such that
$$\tilde{\boldsymbol{\beta}}_{-i} = \arg\min_{\boldsymbol{\beta}} \sum_{j \neq i}^{n} \rho(u_j - \boldsymbol{x}_j'\boldsymbol{\beta}).$$
Also compute the M-estimate $\tilde{\boldsymbol{\beta}}_{-i}(-h)$ of $\boldsymbol{\beta}(-h)$ based on the data $\{(u_1, \boldsymbol{x}_1(-h)), \ldots, (u_{i-1}, \boldsymbol{x}_{i-1}(-h)), (u_{i+1}, \boldsymbol{x}_{i+1}(-h)), \ldots, (u_n, \boldsymbol{x}_n(-h))$ such that

$$\tilde{\boldsymbol{\beta}}_{-i}(-h) = \arg\min_{\boldsymbol{\beta}} \sum_{j \neq i}^{n} \rho(u_j - \boldsymbol{x}_j(-h)'\boldsymbol{\beta}(-h)),$$

for $i = 1, \ldots, n$. Then compute $\Delta_n(h) = D_n(h) - D_{n0}$, $h = 1, \ldots, p$, where $D_n(h) = \sum_{i=1}^n \rho(u_i - \boldsymbol{x}_i(-h)'\tilde{\boldsymbol{\beta}}_{-i}(-h))$ and $D_{n0} = \sum_{i=1}^n \rho(u_i - \boldsymbol{x}_i'\tilde{\boldsymbol{\beta}}_{-i})$.

5. Define

$$C_n^{(R)} = \min\{\Delta_n(h),\ h = 1, \ldots, p\} + \tau. \tag{7}$$

Then choose $C_n^{(R)}$ as the penalty C_n in the criterion RC.

REMARK: κ in (6) is a distinguishability level, which needs to be selected by the practical users. For example, the state treasure department counts money in million dollars, so κ can be chosen as 1,000,000, a store manager counts consumer's income in hundreds dollars, κ can be chosen as 100, and an engineering counts unit cost of a product in milli-cents, κ can be chosen as 0.00001.

How to choose τ in (7) depends on one's tolerance to the overfitting. Large value of τ will reduce the chance of overfitting but it will also increase the chance of underfitting. In the simulation study presented in the next section, τ is set to be 5.

3 A simulation study

In this section, we will study the finite sample performance of the criterion RC. The regression model is assumed to be:

$$y_i = \beta_1 x_{1i} + \beta_2 x_{2i} + \beta_3 x_{3i} + \beta_4 x_{4i} + \beta_5 x_{5i} + e_i, \quad i = 1, \ldots, n,$$

where $x_{1i}, \ldots, x_{5i}$, $i = 1, \ldots, n$, are iid $N(0,1)$ random variables, $\{e_i\}$ are independently and identically distributed from (1) $N(0,1)$; (2) Cauchy(0,1), and $\boldsymbol{\beta} = (\beta_1, \beta_2, \beta_3, \beta_4, \beta_5)'$ is set as $(1.3, 1.5, 0, 0, 1.3)'$. In our simulation, $\rho(u) = 0.5u^2$ if $|u| \leq 1.345$ and $\rho(u) = 1.345|u| - 0.5 \cdot (1.345)^2$ otherwise (Huber ρ), $q(k) = k$, κ is set to be 0.01, C_n is chosen as $\log(n)$, $\sqrt{n}[\log(n)]^{0.1}$, or the date-oriented penalty $C_n^{(R)}$ given in (7) with $\tau = 5$.

The simulation results are reported in Tables 1-2. In both tables, G_n^{CV} denotes the penalized least squares cross-validation criterion proposed in Rao & Wu (2004); RC represents the criterion given in (5); the $Diff$ is equal to $-i$ if the selected model has eliminated i nonzero parameters for $i = 1, 2, 3$; the $Diff$ is equal to j if the selected model has included j more independent variables than the smallest true model for $j = 1, 2, 3$; the $Diff$ is equal to 0 if the smallest true model is selected; $C_n^{(1)} = \log(n)$; $C_n^{(2)} = \sqrt{n}[\log(n)]^{0.1}$. The entries give the actual number of the runs falling into each category based on 1,000 replications. Note that if $C_n = 0$, the penalized cross-validation criteria reduce to ordinary cross-validation criteria. The reasons for selecting $C_n^{(1)} = \log(n)$ and

$C_n^{(2)} = \sqrt{n}[\log(n)]^{0.1}$ in the simulation are due to the well-known BIC criterion and (10), respectively.

From the two tables, it can be seen that when the random errors obey standard normal distribution, both criteria G_n^{CV} and RC with $C_n = \log(n)$ or $\sqrt{n}[\log(n)]^{0.1}$ outperform ordinary cross-validation criteria, i.e., the criteria G_n^{CV} and RC with $C_n = 0$. When the random errors are Cauchy distributed, the criterion RC with $C_n = \log(n)$ or $\sqrt{n}[\log(n)]^{0.1}$ performs significantly better than the criteria RC with $C_n = 0$ and G_n^{CV} with $C_n = 0$, $\log(n)$ or $\sqrt{n}[\log(n)]^{0.1}$. The performance of the criterion RC with use of $C_n^{(R)}$ seems to have improved average performance over these fixed choices of C_n. Since the computation time is no more a problem in today's computing environment, RC with use of $C_n^{(R)}$ is recommended for use in a variable selection problem for small and medium-size samples.

Table 1. Results based on 1,000 simulations with $e_i \sim N(0,1)$

Sample		G_n^{CV}			RC			
Size	Diff	$C_n = 0$	$C_n = C_n^{(1)}$	$C_n = C_n^{(2)}$	$C_n = 0$	$C_n = C_n^{(1)}$	$C_n = C_n^{(2)}$	$C_n = C_n^{(R)}$
	−1	**0**	0	0	0	0	3	1
$n = 30$	0	**679**	935	983	582	987	997	999
	1	**279**	61	16	319	11	0	0
	2	**42**	4	1	99	2	0	0
	−1	**0**	0	0	0	0	0	0
$n = 50$	0	**691**	957	996	614	994	1000	1000
	1	**268**	42	4	317	6	0	0
	2	**41**	1	0	69	0	0	0
	−1	**0**	0	0	0	0	0	0
$n = 100$	0	**698**	981	1000	586	995	1000	1000
	1	**262**	19	0	330	5	0	0
	2	**40**	0	0	84	0	0	0

Appendix

In this appendix, the asymptotic behavior of the model selection criterion proposed in Section 2 is investigated.

Consider the model (1). We assume that the following assumptions hold true, and hence the results in Lemma A3, Corollary A5, Theorem 3.1 and Theorem 4.1 of Wu & Zen (1999) are valid.

(A1) $\rho(\cdot)$ is a convex function on R^1.

Table 2. Results based on 1,000 simulations with $e_i \sim$ Cauchy(0,1)

Sample Size	Diff	G_n^{CV} $C_n = 0$	$C_n = C_n^{(1)}$	$C_n = C_n^{(2)}$	RC $C_n = 0$	$C_n = C_n^{(1)}$	$C_n = C_n^{(2)}$	$C_n = C_n^{(R)}$
$n = 30$	-3	230	235	239	9	4	9	6
	-2	250	261	264	31	22	71	45
	-1	263	265	272	88	150	252	226
	0	177	177	177	531	717	651	702
	1	68	56	44	276	98	17	21
	2	12	6	4	65	9	0	0
$n = 50$	-3	244	246	248	0	0	2	0
	-2	230	236	244	4	0	22	4
	-1	293	294	298	27	52	169	76
	0	171	177	175	631	873	800	888
	1	52	39	29	271	66	7	32
	2	10	8	1	67	9	0	0
$n = 100$	-3	219	220	223	0	0	0	0
	-2	251	255	264	0	0	0	0
	-1	272	274	269	1	1	12	0
	0	185	291	197	604	946	985	969
	1	70	59	46	332	53	3	31
	2	3	1	1	63	0	0	0

(A2) $\mathrm{E}[\rho(e_1)]$ is finite.

(A3) For any $\boldsymbol{\beta}$, $\liminf_{n\to\infty} \frac{1}{n}\sum_{i=1}^{n}\{\mathrm{E}[\rho(e_i - \boldsymbol{x}_i'\boldsymbol{\beta}) - \rho(e_i)]\} \geq g(\boldsymbol{\beta})$, where $g(\cdot)$ is a nonnegative convex function and is strictly convex in a neighborhood of $\mathbf{0}$.

Let $\sigma(\cdot)$ be any choice of the subgradient of $\rho(\cdot)$ and denote by $\mathcal{U}$ the set of discontinuity points of σ, which is the same for all choices of σ.

(B1) The common distribution function F of e_i satisfies $F(\mathcal{U}) = 0$. $\mathrm{E}[\sigma(e_1)] = 0$, and

$$\mathrm{E}[\sigma(e_1 + u)] = au + o(|u|), \quad \text{as } u \to 0,$$

where a is a positive constant.

(B2) There exist positive constants ζ and h_0 such that for any $h \in [0, h_0]$ and any u,

$$\sigma(u+h) - \sigma(u) \leq \zeta.$$

(B3) σ is bounded by $0 < L < \infty$.

Let $S_n = X_n' X_n$ and $d_n^2 = \max_{1\le i\le n} \boldsymbol{x}_i' S_n^{-1} \boldsymbol{x}_i$. Denote the eigenvalues of a symmetric matrix B of order k by $\lambda_1(B) \geq \ldots \geq \lambda_k(B)$.

(X) There are N_0 and constants a_1 and a_2 such that for $n \geq N_0$,

$$0 < a_1 n \leq \lambda_p(S_n) \leq \lambda_1(S_n) \leq a_2 n.$$

(X1) $d_n(\log\log n)^{\frac{1}{2}} \to 0$ as $n \to \infty$.

Assume that $\beta_k = 0$. By the convexity of ρ, Lemma A3, Corollary A5 and Theorem 3.1 of Wu & Zen (1999), it follows that

$$\begin{aligned}
&\sum_{i=1}^{n} \rho(y_i - \boldsymbol{x}_i(-k)'\hat{\boldsymbol{\beta}}_{-i}(-k)) - \sum_{i=1}^{n} \rho(y_i - \boldsymbol{x}_i'\hat{\boldsymbol{\beta}}_{-i}) \\
&\leq \sum_{i=1}^{n} \sigma(y_i - \boldsymbol{x}_i(-k)'\hat{\boldsymbol{\beta}}_{-i}(-k))(\boldsymbol{x}_i'\hat{\boldsymbol{\beta}}_{-i} - \boldsymbol{x}_i(-k)'\hat{\boldsymbol{\beta}}_{-i}(-k)) \\
&\leq L\sqrt{n}\left[\sum_{i=1}^{n} |\boldsymbol{x}_i'\hat{\boldsymbol{\beta}}_{-i} - \boldsymbol{x}_i(-k)'\hat{\boldsymbol{\beta}}_{-i}(-k)|^2\right]^{1/2} = O_{\text{a.s.}}\left(\sqrt{n\log\log n}\right). \quad (8)
\end{aligned}$$

Assume that $\beta_k \neq 0$. Note that for any k,

$$\sum_{i=1}^{n} \rho(y_i - \boldsymbol{x}_i(-k)'\hat{\boldsymbol{\beta}}_{-i}(-k)) \geq \sum_{i=1}^{n} \rho(y_i - \boldsymbol{x}_i(-k)'\hat{\boldsymbol{\beta}}(-k)),$$

where $\hat{\boldsymbol{\beta}}(-k)$ is the M-estimate of $\boldsymbol{\beta}(-k)$ in the model (2) based on the data $\{(y_1, \boldsymbol{x}_1), \ldots, (y_n, \boldsymbol{x}_n)\}$ so that

$$\hat{\boldsymbol{\beta}}(-k) = \arg\min_{\boldsymbol{\beta}(-k)} \sum_{j=1}^{n} \rho(y_j - \boldsymbol{x}_j'\boldsymbol{\beta}(-k)).$$

By the convexity of ρ, (3), (4), Lemma A3, Corollary A5, Theorem 3.1 and Theorem 4.1 of Wu & Zen (1999), it follows that

$$\begin{aligned}
&\sum_{i=1}^{n} \rho(y_i - \boldsymbol{x}_i(-k)'\hat{\boldsymbol{\beta}}_{-i}(-k)) - \sum_{i=1}^{n} \rho(y_i - \boldsymbol{x}_i'\hat{\boldsymbol{\beta}}_{-i}) \\
&\geq \sum_{i=1}^{n} \rho(y_i - \boldsymbol{x}_i(-k)'\hat{\boldsymbol{\beta}}(-k)) - \sum_{i=1}^{n} \rho(y_i - \boldsymbol{x}_i'\hat{\boldsymbol{\beta}}_{-i}) \\
&\geq \sum_{i=1}^{n} \rho(y_i - \boldsymbol{x}_i(-k)'\hat{\boldsymbol{\beta}}(-k)) - \sum_{i=1}^{n} \rho(y_i - \boldsymbol{x}_i'\boldsymbol{\beta}_0) - L\sum_{i=1}^{n} |\boldsymbol{x}_i'\hat{\boldsymbol{\beta}}_{-i} - \boldsymbol{x}_i'\boldsymbol{\beta}_0| \\
&\geq cn - O_{a.s.}\left(\sqrt{n\log\log n}\right). \quad (9)
\end{aligned}$$

where $c > 0$ is a constant.

If the sequence $\{C_n\}$ satisfies that

$$\frac{C_n}{n} \to 0, \qquad \frac{C_n}{\sqrt{n \log\log n}} \to \infty, \tag{10}$$

by (8), and (9), it follows that the model selection criterion eventually selects the smallest true model with probability one. It is noticed that the second condition in (10) may be further weakened.

Acknowledgments

This research was supported in part by a grant from the Natural Sciences and Engineering Research Council of Canada.

References

Bai, Z. D., Rao, C. R. & Wu, Y. (1999). Model selection with data-oriented penalty. *J. Statist. Plann. Inference* **77** 103–117.

Li, K. C. (1987). Asymptotic optimality for C_p, C_L, cross-validation and generalized cross-validation: discrete index set. *Ann. Statist.* **15** 958–975.

McQuarrie, A. & Tsai, C. L. (1998). *Regression And Time Series Model Selection.* World Scientific, Singapore.

Rao, C. R. & Wu, Y. (1989). A strongly consistent procedure for model selection in a regression problem. *Biometrika* **76** 369–374.

Rao, C. R. & Wu, Y. (2001). On model selection (with discussion). IMS Lecture Notes **38** 1–64.

Rao, C. R. & Wu, Y. (2004). Linear model selection by cross-validation. *J. Statist. Plann. Inference.* In press.

Ronchetti, E., Field, C. & Blanchard, W. (1997). Robust linear model selection by cross-validation. *J. Amer. Statist. Assoc.* **92** 1017–1023.

Shao, J. (1993). Linear model selection by cross-validation. *J. Amer. Statist. Assoc.* **88** 486–494.

Shao, J. (1998). Convergence rates of the generalized information criterion. *J. Nonparametr. Statist.* **9** 217–225.

Stone, M. (1974). Cross-validatory choice and assessment of statistical prediction. *J. Roy. Statist. Soc. Ser. B.* **36** 111–133.

Wu, Y. (2001). An M-estimation-based model selection criterion with a data-oriented penalty. *J. Statist. Comput. Simulation* **70** 71–88.

Wu, Y. & Zen, M. M. (1999). A strongly consistent information criterion for linear model selection based on M-estimation. *Probab. Theory relat. Fields* **113** 599–625.

Order-dependent Thresholding with Applications to Regression Splines

Jin-Ting Zhang[1]

Dept of Stat & Appl Prob., National Univ. of Singapore, Singapore
(stazjt@nus.edu.sg)

Summary. Donoho & Johnstone (1994) threshold a sequence of random variables, e.g., wavelet coefficients using a same thresholding parameter. In this paper, we attempt to extend their thresholding technologies to threshold a sequence of random variables using order-dependent thresholding parameters. Some insights about the thresholding estimates are discussed. Applications of the proposed methodologies to regression splines are investigated. A simple extension to nonparametric additive models is briefly discussed. Illustrations are made via applying the methodologies to two real data sets. Simulations are conducted to assess the methodologies empirically.

Key words: Order-dependent thresholding, regression splines, thresholding parameter selection, thresholding rules.

1 Introduction

Given a sample of observations $(t_i, y_i), i = 1, 2, \ldots, n$ generated from the following nonparametric regression model:

$$y = f(t) + \epsilon, \quad t \in [0,1], \quad \epsilon \sim N(0, \sigma^2), \tag{1}$$

it is of great interest in many contexts to recover the unknown smooth function f. Without loss of generality, throughout this article, we assume f has a unit support $[0,1]$. There exist many techniques in the literature that can be used to

Contemporary Multivariate Analysis and Experimental Design—In Honor Celebration of Professor Kai-Tai Fang's 65th birthday. Edited by Jianqing Fan and Gang Li. The World Scientific Publisher, 2005.

recover f, including kernel, local polynomial, smoothing splines, regression splines and wavelet-based methods among others. For the same end, these techniques use somewhat different ideas. The first two use kernel, often taken as a probability density function, and a smoothing parameter, called bandwidth, to specify a neighborhood where a constant function or a polynomial function approximates the unknown function well. The last three use linear combinations of a set of basis functions to estimate the unknown function. Our work is related to the last three techniques.

Smoothing splines essentially use the same number of basis functions as sample size, together with a roughness penalty and a smoothing parameter to balance goodness-of-fit and model complexity. Smoothing splines have a nice Bayesian interpretation. A drawback for smoothing splines is that it is often intensive in computation especially when the sample size is large. Regression splines use fewer basis functions. The balance between goodness-of-fit and model complexity is obtained via proper knot locating. The number of knots is usually small. Penalized splines are combinations of regression splines and smoothing splines, via penalizing the jumps of the truncated power basis functions of regression splines.

Wavelets are a group of orthogonal basis functions, constructed via rescale or shift of a mother wavelet. When wavelets are used to recover f in (1), most of the resulting wavelet coefficients are small and contain just noise. In order to reduce estimation risk, Donoho & Johnstone (1994) propose thresholding methods to truncate small wavelet coefficients to 0. Applications of wavelets are often hampered by some usual requirements of wavelets: the design time points are equally spaced and the sample size must be a power of 2. There exist many techniques for relaxing these requirements, including the recent one by Antoniadis & Fan (2001) where penalized ideas from smoothing splines (Green & Silverman (1994), Wahba (1990) and penalized splines (Ruppert & Carroll (1997) are applied to wavelet smoothing problems.

In this article, we attempt to extend the thresholding ideas of Donoho & Johnstone (1994) to regression splines, i.e., to the coefficients of the induced orthonormal basis functions (i.e.,eigen vectors) resulted from the singular value decompositions of the design matrix. The design matrix is formed via evaluating the regression spline basis functions at design time points. According to the definition of singular value decomposition, the eigen vectors and hence the associated coefficients are sorted in an order that the associated singular values are decreasing so that the lower (higher) order coefficients contain more signal (noise) than noise (signal). To take this into account, we threshold the coefficients using order-depending thresholding parameters for each coefficient. That is, we threshold less (more) those lower (higher) order coefficients. Since our thresholding technique is componentwise, computation is less intensive than that for usual regression splines.

This article is organized as follows. In Sect. 2, we review some basic ideas about thresholding rules. Sect. 3 focuses on how to apply these ideas to regression splines, the basis choice, model approximation, thresholding parameter selections, and constructions of approximate standard deviation bands. Sect. 4 presents some simulation studies where we compare different thresholding rules, different thresholding parameter selections, and different knot distribution rules. In Sect. 5, we illustrate our methods using two real data examples. Extension to nonparametric additive models is briefly discussed in Sect. 6.

2 Thresholding Rules

In this section, to explore more insights about thresholding rules, we first discuss thresholding rules for a single normal random variable, and then extend them for a sequence of normal random variables.

2.1 Thresholding Rules for a Single Normal Random Variable

Let $z \sim N(\mu, \sigma^2)$ be a single observation from a normal distribution with unknown mean μ and unknown variance σ. We want to estimate μ based on the z. From the frequentist view, the best linear unbiased estimate is the least squares estimator z, i.e., $\hat{\mu} = z$ with a risk $\mathrm{E}(\mu - \hat{\mu})^2 = \sigma^2$. However, if we have some prior knowledge about the relationship between μ and σ, we may have better estimates with smaller risks. For example, if we know $|\mu| < \sigma$, we may estimate μ by 0 with a risk $\mathrm{E}(\mu - 0)^2 = \mu^2$ which is smaller than σ^2. Of course, if we know $|\mu| \geq \sigma$, we still use z to estimate μ. The resulting estimate is known as the oracle estimate of μ (Donoho & Johnstone (1994)):

$$\hat{\mu}_O(z) = z 1_{\{|\mu|/\sigma \geq 1\}}, \tag{2}$$

where 1_A is the indicator of A, and $|\mu|/\sigma$ is the signal-to-noise ratio. The rationale of the oracle estimate (2) is that when the signal μ is too small compared to the noise level σ, we better estimate it as 0 to avoid estimating noise.

The oracle estimate improves the ordinary least squares estimate, paying a price losing the latter's linearity and unbiasedness. But its quadratic risk $\sigma^2 \min\left(\frac{|\mu|^2}{\sigma^2}, 1\right)$ is in general not attainable since both μ and σ are often unknown. This risk, however, can be well approached via mimicking the oracle estimate (2) by the following hard-thresholding rule (Donoho & Johnstone (1994)):

$$\hat{\mu}_H(z) = z 1_{\{|z^*| \geq 1\}} = z 1_{\{|z| \geq \sigma\}}, \tag{3}$$

where and throughout $z^* = z/\sigma$ rescales z so that $\mathrm{Var}(z^*) = 1$. When it is unknown, σ is also called thresholding parameter since it thresholds the least squares estimate z at $|z| = \sigma$, i.e., $|z^*| = 1$ (see Fig. 1 (a)). The rationale behind the hard-thresholding rule (3) is that when σ is known, the prior knowledge "$|\mu|/\sigma \geq 1$" is estimated by "$|z^*| \geq 1$ " via estimating μ by z; when it is unknown, σ should be

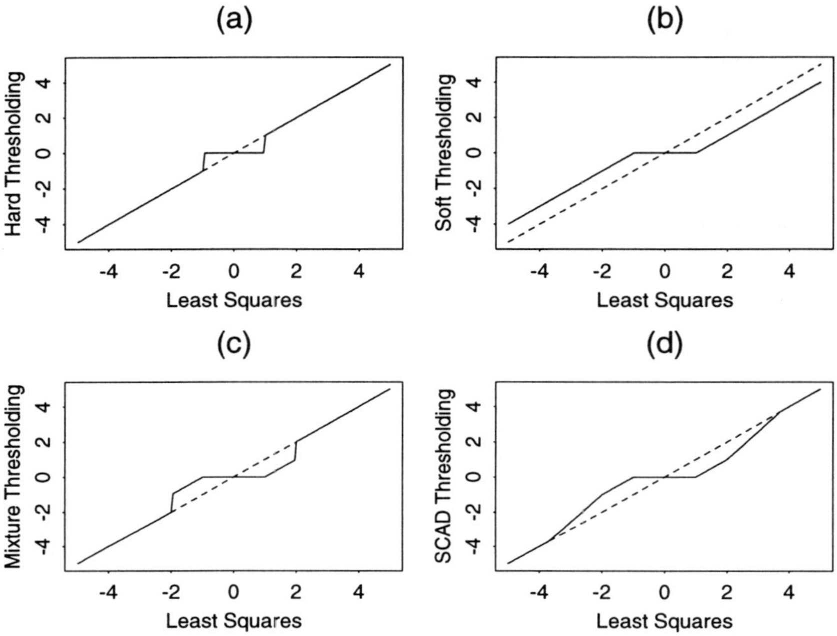

Fig. 1. Thresholding rules against the least squares estimator: (a) Hard-thresholding; (b) Soft-thresholding; (c) Mixture thresholding; and (d) SCAD-thresholding. The dashed diagonal line indicates how different a thresholding rule is from the least squares estimator.

estimated too, together with other information. The hard-thresholding rule (3) is a shrinking estimate of μ in the sense of $|\hat{\mu}_H(z)| \leq |z|$.

The hard-thresholding rule (3) has a strong connection with hypothesis testing. Actually, $|z^*|$ is the z-test score for the following simple hypothesis testing problem:

$$H_0 : \mu = 0 \quad \text{vs} \quad H_1 : \mu \neq 0, \tag{4}$$

based on the observed z. The decision rule for accepting or rejecting H_0 at the significance level of $\alpha = P(|z^*| \geq 1) \approx 30\%$ is exactly the hard-thresholding rule (3). That is, at that significance level, we accept H_0 (i.e. accept $\mu = 0$) when $|z^*| < 1$; otherwise, we reject H_0 (i.e. accept $\mu \neq 0$) so that we may estimate μ by z.

The hard-thresholding rule (3) is discontinuous at $|z^*| = 1$. The continuity of a thresholding rule may be advantageous. For instance, to apply SURE technique (Donoho & Johnstone (1994)) for selecting a thresholding parameter, a rule should be at least weakly differentiable. To overcome this drawback, we may use the following soft-thresholding rule (Donoho & Johnstone (1994)):

$$\hat{\mu}_S(z) = z\left(1 - \frac{\sigma}{|z|}\right)_+ = z\left(1 - \frac{1}{|z^*|}\right)_+, \tag{5}$$

obtained via modifying the hard-thresholding rule (3) by an upward or downward σ-shift according to the sign of z when $|z^*| \geq 1$ (see Fig. 1(b)) where and throughout $u_+ = u1_{\{u \geq 0\}} = \max(u, 0)$. The soft-thresholding rule (5) is also a shrinking estimate of μ.

The risk of the soft-thresholding rule (5) is $2\sigma^2$ when μ is large (see Fig. 2 (b)). This is due to the bias caused by the upward or downward σ-shift for z when $|z^*| > 1$. To reduce this bias and hence reduce the risk, we may do a further modification of the soft-thresholding rule. That is, when $|z^*|$ is very large, we do not shift z upward or downward. This is because when $|z^*|$ is large, say, $|z^*| \geq \omega$ for some $\omega > 1$, we have stronger evidence to believe that μ is not zero. Thus, in this case, it seems more appropriate to estimate μ by the hard-thresholding rule (3) than the soft-thresholding rule (5). This results in the following so-called mixture thresholding rule:

$$\hat{\mu}_M(z) = z\left\{\left(1 - \frac{1}{|z^*|}\right)_+ 1_{\{|z^*|<\omega\}} + 1_{\{|z^*|\geq\omega\}}\right\}. \tag{6}$$

That is, when $1 \leq |z^*| < \omega$, the soft-thresholding rule (5) is applied; when $|z^*| \geq \omega$, the hard-thresholding rule (3) is applied (see Fig. 1 (c)).

The mixture thresholding rule (6) is discontinuous at $|z^*| = \omega$. Modification can be made to make it continuous there. For this end, Fan (1997) proposed the following so-called smoothed clipped absolute deviation (SCAD) rule (see Fig. 1(d)):

$$\begin{aligned}\hat{\mu}_D(z) = z\Bigg\{&\left(1 - \frac{1}{|z^*|}\right)_+ 1_{\{|z^*|<\omega\}} \\ &+ \left(1 + \frac{1 - a|z^*|^{-1}}{a - \omega}\right) 1_{\{\omega\leq|z^*|<a\}} + 1_{\{|z^*|\geq a\}}\Bigg\},\end{aligned} \tag{7}$$

where $1 < \omega < a$. When $\omega = 2$, Fan & Li (2001) suggested $a = 3.7$ based on a Bayesian argument.

A general SCAD-type thresholding rule is:

$$\begin{aligned}\hat{\mu}(z) = z\Bigg\{&\left(1 - \frac{\lambda}{|z^*|}\right)_+ 1_{\{|z^*|<\omega\}} \\ &+ \left(1 + \frac{\lambda(1 - a|z^*|^{-1})}{a - \omega}\right) 1_{\{\omega\leq|z^*|<a\}} + 1_{\{|z^*|\geq a\}}\Bigg\},\end{aligned} \tag{8}$$

where $0 \leq \lambda \leq \omega \leq a$. When λ, ω, and a take proper values, (8) will reduce to one of the foregoing thresholding rules. An example is when $\lambda = \omega = a = 1$, (8) reduces to the hard-thresholding rule (3). Other values taken by λ, ω and a may lead to some other thresholding rules, which are different from those foregoing rules. For instance, when $\lambda = \omega = 1$, (8) reduces to the following thresholding rule:

$$\hat{\mu}(z) = z\left\{\left(1 + \frac{1 - a|z^*|^{-1}}{a-1}\right) 1_{\{1 \le |z^*| < a\}} + 1_{\{|z^*| \ge a\}}\right\}. \tag{9}$$

This rule is simpler than the SCAD rule (7) while keeps the continuity of the SCAD rule.

All the foregoing thresholding rules can be expressed in the following general form:

$$\hat{\mu}(z) = zc(z;\sigma) = c(z;\sigma)z, \tag{10}$$

where $c(z;\sigma)$ denotes the associated coefficient for a particular rule, which is specified by some given parameters λ, ω and a. We emphasize σ in the coefficient $c(z;\sigma)$ since it acts as an unknown thresholding parameter and needs to be estimated based on the data. All the foregoing thresholding rules are shrinking estimates of μ since $|c(z;\sigma)| \le 1$, and are nonlinear since $c(z,\sigma)$ depends on z.

The unified expression (8) allows a unified formula for the estimation risks of the thresholding rules. For this end, we need the following notation. Denote

$$\Phi_r(x) = \int_{-\infty}^{x} u^r \phi(u) du, \quad r = 0, 1, 2, \ldots, \tag{11}$$

where $\phi(\cdot)$ denotes the standard normal density function. Let $\Phi(\cdot)$ denote the standard normal cumulative distribution function. Then it is easy to show that

$$\begin{aligned}
\Phi_0(x) &= \Phi(x), & \Phi_1(x) &= -\phi(x), \\
\Phi_2(x) &= \Phi(x) - x\phi(x), & \Phi_3(x) &= -(x^2+2)\phi(x), \\
\Phi_4(x) &= 3\Phi(x) - x(x^2+3)\phi(x),
\end{aligned}$$

and $\Phi_r(-\infty) = 0, \Phi_{2r+1}(+\infty) = 0, \Phi_{2r+2}(+\infty) = 1 \cdot 3 \ldots \cdot (2r+1) = (2r+1)!!$ for $r = 0, 1, 2, \ldots$. Thus, $\Phi_r(x)$ can be computed using $\Phi(x)$ and $\phi(x)$. Let $\Phi_r(x)|_v^u = \Phi_r(u) - \Phi_r(v)$. Then simple algebra shows that the estimation risk of the general SCAD-type rule (8) is

$$\rho(\hat{\mu}(z), \mu) = \mathrm{E}\left(\hat{\mu}(z) - \mu\right)^2 = c_\mu \mu^2 + c_\sigma \sigma^2, \tag{12}$$

where $c_\mu = \Phi_0(x)|_{a_1}^{b_1}$, indicating the probability when a z is truncated to 0 and causes bias, and

$$\begin{aligned}
c_\sigma = {} & c_0^2 \Phi_0(x)|_{a_3}^{a_2} + \lambda^2 \left(\Phi_0(x)|_{a_2}^{a_1} + \Phi_0(x)|_{b_1}^{b_2}\right) + d_0^2 \Phi_0(x)|_{b_2}^{b_3} \\
& + 2c_0c_1\Phi_1(x)|_{a_3}^{a_2} + 2\lambda\left(\Phi_1(x)|_{a_1}^{a_2} - \Phi_1(x)|_{b_1}^{b_2}\right) + 2d_0d_1\Phi_1(x)|_{b_2}^{b_3} \\
& + \Phi_2(x)|_{-\infty}^{a_3} + c_1^2\Phi_2(x)|_{a_3}^{a_2} + \Phi_2(x)|_{a_2}^{a_1} + \Phi_2(x)|_{b_1}^{b_2} + d_1^2\Phi_2(x)|_{b_2}^{b_3} \\
& + \Phi_2(x)|_{b_3}^{+\infty},
\end{aligned} \tag{13}$$

with

$$
\begin{aligned}
a_1 &= -\lambda - \mu/\sigma,\ b_1 = \lambda - \mu/\sigma,\ c_0 = \lambda(\mu/\sigma + a)/(a - \omega),\\
a_2 &= -\omega - \mu/\sigma,\ b_2 = \omega - \mu/\sigma,\ d_0 = \lambda(\mu/\sigma - a)/(a - \omega), \qquad (14)\\
a_3 &= -a - \mu/\sigma,\ b_3 = a - \mu/\sigma,\ c_1 = d_1 = (a - \omega + \lambda)/(a - \omega).
\end{aligned}
$$

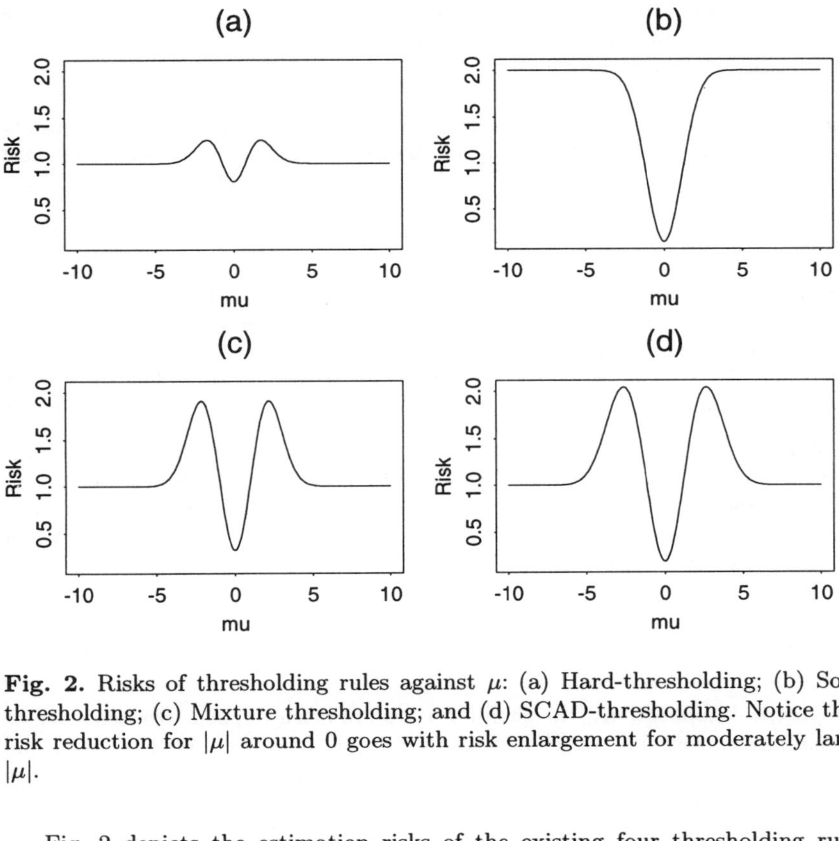

Fig. 2. Risks of thresholding rules against μ: (a) Hard-thresholding; (b) Soft-thresholding; (c) Mixture thresholding; and (d) SCAD-thresholding. Notice that risk reduction for $|\mu|$ around 0 goes with risk enlargement for moderately large $|\mu|$.

Fig. 2 depicts the estimation risks of the existing four thresholding rules against μ when $\sigma = 1$ so that the risk of the ordinary least squares estimator is 1. Some simple observations may be listed as follows: (1) For all the thresholding rules, the risks around $\mu = 0$ are much smaller than 1 while the risks for those moderately large μ are much larger than 1. This indicates that the achievement of the risk reduction around $\mu = 0$ goes with the risk enlargement for moderately large μ. This is beneficial when the thresholding rules are applied to a sequence of normal random variables which are small most of the time; (2) For a large μ, the risk for the soft-thresholding rule is 2. This is due to the bias caused by the upward or downward shift of the soft-thresholding rule. However, when the soft-thresholding rule is applied to a sequence of normal random variables which are small most of the time, then it is still possible that the total risk reduction for the sequence is larger than the total risk enlargement so that the soft-thresholding is beneficial. Moreover, since the risk reduction made by the soft-thresholding rule is much larger than those by other thresholding rules, it is expected that the soft-thresholding rule may outperform other thresholding rules; (3) For large μ, all the rules except the soft-thresholding rule have the same risk 1 as the least squares estimator does. The rationale behind this is that when μ is large, the sample z is generally large so that z will not be thresholded. This is generally true as stated in the following result.

Theorem 1. *Assume $\hat{\mu}(z)$ is a general thresholding rule as defined in (10) which is not thresholded or shrunk to 0 when z is large. That is, there is some constant $c > 0$ such that $\hat{\mu}(z) = z$ when $|z| > c$. Let c_μ and c_σ be defined as in (12). Then as $|\mu| \to \infty$, we have*

$$c_\mu \to 0, \quad \textit{and} \quad c_\sigma \to 1.$$

Proof. Assume the required condition is satisfied. Then as $|\mu| \to \infty$, by (14), we have $a_r, b_r \to -\infty, r = 1, 2, 3, c_0, d_0 \to \infty$, and c_1, d_1 are finite. It follows that $c_\mu \to 0$ since $\Phi_0(x) = \Phi(x)$. Notice that in the expression (13) of c_σ, only the last term $\Phi_2(x)|_{b_3}^{+\infty} \to 1$ as $b_3 \to -\infty$ and other terms all tend to 0. The theorem is proved.

When $\lambda < \omega = a = +\infty$, (10) defines a soft-thresholding rule. It does not satisfy the condition required by the above theorem. Thus, the above theorem is not applicable for the soft-thresholding rule. However, similar calculations show that c_μ has the same asymptotic behavior but c_σ tends to $\lambda^2 + 1$, which is 2 when $\lambda = 1$. The latter defines the soft-thresholding rule (5).

2.2 Thresholding a Sequence of Normal Random Variables

For practical applications, thresholding rules are often applied to a sequence of normal random variables which may result from, say, wavelet applications to some nonparametric smoothing problems (Donoho & Johnstone (1994)). Let a sequence of independent normal random variables be

$$z_i \sim N(\mu_i, \sigma_i^2), \quad i = 1, 2, \ldots, n. \tag{15}$$

Then a direct application of the general thresholding rule (10) will result in following thresholding sequence:

$$\hat{\mu}_i = \hat{\mu}_i(z_i) = c(z_i; \sigma_i) z_i, \quad i = 1, 2, \ldots, n. \tag{16}$$

When $\sigma_i, i = 1, 2, \ldots, n$ are not the same, the associated thresholding parameters are naturally different, resulting in the so-called order-dependent thresholding;

and when they are the same, the associated thresholding parameters are the same, resulting in the case of Donoho & Johnstone (1994) where the same thresholding parameter is used. In some other situations such as the one we shall investigate in next section, we must use different thresholding parameters for different components although all the $\sigma_i, i = 1, 2, \ldots, n$ are the same. This is to take some other information into account for efficient estimation of the underlying quantities.

Due to the independence of the components, the estimation risk is the sum of the componentwise risks. That is,

$$\rho(\hat{\mu}, \mu) = \mathrm{E}_\mu \|\hat{\mu} - \mu\|^2 = \sum_{i=1}^{n} \mathrm{E}_\mu (\hat{\mu}_i - \mu_i)^2 = \sum_{i=1}^{n} \rho(\hat{\mu}_i, \mu_i).$$

Theoretically, we can compute the estimation risk of $\hat{\mu}$ using the formula (12) componentwisely. When most of μ_i's are small, it is expected that the total estimation risk for using any thresholding rule will be small too. This is the basis for thresholding rules' application.

3 Thresholding Regression Splines

In regression spline smoothing, we approximate f in (1) using a linear combination of a set of truncated power basis functions as described below. Given a sequence of K interior knots $0 < \tau_1 < \ldots < \tau_K < 1$, the regression splines basis functions of order q are

$$1, t, t^2, \ldots, t^q, (t - \tau_1)_+^q, \ldots, (t - \tau_K)_+^q. \tag{17}$$

Denote the $K + q + 1$ basis functions as $\sigma_r(t), r = 0, 1, 2, \ldots, K + q$, and let $\mathbf{\Psi}(t)$ be the vector of the basis functions. The regression spline smoothing is to model f in (1) using a linear combination of the basis functions:

$$s(t) = \mathbf{\Psi}(t)^T \beta, \tag{18}$$

where $\beta = (\beta_0, \beta_1, \ldots, \beta_{K+q})^T$ denoting the basis coefficients. The function s defined above is a piecewise polynomial. That is, within any two neighboring knots, s is a polynomial of order q while at any interior knot τ_k, s has up to $(q-1)$-times continuous derivatives. When f has up to $(q-1)$-times continuous derivatives, it can be well approximated by some s of order q.

In practice, the degree of smoothness of f is usually unknown. To form a basis adaptive to various degrees of smoothness of f, we may combine basis functions of various orders, say, from order q up to order p. The resulting basis may be expressed as follows:

$$1, t, t^2, \ldots, t^p, (t - \tau_1)_+^q, \ldots, (t - \tau_K)_+^q, \ldots, (t - \tau_1)_+^p, \ldots, (t - \tau_K)_+^p. \tag{19}$$

The total number of basis functions is $K(p-q+1)+(p+1)$. Any f that has up to $(r-1)$-times continuous derivatives where $q \le r \le p$ can be well approximated by some linear combination of the above basis functions. Therefore, the basis defined in (19) is more flexible than the usual regression spline basis (17).

Let $\mathbf{\Psi}(t)$ denote a basis vector defined either in (17) or (19). Then the model (1) may be further written as:

$$y_i = \mathbf{\Psi}(t_i)^T \boldsymbol{\beta} + \epsilon_i, \quad t_i \in [0,1], \quad \epsilon_i \sim N(0, \sigma^2), \quad i = 1, 2, \ldots, n. \tag{20}$$

Let $\mathbf{X}_i = \mathbf{\Psi}(t_i)$ and $\mathbf{X} = (\mathbf{X}_1, \ldots, \mathbf{X}_n)^T$. The model (20) essentially becomes the following multiple linear regression model:

$$\mathbf{y} = \mathbf{X}\boldsymbol{\beta} + \boldsymbol{\epsilon}, \quad \boldsymbol{\epsilon} \sim N(0, \sigma^2 \mathbf{I}_n), \tag{21}$$

where and throughout, $\mathbf{y} = (y_1, \ldots, y_n)^T, \boldsymbol{\epsilon} = (\epsilon_1, \ldots, \epsilon_n)^T$ and $\mathbf{I}_n$ is the $n \times n$ identity matrix. Unlike the design matrix in a general multiple linear regression model, the design matrix $\mathbf{X}$ here is induced from evaluating the smooth basis function vector $\mathbf{\Psi}(t)$ at the design time points.

Without loss of generality, let the number of basis functions be m, which is smaller than n. The singular value decomposition (SVD) of $\mathbf{X}$ is

$$\mathbf{X} = \mathbf{U}\mathbf{D}\mathbf{V}^T, \tag{22}$$

where $\mathbf{U} = (\mathbf{u}_1, \mathbf{u}_2, \ldots, \mathbf{u}_n)$ and $\mathbf{V} = (\mathbf{v}_1, \mathbf{v}_2, \ldots, \mathbf{v}_m)$ are $n \times n$ and $m \times m$ orthonormal matrices so that $\mathbf{U}^T\mathbf{U} = \mathbf{I}_n$ and $\mathbf{V}^T\mathbf{V} = \mathbf{I}_m$, $\mathbf{D} = (\mathbf{D}_1, \mathbf{D}_2)^T$ where $\mathbf{D}_1 = \mathrm{diag}(d_1, d_2, \ldots, d_m)$ with $d_1 \geq d_2 \geq \ldots \geq d_m$, and $\mathbf{D}_2 : (n-m) \times m$ is a zero matrix. For convenience, let $d_i = 0, i = m+1, m+2, \ldots, n$. Then $\mathbf{u}_1, \ldots, \mathbf{u}_n$ are the n-dimensional orthonormal eigenvectors of $\mathbf{X}\mathbf{X}^T$, associated with eigenvalues $d_1^2, d_2^2, \ldots, d_n^2$ while $\mathbf{v}_1, \ldots, \mathbf{v}_m$ are the m-dimensional orthonormal eigenvectors of $\mathbf{X}^T\mathbf{X}$, associated with eigenvalues $d_1^2, d_2^2, \ldots, d_m^2$.

Let

$$\mathbf{z} = \mathbf{U}^T\mathbf{y}, \quad \mathbf{e} = \mathbf{U}^T\boldsymbol{\epsilon}, \tag{23}$$

be the orthogonal transformation of $\mathbf{y}$ and $\boldsymbol{\epsilon}$ respectively. They are the projection coefficients of $\mathbf{y}$ and $\boldsymbol{\epsilon}$ on the space spanned by the eigen vectors $\mathbf{u}_1, \ldots, \mathbf{u}_n$ respectively. Then

$$z_i = \mathbf{u}_i^T\mathbf{y} = \mu_i + e_i, \quad e_i \sim N(0, \sigma^2), \quad i = 1, 2, \ldots, n. \tag{24}$$

Moreover,

$$\mathbf{y} = \sum_{i=1}^n z_i \mathbf{u}_i, \quad \boldsymbol{\epsilon} = \sum_{i=1}^n e_i \mathbf{u}_i. \tag{25}$$

Based on the model (24), the usual least squares estimators for $\mu_i, i = 1, 2, \ldots, n$ are

$$\hat{\mu}_i = z_i, i = 1, 2, \ldots, n. \tag{26}$$

However, it is obvious that $\mathrm{E}z_i = \mu_i = 0$ for $i = m+1, m+2, \ldots, n$; the associated z_i's are purely noise, normally distributed with mean 0 and variance σ^2. Therefore, the least squares estimators (26) for $\hat{\mu}_i, i = m+1, \ldots, n$ are estimating noise and hence are not favorable.

Based on the model (24), it is seen that $z_1, z_2, \ldots, z_n$ are homogeneous since the noise terms $e_1, e_2, \ldots, e_n$ have the same variance σ^2. A simple application of the general thresholding rules (10) to the projection coefficients $z_i, i = 1, 2, \ldots, n$ using the same thresholding parameter λ will result in the following estimators for $\mu_i, i = 1, 2, \ldots, n$:

$$\hat{\mu}_i = c(z_i; \lambda) z_i, i = 1, 2, \ldots, n. \tag{27}$$

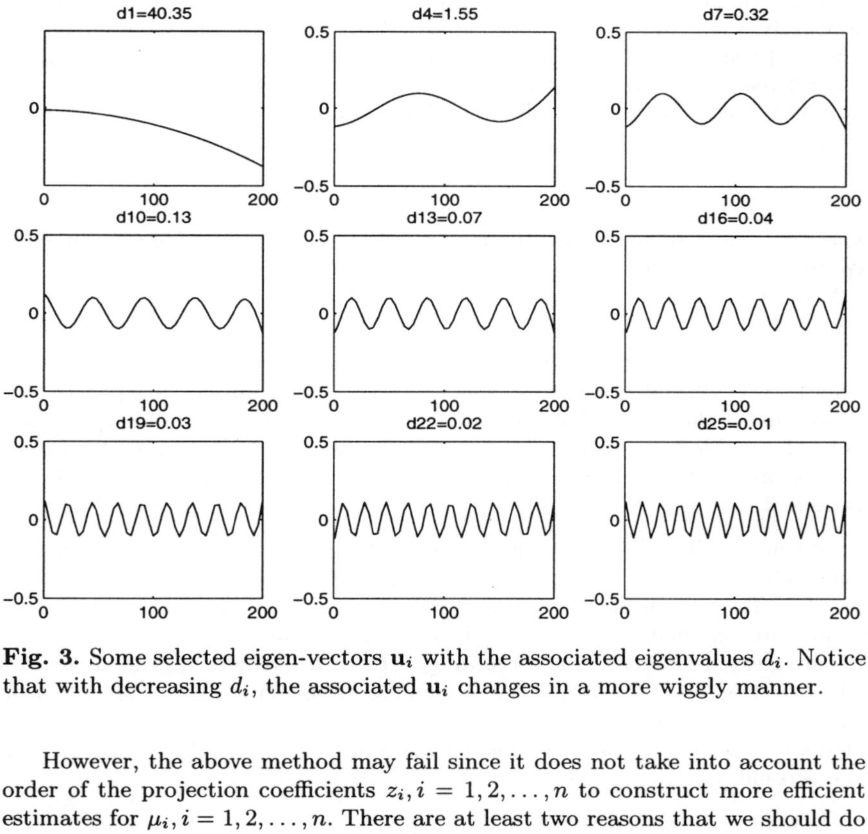

Fig. 3. Some selected eigen-vectors $\mathbf{u}_i$ with the associated eigenvalues d_i. Notice that with decreasing d_i, the associated $\mathbf{u}_i$ changes in a more wiggly manner.

However, the above method may fail since it does not take into account the order of the projection coefficients $z_i, i = 1, 2, \ldots, n$ to construct more efficient estimates for $\mu_i, i = 1, 2, \ldots, n$. There are at least two reasons that we should do so. First of all, the amount of the signal that the projection coefficients $z_i, i = 1, 2, \ldots, n$ contain is mainly determined by their orders.

According to the definition of the singular value decomposition, the eigen vectors $\mathbf{u}_i, i = 1, 2, \ldots, n$ and hence the associated projection coefficients $z_i, i = 1, 2, \ldots, n$ are sorted in the order that the associated singular values $d_1 \leq d_2 \leq$

$\ldots \leq d_n$ so that the lower (higher) order projection coefficients contain more (less) signal than noise out of the response vector $\mathbf{y}$ and the design matrix $\mathbf{X}$. Secondly, the roughness of the eigen vectors $\mathbf{u}_i, i = 1, 2, \cdots, n$ are also closely related to the orders of the associated projection coefficients $z_i, i = 1, 2, \ldots, n$, or equivalently the associated singular values $d_i, i = 1, 2, \ldots, n$. Fig. 3 displays some selected $\mathbf{u}_i$'s (resulted from $n = 200$ equally spaced time points t_i's) with associated d_i's. It is seen that with decreasing d_i, the associated $\mathbf{u}_i$ is generally getting rougher (the curve changes in a higher frequent manner, i.e., with more wiggly waves). Similar situations can be observed for unequally spaced t_i's and/or for other sample sizes n. Therefore, to construct an informative and smooth estimate [see (31) for example] for the underlying function f in (1), we should threshold less (more) the lower (higher) order projection coefficients z_i's. In particular, as mentioned previously, $z_i, i = m+1, \ldots, n$ (associated with $d_i \equiv 0, i = m+1, \ldots, n$) are purely noise and hence they should be thresholded as 0. In other words, we should treat z_i's differently according to their orders, i.e., the values of d_i's. A natural way to reflect this is to use the following thresholding parameters λ_i for different projection coefficients z_i's:

$$\lambda_i = \lambda / d_i^k, i = 1, 2, \ldots, n, \tag{28}$$

where $k > 0$ is some given integer and λ a common thresholding parameter. We then define the following order-dependent thresholding estimators for $\mu_i, i = 1, 2, \cdots, n$:

$$\hat{\mu}_i = c(z_i; \lambda_i) z_i, i = 1, 2, \ldots, n, \tag{29}$$

by applying the general thresholding rule $c(z; \lambda)$ as defined in (10). It is worthwhile to notice that when d_i are larger (smaller), the associated thresholding parameters λ_i are smaller (larger) so that there is a less (more) chance for the associated z_i's to be thresholded; in particular, when $d_i = 0$, the associated $\lambda_i = \infty$ (as usual, we interpret $\lambda/0$ as ∞) so that the associated z_i are definitely thresholded as 0 as desired.

When we allow $k = 0$ in the formula (28), it indicates all the λ_i are the same. This reduces to the case of simple application of the thresholding rules, i.e., (27). This is equivalent to ignoring the order of the z_i's and hence may result in undesired estimates [see Fig. 11 (d) in Sect. 5]. However, when $k > 0$, all the z_i's are treated differently and according to their d_i's values. In general, we can take $k = 1$ or $k = 2$ for simplicity. A small-scale simulation study presented in Sect. 4 shows that $k = 1$ is slightly better than $k = 2$ while they both are better than $k = 0$; see Fig. 7. Thus, it is beneficent to take the order of $z_i, i = 1, 2, \ldots, n$ into account.

Denote

$$\hat{z}_i = \hat{\mu}_i = c(z_i; \lambda / d_i^k) z_i, \quad i = 1, 2, \ldots, n. \tag{30}$$

In matrix notation, that is,

$$\hat{\mathbf{z}} = (\hat{z}_1, \ldots, \hat{z}_n)^T = \mathbf{C}(\mathbf{z}; \lambda, \mathbf{d}, k)\mathbf{z},$$

where $\mathbf{C}(\mathbf{z};\lambda,\mathbf{d},k)$ is a diagonal matrix with diagonal elements $c(z_i;\lambda/d_i^k), i = 1,2,\ldots,n$. The fitted $\mathbf{y}$ is then

$$\hat{\mathbf{y}} = \mathbf{U}\hat{\mathbf{z}} = \mathbf{U}\mathbf{C}(\mathbf{z};\lambda,\mathbf{d},k)\mathbf{z} = \mathbf{U}\mathbf{C}(\mathbf{U}^T\mathbf{y};\lambda,\mathbf{d},k)U^T\mathbf{y} = \mathbf{S}_\lambda(\mathbf{y})\mathbf{y}, \tag{31}$$

where $\mathbf{d} = (d_1, d_2, \ldots, d_n)^T$ and

$$\mathbf{S}_\lambda(\mathbf{y}) = \mathbf{U}\mathbf{C}(\mathbf{U}^T\mathbf{y};\lambda,\mathbf{d},k)\mathbf{U}^T \tag{32}$$

is the associated smoother matrix. Notice that $\hat{\mathbf{y}}$ is a nonlinear shrinking estimator of $\mathbf{f}$ since $\|\hat{\mathbf{y}}\| \le \|\mathbf{y}\|$.

When $\mathbf{S}_\lambda$ does not depend on $\mathbf{y}$, i.e., $\hat{\mathbf{y}}$ is a linear smoother, we have

$$\mathrm{Var}(\hat{y}_i) = \sigma^2 \mathbf{e}_i^T \mathbf{S}_\lambda^2 \mathbf{e}_i, \tag{33}$$

where $\mathbf{e}_i$ denotes an n-dimensional vector with its i-th component to be 1 and 0 otherwise. The above variance expression is in general not true for the nonlinear thresholding smoother (31), and we expect that the variance for $\hat{y}_i$ in (31) should be larger than the one given in (33) due to the dependence of $\mathbf{S}_\lambda$ on $\mathbf{y}$. Nevertheless, it is noticed that the dependence is generally weak since $\mathbf{C}$ is just a diagonal matrix containing some indicator functions specifying the relationship between $\mathbf{y}$ and the thresholding parameters. Thus, we may use the following approximated 2 standard deviation bands to give some basic feeling about the variability of $\hat{y}_i$:

$$\hat{y}_i \pm 2\widehat{\mathrm{Var}}^{1/2}(\hat{y}_i), i = 1,2,\ldots,n, \tag{34}$$

where $\widehat{\mathrm{Var}}(\hat{y}_i)$ is obtained using (33) with σ^2 replaced by a proper estmator, for example,

$$\hat{\sigma}^2 = \sum_{i=1}^n \{1 - c(z_i;\lambda/d_i^k)\}^2 z_i^2 / (n - \sum_{i=1}^n c(z_i;\lambda/d_i^k)); \tag{35}$$

When $c(z_i;\lambda/d_i^k)$ are independent of z_i, it is easy to show that $\hat{\sigma}^2$ is an unbiased and consistent estimator of σ^2.

The f in (1) can also be estimated using the following penalized regression spline: $\hat{f}(t) = \mathbf{\Psi}(t)^T\hat{\boldsymbol{\beta}}$ where $\hat{\boldsymbol{\beta}}$ is the solution to the following penalized likelihood problem:

$$\sum_{i=1}^n (y_i - \mathbf{\Psi}(t_i)^T\boldsymbol{\beta})^2 + \lambda\boldsymbol{\beta}^T\boldsymbol{\beta}. \tag{36}$$

That is,

$$\hat{\boldsymbol{\beta}} = (\mathbf{X}^T\mathbf{X} + \lambda\mathbf{I}_m)^{-1}\mathbf{X}^T\mathbf{y}, \tag{37}$$

where m is the number of basis functions in $\mathbf{\Psi}(t)$. Therefore,

$$\hat{\mathbf{y}} = \mathbf{X}(\mathbf{X}^T\mathbf{X} + \lambda\mathbf{I}_m)^{-1}\mathbf{X}^T\mathbf{y}. \tag{38}$$

Using the singular value decomposition (23) and let $\mathbf{z} = \mathbf{U}^T\mathbf{y}, \hat{\mathbf{z}} = \mathbf{U}^T\hat{\mathbf{y}}$, and $d_i = 0, i = m+1,\ldots,n$, we have

$$\hat{z}_i = \frac{z_i}{1 + \lambda/d_i^2}, i = 1, 2, \ldots, n, \tag{39}$$

so that $\hat{\mathbf{y}} = \mathbf{U}\hat{\mathbf{z}}$. This expression can be unified into (30) using $c(z_i; \lambda/d_i^2) = (1 + \lambda/d_i^2)^{-1}$ but bear in mind that $c(z_i; \lambda/d_i^2) = 0$ when $d_i = 0$.

The regression spline smoother (38) is obtained via penalizing all the coefficients in $\boldsymbol{\beta}$ equally. It is slightly different from the one proposed and studied by Wand and Carroll (1998) where they penalize just those β_i's reflecting jumps of the regression spline (18).

3.1 Locating the Knots

To recover f in (1), at the initial stage of modeling, one often uses a large number of knots, i.e. K is large. In general, we take K such that $m < n$ where m is the total number of basis functions so that all the d_i's are positive.

There are several ways to locate the knots $\tau_k, k = 1, 2, \ldots, K$, often depending on the designs of the time points t_i's or the structure of the regression function f. For instance, when the t_i's are uniformly scattered in $[0, 1]$, the knots can be uniformly scattered in $[0, 1]$, i.e., $\tau_k = k/(K + 1), k = 1, 2, \ldots, K$. This method is widely used due to its simplicity even when the t_i's are not uniformly scattered. To be adaptive to the designs of the t_i's, the k-th knot τ_k's are usually chosen as the k-th sample quantile of $t_i, i = 1, 2, \ldots, n$. That is, let $t_{(i)}, i = 1, 2, \ldots, n$ be the order statistics of $t_i, i = 1, 2, \ldots, n$. Then $\tau_k = t_{[kn/(K+1)]}, k = 1, 2, \ldots, K$ where $[x]$ denotes the largest integer that is smaller than x. The above procedure allows to scatter more (fewer) knots where more (fewer) design time points are around. When f is spatially imhomogeneous, the knots τ_k's may be chosen to be adaptive to the structure of f. For instance, for the "Doppler" function of Donoho & Johnstone (1994) [see Fig. 4(a), the solid curve) in Sect. 4], the left side of f has more wiggly waves than its right side and hence we should put more knots at the left side of the support of f to improve the performance of the estimate. For this end, we may choose τ_k's such that $\log(\tau_k)$'s are uniformly scattered. We term the above three knot location procedures as "uniformly scattered", "quantiles as knots" and "log-uniformly scattered" respectively for easy reference. A small-scale simulation study conducted in Sect. 4 shows that the "quantiles as knots" procedure generally works well while the third procedure works well for the "Doppler" function; see Fig. 9.

3.2 Thresholding Parameter Selection

For each given thresholding parameter λ, the smoother $\hat{\mathbf{y}}$ of f in (1) is given by (31) and (30). Whether $\hat{\mathbf{y}}$ is a good estimate for $\mathbf{f} = (f(t_1), \ldots, f(t_n))^T$ depends strongly on the choice of λ. A measure for the accuracy of $\hat{\mathbf{y}}$ estimating $\mathbf{f}$ is the mean squared error (MSE):

$$E\sum_{i=1}^{n}(\hat{y}_i - f(t_i))^2 = \sum_{i=1}^{n}\text{Bias}^2(\hat{y}_i) + \sum_{i=1}^{n}\text{Var}(\hat{y}_i), \tag{40}$$

where all the expectations involved are taken conditional to the given $\mathbf{f}$. The first (bias) term in the right-hand side represents the goodness of fit of the estimator $\hat{\mathbf{y}}$ and the second (variance) term its total variation. As seen from (30), when λ is large, most of the z_i's are thresholded or shrunk to 0 so that the variance term can be made as small as possible; on the other hand, when λ is small, only few of the z_i's are thresholded or shrunk to 0 so that the bias term can be made as small as possible. Therefore, choosing λ is equivalent to balancing the model complexity and the goodness of fit of the estimator. That is, a good λ should be chosen to minimize the MSE. Unfortunately, the above MSE is not computable since $\mathbf{f}$ is unknown. This motivates proposals of many model selection criteria which mimic the MSE or other accuracy measures for estimates. All these criteria consist of two parts: one represents the goodness of fit and the other the model complexity. These two parts can be balanced by some proper choice of the thresholding parameter λ.

The most popular measure for goodness of fit of the estimator (31) is the sum of squared errors (SSE):

$$\text{SSE}_\lambda = \sum_{i=1}^{n}(y_i - \hat{y}_i)^2 = \sum_{i=1}^{n}\{1 - c(z_i; \lambda/d_i^k)\}^2 z_i^2, \tag{41}$$

using the singular value decomposition (23). As seen from (41), when λ increases, more z_i's will be truncated or shrunk to 0 so that the SSE_λ increases. Thus, use of large λ worsens the goodness of fit due to fewer of the z_i's are used in the model.

A popular measure for model complexity is the trace of the smoother matrix $\mathbf{S}_\lambda$ as defined in (31):

$$\text{tr}(\mathbf{S}_\lambda) = \sum_{i=1}^{n} c(z_i; \lambda/d_i^k). \tag{42}$$

As seen from the above expression, when λ increases, the trace will decrease due to fewer of the z_i's involved so that the model becomes less complicated. In particular, when the smoother is the hard-thresholding rule (3), the trace is exactly the number of the z_i's that are not truncated into 0.

To select good λ, some model selection criterion that tradeoffs the goodness of fit and the model complexity of an estimator should be applied. The existing model selection criteria include cross-validation (CV), generalized cross-validation (GCV), Akaike information criterion (AIC), Bayesian information criterion (BIC), C_p statistics and Stein's unbiased risk estimator (SURE) among others. Although not all of these criteria are constructed based on SSE_λ and $\text{tr}(\mathbf{S}_\lambda)$, they all aim to tradeoff the goodness of fit and the model complexity of an estimator.

Cross-validation is a popular model selection criterion due to its simplicity. The CV score of the smoother $\hat{\mathbf{y}}$ (31) is computed as follows:

$$\mathrm{CV}(\lambda) = \sum_{i=1}^{n} \left(y_i - \hat{y}_i^{(-i)} \right)^2 \approx \sum_{i=1}^{n} \left(\frac{y_i - \hat{y}_i}{1 - s_{ii}} \right)^2, \tag{43}$$

where $\hat{y}_i^{(-i)}$ denotes the prediction of y_i computed using (31) based the data with the i-th observation excluded and s_{ii} the i-th diagonal element of the smoother matrix $\mathbf{S}_\lambda$ as in (32). We expect the second relationship in the expression (43) approximately holds although $\hat{\mathbf{y}}$ is a nonlinear smoother; alternatively, we may directly define the CV score as the right-hand side of the expression. Based on this and for saving computation, we may define the generalized cross-validation (GCV) score as:

$$\mathrm{GCV}(\lambda) = \frac{\mathrm{SSE}_\lambda}{[1 - n^{-1}\mathrm{tr}(\mathbf{S}_\lambda)]^2}. \tag{44}$$

The above expression is actually obtained via replacing all the s_{ii}'s in the CV score (43) by their average $n^{-1}\sum_{i=1}^{n} s_{ii} = n^{-1}\mathrm{tr}(\mathbf{S}_\lambda)$. The CV and GCV ideas are widely used in nonparametric smoothing such as smoothing splines, regression splines, kernel and local polynomial smoothings.

AIC and BIC are two other popular model selection criteria, which, in our context, may be defined respectively as follows:

$$\mathrm{AIC}(\lambda) = \log(\mathrm{SSE}_\lambda) + 2n^{-1}\mathrm{tr}(\mathbf{S}_\lambda), \tag{45}$$

$$\mathrm{BIC}(\lambda) = \log(\mathrm{SSE}_\lambda) + 2n^{-1}\log(n)\mathrm{tr}(\mathbf{S}_\lambda), \tag{46}$$

where $\mathrm{tr}(\mathbf{S}_\lambda)$ is used to represent the model complexity, replacing the number of the parameters as in AIC and BIC for multiple linear regression model selection. It is seen that BIC is more parsimonious than AIC since it puts more weights on the model complexity, $\mathrm{tr}(\mathbf{S}_\lambda)$, than AIC does. Consequently, the λ chosen by BIC is larger than the λ by AIC. That is, BIC yields a smoother $\hat{\mathbf{y}}$.

C_p-statistic is also a popular model selection criterion. It is defined based on an unbiased estimator of the MSE (40):

$$\mathrm{SSE}_\lambda + \sigma^2(2\mathrm{tr}(\mathbf{S}_\lambda) - n).$$

That is, given a good estimate of the σ^2, the C_p statistic is constructed as

$$C(\lambda) = \mathrm{SSE}_\lambda + \hat{\sigma}^2(2\mathrm{tr}(\mathbf{S}_\lambda) - n), \tag{47}$$

where λ replaces the p in the usual C_p statistic for multiple linear model selection.

SURE (Stein's Unbiased Risk Estimate) criterion was proposed in Donoho & Johnstone (1998) for wavelet thresholding parameter selection. Johnstone & Silverman (1997) extended it to correlated data analysis using wavelets. Let $z_i \sim N(\theta_i, \sigma^2)$ be independent, and $\hat{z}_i = z_i + g(z_i)$ be estimates of $\theta_i, i = 1, 2, \ldots, n$ where $g(\cdot)$ is a weakly differentiable function. It is easy to show that

$$\mathrm{E}_\theta\|\hat{\mathbf{z}} - \boldsymbol{\theta}\|^2 = n\sigma^2 + 2\sigma^2\mathrm{E}_{\boldsymbol{\theta}}\sum_{i=1}^{n} g'(z_i) + \mathrm{E}_{\boldsymbol{\theta}}\sum_{i=1}^{n} g^2(z_i),$$

where $\mathbf{z} = (z_1, \ldots, z_n)^T$, $\boldsymbol{\theta} = (\theta_1, \ldots, \theta_n)^T$ and $\mathrm{E}_{\boldsymbol{\theta}}$ indicates the conditional expectation given $\boldsymbol{\theta}$. Applying this rule to the sequence (30), we got the SURE criterion defined as follows:

$$\mathrm{SURE}(\lambda) = \hat{\sigma}^2 + 2\hat{\sigma}^2 n^{-1} \sum_{i=1}^{n} g'(z_i) + n^{-1} \sum_{i=1}^{n} g^2(z_i), \tag{48}$$

where $g(z_i) = \{c(z_i; \lambda/d_i^k) - 1\}z_i$ using (30) so that $g'(z_i) = c'(z_i; \lambda/d_i^k)z_i + c(z_i; \lambda/d_i^k) - 1$, and $\hat{\sigma}^2$ is some consistent estimate of σ^2 such as the one defined in (35). Note that the SURE in Donoho & Johnstone (1998) and Johnstone & Silverman (1997) was defined for the soft-thresholding rule only. However, since $c(z_i; \lambda/d_i^k)$ has at most a finite number of discontinuity points such as the hard-thresholding rule, we expect SURE applies for all thresholding rules and the usual regression splines. Simulations presented in Sect. 4 show that SURE performed quite well for all the thresholding rules including the hard-thresholding rule.

4 Simulation Studies

In the previous sections, we have carefully discussed the methodologies for order-dependent thresholding and its applications to regression splines. These methodologies include the construction of thresholding rules, the selection of the order k, the methods for knot locating, and the methods for thresholding parameter selection. In this section, we aim to investigate their performance via simulations.

The testing functions include Doppler, Heavisine, Bumps and Blocks, which are defined and depicted in Donoho & Johnstone (1994). These functions caricature spatially imhomogeneous functions that arise in imaging, spectroscopy and other scientific signal processing. They have been extensively used for illustrating variable bandwidth selection (Fan and Gijbels (1996)) or wavelet smoothing methodologies (Antoniadis & Fan (2001); Donoho and Johnstone (1994) etc.).

In Fig. 4, four noisy data (dots) sets, which were respectively simulated from the four functions (solid curves), are displayed. The noise level σ was chosen such that the associated signal to noise ratio is about 7. The sample size is $n = 1000$ and the design time points t_i were uniformly sampled over $[0, 1]$. As an example, we illustrate how we fit these data sets using our methodologies. First of all, to fit these highly spatially imhomogeneous functions with small biases, we chose K as large as $n/4 = 250$ knots. We then located these knots using the "quantiles as knots" method except for the Doppler function, we used the "log-uniformly scattered" method to put more knots at the left end of the function support. We then constructed the basis functions using (19) with $q = 1$ and $p = 2$ so that we have as many as $m = K(p - q + 1) + (p + 1) = 502$ basis functions. After singular value decomposition, we applied the soft-thresholding rule (5) to the z_i's computed using (24). The $\hat{z}_i$'s were computed using (30) with $k = 1$. The thresholding parameter λ was chosen by the GCV rule (44).

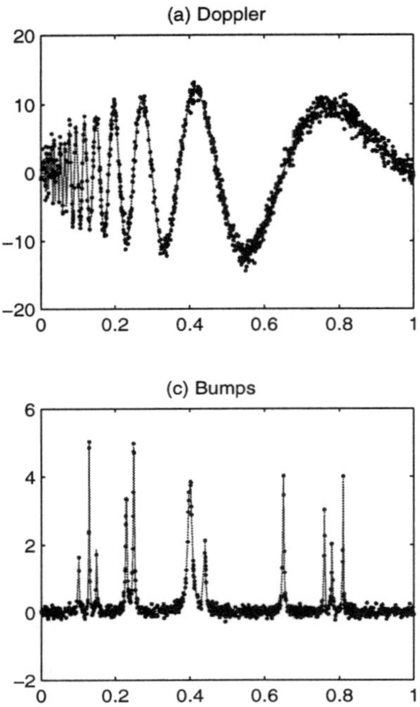

Fig. 4. Noisy simulated data for : (a) Doppler, (b) Heavisine, (c) Bumps, and (d) Blocks. The noise level σ^2 was chosen respectively so that the signal to noise ratio is about 7.

The resulting fits are presented in Fig. 5. These fits are reasonably good although they appeared undersmoothed. This is probably due to the fact that these functions are too spatially imhomogeneous. This difficulty may be overcome via more carefully locating the knots. Nevertheless, for less spatially imhomogeneous functions such as those presented in Sect. 5, our approach performs rather well. The accuracy of a smoother $\hat{f}$ of an underlying function f is often measured using the mean squared error of the smoother, defined as

$$\mathrm{MSE}(\hat{f}) = n^{-1} \sum_{i=1}^{n} (f(t_i) - \hat{f}(t_i))^2.$$

The smaller the MSE is, the more accurate the smoother is. In what follows, MSE is used to compare smoothers based on different thresholding rules, different order k, different methods for thresholding parameter selection, or different knot locating methods. The four testing functions mentioned previously were used. For each testing function, we first generated a sample of size $n = 300$ with a signal-to-noise ratio about 7. We then fit the sample using our methodologies and then computed the MSE using (40). The whole process was repeated $N = 100$ times

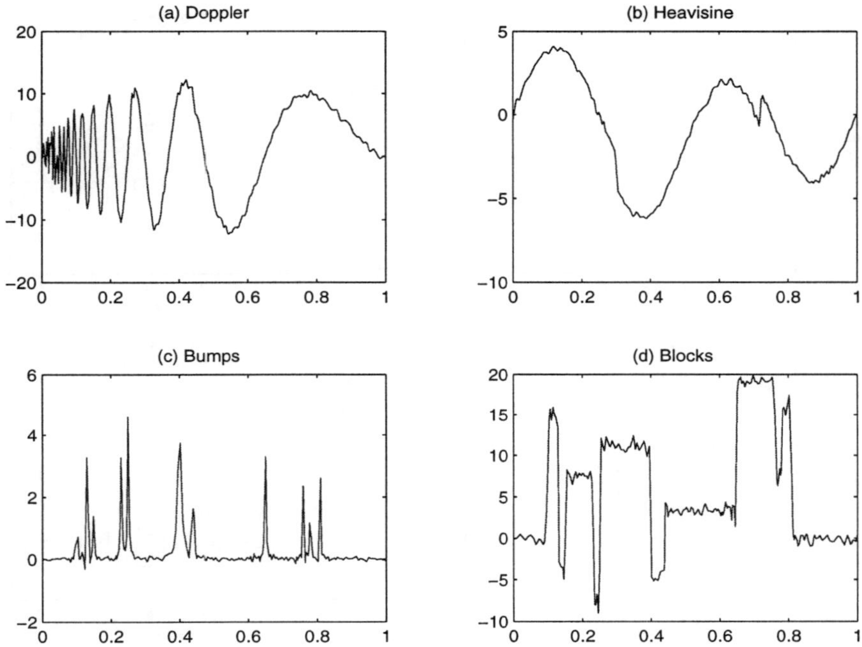

Fig. 5. The fitted curves using the soft-thresholding rule: (a) Doppler, (b) Heavisine, (c) Bumps, and (d) Blocks, where $k = 1$ and the λ's were chosen by GCV.

so that we have $N = 100$ MSEs for each testing function. The performance of a smoother is then summaried by a boxplot of these MSEs; see Fig. 6 for example.

For simplicity, we defaulted the following choice for the parameters involved in simulations, unless they were the targets of the comparison. The number of knots was set to be $K = n/3 = 100$, and they were located using the "quantiles as knots" method except for the Doppler function, the "log-uniformly scattered" method was used. We used the basis (19) with $q = 1$ and $p = 2$. The $k = 1$, and the soft-thresholding rule (5) were used. The thresholding parameter was chosen using the GCV rule (44).

We first present the simulation results for comparing different thresholding rules. Fig. 6 displays the boxplots of the MSEs for the usual regression spline (39), hard (3), soft (5), mixture (6), SCAD-new (9) and SCAD (7) rules as defined in Sect. 2.

For the first three testing functions, it seems that the usual regression spline and the soft-thresholding performed better than the other thresholding rules, while the others performed similarly. However, for the Blocks function, the usual regression spline performed worse than all the thresholding rules significantly.

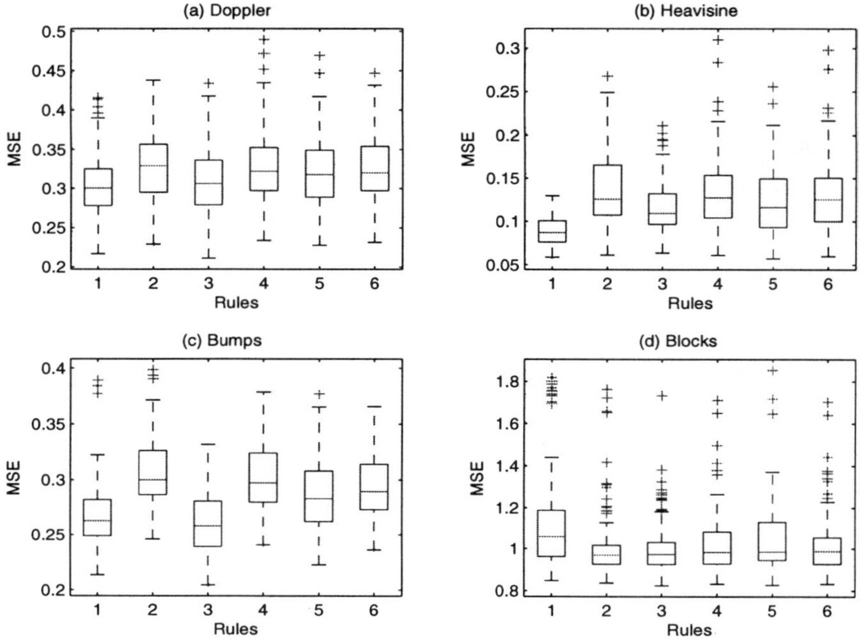

Fig. 6. Boxplots for the MSEs for (a) Doppler, (b) Heavisine, (c) Bumps, and (d) Blocks for thresholding rules (from left to right): usual regression spline, soft, hard, mixed, SCAD-new and SCAD.

This shows that when the testing function is seriously discontinuous, the usual regression spline will usually perform worse than the thresholding rules. Among the thresholding rules, the soft-thresholding rule performed best. This is due to the fact that most of the coefficients are small. We can also see that the soft-thresholding rule is comparable with the usual regression spline: when the testing function is highly discontinuous and highly spatially imhomogeneous, the soft-thresholding performed better, while when the testing function is less spatially imhomogeneous and less discontinuous, the usual regression spline performed better.

We now examine the effect of the different k. For simplicity, we just considered three different k's: $k = 0, 1$, and 2. When $k = 0$, the thresholding is order-independent. That is, all the z_i's in (30) are treated in a same manner. Fig. 7 displays the boxplots of the MSEs for $k = 0, 1$ and 2 respectively for the four testing functions. It is seen that except for the Blocks function which is seriously discontinuous, the order-independent thresholding estimates ($k = 0$) were worse than those order-dependent thresholding estimates ($k > 0$). This says that if the underlying function is reasonably smooth, the order-dependent threshold-

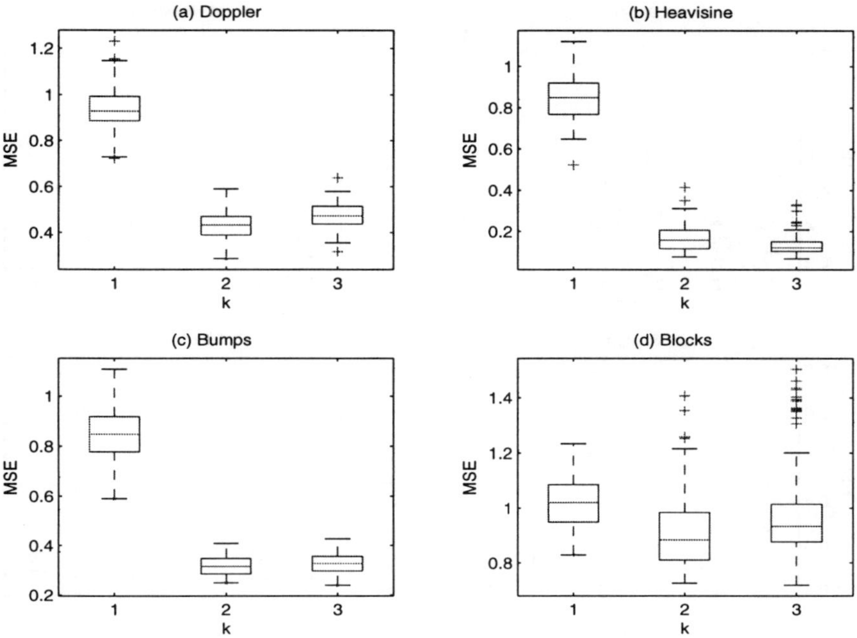

Fig. 7. Boxplots for the MSEs for (a) Doppler, (b) Heavisine, (c) Bumps, and (d) Blocks for (from left to right): $k = 0$, $k = 1$, and $k = 2$.

ing is preferred to order-independent thresholding. Another observation is that thresholding estimates with $k = 1$ generally outperformed those with $k = 2$. This is the reason why in general we used $k = 1$ in the examples and simulation studies.

Which thresholding parameter selection rule is preferred? We used another small-scaled simulation study to answer this question. Fig. 8 presents the box-plots of the MSEs for the six thresholding parameter selection rules: CV, GCV, BIC, AIC, SURE and C_p. The lessons are as follows. First of all, for the four testing functions, the GCV rule performed better or not worse than the CV rule. Second, the SURE and the C_p rules performed similarly; they are both based on the estimates of the noise variance $\hat{\sigma}^2$. Third, the BIC and AIC rules performed much worse than others, probably due to the reason that they generally oversmoothed the testing functions while the testing functions should generally be undersmoothed due to their spatial imhomogeneity. Finally, we can see that among all these criteria, the GCV is definitely preferred due to good performance, less intensive computation, and lack of the need for estimating the noise variance.

Finally, we examined the use of the knot locating methods. We have three possible knot locating approaches: "uniformly scattered", "quantiles as knots"

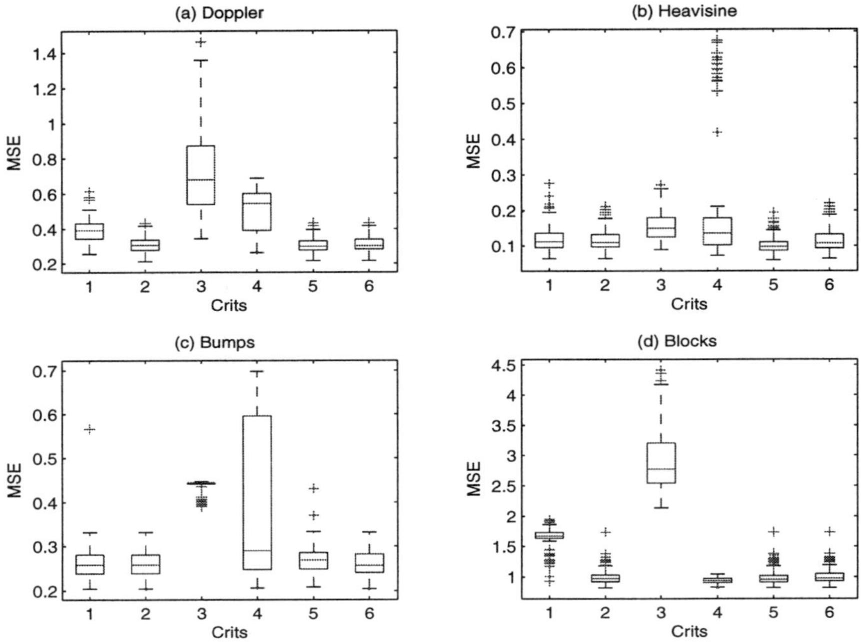

Fig. 8. Boxplots for the MSEs for (a) Doppler, (b) Heavisine, (c) Bumps, and (d) Blocks for the criteria (from left to right): CV, GCV, BIC, AIC, SURE, and C_p.

and "log-uniformly scattered". Fig. 9 displays the boxplots of the associated MSEs for the three methods. For the Doppler function, the third method is definitely better than the other two methods. This is due to the Doppler function is varying more and more rapidly as approaching the left end, and hence needs more knots at the area closer to the left end. For the Heavisine and Bumps functions, all the three methods performed similarly but the third method seems have smaller variation. It is hard to interpret this observation however. For the Blocks function, the first method outperformed the second, and the second outperformed the third. This is probably caused by the so many discontinuous points of the underlying function.

5 Two Real Data Examples

In this section, two real data examples are used to illustrate our methodologies. The first data set is displayed in Fig. 10 (d) as dots. It has been analyzed exten-

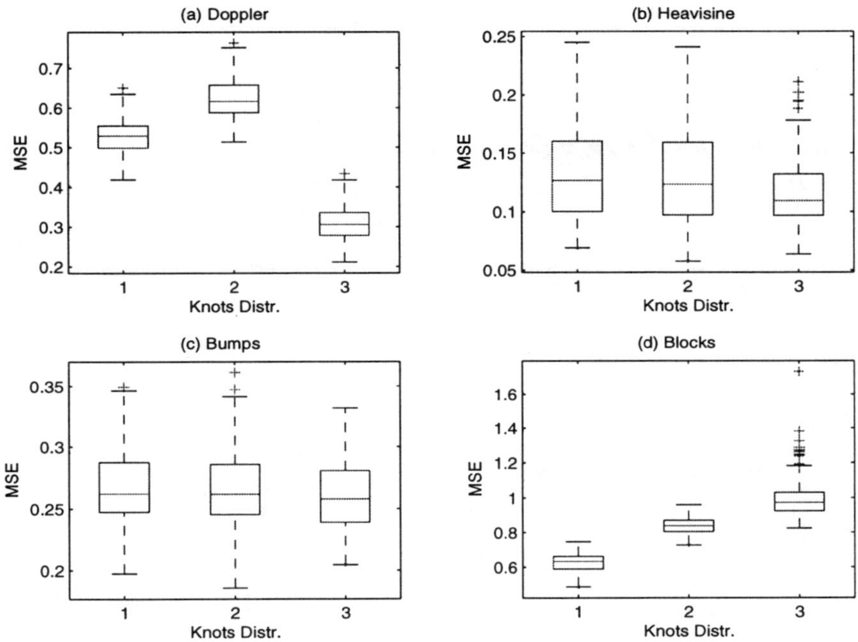

Fig. 9. Boxplots for the MSEs for (a) Doppler, (b) Heavisine, (c) Bumps, and (d) Blocks for the knots' distributions (from left to right): uniformly scattered, quantiles as knots, and log-uniformly scattered.

sively in the area of nonparametric regression smoothing and consists of $n = 133$ observations from a crash test and shows the acceleration of a motorcyclist's head during a crash. Silverman (1985) used it to illustrate his spline smoothing technique while Hall & Turlach (1997), Kovac & Silverman (2000), and Antoniadis & Fan (2001) used it to illustrate their wavelet-based methodologies for smoothing unequally sampled noisy data. As showed in Fig. 7 of Antoniadis & Fan (2001), classical wavelet thresholding or the interpolation method of Hall & Turlach (1997) produce wiggly estimates, while the robust thresholding method of Kovac & Silverman (2000) and the ROSE method of Antoniadis & Fan (2001) gave reasonable estimates. The solid curve in Fig. 10(d) is our estimate and is comparable to the good estimates in the literature. In addition, we also provided the approximate 2 standard deviation bands showed as dashed curves.

As in simulations presented in Sect. 4, our estimate was obtained as follows. We used $K = n/3 \approx 44$ interior knots which were located using the "quantiles as knots" method as described in Sect. 3.1. These knots can be denoted as

$$\tau_r = t_{(1+3*(r-1))}, r = 1, 2, \ldots, K,$$

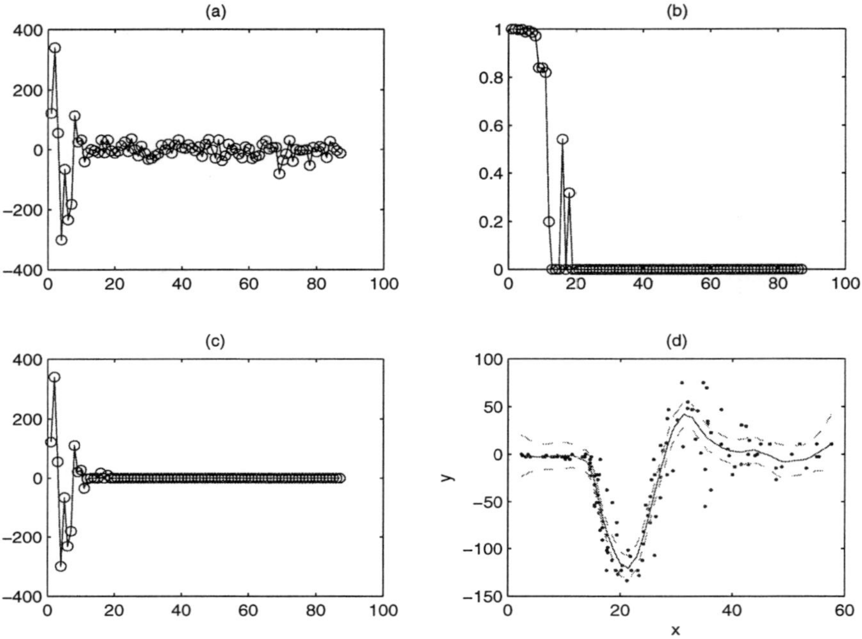

Fig. 10. The motocycle data example: order-dependent soft-thresholding. (a) The z_i's; (b) The $c(z_i; \lambda/d_i^k)$ with $k = 1$; (c) The $\hat{z}_i$'s; (d) The data (dots), fitted curve (solid curve) and approximate 2 standard deviation bands (dashed curves).

where $t_{(i)}, i = 1, 2, \ldots, n$ are the order statistics of the design time points $t_i, i = 1, 2, \cdots, n$.

These knots were then used to specify a design matrix $\mathbf{X} = [\mathbf{\Psi}(t_1), \ldots, \mathbf{\Psi}(t_n)]^T$ using the following basis constructed according to (19) with $q = 1$ and $p = 2$:

$$\mathbf{\Psi}(t) = [1, t, t^2, (t - \tau_1)_+, \ldots, (t - \tau_K)_+, (t - \tau_1)_+^2, \ldots, (t - \tau_K)_+^2]^T.$$

We took $q = 1$ because the left sided data (see Fig. 10 (d), dots) suggest a linear model, and took $p = 2$ because the middle part of the data suggest a quadratic model. The total number of the basis functions in $\mathbf{\Psi}(t)$ is then $m = K(p - q + 1) + (p + 1) = 91$, which is quite large. Denote the singular value decomposition of $\mathbf{X}$ as $\mathbf{UDV}^T$ where $\mathbf{U} = [\mathbf{u}_1, \ldots, \mathbf{u}_n]$ and $\mathbf{V} = [\mathbf{v}_1, \ldots, \mathbf{v}_m]$ and $\mathbf{D} = \text{diag}(d_1, \ldots, d_n)$ defined as before and in particular d_i's are sorted so that $d_1 \geq d_2 \ldots d_n$. We then got the z_i's using (24), i.e. $z_i = \mathbf{u}_i^T \mathbf{y}, i = 1, \ldots, n$. Fig. 10 (a) displays those m z_i's. It is seen that only a few of the $|z_i|$'s are large, say the first 8 in this example, and the rest are quite small. The large $|z_i|$'s are expected to contain signal while those small ones to be noise. To threshold these

z_i's, we used the order-dependent soft-thresholding rule (5) since most of the z_i's are small. The associated $c(z_i; \lambda/d_i^k)$ with $k = 1$ were computed following (30) using the common thresholding parameter $\lambda = .54$ selected by the GCV rule (44) and were displayed in Fig. 10 (b). Only a few of them are nonzero and they are decreasing to 0 from 1 with decreasing the associated d_i's. This well characterizes the soft-thresholding rule (5). The associated $\hat{z}_i$'s as defined in (30) were displayed in Fig. 10 (c). There are about 12 of them being nonzero. The final estimate of the underlying function was then constructed using (31) shown as the solid curve in Fig. 10 (d).

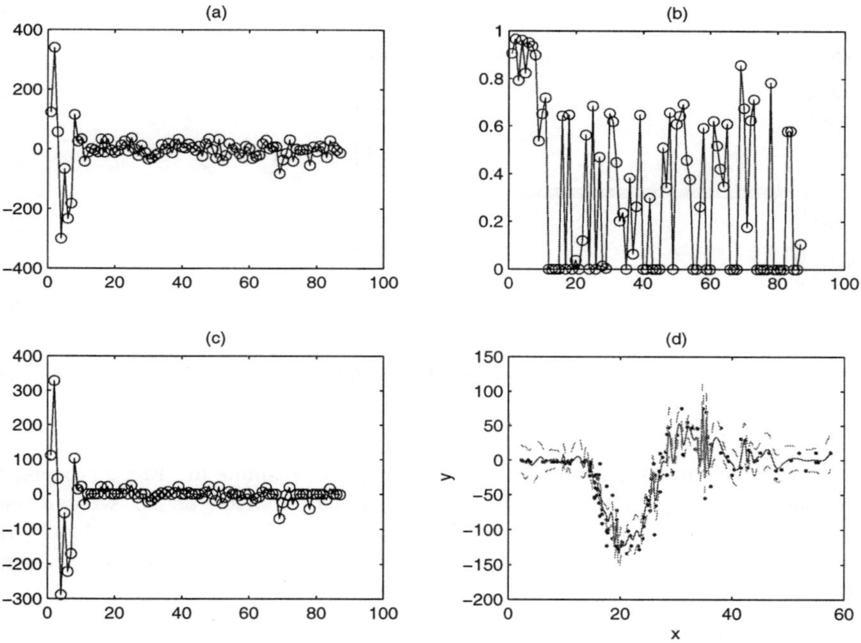

Fig. 11. Similar caption as that of Fig. 10 but now for the order-independent soft-thresholding.

As mentioned in the previous sections, the $k = 0$ means order-independent thresholding with all the z_i's being treated the same. Fig. 11 presents such a kind of thresholding. Although the z_i's (shown in panel (a)) are the same as those presented in Fig. 10 (a), the resulting $c(z_i, \lambda/d_i^k)$'s (with $k = 0$ and λ selected by GCV) are quite different as shown in panel (b) where lots of high frequent noise was not thresholded to 0. Consequently, the resulting fit is very noisy as shown in panel (d) (solid curve). Thus, the order-independent thresholding does not work and the order-dependent thresholding is preferred.

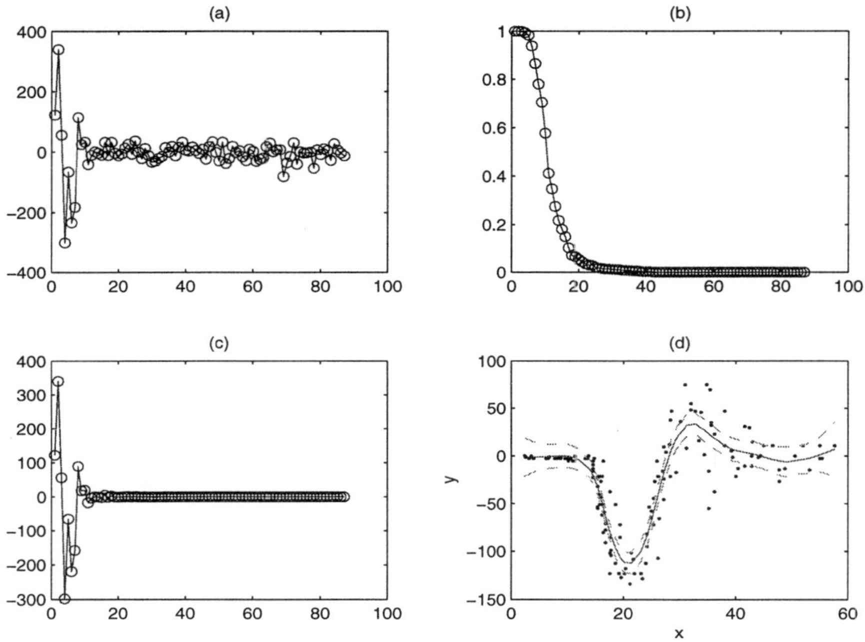

Fig. 12. Similar caption as that of Fig. 10 but now for the usual regression spline smoothing.

How is our fit comparing with the usual regression spline fit? Fig. 12 displays the usual regression spline fitting process. Compared with Fig. 10, we can see that although the $c(z_i, \lambda/d_i^k)$ are slightly different for large $|z_i|$'s (see Panel (b) in both figures), the $\hat{z}_i$'s are quite similar (see Panel (c)) and hence the resulting fits (see Panel (d), solid curve) are almost the same.

As a second example, we now apply our methodologies to a more complicated data set of unequally sampled time series. The response is light magnitude and the predictor is Julian day. The data set is accessible on the World Wide Web at **www.aavso.org**. Due to blockage of the star by sunlights, weather conditions, and availability of telescope time, the magnitudes of the star were measured at irregularly spaced times. Thus, usual wavelet smoothing is not directly applicable. This data set has been used by Sardy, Percival, Bruce, Gao, & Stuelzle (1999) and Antoniadis & Fan (2001) for illustrating their wavelet-based methodologies for denoising unequally sampled noisy signal.

The data set has 295 observations in total but three of them are known to be the upper limits on the star's magnitude and hence were deleted since their error properties are quite different from the remaining observations. We then

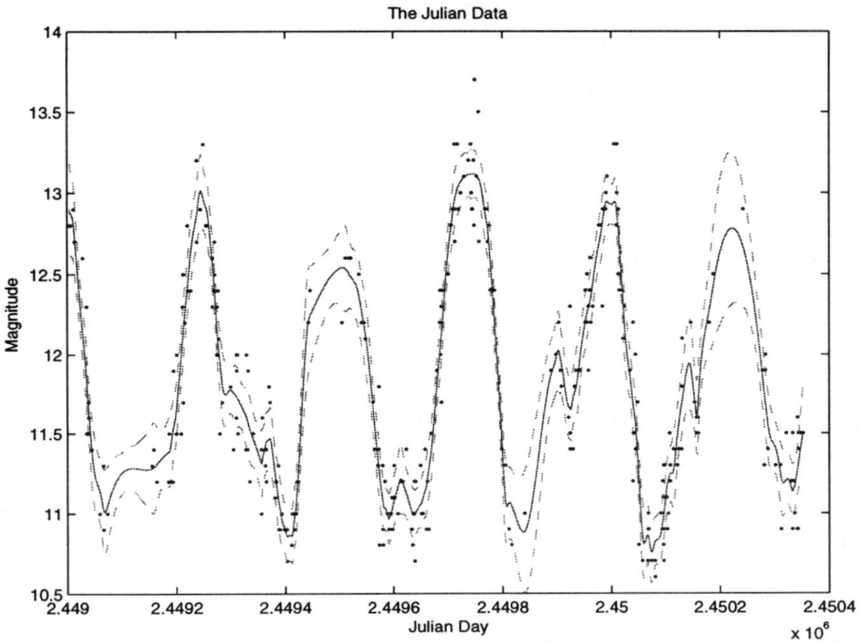

Fig. 13. The Julian data example: order-dependent soft-thresholding. The data are presented as dots, together with the fitted curve (solid curve) and the approximate 2 standard deviation bands (dashed curves).

applied our methodologies to the remaining $n = 292$ observations. As usual, we used defaults for the parameters involved. For example, we used $K = n/3 = 94$ knots which were located using the "quantiles as knots" method. We used the basis (19) with $q = 1$ and $p = 2$. We used $k = 1$ and selected the thresholding parameter λ by the GCV rule (44). The resulting fit is displayed in Fig. 13 (solid curve), together with approximate 2 standard deviation bands (dashed curves). It seems that our fit is comparable with the one given by Antoniadis & Fan (2001). Moreover, we offered approximate 2 standard deviation bands.

6 Extension to Nonparametric Additive Models

It is straightforward to extend our methodologies to some multivariate nonparametric regression models such as additive models. Additive models (Buja, Hastie & Tibshirani (1989)) are useful in data analysis since they allow involving several covariates without suffering the "curse of dimensionality". A general nonparametric additive model may be expressed as:

$$y_i = \alpha + f_1(t_{i1}) + f_2(t_{i2}) + \ldots + f_m(t_{im}) + \epsilon_i, \quad i = 1, 2, \ldots, n, \tag{1}$$

where α is an unknown constant, and $\int f_k(t)dt = 0, k = 1, 2, \ldots, m$ are imposed so that the above additive model is identifiable. Similar to (18), we can model f_k using a basis vector $\boldsymbol{\Psi}_k(t)$ for each k:

$$f_k(t) = \boldsymbol{\Psi}_k(t)^T \boldsymbol{\beta}_k, k = 1, 2, \ldots, m.$$

Then (1) can be approximated by

$$y_i = \alpha + \boldsymbol{\Psi}_1(t_{i1})^T \boldsymbol{\beta}_1 + \ldots + \boldsymbol{\Psi}_m(t_{im})^T \boldsymbol{\beta}_m + \epsilon_i, i = 1, 2, \ldots, n.$$

In matrix form, we have

$$\mathbf{y} = \mathbf{X}\boldsymbol{\beta} + \boldsymbol{\epsilon},$$

where $\boldsymbol{\beta} = (\alpha, \boldsymbol{\beta}_1^T, \ldots, \boldsymbol{\beta}_m^T)^T$ and $\mathbf{X} = (1_n, \mathbf{X}_1, \mathbf{X}_2, \ldots, \mathbf{X}_m)$ with

$$\mathbf{X}_k = (\boldsymbol{\Psi}_k(t_{1k}), \boldsymbol{\Psi}_k(t_{2k}), \ldots, \boldsymbol{\Psi}_k(t_{nk}))^T, k = 1, 2, \ldots, m.$$

This transforms the nonparametric additive model (1) into the general linear model (21) and hence the associated thresholding methodologies can be applied directly.

ACKNOWLEDGEMENTS: Part of the work was done when the author visited Prof. Jianqing Fan at the Department of Statistics, Chinese University of Hong Kong, and Prof. Kai-Tai Fang at the Department of Mathematics, Hong Kong Baptist University in the summer of 2001. The author thanked Profs Fan and Fang for their hospitality, guidance and help. Thanks also go to the Referee, the Editors for their helpful comments and suggestions. The work was partly supported by the National University of Singapore Academic Research grant R-155-000-038-112.

References

Antoniadis, A. & Fan, J. (2001). Regularization of wavelets approximations. *J. Amer. Statist. Assoc.* **96** 939–967.

Buja, A., Hastie, T. & Tibshirani, R. (1989). Linear smoothers and additive models. *Ann. Statis.* **17** 453–555.

Donoho, D. L. & Johnstone, I. M. (1994). Ideal spatial adaptation by wavelet shrinkage. *Biometrika* **85** 425–455.

Donoho, D. L. & Johnstone, I. M. (1998). Minimax estimation via wavelet shrinkage. *Ann. Statist.* **26** 879–921.

Fan, J. (1997). Comments on Wavelets in statistics: a review by A. Antoniadis. *J. Itali. Statist. Assoc.* **6** 131–138.

Fan, J. & Gijbels, I. (1996). *Local Polynomial Modeling and its Applications.* Chapman and Hall, London.

Fan, J. & Li, R. (2001). Variable selection via penalized likelihood. *J. Amer. Statist. Assoc.* 1348–1360.

Green, P. J. & Silverman, B. W. (1994). *Nonparametric Regression and Generalized Linear Models.* Chapman & Hall, London.

Hall, P. & Turlach, B. A. (1997). Interpolation methods for nonlinear wavelet regression with irregularly spaced design. *Ann. Statis.* **25** 1912–1925.

Johnstone, I. M. & Silverman, B. W. (1997). Wavelet threshold estimators for data with correlated noise. *J. Roy. Statist. Soc. B* **59** 319–351.

Kovac, A. & Silverman, B. W. (2000). Extending the scope of wavelet regression methods by coefficient-dependent thresholding. *J. Amer. Statist. Assoc.* **95** 172–183.

Ruppert, D. & Carroll, R. J. (1997). Penalized regression splines. Manuscript.

Sardy, S., Percival, D. B., Bruce, A. G., Gao, H. Y. & Stuelzle, W. (1999). Wavelet shrinkage for unequally spaced data. *Statistics and Computing* **9** 65–75.

Silverman, B. W. (1985). Some aspects of the spline smoothing approach to nonparametric regression curve fitting (with discussion). *J. Roy. Statist. Soc. B* **47** 1–52.

Wahba, G. (1990). *Spline Models for Observational Data.* SIAM, Philadelphia, CBMS-NSF Regional Conference Series in Applied Mathematics, **59**.

A New Transformed Two-Sample t-Test

Hanfeng Chen[1] and Md. Khairul Islam[2]

[1] Bowling Green State University, Department of Mathematics and Statistics, Bowling Green, OH 43403, USA (hchen@math.bgsu.edu)
[2] Bowling Green State University, Department of Mathematics and Statistics, Bowling Green, OH 43403, U.S.A. (mkislam@math.bgsu.edu)

Summary. For a power transformation family $(x^\lambda - 1)/\lambda$, Box & Cox (1964) proposed to use the pseudo maximum likelihood estimation method to determine a value of λ. This approach is intended to transform the non-normal data to normal data and then construct the t-test based on the transformed data in the two-sample problem. In this article, a new procedure of estimating λ, hence a new transformed t-test, is proposed. The idea of the new approach is to select directly a value of λ such that the transformed t-test is as close to the t-distribution as possible. Simulation results show that the new transformed t-test is appropriate to use a t-distribution to determine a threshold and much more powerful than the Box-Cox transformed t-test when the model is nearing a transformed normal model, while two approaches are comparable if the model is far-fetched for transformation to normality.

Key words: Box-Cox power transformation, bootstrap, two-sample problem.

2000 Mathematics Subject Classification: Primary 62G20; secondary 62F03.

1 Introduction

Consider two independent random samples: $X_1, \cdots, X_m$ and $Y_1, \cdots, Y_n$. One wishes to test the null hypothesis that the two populations from which the two

Contemporary Multivariate Analysis and Experimental Design—In Honor Celebration of Professor Kai-Tai Fang's 65th birthday. Edited by Jianqing Fan and Gang Li. The World Scientific Publisher, 2005.

samples come have the identical mean, namely,

$$H_0 : \mu_x = \mu_y$$

where $\mu_x = E(X_1)$ and $\mu_y = E(Y_1)$.

The standard statistical models usually assume that the two population distributions are normal with the same variance. In this case, the Student t-test is often recommended and proved to be the uniformly most powerful unbiased test (see, e.g., Lehmann (1994)). When the normality assumption is weakened or invalid, it is common practice to re-express the data to implement the t-test (Mosteller & Tukey (1977) and Atkinson (1985)). In an oft-cited paper, Box & Cox (1964) suggested to use a power transformation. Suppose that all observations are nonnegative. The Box-Cox power transformation is defined as follows:

$$X_i(\lambda) = \begin{cases} (X_i^\lambda - 1)/\lambda, & \text{if } \lambda \neq 0 \\ \log X_i, & \text{if } \lambda = 0 \end{cases} \tag{1}$$

and similarly $Y_j(\lambda)$. If such a transformation is successful to transform the data to fit a normal model, the profiled log-likelihood function for the transformation parameter λ is

$$l(\lambda) = -\{(m+n)/2\} \log S^2(\lambda) + (\lambda - 1) \left\{ \sum_{i=1}^{m} \log X_i + \sum_{j=1}^{n} \log Y_j \right\},$$

where $S^2(\lambda)$ is the pooled sample variance of the transformed data $X_i(\lambda)$'s and $Y_j(\lambda)$'s defined by (1). Box & Cox (1964) proposed to estimate λ by the maximizer $\hat{\lambda}$ of $l(\lambda)$, i.e., the so-called maximum likelihood estimator (MLE). The two-sided transformed t-test is to reject H_0 if $|T(\hat{\lambda})|$ is greater than the Student t-critical value $t_{\alpha/2,m+n-2}$, where

$$T(\lambda) = \sqrt{mn/(m+n)}\{\bar{X}(\lambda) - \bar{Y}(\lambda)\}/S(\lambda)$$

with $\bar{X}(\lambda)$ and $\bar{Y}(\lambda)$ being the sample means of the transformed data.

Theoretical studies on the Box-Cox transformed data analysis described above have been reported in the literature. Hinkley (1975) and Hernandez & Johnson (1980) investigated the asymptotic properties; Bickel & Doksum (1981) examined critically the behavior of the asymptotic variances of parameter estimates including $\hat{\lambda}$ for regression and analysis of variance situations; Chen & Loh (1992) and Chen (1995) studied the asymptotic testing power. The conclusions are mostly positive. For example, Hernandez & Johnson (1980) discovered that the MLE $\hat{\lambda}$ minimizes the Kullback-Leibler asymptotically; Chen & Loh (1992) and Chen (1995) proved that the Box-Cox transformed t-test is typically more efficient asymptotically than the t-test without transformation.

In this article, we propose an innovative procedure for estimating the transformation parameter λ. The new approach is expected to result in an even more efficient t-test with the transformed data. The simulation results presented in Section

3 show that the new transformed t-test is appropriate to use a t-distribution to determine a threshold and much more powerful than the Box-Cox transformed t-test when the model is nearing a transformed normal model, while two approaches are comparable if the model is far-fetched for transformation to normality.

2 The New Transformed t-Test

The idea of the Box-Cox t-test is to first transform the non-normal model to normal model (or nearing normal model) and then perform the t-test. If the transformation is successful, the transformed t-test is expected to follow a t-distribution under the null model. However, as far as the t-distribution is of concerns, we do not have to require the transformed model to be normal. That is, as long as the transformed t-test statistic follows the t-distribution under the null model, it does the job. Since normality is only a sufficient model condition for the t-test to follow the t-distribution, we expect that transforming the test statistic to t-distribution model directly is an easier task to do than transforming the data to the normal model, so a more efficient procedure than the Box-Cox procedure. For example, the t-test statistic follows the t-distribution under the entire elliptically contoured distribution family that includes multivariate normal distributions. See Anderson & Fang (1987) and Fang, Kotz & Ng (1990).

Viewing the problem of transforming to t-distribution as a fitting problem to the t-distribution, we propose to select the value of λ to minimize the Pearson's χ^2 statistic of goodness-of-fit. In order to calculate the Pearson's χ^2 statistic, we need to have a random sample from the null distribution of the transformed t-test statistic $T(\lambda)$. The bootstrap method can be used to obtain such a random sample.

For a fixed value of λ, let $X_1^*(\lambda), \cdots, X_m^*(\lambda)$ and $Y_1^*(\lambda), \cdots, Y_n^*(\lambda)$ be a bootstrap sample from the $X_i(\lambda)$'s and $Y_j(\lambda)$'s, respectively. Note that the null distribution of $T(\lambda)$ is that of

$$\sqrt{mn/(m+n)}\frac{\{\bar{X}(\lambda) - \bar{Y}(\lambda)\} - \{\mu_x - \mu_y\}}{S(\lambda)}.$$

A bootstrap "observation" on the null distribution of the transformed t-test statistic $T(\lambda)$ is given by

$$T^*(\lambda) = \sqrt{mn/(m+n)}\frac{\{\bar{X}^*(\lambda) - \bar{Y}^*(\lambda)\} - \{\bar{X}(\lambda) - \bar{Y}(\lambda)\}}{S^*(\lambda)}, \tag{2}$$

where $\bar{X}^*(\lambda)$ and $\bar{Y}^*(\lambda)$ are the sample means of the bootstrap samples and $S*(\lambda)$ is the pooled sample standard deviation of the bootstrap samples. Repeat the bootstrap procedure B times and denote the B bootstrap observations on the null distribution of $T(\lambda)$ by $T_1^*(\lambda), \cdots, T_B^*(\lambda)$.

Let F_0 be the t-distribution with $m+n-2$ degrees of freedom. Let $t_1 < t_2 < \cdots < t_k$ be k varieties of F_0 and put $p_i = F_0(t_i) - F_0(t_{i-1})$. For $i = 1, \cdots, k+1$, define

$$n_i(\lambda) = \sum_{j=1}^{B} I\{t_{i-1} < T_j^*(\lambda) \leq t_i\},$$

where $t_0 = -\infty$ and $t_{k+1} = \infty$ and $I(\cdot)$ is the indicator function. Then the Pearson's χ^2 statistic is defined as

$$\chi^2(\lambda) = \sum_{i=1}^{k+1} \frac{(n_i(\lambda) - Bp_i)^2}{Bp_i}. \quad (3)$$

We propose to estimate λ by the minimizer $\tilde{\lambda}$ of $\chi^2(\lambda)$ and reject the null hypothesis against the two-sided alternative if $|T(\tilde{\lambda})|$ is greater than the Student t-critical value $t_{\alpha/2, m+n-2}$.

An algorithm to obtain the estimate $\tilde{\lambda}$ is summarized as follows:

Step 1. Compute p_i, $i = 1, \cdots, k+1$.
Step 2. For a fixed value λ, draw a bootstrap sample of size m from $X_1(\lambda), \cdots, X_m(\lambda)$ and a bootstrap sample of size n from $Y_1(\lambda), \cdots, Y_n(\lambda)$. From the bootstrap two samples, compute the statistic $T^*(\lambda)$ defined by (2).
Step 3. Repeat Step 2 independently B times. Denote the B bootstrap values of $T^*(\lambda)$ by $T_1^*(\lambda), \cdots, T_B^*(\lambda)$.
Step 4. Compute the Pearson's χ^2-statistic $\chi^2(\lambda)$ as defined by (3).
Step 5. The estimate $\tilde{\lambda}$ is the minimizer of $\chi^2(\lambda)$.

In Step 5, if an exhaustive procedure of minimization is utilized, one needs to run Steps 2-4 for a set of λ values, say $\lambda = -1.0(.1)1.0$.

3 Simulation and Discussion

In this section, we present a simulation study to compare the new transformed t-test procedure with the Box-Cox transformed t-test procedure. The simulation settings and arrangements are as follows:

1. Sample sizes considered are $(25, 30)$, $(50, 75)$ and $(100, 150)$. A nominal significance level of 5% is used. With $k = 5$, six varietes of a t-distributions are chosen such that $p_1 = .0250$, $p_2 = p_3 = p_4 = p_5 = .2375$, and $p_6 = .0250$.
2. In all cases, the Monte Carlo size is 4,000 and the bootstrap size is $B = 500$.
3. In each case, the estimate $\tilde{\lambda}$ is obtained by an exhaustive search over $\lambda \in [-1, 1]$ with $\lambda = -1.0(0.1)1.0$.

All simulations are performed by using the MatLab software. One remark we would like to make is that the bootstrap size of 500 was determined not by a bootstrap theory, but by following common practice of a bootstrap size in the literature.

The following null models are considered: log-normal $LN(0, 0.01)$ and $LN(0, 1)$; Gamma $G(1, 1)$; exponential $Exp(1)$. The simulated rejection rates under these models are reported in Table 1.

Alternative models considered are as follows:

A1: $X \sim LN(0, .724)$ and $Y \sim LN(0, .01)$.
A2: $X \sim G(1, 1.2)$ and $Y \sim G(1, 1)$.
A3: $X \sim Exp(1)$ and $Y \sim Exp(1.3)$.
A4: $X \sim N(2, 1|.5)$ and $Y \sim N(2, 1|.5) + .3$,

where $N(2, 1|.5)$ stands for the truncated normal $N(2, 1)$ from below at 0.5. Note that all the alternative models A1, A3 and A4 are chosen to have $\mu_x - \mu_y = .3$ and A2 to have $\mu_x - \mu_y = .2$ to ensure a testing power is well away from 0 and 1 for the comparison purposes. The simulated powers are reported in Table 2.

Table 1. Simulated rejection rates of the null distributions. The nominal signific ance level is 0.05.

Sample size (m, n)	(25,30)		(50, 75)		(100, 150)	
Null model	Box-Cox	New	Box-Cox	New	Box-Cox	New
$LN(0, .01)$	.0462	.0460	.0550	.0550	.0535	.0533
$LN(0, 1)$	.0515	.0460	.0512	.0430	.0468	.0362
$G(1, 1)$	.0495	.0400	.0500	.0410	.0525	.0450
$Exp(1)$	.0545	.0485	.0520	.0500	.0485	.0435

From the simulation results, the new transformed t-test is appropriate to use a t-distribution to determine a threshold and much more powerful than the Box-Cox transformed t-test when the models are nearing a transformed normal model, while the two tests are mostly comparable when the models are far-fetched for transformation to normality. One explanation for this can be that if a model is far-fetched for transformation to normal by a power function, the transformed tests are not very sensitive to choices of λ.

Chen & Loh (1991) pointed out that a transformed t-test in the two-sample problem may not render a correct significance level when the two population distributions have different shapes. They proposed a bootstrap transformed t-test

Table 2. Simulated powers of the Box-Cox transformed t-test and the new transformed t-test. The significance level is 0.05.

Sample size (m,n)	$(25,30)$		$(50,75)$		$(100,150)$	
Alternative model	Box-Cox	New	Box-Cox	New	Box-Cox	New
A1	.0548	.1375	.0755	.2560	.0850	.3907
A2	.0995	.0925	.1450	.1260	.2365	.2200
A3	.1405	.1290	.2290	.2195	.4450	.4185
A4	.8220	.8410	.9065	.9725	.9175	.9995

to achieve robustness for the Box-Cox transformed t-test. The transformed t-test this paper proposes is not intended to improve the robustness against violation of the distribution shape assumption. However, Chen and Loh (1991)'s bootstrap idea can also be used similarly in the new transformed t-test to gain the robustness advantage.

References

Anderson, T. W. & Fang, K. T. (1987). On the Theory of multivariate elliptically contoured distributions. *Sankhya, Ser. A* **49** 305–315.

Atkinson, A. C. (1985). *Plots, Transformations and Regression.* Oxford University Press: London.

Bickel, P. J. & Doksum, K. A. (1981). An analysis of transformations revisited. *J. Amer. Statist. Assoc.* **76** 296–311.

Box, G. E. P. & Cox, D. R. (1964). An analysis of transformation (with discussion). *J. Roy. Statist. Soc. Ser. B* **26** 211–246.

Chen, H. (1995). Tests following transformations. *Ann. Statist.* **23** 1587–1593.

Chen, H. & Loh, W. Y. (1991). Consistency of the bootstrap for the transformed two-sample t-test. *Commun. Statist. -Theory Meth.* **20** 997–1014.

Chen, H. & Loh, W. Y. (1992). Bounds on ARE's of tests following Box-Cox transformations. *Ann. Statist.* **20** 1485–1500.

Fang, K. T., Kotz, S., & Ng, K. W. (1990). *Symmetric Multivariate & Related Distributions.* Chapman and Hall: London and New York.

Hernandez, F. & Johnson, R. A. (1980). The large-sample behavior of transformations to normality. *J. Amer. Statist. Assoc.* **75** 855–861.

Hinkley, D. Y. (1975). On power transformations to symmetry. *Biometrika* **62** 101–111.

Lehmann, E. L. (1994). *Testing Statistical Hypotheses.* Chapman& hall: New York and London.

Mosteller, F. & Tukey, J. W. (1977). *Data Analysis and Regression.* Addison-Wesley: Reading, MA.

On Improved Estimates of Location in the Presence of an Unknown Scale

Jian-Lun Xu

Biometry Research Group, National Cancer Institute, EPN-3131, 6130 Executive Blvd. MSC 7354, Bethesda, MD 20892-7354, U.S.A. (jx2b@nih.gov)

Summary. Suppose that $(\mathbf{X}', \mathbf{U}')'/\sigma$ has a spherically symmetric distribution about $(\boldsymbol{\theta}', \mathbf{0}')'$, where $\mathbf{X}$ and $\mathbf{U}$ are $p \times 1$ and $m \times 1$ random vectors, respectively, $\boldsymbol{\theta}' = (\theta_1, \ldots, \theta_p)$ is an unknown vector and σ is an unknown scale. Under the loss function $f(||\boldsymbol{\delta} - \boldsymbol{\theta}||^2/\sigma^2)$, where $f(t)$ is a nondecreasing concave function of t, Brandwein & Strawderman (1991a) have investigated conditions under which estimators of the form $\mathbf{X} + a\mathbf{U}'\mathbf{U}\mathbf{g}(\mathbf{X})$ dominate $\mathbf{X}$. Their technique requires that $t^\alpha f'(t)$ is a nondecreasing function of t for some $\alpha \in (0, (p-2)/2)$. Because of this assumption, their bound on a depends on α which is related to the loss function f. This paper, without making the monotone assumption on the function $t^\alpha f'(t)$, investigates the dominance conditions of the estimators $\mathbf{X} + a\mathbf{U}'\mathbf{U}\mathbf{g}(\mathbf{X})$ and obtains a bound of a which is independent of the loss function. Examples related to this problem are also considered.

Key words: James-Stein estimation; Location vector; Nondecreasing concave loss; Quadratic loss; Spherical symmetry; Superharmonic functions.

2000 Mathematics Subject Classification: 62C99, 62F10, 62H12.

1 Introduction

Let $\mathbf{X}' = (X_1, \ldots, X_p)$ and $\mathbf{U}' = (U_1, \ldots, U_m)$ be observed random vectors such that $\mathbf{X}'_* = (\mathbf{X}', \mathbf{U}')$ has a spherically symmetric distribution about $\boldsymbol{\theta}'_* = (\boldsymbol{\theta}', \mathbf{0}')$,

Contemporary Multivariate Analysis and Experimental Design—In Honor Celebration of Professor Kai-Tai Fang's 65th birthday. Edited by Jianqing Fan and Gang Li. The World Scientific Publisher, 2005.

$$\mathbf{X}_* \sim \text{s.s.}(\boldsymbol{\theta}_*, \sigma^2 I), \tag{1}$$

where $\boldsymbol{\theta}' = (\theta_1, \ldots, \theta_p)$ and σ are unknown parameters. This paper is concerned with estimating the location vector $\boldsymbol{\theta}' = (\theta_1, \ldots, \theta_p)$ when the unknown scale parameter σ is estimated from the residual vector $\mathbf{U}$. To be specific, we investigate the conditions under which estimators of the form

$$\boldsymbol{\delta}_a(\mathbf{X}_*) = \mathbf{X} + a\mathbf{U}'\mathbf{U}\mathbf{g}(\mathbf{X}) \tag{2}$$

dominate $\mathbf{X}$ under the loss function

$$L(\boldsymbol{\delta}, \boldsymbol{\theta}) = f(||\boldsymbol{\delta} - \boldsymbol{\theta}||^2/\sigma^2), \tag{3}$$

where $f(t)$ is a nondecreasing concave function of t.

The assumptions on $\mathbf{X}_*$ coincide with the canonical form for the general linear model (see Scheffé (1959)). The setup and notation agree with Brandwein & Strawderman (1991a) (also see Section 5 of Brandwein & Strawderman (1991b)). When the loss function is quadratic ($f(t) = t$), Brandwein & Strawderman (1991b) have investigated conditions under which estimators $\boldsymbol{\delta}_a(\mathbf{X}_*)$ defined by (2) dominate $\mathbf{X}$. dominate $\mathbf{X}$. Their result extends elegantly the robustness property of James-Stein type estimators in this setting shown by Cellier, Fourdrinier & Robert (1989) to the class of estimators studied in Stein (1981).

The extension from the quadratic loss to the loss function (3) has been investigated by Brandwein & Strawderman (1991a) and a nice result has been obtained (see Theorem 2.1 of Brandwein & Strawderman (1991a)). Their technique, however, requires that the loss function f in (3) does not flatten out too quickly. Mathematically, they assume that $t^\alpha f'(t)$ is nondecreasing in t for some $\alpha \in (0, (p-2)/2)$. Because of this additional assumption on the loss function $f(t)$, their bound of a also depends on α. Meanwhile, an interesting example of loss function $f(t)$ in which $t^\alpha f'(t)$ is not nondecreasing in t is also given by them. Some other related studies can be found in Fan and Fang (1990a,b,c) and a review article given by Brandwein & Strawderman (1990). The objective of this paper is to investigate the bound for a which doesn't depend on the loss function $f(t)$ such that the estimators $\boldsymbol{\delta}_a(\mathbf{X}_*)$ dominate $\mathbf{X}$.

In Section 2 we present the main result while its proof is deferred to Section 4. To illustrate the performance of the main result, two examples are also presented in Section 2. To shorten the proof of the main result, we prove a lemma in Section 3. Finally, some concluding remarks are given.

2 Main Result

In this section, we present the main result which states conditions under which the estimators $\boldsymbol{\delta}_a(\mathbf{X}_*)$ of $\boldsymbol{\theta}$ defined by (2) dominate $\mathbf{X}$ for loss function (3), and discuss two related examples.

Theorem 1. *Suppose that* $\mathbf{X}$ *is a* $p \times 1$ $(p \geq 5)$ *random vector and* $\mathbf{U}$ *is an* $m \times 1$ *random vector such that* $\mathbf{X}'_* = (\mathbf{X}', \mathbf{U}') \sim \text{s.s.}(\boldsymbol{\theta}_*, \sigma^2 I)$ *as defined by (1). Let the estimators* $\boldsymbol{\delta}_a(\mathbf{X}_*)$ *and the loss function* $L(\boldsymbol{\delta}, \boldsymbol{\theta})$ *be defined by (2) and (3), respectively. Then* $\boldsymbol{\delta}_a(\mathbf{X}_*)$ *dominate* $\mathbf{X}$ *if*

(i) $||\mathbf{g}(\mathbf{X})||^2/2 \leq -h(\mathbf{X}) \leq -\bigtriangledown \circ \mathbf{g}(\mathbf{X})$, *where* $-h(\mathbf{X})$ *is superharmonic and*

$$\bigtriangledown \circ \mathbf{g}(\mathbf{X}) = \sum_{i=1}^{p} \frac{\partial g_i(\mathbf{X})}{\partial x_i},$$

(ii) $E_{\mathbf{W}}[-R^2 h(\mathbf{W})]$ *is a nondecreasing function of* R, *where* $\mathbf{W}$ *has a uniform distribution in a sphere centered at* $\boldsymbol{\theta}$ *with radius* R *and* (iii) $0 < a \leq 1/(pJ)$, *where*

$$J = \frac{2/(p-2)}{p/(m+2) + 2/(p+m+2)}$$

provided

$$\frac{2/(p-2)}{p/(m+2) + 2/(p+m+2)} \leq \frac{p+m-2}{p+2m+2}.$$

To illustrate the performance of this theorem, we present two examples below. Brandwein & Strawderman (1991a) Theorem 2.1 cannot be applied to the first example since the condition that $t^\alpha f'(t)$ be nondecreasing is not satisfied. The loss function in the second example is also studied by Brandwein & Strawderman (1991a).

Example 1. Consider

$$L(\boldsymbol{\delta}, \boldsymbol{\theta}) = 1 - \exp\left(-\frac{||\boldsymbol{\delta} - \boldsymbol{\theta}||^2}{\sigma^2}\right).$$

Thus the loss, $f(t) = 1 - \exp(-t)$, is a nondecreasing concave function of t. However, as mentioned by Brandwein & Strawderman (1991a), the condition that $t^\alpha f'(t)$ be nondecreasing is not satisfied because the derivative is decreasing faster than any power of t at ∞. Therefore, Brandwein and Strawderman's (1991a) Theorem 2.1 is not applicable to this example. However, one can see from theorem above that $\boldsymbol{\delta}_a(\mathbf{X}_*)$ dominate $\mathbf{X}$ for $0 < a \leq /(pJ)$, where J is given by Condition (iii) of theorem 1.

More generally, we can consider the loss function (3) with $f(t) = P[UZ \leq t]$, $t \geq 0$, where the random variable U has a uniform distribution on the interval $(0, 1)$, Z is an arbitrary nonnegative random variable, and U and Z are independent. It is easy to check that $f(t) = 1 - \exp(-t)$ when the random variable Z has a Gamma distribution Gamma$(2, 1)$. Note that Brandwein and Strawderman's (1991a) Theorem 2.1 cannot be applied if the function $t^\alpha \int_t^\infty (1/z) dN(z)$ is not nondecreasing nondecreasing for $\alpha < (p-2)/2$, where $N(z)$ is the distribution function of the random Z. However, one can see that the theorem above provides a positive answer about dominance of $\boldsymbol{\delta}_a(\mathbf{X}_*)$ with respect to $\mathbf{X}$.

Example 2. Consider $L(\boldsymbol{\delta}, \boldsymbol{\theta}) = ||\boldsymbol{\delta} - \boldsymbol{\theta}||^{2\eta}/\sigma^{2\eta}$ for $0 < \eta \leq 1$. Thus, the loss, $f(t) = t^{\eta}$, is a nondecreasing concave function if t. Applying the theorem above yields that $\boldsymbol{\delta}_a$ defined by (2) dominates $\mathbf{X}$ when $0 < a \leq 1/(pJ)$ with $J = [2/(p-2)][p/(m+2) + 2/(p+m+2)]^{-1}$, while Brandwein and Strawderman's (1991a) result gives $0 < a \leq (p-4)/[p(m+2)]$.

3 A Lemma

In this section, we prove a lemma which is needed to shorten the proof of theorem 1.

Lemma 1. *Assume that β has a Beta-distribution $\beta \sim \mathrm{Beta}(A, B)$, where $A > 2$. For $t \in [0,1]$, let the functions $N_1(t)$ and $N_2(t)$ be defined by*

$$\begin{aligned} N_1(t) &= \left(A + B - 1 - \frac{A-2}{t}\right)\frac{1-t}{t}, \\ N_2(t) &= A + B - 1 - \frac{A-1}{t} \end{aligned} \tag{4}$$

with $N_i(0) = \lim_{t \to 0+} N_i(t) = -\infty$ for $i = 1, 2$. Then

$$E_\beta[N_1(\beta) f(\gamma\beta)] \leq J E_\beta[N_2(\beta) f(\gamma\beta)], \tag{5}$$

where γ is a positive number, $f(\cdot)$ is defined by (3), and

$$J = \frac{B(B+1)}{A-1} \frac{A+B+1}{AB(A+B+1) + B(B+1)} \tag{6}$$

provided

$$\frac{B(B+1)}{A-1} \frac{A+B+1}{AB(A+B+1) + B(B+1)} \leq \frac{A+B-1}{A+2B+1}. \tag{7}$$

Proof. For $t \in [0,1]$, let

$$Q(t) = J(A+B)\left(t - \frac{A}{A+B}\right).$$

Then

$$E_\beta[Q(\beta) f(\gamma\beta)] \geq E_\beta[Q(\beta)] E_\beta[f(\gamma\beta)] = 0 \tag{8}$$

for any $\gamma \geq 0$. Here the equality in (8) follows from the fact that $E_\beta[Q(\beta)] = 0$, while the inequality in (8) follows from an application of Wijsman's (1985) theorem 2 with $f_1(t) = Q(t)$, $f_2(t) = 1$, $g_1(z) = f(\gamma t)$, $g_2(t) = 1$ with probability measure $d\mu = f_{A,B}(t)dt$, a pdf of the Beta-distribution $\mathrm{Beta}(A, B)$.

Write $N(t) = N_1(t) - JN_2(t)$ and $M(t) = Q(t) + N(t)$ for $t \in [0,1]$. Then one can see from (8) that

$$E_\beta[N(\beta)f(\gamma\beta)] \leq E_\beta\{[Q(\beta) + N(\beta)]f(\gamma\beta)\} = E_\beta[M(\beta)f(\gamma\beta)]. \tag{9}$$

Note that the function $M(t)$ can be expressed as

$$M(t) = Q(t) + N(t) = \frac{1}{t}K(t), \tag{10}$$

where $K(t) = C_2t^2 + C_1t + C_0 + C_{-1}t^{-1}$, and

$$C_2 = J(A+B),$$

$$C_1 = -[JA + (J+1)(A+B-1)],$$

$$C_0 = 2A + B - 3 + J(A-1),$$

$$C_{-1} = -(A-2).$$

Now we prove that $K(t)$ is first nondecreasing and then nonincreasing in t. In fact,

$$K'(t) = 2C_2t + C_1 - C_{-1}t^{-2} = t^{-2}k(t), \tag{11}$$

where

$$k(t) = 2C_2t^3 + C_1t^2 - C_{-1}.$$

Note that $k'(t) = 2t(3C_2t + C_1) \leq 0$ because

$$3C_2t + C_1 \leq 3C_2 + C_1 = J(A+2B+1) - (A+B-1) \leq 0$$

under conditions (6) and (7). Therefore, $k(t)$ is nonincreasing in t. Since

$$k(0) = -C_1 = A - 2 > 0,$$

$$k(1) = 2C_2 + C_1 - C_{-1} = (B+1)(J-1) < 0,$$

there exists a unique $t_1 \in (0,1)$ such that $k(t) > 0$ for $t \in [0,t_1)$, $k(t) < 0$ for $t \in (t_1,1]$, and $k(t_1) = 0$. Thus, one can see from (11) that $K(t)$ is first nondecreasing and then nonincreasing in t. Because

$$K(0) = \lim_{t\to 0+} L(t) = -\infty,$$

$$K(1) = C_2 + C_1 + C_0 + C_{-1} = 0,$$

there exists a unique $t_2 \in (0, t_1)$ such that $K(t) < 0$ for $t \in [0, t_2)$, $K(t) > 0$ for $t \in (t_2, 1]$, and $K(t_2) = 0$. Since $f(\gamma t)/t$ is nonincreasing in t, one can obtain from (10) that

$$M(t)f(\gamma t) = K(t)\frac{f(\gamma t)}{t} \le K(t)\frac{f(\gamma t_2)}{t_2}$$

which implies that

$$E_\beta[M(\beta)f(\gamma\beta)] \le \frac{f(\gamma t_2)}{t_2} E_\beta[K(\beta)] = 0$$

because

$$E_\beta[K(\beta)] = C_2\frac{A}{A+B}\frac{A+1}{A+B+1} + C_1\frac{A}{A+B} + C_0 + C_{-1}\frac{A+B-1}{A-1}$$

$$= \frac{B(B+1)}{(A-1)(A+B)} - J\left(\frac{A^2B + AB^2 + AB + B^2 + B}{(A+B)(A+B+1)}\right)$$

$$= 0$$

under condition (6).

4 Proof of the Theorem

Let $R[\boldsymbol{\delta}, \boldsymbol{\theta}] = E[L(\delta, \boldsymbol{\theta})] = E[f(||\boldsymbol{\delta} - \boldsymbol{\theta}||^2/\sigma^2)]$ be the risk of $\boldsymbol{\delta}$. Using the argument of Brandwein & Strawderman (1991a) with a verbatim copy of their (2.1)-(2.4) yields that the difference between the risks of two estimators $\boldsymbol{\delta}_a(\mathbf{X}_*)$ and $\mathbf{X}$ is given by

$$\Delta = R[\boldsymbol{\delta}_a(\mathbf{X}_*), \boldsymbol{\theta}] - R[\mathbf{X}, \boldsymbol{\theta}]$$

$$= E_{\mathbf{X}_*}[f(||\boldsymbol{\delta}_a(\mathbf{X}_*) - \boldsymbol{\theta}||^2/\sigma^2) - f(||\mathbf{X} - \boldsymbol{\theta}||^2/\sigma^2)]$$

$$\le E_{\mathbf{X}_*}\left\{f'(||\mathbf{X} - \boldsymbol{\theta}||^2/\sigma^2)\left[||\boldsymbol{\delta}_a(\mathbf{X}_*) - \boldsymbol{\theta}||^2/\sigma^2 - ||\mathbf{X} - \boldsymbol{\theta}||^2/\sigma^2\right]\right\} \tag{12}$$

$$\le \frac{2a}{\sigma^2}E_{(R^2,T^2)}\left\{f'\left(\frac{R^2}{\sigma^2}\right)\left(\frac{a(T^2 - R^2)^2}{R^2} - \frac{T^2 - R^2}{p}\right)G(R^2)\right\}$$

$$= 2a\Delta_1,$$

where $T^2 = R^2 + S^2, R^2 = ||\mathbf{X} - \boldsymbol{\theta}||^2, S^2 = ||\mathbf{U}||^2$, $G(R^2) = -E_{\mathbf{W}}[R^2\, h(\mathbf{W})]$ with $\mathbf{W} \sim \mathcal{U}\{||\mathbf{X} - \boldsymbol{\theta}||^2 \leq R^2\}$, a uniform distribution in a sphere centered at $\boldsymbol{\theta}$ with radius R, and

$$\Delta_1 = E_{(R^2,T^2)} \left\{ \left(\frac{1}{\sigma^2}\right) f'\left(\frac{R^2}{\sigma^2}\right) \left(\frac{a(T^2 - R^2)^2}{R^2} - \frac{T^2 - R^2}{p}\right) G(R^2) \right\}. \quad (13)$$

(See (2.4) of Brandwein & Strawderman (1991a), page 308).

Let $A = p/2, B = m/2$ and let $\beta = R^2/T^2$. Then β is independent of T^2 and $\beta \sim \text{Beta}(A, B)$ (see Kelker (1970) or Fang, Kotz & Ng (1990)). Because of independence of T^2 and β, we evaluate Δ_1 when T^2 is given. Set $\lambda = a + 1/p$ and $\gamma = T^2/\sigma^2$. Then (13) can be expressed by

$$\begin{aligned} \Delta_1 &= E_\beta \left[\gamma f'(\gamma\beta)(a - \lambda\beta)[\beta^{-1}(1 - \beta)]G(T^2\beta)\right] \\ &= [B/(A-1)]E_{\beta_1} \left[\gamma f'(\gamma\beta_1)(a - \lambda\beta_1)G(T^2\beta_1)\right] \\ &\leq [B/(A-1)]\, G(T^2a/\lambda)E_{\beta_1} \left[\gamma f'(\gamma\beta_1)(a - \lambda\beta_1)\right] \\ &= [B/(A-1)]G(T^2a/\lambda)\Delta_2, \end{aligned} \quad (14)$$

where $\beta_1 \sim \text{Beta}(A - 1, B + 1)$ and Δ_2 is defined by

$$\Delta_2 = E_{\beta_1} \left[\gamma f'(\gamma\beta_1)(a - \lambda\beta_1)\right]. \quad (15)$$

The last inequality of (14) follows from the condition (ii) and the fact that $f'(\gamma\beta_1) \geq 0$. To prove $\Delta \leq 0$, one can see from (12), (13), (14) and (15) that it suffices to show that $\Delta_2 \leq 0$.

Let $C(r, s) = \Gamma(r + s)/[\Gamma(r)\Gamma(s)]$ for any $r > 0, s > 0$. Then using the integration-by-parts yields that

$$\begin{aligned} \Delta_2 &= E_{\beta_1}[\gamma f'(\gamma\beta_1)(a - \lambda\beta_1)] \\ &= \int_0^1 (a - \lambda z) f_{A-1,B+1}(z) df(\gamma z) \\ &= -\frac{C(A-1, B+1)}{C(A, B)} E_\beta \left\{ \left(-aN_1(\beta) + \frac{N_2(\beta)}{p}\right) f(\gamma\beta) \right\} \\ &= -\frac{C(A-1, B+1)}{C(A, B)} \Delta_3, \end{aligned} \quad (16)$$

where $\Delta_3 = E_\beta\{[-aN_1(\beta) + N_2(\beta)/p]f(\gamma\beta)\}$, and $N_1(t)$ and $N_2(t)$ are defined by (4). One can see from (16) that $\Delta_2 \leq 0$ is equivalent to $\Delta_3 \geq 0$. In fact, using Lemma 1 yields that

$$\begin{aligned}
\Delta_3 &= E_\beta\left\{\left(-aN_1(\beta) + \frac{N_2(\beta)}{p}\right) f(\gamma\beta)\right\} \\
&\geq \left(-aJ + \frac{1}{p}\right) E_\beta\left[N_2(\beta)f(\gamma\beta)\right] \\
&\geq \left(-aJ + \frac{1}{p}\right) E_\beta[N_2(\beta)]E_\beta[f(\gamma\beta)] \\
&= 0
\end{aligned} \tag{17}$$

if the condition (iii) of theorem 1 holds because $E_\beta[N_2(\beta)] = 0$. Here the last inequality of (17) follows from the condition (iii) of theorem 1 and an application of Wijsman's (1985) theorem 2 with $f_1(t) = N_2(t)$, $f_2(t) = 1$, $g_1(t) = f(\gamma t)$, $g_2(t) = 1$, and probability measure $d\mu = f_{A,B}(t)dt$. The proof is complete.

5 Concluding Remarks

This article presents conditions under which the estimators $\boldsymbol{\delta}_a(\mathbf{X}_*)$ of $\boldsymbol{\theta}$ defined by (2) dominate $\mathbf{X}$ for loss function (3). The range of a given by theorem 1 which is free of the loss function is by no mean optimal. If we know the form of the loss function $f(t)$, then it is possible to get better bounds for a. For example, the bound of a obtained by Brandwein & Strawderman (1991b) for $f(t) = t$ is $(p-2)/[p(m+2)]$ which does not have any constraint between p and m. On the other hand, our method requires $p \geq 5$ ($A > 2$ in Lemma 1) which is necessary. This requirement can also be seen from Example 2.1 of Brandwein & Strawderman (1991a). As a final point, it would be interesting to extend Theorem 1 above to nondecreasing concave loss functions of general quadratic loss $(\mathbf{X} - \boldsymbol{\theta})'D(\mathbf{X} - \boldsymbol{\theta})/\sigma^2$, where D is a known $p \times p$ positive definite matrix. Although this extension looks very natural, we have been unable to prove such a result.

Acknowledgments

The author wishes to thank Philip C. Prorok for valuable comments.

References

Brandwein, A. C. & Strawderman, W. E. (1990). Stein estimation, the spherically symmetric case. *Statist. Sci.* **5** 356–369.

Brandwein, A. C. & Strawderman, W. E. (1991a). Improved estimates of location in the presence of an unknown scale. *J. Multivariate Anal.* **39** 305–314.

Brandwein, A. C. & Strawderman, W. E. (1991b). Generalizations of James-Stein estimators under spherical symmetry. *Ann. Statist.* **19** 1639–1650.

Cellier, D., Fourdrinier, D. & Robert, C. (1988). Robust shrinkage estimators of the location parameter for elliptically symmetric distributions. *J. Multivariate Anal.* **29** 39–52.

Fan, J. & Fang, K. T. (1990a). Inadmissibility of sample mean & sample regression coefficients for elliptically contoured distributions. In *Statistical Inference in Elliptically Contoured and Related Distributions* (K. T. Fang, T. W. Anderson, eds.) 275–290. Allerton Press Inc., New York.

Fan, J. & Fang, K. T. (1990b). Inadmissibility of the usual estimator for the location parameters of spherically symmetric distributions. In *Statistical Inference in Elliptically Contoured and Related Distributions* (K. T. Fang, T. W. Anderson eds.) 291–297. Allerton Press Inc., New York.

Fan, J. & Fang, K. T. (1990c). Shrinkage estimators and ridge regression estimators for elliptically contoured distributions. In *Statistical Inference in Elliptically Contoured and Related Distributions* (K. T. Fang T. W. Anderson, eds.) 313–326. Allerton Press Inc., New York.

Fang, K. T., Kotz, S. & Ng, K. W. (1990). *Symmetric Multivariate and Related Distributions.* Chapman and Hall.

Kelker, D. (1970). Distribution theory of spherical distributions and a location-scale parameter generalization. *Sankhyā Ser. A* **32** 419–430.

Scheffé, H. (1959). *The Analysis of Variance.* Wiley, New York.

Stein, C. (1981). Estimation of the mean of a multivariate normal distribution. *Ann. Statist.* **9** 1135–1151.

Wijsman, R. A. (1985). A useful inequality on ratios of integrals, with application to maximum likelihood estimation. *J. Amer. Statist. Assoc.* **80** 472–475.

Subject Index

Author Index